THE OXFORD BOOK OF
FANTASY STORIES

THE OXFORD BOOK OF FANTASY STORIES

Selected by

TOM SHIPPEY

Oxford New York

OXFORD UNIVERSITY PRESS

1994

Oxford University Press, Walton Street, Oxford OX2 6DP
Oxford New York Toronto
Delhi Bombay Calcutta Madras Karachi
Kuala Lumpur Singapore Hong Kong Tokyo
Nairobi Dar es Salaam Cape Town
Melbourne Auckland Madrid
and associated companies in
Berlin Ibadan

Oxford is a trade mark of Oxford University Press

First published 1994

British Library Cataloguing in Publication Data
Data available

Library of Congress Cataloging in Publication Data
The Oxford book of fantasy stories / selected by Tom Shippey.
p. cm.
Includes bibliographical references.
1. Fantastic fiction, English. 2. Fantastic fiction, American.
I. Shippey, T. A.
PR1309.F3096 1993 823'.0876608—dc20 93–14469
ISBN 0–19–214216–X

1 3 5 7 9 10 8 6 4 2

Typeset by Best-set Typesetter Ltd.
Printed on acid-free paper in
the United States of America

ACKNOWLEDGEMENTS

I owe special thanks to Simon Polley and Debbi Kerr, who provided continual loans and support during prolonged hesitation; to Mike Dickinson and Michael Walshe for their readiness to ransack their own collections; to Phil Stephensen-Payne and Malcolm Edwards for timely advice and the loan of precious items; to David Daniell for scholarly support; and to Pat Green and Joy Day of the Science Fiction Foundation for their swift response and helpfulness in a late crisis.

CONTENTS

INTRODUCTION

The first words of the first story in this collection make it clear that fantasy is an old and venerable literary form, with roots which go back as far as we have records, or further. 'So you won't sell me your soul?' implies: that there are such things as souls; that they can be sold; that Someone is prepared to buy them: and above all that normally such deals are closed. Richard Garnett, in 1888, could count on his readers remembering if nothing else the legend of Faust, or Doctor Faustus. How old is that legend? Faust stories had been written and told for at least four hundred years by Garnett's time, but in any case the deal-with-the-devil motif is pre-Faustian. A perfectly recognizable example is told by Rutebeuf, a French poet of the thirteenth century, in which the Virgin Mary comes to the assistance of a priest who has sold his soul: but Rutebeuf in his turn makes no claims for originality. We do not know where or when this particular motif arose.

Many other elements of the modern fantasy story go back as far or further. The earliest werewolf story surviving is probably Marie de France's lai *Bisclavret*, from around 1170, but once again Marie seems to feel no need to explain the concept, as if it was familiar already; there was at least one man called Werwulf in Anglo-Saxon England, centuries before Marie. Dragons are even longer attested, being found not only in *Beowulf* and the Eddic literature of Scandinavia, but also in Latin poetry now almost two millennia old. The *Odyssey*, composed some would say before literature in the sense of written compositions even began, contains not only Circe the witch and Polyphemus the one-eyed man-eating giant, but developed ideas of sorcery, shape-shifting, and magic potions. From that time to this (and corroborated here and there even by the Bible) it is safe to say that there has also been an immense, world-wide, but largely unrecorded tradition of telling 'wonder-tales', which we now call fairy-tales, on which many authors have at all times felt free to build. The first version of 'Goldilocks' appears to have been composed by Robert Southey (1774–1843); but 'Cinderella' goes back in its hundreds of recorded versions to China more than a thousand years ago.[1]

It would be impossible to attempt to cover this whole narrative impulse (far wider than any literary mode) in a single anthology, and the age of fantasy is stressed here only to make two points. One is that modern authors do have an immense literary hinterland on which to draw, to which they obviously attach great value. The other is that the connecting feature which enables one to talk of fantasy in this wide-ranging way is at least in outline perfectly clear. Fantasy literature, in its broadest definition which includes everything from 'Cinderella' to *Beowulf* to Stephen Donaldson, is literature which makes deliberate use of something known to be impossible.

Other critics have offered rather similar definitions, as for instance C. N. Manlove, who argues that fantasy must contain 'a substantial and irreducible element of supernatural or impossible worlds, beings or objects', or W. R. Irwin, who looks to 'overt violation of what is generally accepted as possible'.[2] Still, though I believe that such definitions are on the right lines, they do all provoke much the same question: 'known *by whom* to be impossible?, who's to say what's "accepted as possible"?, and what do you mean, "generally"?.' Returning to the case of 'The Demon Pope', it is a fair bet that Richard Garnett did *not* believe in a personal Devil who could be negotiated with and taken advantage of; he would not have been so flippant if he had. Nevertheless, many people did and do believe in the Devil, and would not consider that he is 'known to be impossible' at all. There is meanwhile no space left in the world for proper fire-breathing dragons to exist, and they have accordingly moved firmly into the realm of fantasy. But in earlier periods they might well have been regarded as unusual rather than impossible; it is often pointed out that the *Anglo-Saxon Chronicle* (an authority almost invariably referred to in this context as 'sober') recorded fiery dragons flying in Northumbria in the year 793, while a later contributor to the same *Chronicle* not only recorded the passage of the Wild Hunt but insisted that he had it on the witness of *sothfeste men*, 'truthful men'. Beliefs shift from period to period and from person to person. However, it probably remains true that whatever the cultural context, the unseen or the non-material always remains in a separate category from the everyday. If one looks at the

[1] See *Cinderella: A Folklore Casebook*, ed. Alan Dundes (London and New York, 1982). One of the authors in this volume, Jane Yolen, contributed an essay to Dundes's Collection, on 'America's Cinderella'.

[2] See C. N. Manlove, *Modern Fantasy: Five Studies* (Cambridge, 1975), 10–11, and W. R. Irwin, *The Game of the Impossible: A Rhetoric of Fantasy* (Urbana, Ill., 1976), 4.

personnel of the stories in this collection, I suspect that few if any people now believe in dragons, vampires, trolls, or virgin-enchanted unicorns: though larger numbers might be prepared to believe in demons or witches, while belief in magic (the most consistently repeated motif in these thirty-one stories) is still in one way or another quite widespread. Even those who do believe in it are nevertheless aware that they have to provide some explanation or rationalization for it.

The presence of an element 'known to be impossible' may therefore serve as a rule of thumb for identifying fantasy, by which I mean fantasy in its broadest sense, including all the 'ancestor-genres' such as wonder-tale, medieval romance, epic and heroic poetry, and even the medieval Icelandic *fornaldarsaga*, or 'saga of old times'.[3] It is at any rate much more useful as a guide than some current academic definitions, which leave one wondering whether those who produce them ever stray into an ordinary bookshop at all. The most frequently cited definition of 'the fantastic' in academic circles is thus that of Tzvetan Todorov, who wrote (admittedly in 1970, before the present avalanche of fantasy books had gathered momentum) that 'the very heart of the fantastic' is this:

> In a world which is indeed our world, the one we know . . . there occurs an event which cannot be explained by the laws of this same familiar world. The person who experiences the event must opt for one of two possible solutions: either he is the victim of an illusion of the senses . . . or else the event has indeed taken place, it is an integral part of reality . . . The fantastic occupies the duration of this uncertainty.[4]

Frequently though this definition is quoted, from its first words one can see that it does not apply to nine-tenths of what is at present on sale confidently labelled as 'Fantasy'. Most modern fantasy takes place somewhere, or somewhen, which is very definitely *not* our world. In cases where this is not true, it is interesting to see how modern authors deal with the Todorovian 'hesitation', as for instance in the two stories printed here by Phyllis Eisenstein and 'James Tiptree jr.', both of which appeared in the same issue of *Fantasy and Science Fiction*. Both are very firmly, even angrily set in our world, the world of 'mediocrity and malice', as it has been called in discussions of medieval romance.[5] In Eisenstein's 'Subworld' we stay in that frame of reference for over half the story, until

[3] Another of the authors in this volume, Poul Anderson, produced a novelized version of one of these in his *Hrolf Kraki's Saga* (New York, 1973).

[4] Tzvetan Todorov, *The Fantastic: A Structural Approach to a Literary Genre*, trans. Richard Howard (Ithaca, 1975), 25.

[5] See the references in my 'Breton *Lais* and Modern Fantasies', in *Studies in Medieval English Romances*, ed. D. S. Brewer (2nd edn., Cambridge, 1991), 78.

the woman says 'We're mice' (see p. 399). Just as Todorov predicted, the narrator's response is indeed 'Hesitantly, I chuckled'. And from then on, for several pages, the narrator goes through the motions of rejecting the 'fantastic' event—explaining what has happened to himself as joke, hypnosis, or dream, using the dismissive phrase 'fairy-tale'. But then he gives up. He accepts the fantasy explanation, re-enters the fantasy world, and at the end of the story is considering remaining in it, as it were going into the modern 'Elf-hill' analogue for ever and turning his back on humanity. Yet despite the abandonment of hesitation the story remains to all but academic definers a fantasy, a story of the fantastic. The Tiptree story meanwhile steers even closer to the Todorov definition, only to reject it even more firmly. Its first words are 'My informant was, of course, spectacularly unreliable'; all through, both the female narrator and the man who tells her his strange story keep referring to the fact that he is a drunk; a simple dismissive explanation is continually offered. At the same time, the story's complex frame (story embedded in story) is there entirely to present the male narrator as unusually wise and competent. At the end, both male and female narrators are in complete agreement about what has happened. Any 'hesitation' there has been has served only to make plausible a central fantastic event.

It is possible that these arguments over definition are caused simply by reference to different things. 'The fantastic', as academically defined and studied, is just not the same phenomenon as the bestseller genre of 'fantasy' now to be found in every bookshop. But if the Todorovian definition is far too narrow to have any correspondence with current practice, the 'known-to-be-impossible' definition given above is too broad: it takes in, as has been said, all the ancestor-genres of fairy-tale and romance and saga and the rest. And while modern fantasy draws freely and gratefully on all those sources, it is still recognizably not the same thing, even though it may appeal to some similar impulse. What is the nature of fantasy in the modern world?

Here clues lie in the relation of fantasy, not to its ancestor-genres, but to its currently existing relatives: the ghost story, the horror story, the 'strange tale', the Gothic tale, and science fiction. Between all these there is considerable overlap. Most ghost stories are horror stories, and to some extent vice-versa. Both ghost and horror stories arguably count as fantasy within the broad definition given above. A deal-with-the-devil story from 1899, 'Cavalanci's Curse', is printed as one of the *Strange Tales from the Strand Magazine* edited for OUP by Jack Adrian (1991). Even closer, two of the authors selected for this volume (Lovecraft and Carter) appear also in Chris Baldick's collection *The Oxford Book of Gothic Tales* (OUP 1992), while one story originally intended to appear here (Lovecraft's 'The Rats in the Walls') was later rejected just because it is such a textbook example of the qualities identified by Baldick as 'Gothic'.

However, the closest relationship between all the genres cited above is that between modern fantasy and modern science fiction. It is a major premise of this collection that the two genres have as it were grown up together, been conditioned by the same forces, helped to define each other. It is this connection which has made modern, popular, commercial fantasy so sharply distinct at once from the ancestor-genres on which it draws, from the etiolated notions of 'the fantastic' studied by academics, and even from its relatives listed above. Modern fantasy is above all a reaction to scientific advance and the general spread of the scientific imagination.

This claim can be justified in two ways, by looking at science fiction and by looking at 'pre-modern' fantasy. To take the more obvious point first: the overlap of authorship and readership between fantasy and science fiction is very high indeed. Four of the thirty-one authors selected here appear also in my companion collection, *The Oxford Book of Science Fiction Stories*. Of the remaining twenty-seven, all but seven have also written, often extensively, in the science fiction genre. Meanwhile, the major magazine at present printing fantasy is *The Magazine of Fantasy and Science Fiction*; in the UK the subtitle of *Interzone* is 'science fiction and fantasy'; and the Tanith Lee story printed here, though classic fantasy, appeared first in *Isaac Asimov's Science Fiction Magazine*. A twin-track policy has been evident in editorial circles for sixty years. The highly influential but short-lived fantasy magazine *Unknown* had the same editor, John W. Campbell jr., as *Astounding Science Fiction*. For much of its history *Fantastic* (which first printed the Brunner story collected here) ran in partnership with *Amazing*, which only printed science fiction; for a while the two journals came out in alternate months, probably being bought by virtually the same people. Finally, the title of another short-lived but unexpectedly productive journal, from which two stories are taken here, was *Science Fantasy*. What this title meant to the magazine's readership can probably not be recovered, but I would suggest that it was saying in effect: 'the contents will be like science fiction, but will deal not only with material known at present not to be *true* (as in science fiction, so the "science" part of the title), but also with material we are told at present is *impossible* (so the "fantasy" part).' Individual likes and dislikes apart, modern science fiction and fantasy authors are at the very least well aware of each other. The markets they aim at are also much the same.

It is this similarity which marks off modern fantasy writing from its immediate precursors, especially the Victorian/Edwardian 'strange tale'. As is evident from companion anthologies such as *The Oxford Book of Gothic Tales* and *Strange Tales from the Strand*, there was a considerable liking in late nineteenth-century Britain—successfully exported to the USA—for the strange tale, the weird tale, the traveller's tale. H. P. Lovecraft, for instance, pays respectful tribute to these in section 9 of the long essay 'Supernatural Horror in Literature', reprinted in his collection *Dagon* (see 'Select Bibliography' below): even their format is affectionately imitated in Sterling Lanier's 1968 story 'The Kings of the Sea', reprinted here. Yet as one reads through the Edwardian 'strange tales' now, one cannot help feeling that there is something missing. In those days readers were evidently rather easily disturbed. They did not demand much stimulation. In a characteristic work of what I would label 'genteel fantasy', as written by E. F. Benson, E. M. Forster, Arthur Machen, M. P. Shiel, or 'Saki', a respectable picnic or walk in the country is interrupted by some strange or threatening event. This will usually be pooh-poohed by a rationalist commentator within the story, whose beliefs are then undermined or violently overthrown. At the climax of the story a word will be said, an explanation offered: 'siren', 'elemental', 'werewolf', 'panic'. Pan-figures are indeed the most popular irruptions into the respectable world, seen in different ways in Machen's 'The Great God Pan' (1890), Forster's 'The Story of a Panic' (1904), or E. F. Benson's 'The Man who Went Too Far' (1912). Many readers will remember the appearance of Pan in Kenneth Grahame's *The Wind in the Willows* (1908), as 'the piper at the gates of dawn'. He obviously represents an element of life which English readers especially felt was being crowded out.

But in the 'genteel fantasy' this evocation of the supernatural or impossible goes no further. It is merely presented, left for the reader to consider. The new element in fantasy this century, which I see as a parallel development with science fiction and a parallel response to the scientific imagination, is a wish to control, to explore, to discover the rules and absorb the implications of the other world which produces fantastic events. Richard Garnett's 'The Demon Pope' sets an early marker with its celebration of the sorcerer-pope Gerbert's sharp, manipulative, human intelligence. It is interesting also to note the sub-text of John Buchan's 1928 story in this collection. Buchan had started very much in the 'genteel' tradition, and 'The Wind in the Portico' still has several of its tricks, including the clubhouse atmosphere and the characteristic shoulder-shrug of the ending. But what his narrator is obliquely telling the Latin-educated listeners is not just that pagan gods exist, but that there is a procedure for controlling them—a procedure laid down in past ages and now regarded as superstition, but nevertheless practical and authoritative. So why does it not work in the case narrated? Because the squire Dubellay is not enough of a scholar. He makes mistakes in Latin quantities; he 'had never heard of Haverfield' (the major archaeologist of Roman Britain); he cannot handle Sidonius. The story does not say so, but in its gaps one can read the implication that Dubellay called up his god or devil and then mispronounced the spell. The implication *that* leaves behind is that given another chance with the same god—'last year they dug up a temple of his in Wales', remarks the narrator—a better scholar could bring him out of the shadows, make him in a way an object of study (a motif Buchan uses explicitly elsewhere).

Other comparisons with the 'genteel fantasy' include the Lanier and Tiptree stories collected here: one might note the skilful suggestions of history, archaeology, and mythology in the former, the very deliberate running contrast between the garbage-world of *turismo* and the vengeful sea in the latter. Both present not just a 'fantastic event' but also an explanation, a theory. The distinctive feature of modern fantasy as opposed to its pre-modern precursors is in short not just an interest in the supernatural/impossible, but a demand that that element be brought into some accommodation with the rational and the scientific.

One can see that in many ways in this anthology. John Brunner's 'The Wager Lost by Winning' is the third of five in his 'Traveller in Black' sequence, which presents a continuing battle between the forces of Law and Chaos, which Chaos is steadily throughout the sequence losing. The traveller himself, with his magic, is in a sense a product of Chaos, though he acts for Law. The more efficient he is, the sooner he approaches his own destruction; and that anticipated event will create the world we now live in. The sense of a universe bound by laws (including laws of chance, see the scene on pp. 325–31 below), and the strangeness of a universe where laws can be magically set aside, are essential parts of Brunner's fantasy. In a similar way Niven's appropriately titled story 'Not Long Before the End' is set in a magic al universe running down towards its own cancellation; as in a quite different way is the last story in this collection, by Terry Pratchett. Yet another way of handling the science/magic or law/chaos opposition is to do what Poul Anderson and Avram Davidson do in their two stories here, and create a world which is in most ways exactly parallel to the world here, except that in it magic has taken the place of science. So in Davidson we have a story of Presidential inauguration and office-politics embellished with dragons, cockatrices, witches, and industrial alchemy. Anderson gives us a military setting, with air cover, armour, signals, cavalry, and reconnaissance; but all these familiar functions are provided by, respectively, broomsticks, dragons, crystal balls, virginal unicorn-riders, and werewolves. The comedy of the situation—both stories play for comedy—lies in the ingeniousness with which a logical parallel can be extended. On the 'Contents Page' of the August 1940 issue of *Unknown*, the again significantly titled novella 'The Mathematics of Magic', by L. Sprague de Camp and Fletcher Pratt, was advertised as follows: 'What the old magicians lacked was the Scientific Method; what modern science lacks is magic. Blend the two . . .' The advice became a recipe for many authors.

Yet the most obvious expression of conflict or tension between the rational and the fantastic is in the war between 'swordsman' and 'sorcerer'. This has given a name to a whole fantasy sub-genre, 'sword and sorcery', and stories of this type make up perhaps a third of the contents of this book, from Dunsany in 1908 all the way through (if in parody) to Pratchett in 1992. One has to ask what is the inner spring of this continually repeated motif? An obvious clue lies in the fact that in all the early versions of the contest—here in Dunsany, Howard, Moore, and Leiber—the rule is that *sorcerers always lose*. The strategies that defeat them are as various as the stories, but one might observe that what saves Catherine Moore's heroine Jirel, and gives her the power to defeat the sorceress who rules the strange land into which she ventures, is essentially an element of pity for the dying dryad she tries to protect. The same is true (quite unexpectedly) for Howard's normally brutish hero, Conan, faced with an enslaved and tormented alien. By contrast Leiber's pair of heroes, Fafhrd and the Gray Mouser, are saved more directly in their battle on 'the bleak shore' by the Mouser's intelligence and capacity for 'lateral thinking'. In the Dunsany story Leothric conquers through two qualities: his own native doggedness, which keeps him going through the long battle with the almost-invulnerable Tharagavverug, and the very precise and accurate information which he has to be given about his ultimate enemy Gaznac. It is as if the stories are presenting, first, a very evident challenge to the rationalistic world through the sorcerers with their unknown powers and resources; but then in contrast to that a reaffirmation of the ordinary, non-magical, human powers of courage, pity, self-control, common sense.

Even apparent exceptions to the 'sorcerers always lose' rule tend to confirm the thesis that fantasy stories reinforce the powers of Law. Jack Vance's swordsman-hero loses utterly in his contest with Lith, the witch of Thamber Meadow; but then he is not only evidently amoral but worse, unreflective. Like Buchan's squire Dubellay, Liane the Wayfarer does not trouble to get exact information on what he has found, and ignores obvious warning signs from those better educated. In Larry Niven's 'Not Long Before the End', finally, the whole story is presented from the point of view of the sorcerer, while the swordsman-hero who confronts him is recognizably a parodic mixture of Howard's Conan and Michael Moorcock's Elric. But the sorcerer, for all his youth-spells and demon-tattoo, has become something very close to an experimental scientist, who views magic not as being 'magic', immune to calculation, but being a natural resource exactly parallel to (say) geothermal energy. The reason it does not exist now, within a rationalistic universe, is not that it is alien to that universe but that it has been used up. So the sword-and-sorcery contest in that case has been shifted from being one of humanity versus the unknown to being one of intelligence versus force. Whichever line the authors choose to take, it is rare in any of these stories for the alien, irrational, or chaotic to win a victory over the human, sensible, and self-controlled.

The preceding paragraphs may make it sound as if during this century fantasy has survived only by drawing in large elements of the rational/scientific, very probably at the expense of mystery and glamour. The opposition has been a common literary topos from at least the time of Keats's *Lamia* in 1819. It is striking, though, that some branches of the sciences have proved immensely fertile for fantasy writers, giving them entirely new 'imaginative space' for story settings. I have argued elsewhere that it was nineteenth-century philology, with its glimpses of hitherto unknown Dark Ages, which provided the major impulse for writers of extended fantasies such as William Morris and J. R. R. Tolkien: one thing those two writers shared was Mirkwood the Great, rediscovered in Eddic poetry, and then re-imagined as a fictional setting.[6] A parallel science, developing slightly later, was archaeology. In 'The Nameless City' (written apparently in 1921) H. P. Lovecraft's narrator refers repeatedly to ancient cities: to Babylon, to Memnon, to Ur of the Chaldees. Reading these references seventy years later, it is easy not to

[6] See my *The Road to Middle-earth* (2nd edn., London, 1992), and 'Goths and Huns: the rediscovery of the northern cultures in the nineteenth century', in *The Medieval Legacy: A Symposium*, ed. Andreas Haarder (Odense, 1982), 51–69.

realize how up-to-date Lovecraft was. The great excavations which brought Babylon back to light went on from 1899 to 1917. Flinders Petrie's major accounts of the digs at Thebes and Memnon came out in much the same period, but the greatest find of all, the Tutankhamun tomb, was still in the ground as Lovecraft wrote. Meanwhile, Leonard Woolley's excavations of Ur were already famous, though the definitive account of them had not yet appeared. The point is this: the non-human Old Ones in whom Lovecraft's narrator does his best not to believe may indeed be in the realm of the impossible. But in 1921 nameless and undiscovered cities buried in the sand were not. Moreover, if Babylon and Ur, familiar for millennia only as legends, could emerge from legendary status and become matter of fact—then what about Atlantis? Troy after all had been rediscovered (it was thought) by Heinrich Schliemann, while Sir Arthur Evans was digging up Knossos, with finds rapidly labelled as the Temple of Minos and even the Labyrinth. Lovecraft in short had a much more interesting and a much more solidly-based historical time-span to play with than he would have had even thirty years before. There was a certain sense in the alluring thought that maybe there *is* no smoke without fire: behind all the legends of the past, whether Atlantis or Minos or Anubis or King Arthur, there might be a lingering truth (no doubt misrepresented in its passage by the pre-scientific mind). Recovering that truth became a congenial starting-point for the writer of fantasy.

Many authors took advantage of the 'imaginative space' opened up in the past by nineteenth-century scholarship. Tolkien's *Lord of the Rings* is set in far prehistory, though after the fall of Númenor, Tolkien's Atlantis. Long before that work was published (though not before its universe had begun to take shape) Robert E. Howard had created a whole 'Hyborian Age', complete with Picts, Lemurians, Stygians, and Atlanteans, and incorporating a good deal of what was then known or thought about the rise of civilization, while remaining free to add magical and mythical powers now extinct. Several other authors in this volume take the same line (Dunsany, Niven, Brunner). Others colonize the far future, long after the techno-futures of science fiction, in a time when even the sun is fading (Vance or Clark Ashton Smith's story of 'Zothique'). Others take their heroes and heroines through windows or 'gateways' into unknown universes, like Merritt or Moore, or simply set their stories in some unexplained elsewhere, as in Pratchett, or Leiber's tale of 'Nehwon' ('No-when' backwards). A more traditional and semi-Gothic form of marginalization is to set a story on the fringes of our own civilization: this collection includes two 'hillbilly' stories from Kuttner and Wellman, with an interesting dialectal clash from Keith Roberts's Granny Thompson with her defiant East Midlands speech.

Yet fantasy does not have to be forced out on the margins. At least three of the stories in this collection are set not only 'now' (like the dialect stories just mentioned), but also 'here', in the urban environment of most readers: the ones by Peake, Beagle, and Eisenstein. They, and some of the others set 'now' but not exactly 'here', like Lanier's or Tiptree's, raise the question of how far fantasy is an 'escapist' mode, powered by strong rejection of current reality. In a way Eisenstein's story is centrally about 'escapism'. Her protagonist is a paradigm example of modern man—the word 'man' here being chosen deliberately. He is divorced; with limited access to his son; his weekends a drifting up and down, anxious to be with his son but unable to take him home; caught between love, guilt, and fear of further antagonizing the boy's mother. The scenario is familiar, but what the story asks is whether almost anything would not be better—better for the son? Even becoming an animal? With an animal's life-span? The story is hardly 'escapist' in a traditional sense, for it is made clear that the price of this escape is death. At the end of the story even the conventional, prudent, average narrator is considering the choice carefully.

Stories like this could not be written without some deep sense of the dissatisfaction of modern urban life. The theme is taken further in the story by 'James Tiptree' (really Alice Sheldon), a tale which looks forward to revenge for waste and pollution. It is present in Peter Beagle's 'Lila the Werewolf', which ends with werewolves accepted into modern scientific society but prospective mothers-in-law just as psychically dangerous as ever. Finally, in Peake's highly idiosyncratic fable we come upon another inhibited narrator with a fearfully strong sense of the utter tedium of daily life. It is hard not to interpret 'Same Time, Same Place' in a Freudian way: the young male, anxious to escape from the parental home, full of visions of sexuality and maturity, suddenly about to fulfil those visions—and then equally suddenly daunted by a 'true sight' of his bride-to-be. The story appears to be a confession of male fear of women and of sexuality, an explanation which could even apply to the bride's glimpsed companions—the man with the long neck, the goat-hoofed man, the bearded lady, the bald man with the gold teeth and tattoos, respectively a phallic symbol, a satyr-figure, a cross-gender symbol, and perhaps an image (for the middle-class Peake) of social downfall. If that is the case, then the narrator's happy retreat to what left him completely appalled at the start is a kind of ironic reversal/rejection of the structures of traditional fairy-tale, dedicated as they are to the severing of parental ties and the importance of making one's way in the world: a kind of 'escapism' again, but one which, like Eisenstein's, takes that route only after a long cold look at its disadvantages, and a correspondingly strong sense of the horror of the alternative.

One evident fact about the most modern fantasy is that its authors are perfectly capable of working out Freudian interpretations themselves, and of composing deliberate transformations of traditional structures. In their different ways all the stories by Jane Yolen, Angela Carter, and Tanith Lee take up themes and figures from the ancestor-genres—the girl walking through the wood, the dangerous Erl-King, the duke's daughter disguised in rags—and turn them into statements of a kind, about betrayal, authority, hidden dominance, and the price of escape. Before one is tempted to call this an exclusively feminist mode it is worth noting that Thomas Burnett Swann's little-known story 'The Sudden Wings' also edges recognizably towards allegory.

In the end, though, the true appeal of modern fantasy does not lie in what one can get out of it. It lies instead in a tone: appallingly hard to strike truly, correspondingly valued when it is reached. The tone requires a combination of a certain matter-of-fact authorial control with a unpredictable other-worldly glamour. What it does is to suggest the existence of a complete and crowded alternative universe of which the story being told can give only the barest glimpse; a quality resting almost entirely on corner-of-the-eye vision. One sees it strongly in Catherine Moore's tantalizing images of the sorcerers coming from many worlds to observe the trial of Jirel, the spidery thin figure striding across a red desert, the horribly wise black amoeba, the female peacock/bat, and the lily-creature swaying on its stalk and pouring out an ominous light. None of these beings comes *in* to the story at all; they are there only to give a sense of immensity, and to raise unanswerable questions: why they are there, how they are called, how this Sorcerers' Association works across the worlds. In the same way, Vance's short tale printed here raises more questions than it has pages: what is the association between Lith the witch and Chun the Unavoidable?, where is golden Ariventa and how has Lith been exiled from it?, where is the strange blank space into which Liane climbs and why is Chun immune to its spell?, where do the Twk-men come from? All these questions imply that there could be answers, but each answer would only generate another story—and to those stories only the creator of the fantasy-universe has the key. The sense of strong control, of an author prepared to let only relevant bits of information seep out, is established in the genre first by Dunsany with his inimitable ability to suggest a whole deep perspective from a single detail: the stone gargoyles grinning round the Porte Resonant, the vampires flying up and down and giving praise to Satan as they fly, most of all (to my taste) Wong Bongerok the evil, faithful dragon, whose habit it was to 'prophesy quietly to himself' as he lay curled at his master's feet, and who is therefore well aware that he is fated to die even as he exerts all his arts of deception to see that he does not. It is significant that we get exactly the same total control from the otherwise unpretentious voice of Saunk Hogben the hillbilly in Kuttner's story. Saunk is painfully anxious not to put on airs, but his self-corrections cannot help creating the sense of a vast field of unrevealed knowledge behind him: ' 'Course I don't mean the *real* old country. That got sunk.'

'That got sunk' works in its way very like 'So you won't sell me your soul?' They both create a sense of deep vision, of something recondite and unexpected, which nevertheless cannot quite be brought into focus. The recondite, unexpected, sorcerous and magical elements, one might say, represent a reaction to increasingly powerful forces of secularity and rationalism. The never-satisfied but never-abandoned urge to bring them into focus is a product of the rationalist forces themselves.

However that may be, it is quite certain that the modern fantasy genre contains within it a substantial part of the literary effort and literary skill of present-day English-speaking culture, and provides for millions of readers a distinctive literary reward. It is a pity to see this so continually neglected, and even despised, by modern academic criticism. A recent book on fantasy cites on its back cover the names of Dickens, Dostoyevsky, Mary Shelley, James Hogg, E. T. A. Hoffmann, George Eliot, Henry James, Conrad, Stevenson, Kafka, Peake, and Pynchon. Only Peake appears both here and there, while the critical work finds space even in its index for only four more of the authors I have selected. Nor is any more consideration given to the many authors of fantasy who have made their mark through longer works, such as de Camp, Fletcher Pratt, Gene Wolfe, Stephen Donaldson, Marion Zimmer Bradley, or a dozen others. An even more recent study deals with the six hundred plus titles published as fantasy every year by referring vaguely to a 'cult "fantasy" industry' in the United States: 'cult', 'industry', and probably 'United States' are all intended as pejorative terms.

None of the criticism of this type does any justice to the inventiveness, subtlety, and literary skill of scores of authors writing now. It is a major purpose of this anthology to rescue a powerful living tradition from academic marginalization, and to insist that future definitions and discussions of fantasy should pay some attention to common usage and current practice. An equally important goal is to show how fantasy has developed this century, and to give readers of the genre some taste of the many successes that have fallen out of print or missed out on fame.

RICHARD GARNETT

THE DEMON POPE

'So you won't sell me your soul?' said the Devil.

'Thank you,' replied the student, 'I had rather keep it myself, if it's all the same to you.'

'But it's not all the same to me. I want it very particularly. Come, I'll be liberal. I said twenty years. You can have thirty.'

The student shook his head.

'Forty!'

Another shake.

'Fifty!'

As before.

'Now,' said the Devil, 'I know I'm going to do a foolish thing, but I cannot bear to see a clever, spirited young man throw himself away. I'll make you another kind of offer. We won't have any bargain at present, but I will push you on in the world for the next forty years. This day forty years I come back and ask you for a boon; not your soul, mind, or anything not perfectly in your power to grant. If you give it, we are quits; if not, I fly away with you. What say you to this?'

The student reflected for some minutes. 'Agreed,' he said at last.

Scarcely had the Devil disappeared, which he did instantaneously, ere a messenger reined in his smoking steed at the gate of the University of Cordova (the judicious reader will already have remarked that Lucifer could never have been allowed inside a Christian seat of learning), and, enquiring for the student Gerbert, presented him with the Emperor Otho's nomination to the Abbacy of Bobbio, in consideration, said the document, of his virtue and learning, wellnigh miraculous in one so young. Such messengers were frequent visitors during Gerbert's prosperous career. Abbot, Bishop, Archbishop, Cardinal, he was ultimately enthroned Pope on 2 April 999, and assumed the appellation of Silvester the Second. It was then a general belief that the world would come to an end in the following year, a catastrophe which to many seemed the more imminent from the election of a chief pastor whose celebrity as a theologian, though not inconsiderable, by no means equalled his reputation as a necromancer.

The world, notwithstanding, revolved scathless through the dreaded twelvemonth, and early in the first year of the eleventh century Gerbert was sitting peacefully in his study, perusing a book of magic. Volumes of algebra, astrology, alchemy, Aristotelian philosophy, and other such light reading filled his bookcase; and on a table stood an improved clock of his invention, next to his introduction of the Arabic numerals his chief legacy to posterity. Suddenly a sound of wings was heard, and Lucifer stood by his side.

'It is a long time,' said the fiend, 'since I have had the pleasure of seeing you. I have now called to remind you of our little contract, concluded this day forty years.'

'You remember,' said Silvester, 'that you are not to ask anything exceeding my power to perform.'

'I have no such intention,' said Lucifer. 'On the contrary, I am about to solicit a favour which can be bestowed by you alone. You are Pope, I desire that you would make me a Cardinal.'

'In the expectation, I presume,' returned Gerbert, 'of becoming Pope on the next vacancy.'

'An expectation,' replied Lucifer, 'which I may most reasonably entertain, considering my enormous wealth, my proficiency in intrigue, and the present condition of the Sacred College.'

'You would doubtless,' said Gerbert, 'endeavour to subvert the foundations of the Faith, and, by a course of profligacy and licentiousness, render the Holy See odious and contemptible.'

'On the contrary,' said the fiend, 'I would extirpate heresy, and all learning and knowledge as inevitably tending thereunto. I would suffer no man to read but the priest, and confine his reading to his breviary. I would burn your books together with your bones on the first convenient opportunity. I would observe an austere propriety of conduct, and be especially careful not to loosen one rivet in the tremendous yoke I was forging for the minds and consciences of mankind.'

'If it be so,' said Gerbert, 'let's be off!'

'What!' exclaimed Lucifer, 'you are willing to accompany me to the infernal regions?'

'Assuredly, rather than be accessory to the burning of Plato and Aristotle, and give place to the darkness against which I have been contending all my life.'

'Gerbert,' replied the demon, 'this is arrant trifling. Know you not that no good man can enter my dominions? That, were such a thing possible, my empire would become intolerable to me, and I should be compelled to abdicate?'

'I do know it,' said Gerbert, 'and hence I have been able to receive your visit with composure.'

'Gerbert,' said the Devil, with tears in his eyes, 'I put it to you—is this fair, is this honest? I undertake to promote your interests in the world; I fulfil my promise abundantly. You obtain through my instrumentality a position to which you could never otherwise have aspired. Often have I had a hand in the election of a Pope, but never before have I contributed to confer the tiara on one eminent for virtue and learning. You profit by my assistance to the full, and now take advantage of an adventitious circumstance to deprive me of my reasonable guerdon. It is my constant experience that the good people are much more slippery than the sinners, and drive much harder bargains.'

'Lucifer,' answered Gerbert, 'I have always sought to treat you as a gentleman, hoping that you would approve yourself such in return. I will not enquire whether it was entirely in harmony with this character to seek to intimidate me into compliance with your demand by threatening me with a penalty which you well knew could not be enforced. I will overlook this little irregularity, and concede even more than you have requested. You have asked to be a Cardinal. I will make you Pope—'

'Ha!' exclaimed Lucifer, and an internal glow suffused his sooty hide, as the light of a fading ember is revived by breathing upon it.

'—for twelve hours,' continued Gerbert. 'At the expiration of that time we will consider the matter further; and if, as I anticipate, you are more anxious to divest yourself of the Papal dignity than you were to assume it, I promise to bestow upon you any boon you may ask within my power to grant, and not plainly inconsistent with religion or morals.'

'Done!' cried the demon. Gerbert uttered some cabalistic words, and in a moment the apartment held two Pope Silvesters, entirely indistinguishable save by their attire, and the fact that one limped slightly with the left foot.

'You will find the Pontifical apparel in this cupboard,' said Gerbert, and, taking his book of magic with him, he retreated through a masked door to a secret chamber. As the door closed behind him he chuckled, and muttered to himself, 'Poor old Lucifer! Sold again!'

If Lucifer was sold he did not seem to know it. He approached a large slab of silver which did duty as a mirror, and contemplated his personal appearance with some dissatisfaction.

'I certainly don't look half so well without my horns,' he soliloquized, 'and I am sure I shall miss my tail most grievously.'

A tiara and a train, however, made fair amends for the deficient appendages, and Lucifer now looked every inch a Pope. He was about to call the master of the ceremonies, and summon a consistory, when the door was burst open, and seven Cardinals, brandishing poniards, rushed into the room.

'Down with the sorcerer!' they cried, as they seized and gagged him.

'Death to the Saracen!'

'Practises algebra, and other devilish arts!'

'Knows Greek!'

'Talks Arabic!'

'Reads Hebrew!'

'Burn him!'

'Smother him!'

'Let him be deposed by a general council,' said a young and inexperienced Cardinal.

'Heaven forbid!' said an old and wary one, *sotto voce.*

Lucifer struggled frantically, but the feeble frame he was doomed to inhabit for the next eleven hours was speedily exhausted. Bound and helpless, he swooned away.

'Brethren,' said one of the senior Cardinals, 'it hath been delivered by the exorcists that a sorcerer or other individual in league with the demon doth usually bear upon his person some visible token of his infernal compact. I propose that we forthwith institute a search for this stigma, the discovery of which may contribute to justify our proceedings in the eyes of the world.'

'I heartily approve of our brother Anno's proposition,' said another, 'the rather as we cannot possibly fail to discover such a mark, if, indeed, we desire to find it.'

The search was accordingly instituted, and had not proceeded far ere a simultaneous yell from all the seven Cardinals indicated that their investigation had brought more to light than they had ventured to expect.

The Holy Father had a cloven foot!

For the next five minutes the Cardinals remained utterly stunned, silent, and stupefied with amazement. As they gradually recovered their faculties it would have become manifest to a nice observer that the Pope had risen very considerably in their good opinion.

'This is an affair requiring very mature deliberation,' said one.

'I always feared that we might be proceeding too precipitately,' said another.

'It is written, "the devils believe,"' said a third: 'the Holy Father, therefore, is not a heretic at any rate.'

'Brethren,' said Anno, 'this affair, as our brother Benno well remarks, doth indeed call for mature deliberation. I therefore propose that, instead of smothering his Holiness with cushions, as originally contemplated, we immure him for the present in the dungeon adjoining hereunto, and, after spending the night in meditation and prayer, resume the consideration of the business tomorrow morning.'

'Informing the officials of the palace,' said Benno, 'that his Holiness has retired for his devotions, and desires on no account to be disturbed.'

'A pious fraud,' said Anno, 'which not one of the Fathers would for a moment have scrupled to commit.'

The Cardinals accordingly lifted the still insensible Lucifer, and bore him carefully, almost tenderly, to the apartment appointed for his detention. Each would fain have lingered in hopes of his recovery, but each felt that the eyes of his six brethren were upon him: and all, therefore, retired simultaneously, each taking a key of the cell.

Lucifer regained consciousness almost immediately afterwards. He had the most confused idea of the circumstances which had involved him in his present scrape, and could only say to himself that if they were the usual concomitants of the Papal dignity, these were by no means to his taste, and he wished he had been made acquainted with them sooner. The dungeon was not only perfectly dark, but horribly cold, and the poor devil in his present form had no latent store of infernal heat to draw upon. His teeth chattered, he shivered in every limb, and felt devoured with hunger and thirst. There is much probability in the assertion of some of his biographers that it was on this occasion that he invented ardent spirits; but, even if he did, the mere conception of a glass of brandy could only increase his sufferings. So the long January night wore wearily on, and Lucifer seemed likely to expire from inanition, when a key turned in the lock, and Cardinal Anno cautiously glided in, bearing a lamp, a loaf, half a cold roast kid, and a bottle of wine.

'I trust,' he said, bowing courteously, 'that I may be excused any slight breach of etiquette of which I may render myself culpable from the difficulty under which I labour of determining whether, under present circumstances, "Your Holiness," or "Your Infernal Majesty" be the form of address most befitting me to employ.'

'Bub-ub-bub-boo,' went Lucifer, who still had the gag in his mouth.

'Heavens!' exclaimed the Cardinal, 'I crave your Infernal Holiness's forgiveness. What a lamentable oversight!'

And, relieving Lucifer from his gag and bonds, he set out the refection, upon which the demon fell voraciously.

'Why the devil, if I may so express myself,' pursued Anno, 'did not your Holiness inform us that you *were* the devil? Not a hand would then have been raised against you. I have myself been seeking all my life for the audience now happily vouchsafed me. Whence this mistrust of your faithful Anno, who has served you so loyally and zealously these many years?'

Lucifer pointed significantly to the gag and fetters.

'I shall never forgive myself,' protested the Cardinal, 'for the part I have borne in this unfortunate transaction. Next to ministering to your Majesty's bodily necessities, there is nothing I have so much at heart as to express my penitence. But I entreat your Majesty to remember that I believed myself to be acting in your Majesty's interest by overthrowing a magician who was accustomed to send your Majesty upon errands, and who might at any time enclose you in a box, and cast you into the sea. It is deplorable that your Majesty's most devoted servants should have been thus misled.'

'Reasons of State,' suggested Lucifer.

'I trust that they no longer operate,' said the Cardinal. 'However, the Sacred College is now fully possessed of the whole matter: it is therefore unnecessary to pursue this department of the subject further. I would now humbly crave leave to confer with your Majesty, or rather, perhaps, your Holiness, since I am about to speak of spiritual things, on the important and delicate point of your Holiness's successor. I am ignorant how long your Holiness proposes to occupy the Apostolic chair; but of course you are aware that public opinion will not suffer you to hold it for a term exceeding that of the pontificate of Peter. A vacancy, therefore, must one day occur; and I am humbly to represent that the office could not be filled by one more congenial than myself to the present incumbent, or on whom he could more fully rely to carry out in every respect his views and intentions.'

And the Cardinal proceeded to detail various circumstances of his past life, which certainly seemed to corroborate his assertion. He had not, however, proceeded far ere he was disturbed by the grating of another key in the lock, and had just time to whisper impressively, 'Beware of Benno,' ere he dived under a table.

Benno was also provided with a lamp, wine, and cold viands. Warned by the other lamp and the remains of Lucifer's repast that some colleague had been beforehand with him, and not knowing how many more might be in the field, he came briefly to the point as regarded the Papacy, and preferred his claim in much the same manner as Anno. While he was earnestly cautioning Lucifer against this Cardinal as one who could and would cheat the very Devil himself, another key turned in the lock, and Benno escaped under the table, where Anno immediately inserted his finger into his right eye. The little squeal consequent upon this occurrence Lucifer successfully smothered by a fit of coughing.

Cardinal No. 3, a Frenchman, bore a Bayonne ham, and exhibited the same disgust as Benno on seeing himself forestalled. So far as his requests transpired they were moderate, but no one knows where he would have stopped if he had not been scared by the advent of Cardinal No. 4. Up to this time he had only asked for an inexhaustible purse, power to call up the Devil *ad libitum*, and a ring of invisibility to allow him free access to his mistress, who was unfortunately a married woman.

Cardinal No. 4 chiefly wanted to be put into the way of poisoning Cardinal No. 5; and Cardinal No. 5 preferred the same petition as respected Cardinal No. 4.

Cardinal No. 6, an Englishman, demanded the reversion of the Archbishoprics of Canterbury and York, with the faculty of holding them together, and of unlimited non-residence. In the course of his harangue he made use of the phrase *non obstantibus*, of which Lucifer immediately took a note.

What the seventh Cardinal would have solicited is not known, for he had hardly opened his mouth when the twelfth hour expired, and Lucifer, regaining his vigour with his shape, sent the Prince of the Church spinning to the other end of the room, and split the table with a single stroke of his tail. The six crouched and huddling Cardinals cowered revealed to one another, and at the same time enjoyed the spectacle of his Holiness darting through the stone ceiling, which yielded like a film to his passage, and closed up afterwards as if nothing had happened. After the first shock of dismay they unanimously rushed to the door, but found it bolted on the outside. There was no other exit, and no means of giving an alarm. In this emergency the demeanour of the Italian Cardinals set a bright example to their ultra-montane colleagues. '*Bisogna pazienzia*,' they said, as they shrugged their shoulders. Nothing could exceed the mutual politeness of Cardinals Anno and Benno, unless that of the two who had sought to poison each other. The Frenchman was held to have gravely derogated from good manners by alluding to this circumstance, which had reached his ears while he was under the table: and the Englishman swore so outrageously at the plight in which he found himself that the Italians then and there silently registered a vow that none of his nation should ever be Pope, a maxim which, with one exception, has been observed to this day.

Lucifer, meanwhile, had repaired to Silvester, whom he found arrayed in all the insignia of his dignity; of which, as he remarked, he thought his visitor had probably had enough.

'I should think so indeed,' replied Lucifer. 'But at the same time I feel myself fully repaid for all I have undergone by the assurance of the loyalty of my friends and admirers, and the conviction that it is needless for me to devote any considerable amount of personal attention to ecclesiastical affairs. I now claim the promised boon, which it will be in no way inconsistent with thy functions to grant, seeing that it is a work of mercy. I demand that the Cardinals be released, and that their conspiracy against thee, by which I alone suffered, be buried in oblivion.'

'I hoped you would carry them all off,' said Gerbert, with an expression of disappointment.

'Thank you,' said the Devil. 'It is more to my interest to leave them where they are.'

So the dungeon door was unbolted, and the Cardinals came forth, sheepish and crestfallen. If, after all, they did less mischief than Lucifer had expected from them, the cause was their entire bewilderment by what had passed, and their utter inability to penetrate the policy of Gerbert, who henceforth devoted himself even with ostentation to good works. They could never quite satisfy themselves whether they were speaking to the Pope or to the Devil, and when under the latter impression habitually emitted propositions which Gerbert justly stigmatized as rash, temerarious, and scandalous. They plagued him with allusions to certain matters mentioned in their interviews with Lucifer, with which they naturally but erroneously supposed him to be conversant, and worried him by continual nods and titterings as they glanced at his nether extremities. To abolish this nuisance, and at the same time silence sundry unpleasant rumours which had somehow got abroad, Gerbert devised the ceremony of kissing the Pope's feet, which, in a grievously mutilated form, endures to this day. The stupefaction of the Cardinals on discovering that the Holy Father had lost his hoof surpasses all description, and they went to their graves without having obtained the least insight into the mystery.

LORD DUNSANY

THE FORTRESS UNVANQUISHABLE, SAVE FOR SACNOTH

In a wood older than record, a foster brother of the hills, stood the village of Allathurion; and there was peace between the people of that village and all the folk who walked in the dark ways of the wood, whether they were human or of the tribes of the beasts or of the race of the fairies and the elves and the little sacred spirits of trees and streams. Moreover, the village people had peace among themselves and between them and their lord, Lorendiac. In front of the village was a wide and grassy space, and beyond this the great wood again, but at the back the trees came right up to the houses, which, with their great beams and wooden framework and thatched roofs, green with moss, seemed almost to be a part of the forest.

Now in the time I tell of, there was trouble in Allathurion, for of an evening fell dreams were wont to come slipping through the tree trunks and into the peaceful village; and they assumed dominion of men's minds and led them in watches of the night through the cindery plains of Hell. Then the magician of that village made spells against those fell dreams; yet still the dreams came flitting through the trees as soon as the dark had fallen, and led men's minds by night into terrible places and caused them to praise Satan openly with their lips.

And men grew afraid of sleep in Allathurion. And they grew worn and pale, some through the want of rest, and others from fear of the things they saw on the cindery plains of Hell.

Then the magician of the village went up into the tower of his house, and all night long those whom fear kept awake could see his window high up in the night glowing softly alone. The next day, when the twilight was far gone and night was gathering fast, the magician went away to the forest's edge, and uttered there the spell that he had made. And the spell was a compulsive, terrible thing, having a power over evil dreams and over spirits of ill; for it was a verse of forty lines in many languages, both living and dead, and had in it the word wherewith the people of the plains are wont to curse their camels, and the shout wherewith the whalers of the north lure the whales shoreward to be killed, and a word that causes elephants to trumpet; and every one of the forty lines closed with a rhyme for 'wasp'.

And still the dreams came flitting through the forest, and led men's souls into the plains of Hell. Then the magician knew that the dreams were from Gaznak. Therefore he gathered the people of the village, and told them that he had uttered his mightiest spell—a spell having power over all that were human or of the tribes of the beasts; and that since it had not availed the dreams must come from Gaznak, the greatest magician among the spaces of the stars. And he read to the people out of the Book of Magicians, which tells the comings of the comet and foretells his coming again. And he told them how Gaznak rides upon the comet, and how he visits Earth once in every two hundred and thirty years, and makes for himself a vast, invincible fortress and sends out dreams to feed on the minds of men, and may never be vanquished but by the sword Sacnoth.

And a cold fear fell on the hearts of the villagers when they found that their magician had failed them.

Then spake Leothric, son of the Lord Lorendiac, and twenty years old was he: 'Good Master, what of the sword Sacnoth?'

And the village magician answered: 'Fair Lord, no such sword as yet is wrought, for it lies as yet in the hide of Tharagavverug, protecting his spine.'

Then said Leothric: 'Who is Tharagavverug, and where may he be encountered?'

And the magician of Allathurion answered: 'He is the dragon-crocodile who haunts the Northern marshes and ravages the homesteads by their marge. And the hide of his back is of steel, and his under parts are of iron; but along the midst of his back, over his spine, there lies a narrow strip of unearthly steel. This strip of steel is Sacnoth, and it may be neither cleft nor molten, and there is nothing in the world that may avail to break it, nor even leave a scratch upon its surface. It is of the length of a good sword, and of the breadth thereof. Shouldst thou prevail against Tharagavverug, his hide may be melted away from Sacnoth in a furnace; but there is only one thing that may sharpen Sacnoth's edge, and this is one of Tharagavverug's own steel eyes; and the other eye thou must fasten to Sacnoth's hilt, and it will watch for thee. But it is a hard task to vanquish Tharagavverug, for no sword can pierce his hide; his back cannot be broken, and he can neither burn nor drown. In one way only can Tharagavverug die, and that is by starving.'

Then sorrow fell upon Leothric, but the magician spoke on:

'If a man drive Tharagavverug away from his food with a stick for three days, he will starve on the third day at sunset. And though he is not vulnerable, yet in one spot he may take hurt, for his nose is only of lead. A sword would merely lay bare the uncleavable bronze beneath, but if his nose be smitten constantly with a stick he will always recoil from the pain, and thus may Tharagavverug, to left and right, be driven away from his food.'

Then Leothric said: 'What is Tharagavverug's food?'

And the magician of Allathurion said: 'His food is men.'

But Leothric went straightway thence, and cut a great staff from a hazel tree, and slept early that evening. But the next morning, awaking from troubled dreams, he arose before the dawn, and, taking with him provisions for five days, set out through the forest northwards towards the marshes. For some hours he moved through the gloom of the forest, and when he emerged from it the sun was above the horizon shining on pools of water in the waste land. Presently he saw the claw-marks of Tharagavverug deep in the soil, and the track of his tail between them like a furrow in a field. Then Leothric followed the tracks till he heard the bronze heart of Tharagavverug before him, booming like a bell.

And Tharagavverug, it being the hour when he took the first meal of the day, was moving towards a village with his heart tolling. And all the people of the village were come out to meet him, as it was their wont to do; for they abode not the suspense of awaiting Tharagavverug and of hearing him sniffing brazenly as he went from door to door, pondering slowly in his metal mind what habitant he should choose. And none dared to flee, for in the days when the villagers fled from Tharagavverug, he, having chosen his victim, would track him tirelessly, like a doom. Nothing availed them against Tharagavverug. Once they climbed the trees when he came, but Tharagavverug went up to one, arching his back and leaning over slightly, and rasped against the trunk until it fell. And when Leothric came near, Tharagavverug saw him out of one of his small steel eyes and came towards him leisurely, and the echoes of his heart swirled up through his open mouth. And Leothric stepped sideways from his onset, and came between him and the village and smote him on the nose, and the blow of the stick made a dint in the soft lead. And Tharagavverug swung clumsily away, uttering one fearful cry like the sound of a great church bell that had become possessed of a soul that fluttered upward from the tombs at night—an evil soul, giving the bell a voice. Then he attacked Leothric, snarling, and again Leothric leapt aside, and smote him on the nose with his stick. Tharagavverug sounded like a bell howling. And whenever the dragon-crocodile attacked him, or turned towards the village, Leothric smote him again.

So all day long Leothric drove the monster with a stick and he drove him further and further from his prey, with his heart tolling angrily and his voice crying out for pain.

Towards evening Tharagavverug ceased to snap at Leothric, but ran before him to avoid the stick, for his nose was sore and shining; and in the gloaming the villagers came out and danced to cymbal and psaltery. When Tharagavverug heard the cymbal and psaltery, hunger and anger came upon him, and he felt as some lord might feel who was held by force from the banquet in his own castle and heard the creaking spit go round and round and the good meat crackling on it. And all that night he attacked Leothric fiercely, and oft-times nearly caught him in the darkness; for his gleaming eyes of steel could see as well by night as by day. And Leothric gave ground slowly till the dawn, and when the light came they were near the village again; yet not so near to it as they had been when they encountered, for Leothric drove Tharagavverug further in the day than Tharagavverug had forced him back in the night. Then Leothric drove him again with his stick till the hour came when it was the custom of the dragon-crocodile to find his man. One third of his man he would eat at the time he found him, and the rest at noon and evening. But when the hour came for finding his man a great fierceness came on Tharagavverug, and he grabbed rapidly at Leothric, but could not seize him, and for a long while neither of them would retire. But at last the pain of the stick on his leaden nose overcame the hunger of the dragon-crocodile, and he turned from it howling. From that moment Tharagavverug weakened. All that day Leothric drove him with his stick, and at night both held their ground; and when the dawn of the third day was come the heart of Tharagavverug beat slower and fainter. It was as though a tired man was ringing a bell. Once Tharagavverug nearly seized a frog, but Leothric snatched it away just in time. Towards noon the dragon-crocodile lay still for a long while, and Leothric stood near him and leaned on his trusty stick. He was very tired and sleepless, but had more leisure now for eating his provisions. With Tharagavverug the end was coming fast, and in the afternoon his breath came hoarsely, rasping in his throat. It was as the sound of many huntsmen blowing blasts on horns, and towards evening his breath came faster but fainter, like the sound of a hunt going furious to the distance and dying away, and he made desperate rushes towards the village; but Leothric still leapt about him, battering his leaden nose. Scarce audible now at all was the sound of his heart: it was like a church bell tolling beyond hills for the death of some one unknown and far away. Then the sun set and flamed in the village windows, and a chill went over the world, and in some small garden a woman sang; and Tharagavverug lifted up his head and starved, and his life went from his invulnerable body, and Leothric lay down beside him and slept. And later in the starlight the villagers came out and carried Leothric, sleeping, to the village, all praising him in whispers as they went. They laid him down upon a couch in a house, and danced outside in silence, without psaltery or cymbal. And the next day, rejoicing, to Allathurion they hauled the dragon-crocodile. And Leothric went with them, holding his battered staff; and a tall, broad man, who was smith of Allathurion, made a great furnace, and melted Tharagavverug away till only Sacnoth was left, gleaming among the ashes. Then he took one of the small eyes that had been chiselled out, and filed an edge on Sacnoth, and gradually the steel eye wore away facet by facet, but ere it was quite gone it had sharpened redoubtable Sacnoth. But the other eye they set in the butt of the hilt, and it gleamed there bluely.

And that night Leothric arose in the dark and took the sword, and went westwards to find Gaznak; and he went through the dark forest till the dawn, and all the morning and till the afternoon. But in the afternoon he came into the open and saw in the midst of The Land Where No Man Goeth the fortress of Gaznak, mountainous before him, little more than a mile away.

And Leothric saw that the land was marsh and desolate. And the fortress went up all white out of it, with many buttresses, and was broad below but narrowed higher up, and was full of gleaming windows with the light upon them. And near the top of it a few white clouds were floating, but above them some of its pinnacles reappeared. Then Leothric advanced into the marshes, and the eye of Tharagavverug looked out warily from the hilt of Sacnoth; for Tharagavverug had known the marshes well, and the sword nudged Leothric to the right or pulled him to the left away from the dangerous places, and so brought him safely to the fortress walls.

And in the wall stood doors like precipices of steel, all studded with boulders of iron, and above every window were terrible gargoyles of stone; and the name of the fortress shone on the wall, writ large in letters of brass: 'The Fortress Unvanquishable, Save For Sacnoth.'

Then Leothric drew and revealed Sacnoth, and all the gargoyles grinned, and the grin went flickering from face to face right up into the cloud-abiding gables.

And when Sacnoth was revealed and all the gargoyles grinned, it was like the moonlight emerging from a cloud to look for the first time upon a field of blood, and passing swiftly over the wet faces of the slain that lie together in the horrible night. Then Leothric advanced towards a door, and it was mightier than the marble quarry, Sacremona, from which old men cut enormous slabs to build the Abbey of the Holy Tears. Day after day they wrenched out the very ribs of the hill until the Abbey was builded, and it was more beautiful than anything in stone. Then the priests blessed Sacremona, and it had rest, and no more stone was ever taken from it to build the houses of men. And the hill stood looking southwards lonely in the sunlight, defaced by that mighty scar. So vast was the door of steel. And the name of the door was The Porte Resonant, the Way of Egress for War.

Then Leothric smote upon the Porte Resonant with Sacnoth, and the echo of Sacnoth went ringing through the halls, and all the dragons in the fortress barked. And when the baying of the remotest dragon had faintly joined in the tumult, a window opened far up among the clouds below the twilit gables, and a woman screamed, and far away in Hell her father heard her and knew that her doom was come.

And Leothric went on smiting terribly with Sacnoth, and the grey steel of the Porte Resonant, the Way of Egress for War, that was tempered to resist the swords of the world, came away in ringing slices.

Then Leothric, holding Sacnoth in his hand, went in through the hole that he had hewn in the door, and came into the unlit, cavernous hall.

An elephant fled trumpeting. And Leothric stood still, holding Sacnoth. When the sound of the feet of the elephant had died away in remoter corridors, nothing more stirred, and the cavernous hall was still.

Presently the darkness of the distant halls became musical with the sound of bells, all coming nearer and nearer.

Still Leothric waited in the dark, and the bells rang louder and louder, echoing through the halls, and there appeared a procession of men on camels riding two by two from the interior of the fortress, and they were armed with scimitars of Assyrian make and were all clad with mail, and chain-mail hung from their helmets about their faces, and flapped as the camels moved. And they all halted before Leothric in the cavernous hall, and the camel bells clanged and stopped. And the leader said to Leothric:

'The Lord Gaznak has desired to see you die before him. Be pleased to come with us, and we can discourse by the way of the manner in which the Lord Gaznak has desired to see you die.'

And as he said this he unwound a chain of iron that was coiled upon his saddle, and Leothric answered:

'I would fain go with you, for I am come to slay Gaznak.'

Then all the camel-guard of Gaznak laughed hideously, disturbing the vampires that were asleep in the measureless vault of the roof. And the leader said:

'The Lord Gaznak is immortal, save for Sacnoth, and weareth armour that is proof even against Sacnoth himself, and hath a sword the second most terrible in the world.'

Then Leothric said: 'I am the Lord of the sword Sacnoth.'

And he advanced towards the camel-guard of Gaznak, and Sacnoth lifted up and down in his hand as though stirred by an exultant pulse. Then the camel-guard of Gaznak fled, and the riders leaned forward and smote their camels with whips, and they went away with a great clamour of bells through colonnades and corridors and vaulted halls, and scattered into the inner darknesses of the fortress. When the last sound of them had died away, Leothric was in doubt which way to go, for the camel-guard was dispersed in many directions, so he went straight on till he came to a great stairway in the midst of the hall. Then Leothric set his foot in the middle of a wide step, and climbed steadily up the stairway for five minutes. Little light was there in the great hall through which Leothric ascended, for it only entered through arrow slits here and there, and in the world outside evening was waning fast. The stairway led up to two folding doors, and they stood a little ajar, and through the crack Leothric entered and tried to continue straight on, but could get no further, for the whole room seemed to be full of festoons of ropes which swung from wall to wall and were looped and draped from the ceiling. The whole chamber was thick and black with them. They were soft and light to the touch, like fine silk, but Leothric was unable to break any one of them, and though they swung away from him as he pressed forward, yet by the time he had gone three yards they were all about him like a heavy cloak. Then Leothric stepped back and drew Sacnoth, and Sacnoth divided the ropes without a sound, and without a sound the severed pieces fell to the floor. Leothric went forward slowly, moving Sacnoth in front of him up and down as he went. When he was come into the middle of the chamber, suddenly, as he parted with Sacnoth a great hammock of strands, he saw a spider before him that was larger than a ram, and the spider looked at him with eyes that were little, but in which there was much sin, and said:

'Who are you that spoil the labour of years all done to the honour of Satan?'

And Leothric answered: 'I am Leothric, son of Lorendiac.'

And the spider said: 'I will make a rope at once to hang you with.'

Then Leothric parted another bunch of strands, and came nearer to the spider as he sat making his rope, and the spider, looking up from his work, said: 'What is that sword which is able to sever my ropes?'

And Leothric said: 'It is Sacnoth.'

Thereat the black hair that hung over the face of the spider parted to left and right, and the spider frowned: then the hair fell back into its place, and hid everything except the sin of the little eyes which went on gleaming lustfully in the dark. But before Leothric could reach him, he climbed away with his hands, going up by one of his ropes to a lofty rafter, and there sat, growling. But clearing his way with Sacnoth, Leothric passed through the chamber, and came to the further door; and the door being shut, and the handle far up out of his reach, he hewed his way through it with Sacnoth in the same way as he had through the Porte Resonant, the Way of Egress for War. And so Leothric came into a well-lit chamber, where Queens and Princes were banqueting together, all at a great table; and thousands of candles were glowing all about, and their light shone in the wine that the Princes drank and on the huge gold candelabra, and the royal faces were irradiant with the glow, and the white table-cloth and the silver plates and the jewels in the hair of the Queens, each jewel having a historian all to itself, who wrote no other chronicles all his days. Between the table and the door there stood two hundred footmen in two rows of one hundred facing one another. Nobody looked at Leothric as he entered through the hole in the door, but one of the Princes asked a question of a footman, and the question was passed from mouth to mouth by all the hundred footmen till it came to the last one nearest Leothric; and he said to Leothric, without looking at him:

'What do you seek here?'

And Leothric answered: 'I seek to slay Gaznak.'

And footman to footman repeated all the way to the table: 'He seeks to slay Gaznak.'

And another question came down the line of footmen: 'What is your name?'

And the line that stood opposite took his answer back.

Then one of the Princes said: 'Take him away where we shall not hear his screams.'

And footman repeated it to footman till it came to the last two, and they advanced to seize Leothric.

Then Leothric showed to them his sword, saying, 'This is Sacnoth,' and both of them said to the man nearest: 'It is Sacnoth,' then screamed and fled away.

And two by two, all up the double line, footman to footman repeated: 'It is Sacnoth,' then screamed and fled, till the last two gave the message to the table, and all the rest had gone. Hurriedly then arose the Queens and Princes, and fled out of the chamber. And the goodly table, when they were all gone, looked small and disorderly and awry. And to Leothric, pondering in the desolate chamber by what door he should pass onwards, there came from far away the sounds of music, and he knew that it was the magical musicians playing to Gaznak while he slept.

Then Leothric, walking towards the distant music, passed out by the door opposite to the one through which he had cloven his entrance, and so passed into a chamber vast as the other, in which were many women, weirdly beautiful. And they all asked him of his quest, and when they heard that it was to slay Gaznak, they all besought him to tarry among them, saying that Gaznak was immortal, save for Sacnoth, and also that they had need of a knight to protect them from the wolves that rushed round and round the wainscot all the night and sometimes broke in upon them through the mouldering oak. Perhaps Leothric had been tempted to tarry had they been human women, for theirs was a strange beauty, but he perceived that instead of eyes they had little flames that flickered in their sockets, and knew them to be the fevered dreams of Gaznak. Therefore he said:

'I have a business with Gaznak and with Sacnoth,' and passed on through the chamber.

And at the name of Sacnoth those women screamed, and the flames of their eyes sank low and dwindled to sparks.

And Leothric left them, and, hewing with Sacnoth, passed through the further door.

Outside he felt the night air on his face, and found that he stood upon a narrow way between two abysses. To left and right of him, as far as he could see, the walls of the fortress ended in a profound precipice, though the roof still stretched above him; and before him lay the two abysses full of stars, for they cut their way through the whole Earth and revealed the under sky; and threading its course between them went the way, and it sloped upward and its sides were sheer. And beyond the abysses, where the way led up to the further chambers of the fortress, Leothric heard the musicians playing their magical tune. So he stepped on to the way, which was scarcely a stride in width, and moved along it holding Sacnoth naked. And to and fro beneath him in each abyss whirred the wings of vampires passing up and down, all giving praise to Satan as they flew. Presently he perceived the dragon Thok lying upon the way, pretending to sleep, and his tail hung down into one of the abysses.

And Leothric went towards him, and when he was quite close Thok rushed at Leothric.

And he smote deep with Sacnoth, and Thok tumbled into the abyss, screaming, and his limbs made a whirring in the darkness as he fell, and he fell till his scream sounded no louder than a whistle and then could be heard no more. Once or twice Leothric saw a star blink for an instant and reappear again, and this momentary eclipse of a few stars was all that remained in the world of the body of Thok. And Lunk, the brother of Thok, who had lain a little behind him, saw that this must be Sacnoth and fled lumbering away. And all the while that he walked between the abysses, the mighty vault of the roof of the fortress still stretched over Leothric's head, all filled with gloom. Now, when the further side of the abyss came into view, Leothric saw a chamber that opened with innumerable arches upon the twin abysses, and the pillars of the arches went away into the distance and vanished in the gloom to left and right.

Far down the dim precipice on which the pillars stood he could see windows small and closely barred, and between the bars there showed at moments, and disappeared again, things that I shall not speak of.

There was no light here except for the great Southern stars that shone below the abysses, and here and there in the chamber through the arches lights that moved furtively without the sound of footfall.

Then Leothric stepped from the way, and entered the great chamber.

Even to himself he seemed but a tiny dwarf as he walked under one of those colossal arches.

The last faint light of evening flickered through a window painted in sombre colours commemorating the achievements of Satan upon Earth. High up in the wall the window stood, and the streaming lights of candles lower down moved stealthily away.

Other light there was none, save for a faint blue glow from the steel eye of Tharagavverug that peered restlessly about it from the hilt of Sacnoth.

Heavily in the chamber hung the clammy odour of a large and deadly beast.

Leothric moved forward slowly with the blade of Sacnoth in front of him feeling for a foe, and the eye in the hilt of it looking out behind.

Nothing stirred.

If anything lurked behind the pillars of the colonnade that held aloft the roof, it neither breathed nor moved.

The music of the magical musicians sounded from very near.

Suddenly the great doors on the far side of the chamber opened to left and right. For some moments Leothric saw nothing move, and waited clutching Sacnoth. Then Wong Bongerok came towards him, breathing.

This was the last and faithfullest guard of Gaznak, and came from slobbering just now his master's hand.

More as a child than a dragon was Gaznak wont to treat him, giving him often in his fingers tender pieces of man all smoking from his table.

Long and low was Wong Bongerok, and subtle about the eyes, and he came breathing malice against Leothric out of his faithful breast, and behind him roared the armoury of his tail, as when sailors drag the cable of the anchor all rattling down the deck.

And well Wong Bongerok knew that he now faced Sacnoth, for it had been his wont to prophesy quietly to himself for many years as he lay curled at the feet of Gaznak.

And Leothric stepped forward into the blast of his breath, and lifted Sacnoth to strike.

But when Sacnoth was lifted up, the eye of Tharagavverug in the butt of the hilt beheld the dragon and perceived his subtlety.

For he opened his mouth wide, and revealed to Leothric the ranks of his sabre teeth, and his leather gums flapped upwards. But while Leothric made to smite at his head, he shot forward scorpion-wise over his head the length of his armoured tail. All this the eye perceived in the hilt of Sacnoth, who smote suddenly sideways. Not with the edge smote Sacnoth, for, had he done so, the severed end of the tail had still come hurtling on, as some pine tree that the avalanche has hurled point foremost from the cliff right through the broad breast of some mountaineer. So had Leothric been transfixed; but Sacnoth smote sideways with the flat of his blade, and sent the tail whizzing over Leothric's left shoulder; and it rasped upon his armour as it went, and left a groove upon it. Sideways then Leothric smote the foiled tail of Wong Bongerok, and Sacnoth parried, and the tail went shrieking up the blade and over Leothric's head. Then Leothric and Wong Bongerok fought sword to tooth, and the sword smote as only Sacnoth can, and the evil faithful life of Wong Bongerok the dragon went out through the wide wound.

Then Leothric walked on past that dead monster, and the armoured body still quivered a little. And for a while it was like all the ploughshares in a county working together in one field behind tired and struggling horses; then the quivering ceased, and Wong Bongerok lay still to rust.

And Leothric went on to the open gates, and Sacnoth dripped quietly along the floor.

By the open gates through which Wong Bongerok had entered, Leothric came into a corridor echoing with music. This was the first place from which Leothric could see anything above his head, for hitherto the roof had ascended to mountainous heights and had stretched indistinct in the gloom. But along the narrow corridor hung huge bells low and near to his head, and the width of each brazen bell was from wall to wall, and they were one behind the other. And as he passed under each the bell uttered, and its voice was mournful and deep, like to the voice of a bell speaking to a man for the last time when he is newly dead. Each bell uttered once as Leothric came under it, and their voices sounded solemnly and wide apart at ceremonious intervals. For if he walked slow, these bells came closer together, and when he walked swiftly they moved further apart. And the echoes of each bell tolling above his head went on before him whispering to the others. Once when he stopped they all jangled angrily till he went on again.

Between these slow and boding notes came the sound of the magical musicians. They were playing a dirge now very mournfully.

And at last Leothric came to the end of the Corridor of the Bells, and beheld there a small black door. And all the corridor behind him was full of the echoes of the tolling, and they all muttered to one another about the ceremony; and the dirge of the musicians came floating slowly through them like a procession of foreign elaborate guests, and all of them boded ill to Leothric.

The black door opened at once to the hand of Leothric, and he found himself in the open air in a wide court paved with marble. High over it shone the moon, summoned there by the hand of Gaznak.

There Gaznak slept, and around him sat his magical musicians, all playing upon strings. And even sleeping Gaznak was clad in armour, and only his wrists and face and neck were bare.

But the marvel of that place was the dreams of Gaznak; for beyond the wide court slept a dark abyss, and into the abyss there poured a white cascade of marble stairways, and widened out below into terraces and balconies with fair white statues on them, and descended again in a wide stairway, and came to lower terraces in the dark, where swart uncertain shapes went to and fro. All these were the dreams of Gaznak, and issued from his mind, and, becoming marble, passed over the edge of the abyss as the musicians played. And all the while out of the mind of Gaznak, lulled by that strange music, went spires and pinnacles beautiful and slender, ever ascending skywards. And the marble dreams moved slow in time to the music. When the bells tolled and the musicians played their dirge, ugly gargoyles came out suddenly all over the spires and pinnacles, and great shadows passed swiftly down the steps and terraces, and there was hurried whispering in the abyss.

When Leothric stepped from the black door, Gaznak opened his eyes. He looked neither to left nor right, but stood up at once facing Leothric.

Then the magicians played a deathspell on their strings, and there arose a humming along the blade of Sacnoth as he turned the spell aside. When Leothric dropped not down, and they heard the humming of Sacnoth, the magicians arose and fled, all wailing, as they went, upon their strings.

Then Gaznak drew out screaming from its sheath the sword that was the mightiest in the world except for Sacnoth, and slowly walked towards Leothric; and he smiled as he walked, although his own dreams had foretold his doom. And when Leothric and Gaznak came together, each looked at each, and neither spoke a word; but they smote both at once, and their swords met, and each sword knew the other and from whence he came. And whenever the sword of Gaznak smote on the blade of Sacnoth it rebounded gleaming, as hail from off slated roofs; but whenever it fell upon the armour of Leothric, it stripped it off in sheets. And upon Gaznak's armour Sacnoth fell oft and furiously, but ever he came back snarling, leaving no mark behind, and as Gaznak fought he held his left hand hovering close over his head. Presently Leothric smote fair and fiercely at his enemy's neck, but Gaznak, clutching his own head by the hair, lifted it high aloft, and Sacnoth went cleaving through an empty space. Then Gaznak replaced his head upon his neck, and all the while fought nimbly with his sword; and again and again Leothric swept with Sacnoth at Gaznak's bearded neck, and ever the left hand of Gaznak was quicker than the stroke, and the head went up and the sword rushed vainly under it.

And the ringing fight went on till Leothric's armour lay all round him on the floor and the marble was splashed with his blood, and the sword of Gaznak was notched like a saw from meeting the blade of Sacnoth. Still Gaznak stood unwounded and smiling still.

At last Leothric looked at the throat of Gaznak and aimed with Sacnoth, and again Gaznak lifted his head by the hair; but not at his throat flew Sacnoth, for Leothric struck instead at the lifted hand, and through the wrist of it went Sacnoth whirring, as a scythe goes through the stem of a single flower.

And bleeding, the severed hand fell to the floor; and at once blood spurted from the shoulders of Gaznak and dripped from the fallen head, and the tall pinnacles went down into the earth, and the wide fair terraces all rolled away, and the court was gone like the dew, and a wind came and the colonnades drifted thence, and all the colossal halls of Gaznak fell. And the abysses closed up suddenly as the mouth of a man who, having told a tale, will for ever speak no more.

Then Leothric looked around him in the marshes where the night mist was passing away, and there was no fortress nor sound of dragon or mortal, only beside him lay an old man, wizened and evil and dead, whose head and hand were severed from his body.

And gradually over the wide lands the dawn was coming up, and ever growing in beauty as it came, like to the peal of an organ played by a master's hand, growing louder and lovelier as the soul of the master warms, and at last giving praise with all its mighty voice.

Then the birds sang, and Leothric went homeward, and left the marshes and came to the dark wood, and the light of the dawn ascending lit him upon his way. And into Allathurion he came ere noon, and with him brought the evil wizened head, and the people rejoiced, and their nights of trouble ceased.

This is the tale of the vanquishing of The Fortress Unvanquishable, Save For Sacnoth, and of its passing away, as it is told and believed by those who love the mystic days of old.

Others have said, and vainly claim to prove, that a fever came to Allathurion, and went away; and that this same fever drove Leothric into the marshes by night, and made him dream there and act violently with a sword.

And others again say that there hath been no town of Allathurion, and that Leothric never lived.

Peace to them. The gardener hath gathered up this autumn's leaves. Who shall see them again, or who wot of them? And who shall say what hath befallen in the days of long ago?

THROUGH THE DRAGON GLASS

Herndon helped loot the Forbidden City when the Allies turned the suppression of the Boxers into the most gorgeous burglar-party since the days of Tamerlane. Six of his sailormen followed faithfully his buccaneering fancy. A sympathetic Russian highness whom he had entertained in New York saw to it that he got to the coast and his yacht. That is why Herndon was able to sail through the Narrows with as much of the Son of Heaven's treasures as the most accomplished labourer in Peking's mission vineyard.

Some of the loot he gave to charming ladies who had dwelt or were still dwelling on the sunny side of his heart. Most of it he used to fit up those two astonishing Chinese rooms in his Fifth Avenue house. And a little of it, following a vague religious impulse, he presented to the Metropolitan Museum. This, somehow, seemed to put the stamp of legitimacy on his part of the pillage—like offerings to the gods and building hospitals and peace palaces and such things.

But the Dragon Glass, because he had never seen anything quite so wonderful, he set up in his bedroom where he could look at it the first thing in the morning, and he placed shaded lights about it so that he could wake up in the night and look at it! Wonderful? It is more than wonderful, the Dragon Glass! Whoever made it lived when the gods walked about the earth creating something new every day. Only a man who lived in that sort of atmosphere could have wrought it. There was never anything like it.

I was in Hawaii when the cables told of Herndon's first disappearance. There wasn't much to tell. This man had gone to his room to awaken him one morning—and Herndon wasn't there. All his clothes were, though. Everything was just as if Herndon ought to be somewhere in the house—only he wasn't.

A man worth ten millions can't step out into thin air and vanish without leaving behind him the probability of some commotion, naturally. The newspapers attended to the commotion, but the columns of type boiled down to essentials contained just two facts—that Herndon had come home the night before, and in the morning he was undiscoverable.

I was on the high seas, homeward bound to help in the search, when the wireless told the story of his reappearance. They had found him on the floor of his bedroom, shreds of a silken robe on him, and his body mauled as though by a tiger. But there was no more explanation of his return than there had been of his disappearance.

The night before he hadn't been there—and in the morning there he was. Herndon, when he was able to talk, utterly refused to confide in his doctors. I went straight through to New York, and waited until the men of medicine decided that it was better to let him see me than to have him worry any longer about not seeing me.

Herndon got up from a big invalid chair when I entered. His eyes were clear and bright, and there was no weakness in the way he greeted me, nor in the grip of his hand. A nurse slipped from the room.

'What was it, Jim?' I cried. 'What on earth happened to you?'

'Not so sure it was on earth,' he said. He pointed to what looked like a tall easel hooded with a heavy piece of silk covered with embroidered Chinese characters. He hesitated for a moment and then walked over to the closet. He drew out two heavy bore guns, the very ones, I remembered, that he had used in his last elephant hunt.

'You won't think me crazy if I ask you to keep one of these handy while I talk, will you, Ward?' he asked rather apologetically. 'This looks pretty real, doesn't it?'

He opened his dressing gown and showed me his chest swathed in bandages. He gripped my shoulder as I took without question one of the guns. He walked to the easel and drew off the hood.

'There it is,' said Herndon.

And then, for the first time, I saw the Dragon Glass!

There has never been anything like that thing! Never! At first all you saw was a cool, green, glimmering translucence, like the sea when you are swimming under water on a still summer day and look up through it. Around its edges ran flickers of scarlet and gold, flashes of emerald, shimmers of silver and ivory. At its base a disk of topaz rimmed with red fire shot up dusky little vaporous yellow flames.

Afterward you were aware that the green translucence was an oval slice of polished stones. The flashes and flickers became dragons. There were twelve of them. Their eyes were emeralds, their fangs were ivory, their claws were gold. They were scaled dragons and each scale was so inlaid that the base, green as the primeval jungle, shaded off into a vivid scarlet, and the scarlet into tips of gold. Their wings were of silver and vermilion, and were folded close to their bodies.

But they were alive, those dragons. There was never so much life in metal and wood since Al-Akram, the sculptor of ancient Ad, carved the first crocodile, and the jealous Almighty breathed life into it for a punishment!

And last you saw that the topaz disc that sent up little yellow flames was the top of a metal sphere around which coiled a thirteenth dragon, thin and red, and biting its scorpion-tipped tail.

It took your breath away, the first glimpse of the Dragon Glass. Yes, and the second and third glimpse, too—and every other time you looked at it.

'Where did you get it?' I asked, a little shakily.

Herndon said evenly: 'It was in a small hidden crypt in the Imperial Palace. We broke into the crypt quite by'—he hesitated—'well, call it by accident. As soon as I saw it I knew I must have it. What do you think of it?'

'Think!' I cried. 'Think! Why, it's the most marvellous thing that the hands of man ever made! What is that stone? Jade?'

'I'm not sure,' said Herndon. 'But come here. Stand just in front of me.'

He switched out the lights in the room. He turned another switch, and on the glass opposite me three shaded electrics threw their rays into its mirror-like oval.

'Watch!' said Herndon. 'Tell me what you see!'

I looked into the glass. At first I could see nothing but the rays shining further, further—into infinite distances, it seemed. And then—

'Good God!' I cried, stiffening with horror. 'Jim, what hellish thing is this?'

'Steady, old man,' came Herndon's voice. There was relief and a curious sort of joy in it. 'Steady; tell me just what you see.'

I said: 'I seem to see through infinite distances—and yet what I see is as close to me as though it were just on the other side of the glass. I see a cleft that cuts through two masses of darker green. I see a claw, a gigantic, hideous claw that stretches out through the cleft. The claw has seven talons that open and close—open and close. Good God, such a claw, Jim! It is like the claws that reach out from the holes in the lama's hell to grip the blind souls as they shudder by!'

'Look, look further, up through the cleft, above the claw. It widens. What do you see?'

I said: 'I see a peak rising enormously high and cutting the sky like a pyramid. There are flashes of flame that dart from behind and outline it. I see a great globe of light like a moon that moves slowly out of the flashes: there is another moving across the breast of the peak; there is a third that swims into the flame at the furthest edge—'

'The seven moons of Rak,' whispered Herndon, as though to himself. 'The seven moons that bathe in the rose flames of Rak which are the fires of life and that circle Lalil like a diadem. He upon whom the seven moons of Rak have shone is bound to Lalil for this life, and for ten thousand lives.'

He reached over and turned the switch again. The lights of the room sprang up.

'Jim,' I said, 'it can't be real! What is it? Some devilish illusion in the glass?'

He unfastened the bandages about his chest.

'The claw you saw had seven talons,' he answered quietly. 'Well, look at this.'

Across the white flesh of his breast, from left shoulder to the lower ribs on the right, ran seven healing furrows. They looked as though they had been made by a gigantic steel comb that had been drawn across him. They gave one the thought they had been ploughed.

'The claw made these,' he said as quietly as before.

'Ward,' he went on, before I could speak, 'I wanted you to see—what you've seen. I didn't know whether you would see it. I don't know whether you'll believe me even now. I don't suppose I would if I were in your place—still—'

He walked over and threw the hood upon the Dragon Glass.

'I'm going to tell you,' he said. 'I'd like to go through it—uninterrupted. That's why I cover it.

'I don't suppose,' he began slowly—'I don't suppose, Ward, that you've ever heard of Rak the Wonder-Worker, who lived somewhere back at the beginning of things, nor how the Greatest Wonder-Worker banished him somewhere outside the world?'

'No,' I said shortly, still shaken by what I had seen.

'It's a big part of what I've got to tell you,' he went on. 'Of course you'll think it rot, but—I came across the legend in Tibet first. Then I ran across it again—with the names changed, of course—when I was getting away from China.

'I take it that the gods were still fussing around close to man when Rak was born. The story of his parentage is somewhat scandalous. When he grew older, Rak wasn't satisfied with just seeing wonderful things being done. He wanted to do them himself, and he—well, he studied the method. After a while the Greatest Wonder-Worker ran across some of the things Rak had made, and he found them admirable—a little too admirable. He didn't like to destroy the lesser wonder-worker because, so the gossip ran, he felt a sort of responsibility. So he gave Rak a place somewhere—outside the world—and he gave him power over every one out of so many millions of births to lead or lure or sweep that soul into his domain so that he might build up a people—and over his people Rak was given the high, the low, and the middle justice.

'And outside the world Rak went. He fenced his domain about with clouds. He raised a great mountain, and on its flank he built a city for the men and women who were to be his. He circled the city with wonderful gardens, and he placed in the gardens many things, some good and some very—terrible. He set around the mountain's brow seven moons for a diadem, and he fanned behind the mountain a fire which is the fire of life, and through which the moons pass eternally to be born again.'

Herndon's voice sank to a whisper.

'Through which the moons pass,' he said. 'And with them the souls of the people of Rak. They pass through the fires and are born again—and again—for ten thousand lives. I have seen the moons of Rak and the souls that march with them into the fires. There is no sun in the land—only the new-born moons that shine green on the city and on the gardens.'

'Jim,' I cried impatiently. 'What in the world are you talking about? Wake up, man! What's all that nonsense got to do with this?'

I pointed to the hooded Dragon Glass.

'That,' he said. 'Why, through that lies the road to the gardens of Rak!'

The heavy gun dropped from my hand as I stared at him, and from him to the glass and back again. He smiled and pointed to his bandaged breast.

He said: 'I went straight through to Peking with the Allies. I had an idea what was coming, and I wanted to be in at the death. I was among the first to enter the Forbidden City. I was as mad for loot as any of them. It was a maddening sight, Ward. Soldiers with their arms full of precious stuff even Morgan couldn't buy; soldiers with wonderful necklaces around their hairy throats and their pockets stuffed with jewels; soldiers with their shirts bulging treasures the Sons of Heaven had been hoarding for centuries! We were Goths sacking imperial Rome. Alexander's hosts pillaging that ancient gemmed courtesan of cities, royal Tyre! Thieves in the great ancient scale, a scale so great that it raised even thievery up to something heroic.

'We reached the throne-room. There was a little passage leading off to the left, and my men and I took it. We came into a small octagonal room. There was nothing in it except a very extraordinary squatting figure of jade. It squatted on the floor, its back turned toward us. One of my men stooped to pick it up. He slipped. The figure flew from his hand and smashed into the wall. A slab swung outward. By a—well, call it a fluke, we had struck the secret of the little octagonal room!

'I shoved a light through the aperture. It showed a crypt shaped like a cylinder. The circle of the floor was about ten feet in diameter. The walls were covered with paintings, Chinese characters, queer-looking animals, and things I can't well describe. Around the room, about seven feet up, ran a picture. It showed a sort of island floating off into space. The clouds lapped its edges like frozen seas full of rainbows. There was a big pyramid of a mountain rising out of the side of it. Around its peak were seven moons, and over the peak—a face!

'I couldn't place that face and I couldn't take my eyes off it. It wasn't Chinese, and it wasn't of any other race I'd ever seen. It was as old as the world and as young as tomorrow. It was benevolent and malicious, cruel and kindly, merciful and merciless, saturnine as Satan and as joyous as Apollo. The eyes were as yellow as buttercups, or as the sunstone on the crest of the Feathered Serpent they worship down in the Hidden Temple of Tuloon. And they were as wise as Fate.

'"There's something else here, sir," said Martin—you remember Martin, my first officer. He pointed to a shrouded thing on the side. I entered, and took from the thing a covering that fitted over it like a hood. It was the Dragon Glass!

'The moment I saw it I knew I had to have it—and I knew I would have it. I felt I did not want to get the thing away any more than the thing itself wanted to get away. From the first I thought of the Dragon Glass as something alive. Just as much alive as you and I are. Well, I did get it away. I got it down to the yacht, and then the first odd thing happened.

'You remember Wu-Sing, my boat steward? You know the English Wu-Sing talks. Atrocious! I had the Dragon Glass in my stateroom. I'd forgotten to lock the door. I heard a whistle of sharply indrawn breath. I turned, and there was Wu-Sing. Now, you know that Wu-Sing isn't what you'd call intelligent-looking. Yet as he stood there something seemed to pass over his face, and very subtly change it. The stupidity was wiped out as though a sponge had been passed over it. He did not raise his eyes, but he said, in perfect English, mind you: "Has the master augustly counted the cost of his possession?"

'I simply gaped at him.

'"Perhaps,' he continued, "the master has never heard of the illustrious Hao-Tzan? Well, he shall hear."

'Ward, I couldn't move or speak. But I know now it wasn't sheer astonishment that held me. I listened while Wu-Sing went on to tell in polished phrases the same story that I had heard in Tibet, only there they called him Rak instead of Hao-Tzan. But it was the same story.

'"And," he finished, "before he journeyed afar, the illustrious Hao-Tzan caused a great marvel to be wrought. He called it the Gateway!" Wu-Sing waved his hand at the Dragon Glass. "The master has it. But what shall he who has a Gateway do but pass through it? Is it not better to leave the Gateway behind—unless he dare go through it?"

'He was silent. I was silent, too. All I could do was wonder where the fellow had so suddenly got his command of English. And then Wu-Sing straightened. For a moment his eyes looked into mine. They were as yellow as buttercups, Ward, and wise, wise! My mind rushed back to the little room behind the panel. Ward—the eyes of Wu-Sing were the eyes of the face that brooded over the peak of the seven moons!

'And all in a moment, the face of Wu-Sing dropped back into its old familiar stupid lines. The eyes he turned to me were black and clouded. I jumped from my chair.

'"What do you mean, you yellow fraud!" I shouted. "What do you mean by pretending all this time that you couldn't talk English?"

'He looked at me stupidly, as usual. He whined in his pidgin that he didn't understand; that he hadn't spoken a word to me until then. I couldn't get anything else out of him, although I nearly frightened his wits out. I had to believe him. Besides, I had seen his eyes. Well, I was fair curious by this time, and I was more anxious to get the glass home safely than ever.

'I got it home. I set it up here, and I fixed those lights as you saw them. I had a sort of feeling that the glass was waiting—for something. I couldn't tell just what. But that it was going to be rather important, I knew—'

He suddenly thrust his head into his hands, and rocked to and fro.

'How long, how long,' he moaned, 'how long, Santhu?'

'Jim!' I cried. 'Jim! What's the matter with you?'

He straightened. 'In a moment you'll understand,' he said.

And then, as quietly as before: 'I felt that the glass was waiting. The night I disappeared I couldn't sleep. I turned out the lights in the room; turned them on around the glass and sat before it. I don't know how long I sat but all at once I jumped to my feet. The dragons seemed to be moving! They were moving! They were crawling round and round the glass. They moved faster and faster. The thirteenth dragon spun about the topaz globe. They circled faster and faster until they were nothing but a halo of crimson and gold flashes. As they spun, the glass itself grew misty, mistier, mistier still, until it was nothing but a green haze. I stepped over to touch it. My hand went straight on through it as though nothing were there.

'I reached in—up to the elbow, up to the shoulder. I felt my hand grasped by warm little fingers. I stepped through—'

'Stepped through the glass?' I cried.

'Through it,' he said, 'and then—I felt another little hand touch my face. I saw Santhu!

'Her eyes were as blue as the cornflowers, as blue as the big sapphire that shines in the forehead of Vishnu, in his temple at Benares. And they were set wide, wide apart. Her hair was blue-black, and fell in two long braids between her little breasts. A golden dragon crowned her, and through its paws slipped the braids. Another golden dragon girded her. She laughed into my eyes, and drew my head down until my lips touched hers. She was lithe and slender and yielding as the reeds that grow before the Shrine of Hathor that stands on the edge of the Pool of Djeeba. Who Santhu is, or where she came from—how do I know? But this I know—she is lovelier than any woman who ever lived on earth. And she is a woman!

'Her arms slipped from about my neck and she drew me forward. I looked about me. We stood in a cleft between two great rocks. The rocks were a soft green, like the green of the Dragon Glass. Behind us was a green mistiness. Before us the cleft ran only a little distance. Through it I saw an enormous peak jutting up like a pyramid, high, high into a sky of chrysoprase. A soft rose radiance pulsed at its sides, and swimming slowly over its breast was a huge globe of green fire. The girl pulled me gently towards the opening. We walked on silently, hand in hand. Quickly it came to me—Ward, I was in the place whose pictures had been painted in the room of the Dragon Glass!

'We came out of the cleft and into a garden. The Gardens of Many-Columned Iram, lost in the desert because they were too beautiful, must have been like that place. There were strange, immense trees whose branches were like feathery plumes and whose plumes shone with fires like those that clothe the feet of Indra's dancers. Strange flowers raised themselves along our path, and their hearts glowed like the glow-worms that are fastened to the rainbow bridge to Asgard. A wind sighed through the plumed trees, and luminous shadows drifted past their trunks. I heard a girl laugh, and the voice of a man singing.

'We went on. Once there was a low wailing far in the garden, and the girl threw herself before me, her arms outstretched. The wailing ceased, and we went on. The mountain grew plainer. I saw another globe of green fire swing out of the rose flashes at the right of the peak. I saw another shining into the glow at the left. There was a curious trail of mist behind it. It was a mist that had tangled in it a multitude of little stars. Everything was bathed in a soft green light—such a light as you would have if you lived within a pale emerald.

'We turned and went along another little trail. The little trail ran up a little hill, and on the hill was a little house. It looked as though it was made of ivory. It was a very odd little house. It was more like the Jain pagodas at Brahmaputra than anything else. The walls glowed as though they were full of light. The girl touched the wall, and a panel slid away. We entered, and the panel closed after us.

'The room was filled with a whispering yellow light. I say whispering because that is how one felt about it. It was gentle and alive. A stairway of ivory ran up to another room above. The girl pressed me towards it. Neither of us had uttered a word. There was a spell of silence upon me. I could not speak. There seemed to be nothing to say. I felt a great rest and a great peace—as though I had come home. I walked up the stairway and into the room above. It was dark except for a bar of green light that came through the long and narrow window. Through it I saw the mountain and its moons. On the floor was an ivory head-rest and some silken cloths. I felt suddenly very sleepy. I dropped to the cloths, and at once was asleep.

'When I awoke the girl with the cornflower eyes was beside me! She was sleeping. As I watched, her eyes opened. She smiled and drew me to her—

'I do not know why, but a name came to me. "Santhu!" I cried. She smiled again, and I knew that I had called her name. It seemed to me that I remembered her, too, out of immeasurable ages. I arose and walked to the window. I looked toward the mountain. There were now two moons on its breast. And then I saw the city that lay on the mountain's flank. It was such a city as you see in dreams, or as the tale-tellers of El-Bahara fashion out of the mirage. It was all of ivory and shining greens and flashing blues and crimsons. I could see people walking about its streets. There came the sound of little golden bells chiming.

'I turned towards the girl. She was sitting up, her hands clasped about her knees, watching me. Love came, swift and compelling. She arose—I took her in my arms—

'Many times the moons circled the mountain, and the mist held the little tangled stars passing with them. I saw no one but Santhu; no thing came near us. The trees fed us with fruits that had in them the very essence of life. Yes, the fruit of the Tree of Life that stood in Eden must have been like the fruit of those trees. We drank of green water that sparkled with green fires, and tasted like the wine Osiris gives the hungry souls in Amenti to strengthen them. We bathed in pools of carved stone that welled with water yellow as amber. Mostly we wandered in the gardens. There were many wonderful things in the gardens. They were very unearthly. There was no day or night. Only the green glow of the ever-circling moons. We never talked to each other. I don't know why. Always there seemed nothing to say.

'Then Santhu began to sing to me. Her songs were strange songs. I could not tell what the words were. But they built up pictures in my brain. I saw Rak the Wonder-Worker fashioning his gardens, and filling them with things beautiful and things—evil. I saw him raise the peak, and knew that it was Lalil; saw him fashion the seven moons and kindle the fires that are the fires of life. I saw him build his city, and I saw men and women pass into it from the world through many gateways.

'Santhu sang—and I knew that the marching stars in the mist were the souls of the people of Rak which sought rebirth. She sang, and I saw myself ages past walking in the city of Rak with Santhu beside me. Her song wailed, and I felt myself one of the mist-entangled stars. Her song wept, and I felt myself a star that fought against the mist, and, fighting, break away—a star that fled out and out through immeasurable green space—

'A man stood before us. He was very tall. His face was both cruel and kind, saturnine as Satan and joyous as Apollo. He raised his eyes to us, and they were yellow as buttercups, and wise, so wise! Ward, it was the face above the peak in the room of the Dragon Glass! The eyes that had looked at me out of Wu-Sing's face! He smiled on us for a moment and then—he was gone!

'I took Santhu by the hand and began to run. Quite suddenly it came to me that I had enough of the haunted gardens of Rak; that I wanted to get back to my own land. But not without Santhu. I tried to remember the road to the cleft. I felt that there lay the path back. We ran. From far behind came a wailing. Santhu screamed—but I knew the fear in her cry was not for herself. It was for me. None of the creatures of that place could harm her who was herself one of its creatures. The wailing drew closer. I turned.

'Winging down through the green air was a beast, an unthinkable beast, Ward! It was like the winged beast of the Apocalypse that is to bear the woman arrayed in purple and scarlet. It was beautiful even in its horror. It closed its scarlet and golden wings, and its long gleaming body shot at me like a monstrous spear.

'And then—just as it was about to strike—a mist threw itself between us! It was a rainbow mist, and it was—cast. It was cast as though a hand had held it and thrown it like a net. I heard the winged beast shriek its disappointment. Santhu's hand gripped mine tighter. We ran through the mist.

'Before us was the cleft between the two green rocks. Time and time again we raced for it, and time and time again that beautiful shining horror struck at me—and each time came the thrown mist to baffle it. It was a game! Once I heard a laugh, and then I knew who was my hunter. The master of the beast and the caster of the mist. It was he of the yellow eyes—and he was playing with me—playing with me as a child plays with a cat when he tempts it with a piece of meat and snatches the meat away again and again from the hungry jaws!

'The mist cleared away from its last throw, and the mouth of the cleft was just before us. Once more the thing swooped—and this time there was no mist. The player had tired of the game! As it struck, Santhu raised herself before it. The beast swerved—and the claw that had been stretched to rip me from throat to waist struck me a glancing blow. I fell—fell through leagues and leagues of green space.

'When I awoke I was here in this bed with the doctor men around me and this—' He pointed to his bandaged breast again.

'That night when the nurse was asleep I got up and looked into the Dragon Glass, and I saw—the claw, even as you did. The beast is there. It is waiting for me!'

Herndon was silent for a moment.

'If he tires of the waiting he may send the beast through for me,' he said. 'I mean the man with the yellow eyes. I've a desire to try one of these guns on it. It's real, you know, the beast is—and these guns have stopped elephants.'

'But the man with the yellow eyes, Jim,' I whispered— 'who is he?'

'He,' said Herndon—'why, he's the Wonder-Worker himself!'

'You don't believe such a story as that!' I cried. 'Why, it's—it's lunacy! It's some devilish illusion in the glass. It's like the—the crystal globe that makes you hypnotize yourself and think the things your own mind creates are real. Break it, Jim! It's devilish! Break it!'

'Break it!' he said incredulously. 'Break it? Not for the ten thousand lives that are the toll of Rak! Not real? Aren't these wounds real? Wasn't Santhu real? Break it! Good God, man, you don't know what you say! Why, it's my only road back to her! If that yellow-eyed devil back there were only as wise as he looks, he would know he didn't have to keep his beast watching there. I want to go, Ward; I want to go and bring her back with me. I've an idea somehow, that he hasn't—well, full control of things. I've an idea that the Greatest Wonder-Worker wouldn't put wholly in Rak's hands the souls that wander through the many gateways into his kingdom. There's a way out, Ward; there's a way to escape him. I won away from him once, Ward. I'm sure of it. But then I left Santhu behind. I have to go back for her. That's why I found the little passage that led from the throne-room. And he knows it, too. That's why he had to turn his beast on me.

'And I'll go through again, Ward. And I'll come back again—with Santhu!'

But he has not returned. It is six months now since he disappeared for the second time. And from his bedroom, as he had done before. By the will that they found—the will that commanded that in the event of his disappearing as he had done before and not returning within a week, I was to have his house and all that was within it—I came into possession of the Dragon Glass. The dragons had spun again for Herndon, and he had gone through the gateway once more. I found only one of the elephant guns, and I knew that he had had time to take the other with him.

I sit night after night before the glass, waiting for him to come back through it—with Santhu. Sooner or later they will come. That I know.

H. P. LOVECRAFT

THE NAMELESS CITY

When I drew nigh the nameless city I knew it was accursed. I was travelling in a parched and terrible valley under the moon, and afar I saw it protruding uncannily above the sands as parts of a corpse may protrude from an ill-made grave. Fear spoke from the age-worn stones of this hoary survivor of the deluge, this great-grandmother of the oldest pyramid; and a viewless aura repelled me and bade me retreat from antique and sinister secrets that no man should see, and no man else had ever dared to see.

Remote in the desert of Araby lies the nameless city, crumbling and inarticulate, its low walls nearly hidden by the sands of uncounted ages. It must have been thus before the first stones of Memphis were laid, and while the bricks of Babylon were yet unbaked. There is no legend so old as to give it a name, or to recall that it was ever alive; but it is told of in whispers around campfires and muttered about by grandmas in the tents of sheiks so that all the tribes shun it without wholly knowing why. It was of this place that Abdul Alhazred the mad poet dreamed on the night before he sang his unexplainable couplet:

> That is not dead which can eternal lie,
> And with strange aeons even death may die.

I should have known that the Arabs had good reason for shunning the nameless city, the city told of in strange tales but seen by no living man, yet I defied them and went into the untrodden waste with my camel. I alone have seen it, and that is why no other face bears such hideous lines of fear as mine; why no other man shivers so horribly when the night wind rattles the windows. When I came upon it in the ghastly stillness of unending sleep it looked at me, chilly from the rays of a cold moon amidst the desert's heat. And as I returned its look I forgot my triumph at finding it, and stopped still with my camel to wait for the dawn.

For hours I waited, till the east grew grey and the stars faded, and the grey turned to roseate light edged with gold. I heard a moaning and saw a storm of sand stirring among the antique stones though the sky was clear and the vast reaches of the desert still. Then suddenly above the desert's far rim came the blazing edge of the sun, seen through the tiny sandstorm which was passing away, and in my fevered state I fancied that from some remote depth there came a crash of musical metal to hail the fiery disc as Memnon hails it from the banks of the Nile. My ears rang and my imagination seethed as I led my camel slowly across the sand to that unvocal stone place; that place too old for Egypt and Meroe to remember; that place which I alone of living men had seen.

In and out amongst the shapeless foundations of houses and places I wandered, finding never a carving or inscription to tell of these men, if men they were, who built this city and dwelt therein so long ago. The antiquity of the spot was unwholesome, and I longed to encounter some sign or device to prove that the city was indeed fashioned by mankind. There were certain *proportions* and *dimensions* in the ruins which I did not like. I had with me many tools, and dug much within the walls of the obliterated edifices; but progress was slow, and nothing significant was revealed. When night and the moon returned I felt a chill wind which brought new fear, so that I did not dare to remain in the city. And as I went outside the antique walls to sleep, a small sighing sandstorm gathered behind me, blowing over the grey stones though the moon was bright and most of the desert still.

I awaked just at dawn from a pageant of horrible dreams, my ears ringing as from some metallic peal. I saw the sun peering redly through the last gusts of a little sandstorm that hovered over the nameless city, and marked the quietness of the rest of the landscape. Once more I ventured within those brooding ruins that swelled beneath the sand like an ogre under a coverlet, and again dug vainly for relics of the forgotten race. At noon I rested, and in the afternoon I spent much time tracing the walls and bygone streets, and the outlines of the nearly vanished buildings. I saw that the city had been mighty indeed, and wondered at the sources of its greatness. To myself I pictured all the splendours of an age so distant that Chaldea could not recall it, and thought of Sarnath the Doomed, that stood in the land Mnar when mankind was young, and of Ib, that was carven of grey stone before mankind existed.

All at once I came upon a place where the bedrock rose stark through the sand and formed a low cliff; and here I saw with joy what seemed to promise further traces of the antediluvian people. Hewn rudely on the face of the cliff were the unmistakable façades of several small, squat rock houses or temples; whose interiors might preserve many secrets of ages too remote for calculation, though sandstorms had long since effaced any carvings which may have been outside.

Very low and sand-choked were all of the dark apertures near me, but I cleared one with my spade and crawled through it, carrying a torch to reveal whatever mysteries it might hold. When I was inside I saw that the cavern was indeed a temple, and beheld plain signs of the race that had lived and worshipped before the desert was a desert. Primitive altars, pillars, and niches, all curiously low, were not absent; and though I saw no sculptures nor frescos, there were many singular stones clearly shaped into symbols by artificial means. The lowness of the chiselled chamber was very strange, for I could hardly kneel upright; but the area was so great that my torch showed only part of it at a time. I shuddered oddly in some of the far corners; for certain altars and stones suggested forgotten rites of terrible, revolting, and inexplicable nature and made me wonder what manner of men could have made and frequented such a temple. When I had seen all that the place contained, I crawled out again, avid to find what the other temples might yield.

Night had now approached, yet the tangible things I had seen made curiosity stronger than fear, so that I did not flee from the long moon-cast shadows that had daunted me when first I saw the nameless city. In the twilight I cleared another aperture and with a new torch crawled into it, finding more vague stones and symbols, though nothing more definite than the other temple had contained. The room was just as low, but much less broad, ending in a very narrow passage crowded with obscure and cryptical shrines. About these shrines I was prying when the noise of a wind and my camel outside broke through the stillness and drew me forth to see what could have frightened the beast.

The moon was gleaming vividly over the primitive ruins, lighting a dense cloud of sand that seemed blown by a strong but decreasing wind from some point along the cliff ahead of me. I knew it was this chilly, sandy wind which had disturbed the camel and was about to lead him to a place of better shelter when I chanced to glance up and saw that there was no wind atop the cliff. This astonished me and made me fearful again, but I immediately recalled the sudden local winds that I had seen and heard before at sunrise and sunset, and judged it was a normal thing. I decided it came from some rock fissure leading to a cave, and watched the troubled sand to trace it to its source; soon perceiving that it came from the black orifice of a temple a long distance south of me, almost out of sight. Against the choking sand-cloud I plodded toward this temple, which as I neared it loomed larger than the rest, and showed a doorway far less clogged with caked sand. I would have entered had not the terrific force of the icy wind almost quenched my torch. It poured madly out of the dark door, sighing uncannily as it ruffled the sand and spread among the weird ruins. Soon it grew fainter and the sand grew more and more still, till finally all was at rest again; but a presence seemed stalking among the spectral stones of the city, and when I glanced at the moon it seemed to quiver as though mirrored in unquiet waters. I was more afraid than I could explain, but not enough to dull my thirst for wonder; so as soon as the wind was quite gone I crossed into the dark chamber from which it had come.

This temple, as I had fancied from the outside, was larger than either of those I had visited before; and was presumably a natural cavern since it bore winds from some region beyond. Here I could stand quite upright, but saw that the stones and altars were as low as those in the other temples. On the walls and roof I beheld for the first time some traces of the pictorial art of the ancient race, curious curling streaks of paint that had almost faded or crumbled away; and on two of the altars I saw with rising excitement a maze of well-fashioned curvilinear carvings. As I held my torch aloft it seemed to me that the shape of the roof was too regular to be natural, and I wondered what the prehistoric cutters of stone had first worked upon. Their engineering skill must have been vast.

Then a brighter flare of the fantastic flame showed that for which I had been seeking, the opening to those remoter abysses whence the sudden wind had blown; and I grew faint when I saw that it was a small and plainly artificial door chiselled in the solid rock. I thrust my torch within, beholding a black tunnel with the roof arching low over a rough flight of very small, numerous and steeply descending steps. I shall always see those steps in my dreams, for I came to learn what they meant. At the time I hardly knew whether to call them steps or mere footholds in a precipitous descent. My mind was whirling with mad thoughts, and the words and warnings of Arab prophets seemed to float across the desert from the lands that men know to the nameless city that men dare not know. Yet I hesitated only a moment before advancing through the portal and commencing to climb cautiously down the steep passage, feet first, as though on a ladder.

It is only in the terrible phantasms of drugs of delirium that any other man can have such a descent as mine. The narrow passage led infinitely down like some hideous haunted well, and the torch I held about my head could not light the unknown depths towards which I was crawling.

I lost track of the hours and forgot to consult my watch, though I was frightened when I thought of the distance I must be traversing. There were chänges of direction and of steepness; and once I came to a long, low, level passage where I had to wriggle feet first along the rocky floor, holding the torch at arm's length beyond my head. The place was not high enough for kneeling. After that were more of the steep steps, and I was still scrambling down interminably when my failing torch died out. I do not think I noticed it at the time for when I did notice it I was still holding it above me as if it were ablaze. I was quite unbalanced with that instinct for the strange and the unknown which had made me a wanderer upon earth and a haunter of far, ancient, and forbidden places.

In the darkness there flashed before my mind fragments of my cherished treasury of daemoniac lore; sentences from Alhazred the mad Arab, paragraphs from the apocryphal nightmares of Damascius, and infamous lines from the delirious 'Image du Monde' of Gauthier de Metz. I repeated queer extracts, and muttered of Afrasiab and the daemons that floated with him down the Oxus; later chanting over and over again a phrase from one of Lord Dunsany's tales— 'The unreverberate blackness of the abyss.' Once when the descent grew amazingly steep I recited something in sing-song from Thomas Moore until I feared to recite more:

> A reservoir of darkness, black
> As witches' cauldrons are, when fill'd
> With moon-drugs in th' eclipse distill'd.
> Leaning to look if foot might pass
> Down thro' that chasm, I saw, beneath,
> As far as vision could explore,
> The jetty sides as smooth as glass,
> Looking as if just varnish'd o'er
> With that dark pitch the Sea of Death
> Throws out upon its slimy shore.

Time had quite ceased to exist when my feet again felt a level floor, and I found myself in a place slightly higher than the rooms in the two smaller temples now so incalculably far above my head. I could not quite stand, but could kneel upright, and in the dark I shuffled and crept hither and thither at random. I soon knew that I was in a narrow passage whose walls were lined with cases of wood having glass fronts. As in that Palaeozoic and abysmal place I felt of such things as polished wood and glass I shuddered at the possible implications. The cases were apparently ranged along each side of the passage at regular intervals, and were oblong and horizontal, hideously like coffins in shape and size. When I tried to move two or three for further examination, I found that they were firmly fastened.

I saw that the passage was a long one, so floundered ahead rapidly in a creeping run that would have seemed horrible had any eye watched me in the blackness; crossing from side to side occasionally to feel of my surroundings and be sure the walls and rows of cases still stretched on. Man is so used to thinking visually that I almost forgot the darkness and pictured the endless corridor of wood and glass in its low-studded monotony as though I saw it. And then in a moment of indescribable emotion I did see it.

Just when my fancy merged into real sight I cannot tell: but there came a gradual glow ahead, and all at once I knew that I saw the dim outlines of the corridor and the cases, revealed by some unknown subterranean phosphorescence. For a little while all was exactly as I had imagined it, since the glow was very faint; but as I mechanically kept stumbling ahead into the stronger light I realized that my fancy had been but feeble. This hall was no relic of crudity like the temples in the city above, but a monument of the most magnificent and exotic art. Rich, vivid, and daringly fantastic designs and pictures formed a continuous scheme of mural painting whose lines and colours were beyond description. The cases were of a strange golden wood, with fronts of exquisite glass, and containing the mummified forms of creatures outreaching in grotesqueness the most chaotic dreams of man.

To convey any idea of these monstrosities is impossible. They were of the reptile kind, with body lines suggesting sometimes the crocodile, sometimes the seal, but more often nothing of which either the naturalist or the palaeontologist ever heard. In size they approximated a small man, and their forelegs bore delicate and evident feet curiously like human hands and fingers. But strangest of all were their heads, which presented a contour violating all known biological principles. To nothing can such things be well compared—in one flash I thought of comparisons as varied as the cat, the bulldog, the mythic Satyr, and the human being. Not Jove himself had had so colossal and protuberant a forehead, yet the horns and the noselessness and the alligator-like jaw placed the things outside all established categories. I debated for a time on the reality of the mummies, half suspecting they were artificial idols; but soon decided they were indeed some palaeogean species which had lived when the nameless city was alive. To crown their grotesqueness, most of them were gorgeously enrobed in the costliest of fabrics, and lavishly laden with ornaments of gold, jewels, and unknown shining metals.

The importance of these crawling creatures must have been vast, for they held first place among the wild designs on the frescoed walls and ceiling. With matchless skill had the artist drawn them in a world of their own, wherein they had cities and gardens fashioned to suit their dimensions; and I could not help but think that their pictured history was allegorical, perhaps showing the progress of the race that worshipped them. These creatures, I said to myself, were to the men of the nameless city what the she-wolf was to Rome, or some totem-beast is to a tribe of Indians.

Holding this view, I could trace roughly a wonderful epic of the nameless city; the tale of a mighty seacoast metropolis that ruled the world before Africa rose out of the waves, and of its struggles as the sea shrank away, and the desert crept into the fertile valley that held it. I saw its wars and triumphs, its troubles and defeats, and afterward its terrible fight against the desert when thousands of its people—here represented in allegory by the grotesque reptiles—were driven to chisel their way down through the rocks in some marvellous manner to another world whereof their prophets had told them. It was all vividly weird and realistic, and its connection with the awesome descent I had made was unmistakable. I even recognized the passages.

As I crept along the corridor towards the brighter light I saw later stages of the painted epic—the leave-taking of the race that had dwelt in the nameless city and the valley around for ten million years; the race whose souls shrank from quitting scenes their bodies had known so long where they had settled as nomads in the earth's youth, hewing in the virgin rock those primal shrines at which they had never ceased to worship. Now that the light was better I studied the pictures more closely and, remembering that the strange reptiles must represent the unknown men, pondered upon the customs of the nameless city. Many things were peculiar and inexplicable. The civilization, which included a written alphabet, had seemingly risen to a higher order than those immeasurably later civilizations of Egypt and Chaldea, yet there were curious omissions. I could, for example, find no pictures to represent deaths or funeral customs, save such as were related to wars, violence, and plagues; and I wondered at the reticence shown concerning natural death. It was as though an ideal of immortality had been fostered as a cheering illusion.

Still nearer the end of the passage were painted scenes of the utmost picturesqueness and extravagance: contrasted views of the nameless city in its desertion and growing ruin, and of the strange new realm of paradise to which the race had hewed its way through the stone. In these views the city and the desert valley were shown always by moonlight, a golden nimbus hovering over the fallen walls and half-revealing the splendid perfection of former times, shown spectrally and elusively by the artist. The paradisal scenes were almost too extravagant to be believed; portraying a hidden world of eternal day filled with glorious cities and ethereal hills and valleys. At the very last I thought I saw signs of an artistic anticlimax. The paintings were less skilful, and much more bizarre than even the wildest of the earlier scenes. They seemed to record a slow decadence of the ancient stock, coupled with a growing ferocity towards the outside world from which it was driven by the desert. The forms of the people—always represented by the sacred reptiles—appeared to be gradually wasting away, though their spirits shown hovering about the ruins by moonlight gained in proportion. Emaciated priests, displayed as reptiles in ornate robes, cursed the upper air and all who breathed it; and one terrible final scene showed a primitive-looking man, perhaps a pioneer of ancient Irem, the City of Pillars, torn to pieces by members of the elder race. I remembered how the Arabs fear the nameless city, and was glad that beyond this place the grey walls and ceiling were bare.

As I viewed the pageant of mural history I had approached very closely the end of the low-ceilinged hall, and was aware of a gate through which came all of the illuminating phosphorescence. Creeping up to it, I cried aloud in transcendent amazement at what lay beyond; for instead of other and brighter chambers there was only an illimitable void of uniform radiance, such as one might fancy when gazing down from the peak of Mount Everest upon a sea of sunlit mist. Behind me was a passage so cramped that I could not stand upright in it; before me was an infinity of subterranean effulgence.

Reaching down from the passage into the abyss was the head of a steep flight of steps—small numerous steps like those of the black passages I had traversed—but after a few feet the glowing vapours concealed everything. Swung back open against the left-hand wall of the passage was a massive door of brass, incredibly thick and decorated with fantastic bas-reliefs, which could if closed shut the whole inner world of light away from the vaults and passages of rock. I looked at the steps, and for the nonce dared not try them. I touched the open brass door, and could not move it. Then I sank prone to the stone floor, my mind aflame with prodigious reflections which not even a death-like exhaustion could banish.

As I lay still with closed eyes, free to ponder, many things I had lightly noted in the frescos came back to me with new and terrible significance—scenes representing the nameless city in its heyday—the vegetation of the valley around it, and the distant lands with which its merchants traded. The allegory of the crawling creatures puzzled me by its universal prominence, and I wondered that it would be so closely followed in a pictured history of such importance. In the frescos the nameless city had been shown in proportions fitted to the reptiles. I wondered what its real proportions and magnificence had been, and reflected a moment on certain oddities I had noticed in the ruins. I thought curiously of the lowness of the primal temples and of the underground corridor, which were doubtless hewn thus out of deference to the reptile deities there honoured; though it perforce reduced the worshippers to crawling. Perhaps the very rites here involved a crawling in imitation of the creatures. No religious theory, however, could easily explain why the level passages in that awesome descent should be as low as the temples—or lower, since one could not even kneel in it. As I thought of the crawling creatures, whose hideous mummified forms were so close to me, I felt a new throb of fear. Mental associations are curious, and I shrank from the idea that except for the poor primitive man torn to pieces in the last painting, mine was the only human form amidst the many relics and symbols of primordial life.

But as always in my strange and roving existence, wonder soon drove out fear, for the luminous abyss and what it might contain presented a problem worthy of the greatest explorer. That a weird world of mystery lay far down that flight of peculiarly small steps I could not doubt, and I hoped to find there those human memorials which the painted corridor had failed to give. The frescos had pictured unbelievable cities and valleys in this lower realm, and my fancy dwelt on the rich and colossal ruins that awaited me.

My fears, indeed, concerned the past rather than the future. Not even the physical horror of my position in that cramped corridor of dead reptiles and antediluvian frescos, miles below the world I knew and faced by another world of eerie light and mist, could match the lethal dread I felt at the abysmal antiquity of the scene and its soul. An ancientness so vast that measurement is feeble seemed to leer down from the primal stones and rock-hewn temples of the nameless city, while the very latest of the astounding maps in the frescos showed oceans and continents that man has forgotten, with only here and there some vaguely familiar outline. Of what could have happened in the geological ages since the paintings ceased and the death-hating race resentfully succumbed to decay, no man might say. Life had once teemed in these caverns and in the luminous realm beyond; now I was alone with vivid relics, and I trembled to think of the countless ages through which these relics had kept a silent deserted vigil.

Suddenly there came another burst of that acute fear which had intermittently seized me ever since I first saw the terrible valley and the nameless city under a cold moon, and despite my exhaustion I found myself starting frantically to a sitting posture and gazing back along the black corridor towards the tunnels that rose to the outer world. My sensations were like those which had made me shun the nameless city at night, and were as inexplicable as they were poignant. In another moment, however, I received a still greater shock in the form of a definite sound—the first which had broken the utter silence of these tomb-like depths. It was a deep, low moaning, as of a distant throng of condemned spirits, and came from the direction in which I was staring. Its volume rapidly grew, till soon it reverberated frightfully through the low passage, and at the same time I became conscious of an increasing draught of cold air, likewise flowing from the tunnels and the city above. The touch of this air seemed to restore my balance, for I instantly recalled the sudden gusts which had risen around the mouth of the abyss each sunset and sunrise, one of which had indeed revealed the hidden tunnels to me. I looked at my watch and saw that sunrise was near, so braced myself to resist the gale that was sweeping down to its cavern home as it had swept forth at evening. My fear again waned low, since a natural phenomenon tends to dispel broodings over the unknown.

More and more madly poured the shrieking, moaning night wind into that gulf of the inner earth. I dropped prone again and clutched vainly at the floor for fear of being swept bodily through the open gate into the phosphorescent abyss. Such fury I had not expected, and as I grew aware of an actual slipping of my form toward the abyss I was beset by a thousand new terrors of apprehension and imagination. The malignancy of the blast awakened incredible fancies; once more I compared myself shudderingly to the only human image in that frightful corridor, the man who was torn to pieces by the nameless race, for in the fiendish clawing of the swirling currents there seemed to abide a vindictive rage all the stronger because it was largely impotent. I think I screamed frantically near the last—I was almost mad—but if I did so my cries were lost in the hell-born babel of the howling wind-wraiths. I tried to crawl against the murderous invisible torrent, but I could not even hold my own as I was pushed slowly and inexorably towards the unknown world. Finally, reason must have wholly snapped; for I fell to babbling over and over that unexplainable couplet of the mad Arab Alhazred, who dreamed of the nameless city:

> That is not dead which can eternal lie,
> And with strange aeons even death may die.

Only the grim brooding desert gods know what really took place—what indescribable struggles and scrambles in the dark I endured or what Abaddon guided me back to life, where I must always remember and shiver in the night wind till oblivion—or worse—claims me. Monstrous, unnatural, colossal, was the thing too far beyond all the ideas of man to be believed except in the silent damnable small hours of the morning when one cannot sleep.

I have said that the fury of the rushing blast was infernal—cacodaemoniacal—and that its voices were hideous with the pent-up viciousness of desolate eternities. Presently these voices, while still chaotic before me, seemed to my beating brain to take articulate form behind me; and down there in the grave of unnumbered aeon-dead antiquities, leagues below the dawn-lit world of men, I heard the ghastly cursing and snarling of strange-tongued fiends. Turning, I saw outlined against the luminous aether of the abyss that could not be seen against the dusk of the corridor—a nightmare horde of rushing devils; hate-distorted, grotesquely panoplied, half transparent devils of a race no man might mistake—the crawling reptiles of the nameless city.

And as the wind died away I was plunged into the ghoul-pooled darkness of earth's bowels; for behind the last of the creatures the great brazen door clanged shut with a deafening peal of metallic music whose reverberations swelled out to the distant world to hail the rising sun as Memnon hails it from the banks of the Nile.

JOHN BUCHAN

THE WIND IN THE PORTICO

A dry wind of the high places . . . not to fan nor to cleanse, even a full wind from those places shall come unto me.

Jeremiah iv: 11–12

Nightingale was a hard man to draw. His doings with the Bedawin had become a legend, but he would as soon have talked about them as claimed to have won the war. He was a slim dark fellow about thirty-five years of age, very short-sighted, and wearing such high-powered double glasses that it was impossible to tell the colour of his eyes. This weakness made him stoop a little and peer, so that he was the strangest figure to picture in a burnous leading an army of desert tribesmen. I fancy his power came partly from his oddness, for his followers thought that the hand of Allah had been laid on him, and partly from his quick imagination and his flawless courage. After the war he had gone back to his Cambridge fellowship, declaring that, thank God, that chapter in his life was over.

As I say, he never mentioned the deeds which had made him famous. He knew his own business, and probably realized that to keep his mental balance he had to drop the curtain on what must have been the most nerve-racking four years ever spent by man. We respected his decision and kept off Arabia. It was a remark of Hannay's that drew from him the following story. Hannay was talking about his Cotswold house, which was on the Fosse Way, and saying that it always puzzled him how so elaborate a civilization as Roman Britain could have been destroyed utterly and left no mark on the national history beyond a few roads and ruins and place-names. Peckwether, the historian, demurred, and had a good deal to say about how much the Roman tradition was woven into the Saxon culture. 'Rome only sleeps,' he said; 'she never dies.'

Nightingale nodded. 'Sometimes she dreams in her sleep and talks. Once she scared me out of my senses.'

After a good deal of pressing he produced this story. He was not much of a talker, so he wrote it out and read it to us.

There is a place in Shropshire which I do not propose to visit again. It lies between Ludlow and the hills, in a shallow valley full of woods. Its name is St Sant, a village with a big house and park adjoining, on a stream called the Vaun, about five miles from the little town of Faxeter. They have queer names in those parts, and other things queerer than the names.

I was motoring from Wales to Cambridge at the close of the long vacation. All this happened before the war, when I had just got my fellowship and was settling down to academic work. It was a fine night in early October, with a full moon, and I intended to push on to Ludlow for supper and bed. The time was about half-past eight, the road was empty and good going, and I was trundling pleasantly along when something went wrong with my headlights. It was a small thing, and I stopped to remedy it beyond a village and just at the lodge-gates of a house.

On the opposite side of the road a carrier's cart had drawn up, and two men, who looked like indoor servants, were lifting some packages from it on to a big barrow. The moon was up, so I didn't need the feeble light of the carrier's lamp to see what they were doing. I suppose I wanted to stretch my legs for a moment, for when I had finished my job I strolled over to them. They did not hear me coming, and the carrier on his perch seemed to be asleep.

The packages were the ordinary consignments from some big shop in town. But I noticed that the two men handled them very gingerly, and that, as each was laid in the barrow, they clipped off the shop label and affixed one of their own. The new labels were odd things, large and square, with some address written on them in very black capital letters. There was nothing in that, but the men's faces puzzled me. For they seemed to do their job in a fever, longing to get it over and yet in a sweat lest they should make some mistake. Their commonplace task seemed to be for them a matter of tremendous importance. I moved so as to get a view of their faces, and I saw that they were white and strained. The two were of the butler or valet class, both elderly, and I could have sworn that they were labouring under something like fear.

I shuffled my feet to let them know of my presence and remarked that it was a fine night. They started as if they had been robbing a corpse. One of them mumbled something in reply, but the other caught a package which was slipping, and in a tone of violent alarm growled to his mate to be careful. I had a notion that they were handling explosives.

I had no time to waste, so I pushed on. That night, in my room at Ludlow, I had the curiosity to look up my map and identify the place where I had seen the men. The village was St Sant, and it appeared that the gate I had stopped at belonged to a considerable demesne called Vauncastle. That was my first visit.

At that time I was busy on a critical edition of Theocritus, for which I was making a new collation of the manuscripts. There was a variant of the Medicean Codex in England, which nobody had seen since Gaisford, and after a good deal of trouble I found that it was in the library of a man called Dubellay. I wrote to him at his London club, and got a reply to my surprise from Vauncastle Hall, Faxeter. It was an odd letter, for you could see that he longed to tell me to go to the devil, but couldn't quite reconcile it with his conscience. We exchanged several letters, and the upshot was that he gave me permission to examine his manuscript. He did not ask me to stay, but mentioned that there was a comfortable little inn in St Sant.

My second visit began on the 27th of December, after I had been home for Christmas. We had had a week of severe frost, and then it had thawed a little; but it remained bitterly cold, with leaden skies that threatened snow. I drove from Faxeter, and as we ascended the valley I remember thinking that it was a curiously sad country. The hills were too low to be impressive, and their outlines were mostly blurred with woods; but the tops showed clear, funny little knolls of grey bent that suggested a volcanic origin. It might have been one of those backgrounds you find in Italian primitives, with all the light and colour left out. When I got a glimpse of the Vaun in the bleached meadows it looked like the 'wan water' of the Border ballads. The woods, too, had not the friendly bareness of English copses in wintertime. They remained dark and cloudy, as if they were hiding secrets. Before I reached St Sant, I decided that the landscape was not only sad, but ominous.

I was fortunate in my inn. In the single street of one-storeyed cottages it rose like a lighthouse, with a cheery glow from behind the red curtains of the bar parlour. The inside proved as good as the outside. I found a bedroom with a bright fire, and I dined in a wainscoted room full of preposterous old pictures of lanky hounds and hollow-backed horses. I had been rather depressed on my journey, but my spirits were raised by this comfort, and when the house produced a most respectable bottle of port I had the landlord in to drink a glass. He was an ancient man who had been a gamekeeper, with a much younger wife, who was responsible for the management. I was curious to hear something about the owner of my manuscript, but I got little from the landlord. He had been with the old squire, and had never served the present one. I heard of Dubellays in plenty—the landlord's master, who had hunted his own hounds for forty years, the Major his brother, who had fallen at Abu Klea; Parson Jack, who had had the living till he died, and of all kinds of collaterals. The 'Deblays' had been a high-spirited, open-handed stock, and much liked in the place. But of the present master of the Hall he could or would tell me nothing. The Squire was a 'great scholard', but I gathered that he followed no sport and was not a convivial soul like his predecessors. He had spent a mint of money on the house, but not many people went there. He, the landlord, had never been inside the grounds in the new master's time, though in the old days there had been hunt breakfasts on the lawn for the whole countryside, and mighty tenantry dinners. I went to bed with a clear picture in my mind of the man I was to interview on the morrow. A scholarly and autocratic recluse, who collected treasures and beautified his dwelling and probably lived in his library. I rather looked forward to meeting him, for the bonhomous sporting squire was not much in my line.

After breakfast next morning I made my way to the Hall. It was the same leaden weather, and when I entered the gates the air seemed to grow bitterer and the skies darker. The place was muffled in great trees which even in their winter bareness made a pall about it. There was a long avenue of ancient sycamores, through which one caught only rare glimpses of the frozen park. I took my bearings, and realized that I was walking nearly due south, and was gradually descending. The house must be in a hollow. Presently the trees thinned, I passed through an iron gate, came out on a big untended lawn, untidily studded with laurels and rhododendrons, and there before me was the house front.

I had expected something beautiful—an old Tudor or Queen Anne façade or a dignified Georgian portico. I was disappointed, for the front was simply mean. It was low and irregular, more like the back parts of a house, and I guessed that at some time or another the building had been turned round, and the old kitchen door made the chief entrance. I was confirmed in my conclusion by observing that the roofs rose in tiers, like one of those recessed New York skyscrapers, so that the present back parts of the building were of an impressive height.

The oddity of the place interested me, and still more its dilapidation. What on earth could the owner have spent his money on? Everything—

lawn, flower-beds, paths—was neglected. There was a new stone doorway, but the walls badly needed pointing, the window woodwork had not been painted for ages, and there were several broken panes. The bell did not ring, so I was reduced to hammering on the knocker, and it must have been ten minutes before the door opened. A pale butler, one of the men I had seen at the carrier's cart the October before, stood blinking in the entrance.

He led me in without question, when I gave my name, so I was evidently expected. The hall was my second surprise. What had become of my picture of the collector? The place was small and poky, and furnished as barely as the lobby of a farmhouse. The only thing I approved was its warmth. Unlike most English country houses there seemed to be excellent heating arrangements.

I was taken into a little dark room with one window that looked out on a shrubbery, while the man went to fetch his master. My chief feeling was of gratitude that I had not been asked to stay, for the inn was paradise compared with this sepulchre. I was examining the prints on the wall, when I heard my name spoken and turned round to greet Mr Dubellay.

He was my third surprise. I had made a portrait in my mind of a fastidious old scholar, with eye-glasses on a black cord, and a finical *weltkind*-ish manner. Instead I found a man still in early middle age, a heavy fellow dressed in the roughest country tweeds. He was as untidy as his demesne, for he had not shaved that morning, his flannel collar was badly frayed, and his fingernails would have been the better for a scrubbing brush. His face was hard to describe. It was high-coloured, but the colour was not healthy; it was friendly, but it was also wary; above all, it was *unquiet*. He gave me the impression of a man whose nerves were all wrong, and who was perpetually on his guard.

He said a few civil words, and thrust a badly tied brown paper parcel at me.

'That's your manuscript,' he said jauntily.

I was staggered. I had expected to be permitted to collate the codex in his library, and in the last few minutes had realized that the prospect was distasteful. But here was this casual owner offering me the priceless thing to take away.

I stammered my thanks, and added that it was very good of him to trust a stranger with such a treasure.

'Only as far as the inn,' he said. 'I wouldn't like to send it by post. But there's no harm in your working at it at the inn. There should be confidence among scholars.' And he gave an odd cackle of a laugh.

'I greatly prefer your plan,' I said. 'But I thought you would insist on my working at it here.'

'No, indeed,' he said earnestly. 'I shouldn't think of such a thing. . . . Wouldn't do at all. . . . An insult to our freemasonry. . . . That's how I should regard it.'

We had a few minutes' further talk. I learned that he had inherited under the entail from a cousin, and had been just over ten years at Vauncastle. Before that he had been a London solicitor. He asked me a question or two about Cambridge—wished he had been at the University—much hampered in his work by a defective education. I was a Greek scholar?—Latin, too, he presumed. Wonderful people the Romans. . . . He spoke quite freely, but all the time his queer restless eyes were darting about, and I had a strong impression that he would have liked to say something to me very different from these commonplaces—that he was longing to broach some subject but was held back by shyness or fear. He had such an odd appraising way of looking at me.

I left without his having asked me to a meal, for which I was not sorry, for I did not like the atmosphere of the place. I took a short cut over the ragged lawn, and turned at the top of the slope to look back. The house was in reality a huge pile, and I saw that I had been right and that the main building was all at the back. Was it, I wondered, like the Alhambra, which behind a front like a factory concealed a treasure-house? I saw, too, that the woodland hollow was more spacious than I had fancied. The house, as at present arranged, faced due north, and behind the south front was an open space in which I guessed that a lake might lie. Far beyond I could see in the December dimness the lift of high dark hills.

That evening the snow came in earnest, and fell continuously for the better part of two days. I banked up the fire in my bedroom and spent a happy time with the codex. I had brought only my working books with me and the inn boasted no library, so when I wanted to relax I went down to the tap-room, or gossiped with the landlady in the bar parlour. The yokels who congregated in the former were pleasant fellows, but, like all the folk on the Marches, they did not talk readily to a stranger and I heard little from them of the Hall. The old squire had reared every year three thousand pheasants, but the present squire would not allow a gun to be fired on his land and there were only a few wild birds left. For the same reason the woods were thick with vermin. This they told me when I professed an interest in shooting. But of Mr Dubellay they would not speak, declaring that they never saw him. I daresay they gossiped wildly about him, and their public reticence struck me as having in it a touch of fear.

The landlady, who came from a different part of the shire, was more communicative. She had not known the former Dubellays and so had no standard of comparison, but she was inclined to regard the present squire as not quite right in the head. 'They do say,' she would begin, but she, too, suffered from some inhibition, and what promised to be sensational would tail off into the commonplace. One thing apparently puzzled the neighbourhood above others, and that was his rearrangement of the house. 'They do say,' she said in an awed voice, 'that he have built a great church.' She had never visited it—no one in the parish had, for Squire Dubellay did not allow intruders—but from Lyne Hill you could see it through a gap in the woods. 'He's no good Christian,' she told me, 'and him and Vicar has quarrelled this many a day. But they do say as he worships summat there.' I learned that there were no women servants in the house, only the men he had brought from London. 'Poor benighted souls, they must live in a sad hobble,' and the buxom lady shrugged her shoulders and giggled.

On the last day of December I decided that I needed exercise and must go for a long stride. The snow had ceased that morning, and the dull skies had changed to a clear blue. It was still very cold, but the sun was shining, the snow was firm and crisp underfoot, and I proposed to survey the country. So after luncheon I put on thick boots and gaiters, and made for Lyne Hill. This meant a considerable circuit, for the place lay south of the Vauncastle park. From it I hoped to get a view of the other side of the house.

I was not disappointed. There was a rift in the thick woodlands, and below me, two miles off, I suddenly saw a strange building, like a classical temple. Only the entablature and the tops of the pillars showed above the trees, but they stood out vivid and dark against the background of snow. The spectacle in that lonely place was so startling that for a little I could only stare. I remember that I glanced behind me to the snowy line of the Welsh mountains, and felt that I might have been looking at a winter view of the Apennines two thousand years ago.

My curiosity was now alert, and I determined to get a nearer view of this marvel. I left the track and ploughed through the snowy fields down to the skirts of the woods. After that my troubles began. I found myself in a very good imitation of a primeval forest, where the undergrowth had been unchecked and the rides uncut for years. I sank into deep pits, I was savagely torn by briars and brambles, but I struggled on, keeping a line as best I could. At last the trees stopped. Before me was a flat expanse which I knew must be a lake, and beyond rose the temple.

It ran the whole length of the house, and from where I stood it was hard to believe that there were buildings at its back where men dwelt. It was a fine piece of work—the first glance told me that—admirably proportioned, classical, yet not following exactly any of the classical models. One could imagine a great echoing interior dim with the smoke of sacrifice, and it was only by reflecting that I realized that the peristyle could not be continued down the two sides, that there was no interior, and that what I was looking at was only a portico.

The thing was at once impressive and preposterous. What madness had been in Dubellay when he embellished his house with such a grandiose garden front? The sun was setting and the shadow of the wooded hills darkened the interior, so I could not even make out the back wall of the porch. I wanted a nearer view, so I embarked on the frozen lake.

Then I had an odd experience. I was not tired, the snow lay level and firm, but I was conscious of extreme weariness. The biting air had become warm and oppressive. I had to drag boots that seemed to weigh tons across that lake. The place was utterly silent in the stricture of the frost, and from the pile in front no sign of life came.

I reached the other side at last and found myself in a frozen shallow of bulrushes and skeleton willow-herbs. They were taller than my head, and to see the house I had to look upward through their snowy traceries. It was perhaps eighty feet above me and a hundred yards distant, and, since I was below it, the delicate pillars seemed to spring to a great height. But it was still dusky, and the only detail I could see was on the ceiling, which seemed either to be carved or painted with deeply-shaded monochrome figures.

Suddenly the dying sun came slanting through the gap in the hills, and for an instant the whole portico to its furthest recesses was washed in clear gold and scarlet. That was wonderful enough, but there was something more. The air was utterly still with not the faintest breath of wind—so still that when I had lit a cigarette half an hour before the flame of the match had burned steadily upward like a candle in a room. As I stood among the sedges not a single frost crystal stirred. . . . But there was a wind blowing in the portico.

I could see it lifting feathers of snow from the base of the pillars and fluffing the cornices. The floor had already been swept clean, but tiny flakes drifted on to it from the exposed edges. The interior was filled with a furious movement, though a yard from it was frozen peace. I felt nothing of the action of the wind, but I knew that it was hot, hot as the breath of a furnace.

I had only one thought, dread of being overtaken by night near that place. I turned and ran. Ran with labouring steps across the lake, panting and stifling with a deadly hot oppression, ran blindly by a sort of instinct in the direction of the village. I did not stop till I had wrestled through the big wood, and come out on some rough pasture above the highway. Then I dropped on the ground, and felt again the comforting chill of the December air.

The adventure left me in an uncomfortable mood. I was ashamed of myself for playing the fool, and at the same time hopelessly puzzled, for the oftener I went over in my mind the incidents of that afternoon the more I was at a loss for an explanation. One feeling was uppermost, that I did not like this place and wanted to be out of it. I had already broken the back of my task, and by shutting myself up for two days I completed it; that is to say, I made my collation as far as I had advanced myself in my commentary on the text. I did not want to go back to the Hall, so I wrote a civil note to Dubellay, expressing my gratitude and saying that I was sending up the manuscript by the landlord's son, as I scrupled to trouble him with another visit.

I got a reply at once, saying that Mr Dubellay would like to give himself the pleasure of dining with me at the inn before I went, and would receive the manuscript in person.

It was the last night of my stay in St Sant, so I ordered the best dinner the place could provide, and a magnum of claret, of which I discovered a bin in the cellar. Dubellay appeared promptly at eight o'clock, arriving to my surprise in a car. He had tidied himself up and put on a dinner jacket, and he looked exactly like the city solicitors you see dining in the Junior Carlton.

He was in excellent spirits, and his eyes had lost their air of being on guard. He seemed to have reached some conclusion about me, or decided that I was harmless. More, he seemed to be burning to talk to me. After my adventure I was prepared to find fear in him, the fear I had seen in the faces of the menservants. But there was none; instead there was excitement, overpowering excitement.

He neglected the courses in his verbosity. His coming to dinner had considerably startled the inn, and instead of a maid the landlady herself waited on us. She seemed to want to get the meal over, and hustled the biscuits and the port on to the table as soon as she decently could. Then Dubellay became confidential.

He was an enthusiast, it appeared, an enthusiast with a single hobby. All his life he had pottered among antiquities, and when he succeeded to Vauncastle he had the leisure and money to indulge himself. The place, it seemed, had been famous in Roman Britain—Vauni Castra—and Faxeter was a corruption of the same. 'Who was Vaunus?' I asked. He grinned, and told me to wait.

There had been an old temple up in the high woods. There had always been a local legend about it, and the place was supposed to be haunted. Well, he had had the site excavated and he had found —— Here he became the cautious solicitor, and explained to me the law of treasure trove. As long as the objects found were not intrinsically valuable, not gold or jewels, the finder was entitled to keep them. He had done so—had not published the results of his excavations in the proceedings of any learned society—did not want to be bothered by tourists. I was different, for I was a scholar.

What had he found? It was really rather hard to follow his babbling talk, but I gathered that he had found certain carvings and sacrificial implements. And—he sunk his voice—most important of all, an altar, an altar of Vaunus, the tutelary deity of the vale.

When he mentioned this word his face took on a new look—not of fear but of secrecy, a kind of secret excitement. I have seen the same look on the face of a street-preaching Salvationist.

Vaunus had been a British god of the hills, whom the Romans in their liberal way appear to have identified with Apollo. He gave me a long confused account of him, from which it appeared that Mr Dubellay was not an exact scholar. Some of his derivations of place-names were absurd—like St Sant from Sancta Sanctorum—and in quoting a line of Ausonius he made two false quantities. He seemed to hope that I could tell him something more about Vaunus, but I said that my subject was Greek, and that I was deeply ignorant about Roman Britain. I mentioned several books, and found that he had never heard of Haverfield.

One word he used, 'hypocaust', which suddenly gave me a clue. He must have heated the temple, as he heated his house, by some very efficient system of hot air. I know little about science, but I imagined that the artificial heat of the portico, as contrasted with the cold outside, might create an air current. At any rate that explanation satisfied me, and my afternoon's adventure lost its uncanniness. The reaction made me feel friendly towards him, and I listened to his talk with sympathy, but I decided not to mention that I had visited his temple.

He told me about it himself in the most open way. 'I couldn't leave the altar on the hillside,' he said. 'I had to make a place for it, so I turned the old front of the house into a sort of temple. I got the best advice, but architects are ignorant people, and I often wished I had been a better scholar. Still the place satisfies me.'

'I hope it satisfies Vaunus,' I said jocularly.

'I think so,' he replied quite seriously, and then his thoughts seemed to go wandering, and for a minute or so he looked through me with a queer abstraction in his eyes.

'What do you do with it now you've got it?' I asked.

He didn't reply, but smiled to himself.

'I don't know if you remember a passage in Sidonius Apollinaris,' I said, 'a formula for consecrating pagan altars to Christian uses. You begin by sacrificing a white cock or something suitable, and tell Apollo with all friendliness that the old dedication is off for the present. Then you have a Christian invocation—'

He nearly jumped out of his chair.

'That wouldn't do—wouldn't do at all! . . . Oh Lord, no! . . . Couldn't think of it for one moment!'

It was as if I had offended his ears by some horrid blasphemy, and the odd thing was that he never recovered his composure. He tried, for he had good manners, but his ease and friendliness had gone. We talked stiffly for another half-hour about trifles, and then he rose to leave. I returned him his manuscript neatly parcelled up, and expanded in thanks, but he scarcely seemed to heed me. He stuck the thing in his pocket, and departed with the same air of shocked absorption.

After he had gone I sat before the fire and reviewed the situation. I was satisfied with my hypocaust theory, and had no more perturbation in my memory about my afternoon's adventure. Yet a slight flavour of unpleasantness hung about it, and I felt that I did not quite like Dubellay. I set him down as a crank who had tangled himself up with a half-witted hobby, like an old maid with her cats, and I was not sorry to be leaving the place.

My third and last visit to St Sant was in the following June—the midsummer of 1914. I had all but finished my Theocritus, but I needed another day or two with the Vauncastle manuscript, and, as I wanted to clear the whole thing off before I went to Italy in July, I wrote to Dubellay and asked if I might have another sight of it. The thing was a bore, but it had to be faced, and I fancied that the valley would be a pleasant place in that hot summer.

I got a reply at once, inviting, almost begging me to come, and insisting that I should stay at the Hall. I couldn't very well refuse, though I would have preferred the inn. He wired about my train, and wired again saying he would meet me. This time I seemed to be a particularly welcome guest.

I reached Faxeter in the evening, and was met by a car from a Faxeter garage. The driver was a talkative young man, and, as the car was a closed one, I sat beside him for the sake of fresh air. The term had tired me, and I was glad to get out of stuffy Cambridge, but I cannot say that I found it much cooler as we ascended the Vaun valley. The woods were in their summer magnificence but a little dulled and tarnished by the heat, the river was shrunk to a trickle, and the curious hilltops were so scorched by the sun that they seemed almost yellow above the green of the trees. Once again I had the feeling of a landscape fantastically un-English.

'Squire Dubellay's been in a great way about your coming, sir,' the driver informed me. 'Sent down three times to the boss to make sure it was all right. He's got a car of his own, too, a nice little Daimler, but he don't seem to use it much. Haven't seen him about in it for a month of Sundays.'

As we turned in at the Hall gates he looked curiously about him. 'Never been here before, though I've been in most gentlemen's parks for fifty miles round. Rum old-fashioned spot, isn't it, sir?'

If it had seemed a shuttered sanctuary in mid-winter, in that June twilight it was more than ever a place enclosed and guarded. There was almost an autumn smell of decay, a dry decay like touchwood. We seemed to be descending through layers of ever-thickening woods. When at last we turned through the iron gate I saw that the lawns had reached a further stage of neglect, for they were as shaggy as a hayfield.

The white-faced butler let me in, and there, waiting at his back, was Dubellay. But he was not the man whom I had seen in December. He was dressed in an old baggy suit of flannels, and his unwholesome red face was painfully drawn and sunken. There were dark pouches under his eyes, and these eyes were no longer excited, but dull and pained. Yes, and there was more than pain in them—there was fear. I wondered if his hobby were becoming too much for him.

He greeted me like a long-lost brother. Considering that I scarcely knew him, I was a little embarrassed by his warmth. 'Bless you for coming, my dear fellow,' he cried. 'You want a wash and then we'll have dinner. Don't bother to change, unless you want to. I never do.' He led me to my bedroom, which was clean enough but small and shabby like a servant's room. I guessed that he had gutted the house to build his absurd temple.

We dined in a fair-sized room which was a kind of library. It was lined with old books, but they did not look as if they had been there long; rather it seemed like a lumber room in which a fine collection had been stored. Once no doubt they had lived in a dignified Georgian chamber. There was nothing else, none of the antiques which I had expected.

'You have come just in time,' he told me. 'I fairly jumped when I got your letter, for I had been thinking of running up to Cambridge to insist on your coming down here. I hope you're in no hurry to leave.'

'As it happens,' I said, 'I *am* rather pressed for time, for I hope to go abroad next week. I ought to finish my work here in a couple of days. I can't tell you how much I'm in your debt for your kindness.'

'Two days,' he said. 'That will get us over midsummer. That should be enough.' I hadn't a notion what he meant.

I told him that I was looking forward to examining his collection. He opened his eyes. 'Your discoveries, I mean,' I said, 'the altar of Vaunus . . .'

As I spoke the words his face suddenly contorted in a spasm of what looked like terror. He choked and then recovered himself. 'Yes, yes,' he said rapidly. 'You shall see it—you shall see everything—but not now—not tonight. Tomorrow—in broad daylight—that's the time.'

After that the evening became a bad dream. Small talk deserted him, and he could only reply with an effort to my commonplaces. I caught him often looking at me furtively, as if he were sizing me up and wondering how far he could go with me. The thing fairly got on my nerves, and to crown all it was abominably stuffy. The windows of the room gave on a little paved court with a background of laurels, and I might have been in Seven Dials for all the air there was.

When coffee was served I could stand it no longer. 'What about smoking in the temple?' I said. 'It should be cool there with the air from the lake.'

I might have been proposing the assassination of his mother. He simply gibbered at me. 'No, no,' he stammered. 'My God, no!' It was half an hour before he could properly collect himself. A servant lit two oil lamps, and we sat on in the frowsty room.

'You said something when we last met,' he ventured at last, after many a sidelong glance at me. 'Something about a ritual for re-dedicating an altar.'

I remembered my remark about Sidonius Apollinaris.

'Could you show me the passage? There is a good classical library here, collected by my great-grandfather. Unfortunately my scholarship is not equal to using it properly.'

I got up and hunted along the shelves, and presently found a copy of Sidonius, the Plantin edition of 1609. I turned up the passage, and roughly translated it for him. He listened hungrily and made me repeat it twice.

'He says a cock,' he hesitated. 'Is that essential?'

'I don't think so. I fancy any of the recognized ritual stuff would do.'

'I am glad,' he said simply. 'I am afraid of blood.'

'Good God, man,' I cried out, 'are you taking my nonsense seriously? I was only chaffing. Let old Vaunus stick to his altar!'

He looked at me like a puzzled and rather offended dog.

'Sidonius was in earnest . . .'

'Well, I'm not,' I said rudely. 'We're in the twentieth century and not in the third. Isn't it about time we went to bed?'

He made no objection, and found me a candle in the hall. As I undressed I wondered into what kind of lunatic asylum I had strayed. I felt the strongest distaste for the place, and longed to go straight off to the inn; only I couldn't make use of a man's manuscripts and insult his hospitality. It was fairly clear to me that Dubellay was mad. He had ridden his hobby to the death of his wits and was now in its bondage. Good Lord! he had talked of his precious Vaunus as a votary talks of a god. I believed he had come to worship some figment of his half-educated fancy.

I think I must have slept for a couple of hours. Then I woke dripping with perspiration, for the place was simply an oven. My window was as wide open as it would go, and, though it was a warm night, when I stuck my head out the air was fresh. The heat came from indoors. The room was on the first floor near the entrance and I was looking on to the overgrown lawns. The night was very dark and utterly still, but I could have sworn that I heard wind. The trees were as motionless as marble, but somewhere close at hand I heard a strong gust blowing. Also, though there was no moon, there was somewhere near me a steady glow of light; I could see the reflection of it round the end of the house. That meant that it came from the temple. What kind of saturnalia was Dubellay conducting at such an hour?

When I drew in my head I felt that if I was to get any sleep something must be done. There could be no question about it; some fool had turned on the steam heat, for the room was a furnace. My temper was rising.

There was no bell to be found, so I lit my candle and set out to find a servant.

I tried a cast downstairs and discovered the room where we had dined. Then I explored a passage at right angles, which brought me up against a great oak door. The light showed me that it was a new door, and that there was no apparent way of opening it. I guessed that it led into the temple, and, though it fitted close and there seemed to be no keyhole, I could hear through it a sound like a rushing wind. . . . Next I opened a door on my right and found myself in a big store cupboard. It had a funny, exotic, spicy smell, and, arranged very neatly on the floor and shelves, was a number of small sacks and coffers. Each bore a label, a square of stout paper with very black lettering. I read '*Pro servitio Vauni.*'

I had seen them before, for my memory betrayed me if they were not the very labels that Dubellay's servants had been attaching to the packages from the carrier's cart that evening in the past autumn. The discovery made my suspicions an unpleasant certainty. Dubellay evidently meant the labels to read 'For the service of Vaunus.' He was no scholar, for it was an impossible use of the word 'servitium', but he was very patently a madman.

However, it was my immediate business to find some way to sleep, so I continued my quest for a servant. I followed another corridor, and discovered a second staircase. At the top of it I saw an open door and looked in. It must have been Dubellay's, for his flannels were tumbled untidily on a chair, but Dubellay himself was not there and the bed had not been slept in.

I suppose my irritation was greater than my alarm—though I must say I was getting a little scared—for I still pursued the evasive servant. There was another stair which apparently led to attics, and in going up it I slipped and made a great clatter. When I looked up the butler in his nightgown was staring down at me, and if ever a mortal face held fear it was his. When he saw who it was he seemed to recover a little.

'Look here,' I said, 'for God's sake turn off that infernal hot air. I can't get a wink of sleep. What idiot set it going?'

He looked at me owlishly, but he managed to find his tongue.

'I beg your pardon, sir,' he said, 'but there is no heating apparatus in this house.'

There was nothing more to be said. I returned to my bedroom and it seemed to me that it had grown cooler. As I leaned out of the window, too, the mysterious wind seemed to have died away, and the glow no longer showed from beyond the corner of the house. I got into bed and slept heavily till I was roused by the appearance of my shaving water about half-past nine. There was no bathroom, so I bathed in a tin pannikin.

It was a hazy morning which promised a day of blistering heat. When I went down to breakfast I found Dubellay in the dining-room. In the daylight he looked a very sick man, but he seemed to have taken a pull on himself, for his manner was considerably less nervy than the night before. Indeed, he appeared almost normal, and I might have reconsidered my view but for the look in his eyes.

I told him that I proposed to sit tight all day over the manuscript, and get the thing finished. He nodded. 'That's all right. I've a lot to do myself, and I won't disturb you.'

'But first,' I said, 'you promised to show me your discoveries.'

He looked at the window where the sun was shining on the laurels and on a segment of the paved court.

'The light is good,' he said—an odd remark. 'Let us go there now. There are times and seasons for the temple.'

He led me down the passage I had explored the previous night. The door opened not by a key but by some lever in the wall. I found myself looking suddenly at a bath of sunshine with the lake below as blue as a turquoise.

It is not easy to describe my impressions of that place. It was unbelievably light and airy, as brilliant as an Italian colonnade in midsummer. The proportions must have been good, for the columns soared and swam, and the roof (which looked like cedar) floated as delicately as a flower on its stalk. The stone was some local limestone, which on the floor took a polish like marble. All around was a vista of sparkling water and summer woods and far blue mountains. It should have been as wholesome as the top of a hill.

And yet I had scarcely entered before I knew that it was a prison. I am not an imaginative man, and I believe my nerves are fairly good, but I could scarcely put one foot before the other, so strong was my distaste. I felt shut off from the world, as if I were in a dungeon or on an ice-floe. And I felt, too, that though far enough from humanity, we were not alone.

On the inner wall there were three carvings. Two were imperfect friezes sculptured in low-relief, dealing apparently with the same subject. It was a ritual procession, priests bearing branches, the ordinary *dendrophori* business. The faces were only half-human, and that was from no lack of skill, for the artist had been a master. The striking thing was that the branches and the hair of the hierophants were being tossed by a violent wind, and the expression of each was of a being in the last stage of endurance, shaken to the core by terror and pain.

Between the friezes was a great roundel of a Gorgon's head. It was not a female head, such as you commonly find, but a male head, with the viperous hair sprouting from chin and lip. It had once been coloured, and fragments of a green pigment remained in the locks. It was an awful thing, the ultimate horror of fear, the last dementia of cruelty made manifest in stone. I hurriedly averted my eyes and looked at the altar.

That stood at the west end on a pediment with three steps. It was a beautiful piece of work, scarcely harmed by the centuries, with two words inscribed on its face—APOLL. VAUN. It was made of some foreign marble, and the hollow top was dark with ancient sacrifices. Not so ancient either, for I could have sworn that I saw there the mark of recent flame.

I do not suppose I was more than five minutes in the place. I wanted to get out, and Dubellay wanted to get me out. We did not speak a word till we were back in the library.

'For God's sake give it up!' I said. 'You're playing with fire, Mr Dubellay. You're driving yourself into Bedlam. Send these damned things to a museum and leave this place. Now, now, I tell you. You have no time to lose. Come down with me to the inn straight off and shut up this house.'

He looked at me with his lip quivering like a child about to cry.

'I will. I promise you I will. . . . But not yet. . . . After tonight. . . . Tomorrow I'll do whatever you tell me. . . . You won't leave me?'

'I won't leave you, but what earthly good am I to you if you won't take my advice?'

'Sidonius . . .' he began.

'Oh, damn Sidonius! I wish I had never mentioned him. The whole thing is arrant nonsense, but it's killing you. You've got it on the brain. Don't you know you're a sick man?'

'I'm not feeling very grand. It's so warm today. I think I'll lie down.'

It was no good arguing with him, for he had the appalling obstinacy of very weak things. I went off to my work in a shocking bad temper.

The day was what it had promised to be, blisteringly hot. Before midday the sun was hidden by a coppery haze, and there was not the faintest stirring of wind. Dubellay did not appear at luncheon—it was not a meal he ever ate, the butler told me. I slogged away all the afternoon, and had pretty well finished my job by six o'clock. That would enable me to leave next morning, and I hoped to be able to persuade my host to come with me.

The conclusion of my task put me into a better humour, and I went for a walk before dinner. It was a very close evening, for the heat haze had not lifted; the woods were as silent as a grave, not a bird spoke, and when I came out of the cover to the burnt pastures the sheep seemed too languid to graze. During my walk I prospected the environs of the house, and saw that it would be very hard to get access to the temple except by a long circuit. On one side was a mass of outbuildings, and then a high wall, and on the other the very closest and highest quickset hedge I have ever seen, which ended in a wood with savage spikes on its containing wall. I returned to my room, had a cold bath in the exiguous tub, and changed.

Dubellay was not at dinner. The butler said that his master was feeling unwell and had gone to bed. The news pleased me, for bed was the best place for him. After that I settled myself down to a lonely evening in the library. I browsed among the shelves and found a number of rare editions which served to pass the time. I noticed that the copy of Sidonius was absent from its place.

I think it was about ten o'clock when I went to bed, for I was unaccountably tired. I remember wondering whether I oughtn't to go and visit Dubellay, but decided that it was better to leave him alone. I still reproach myself for that decision. I know now I ought to have taken him by force and haled him to the inn.

Suddenly I came out of heavy sleep with a start. A human cry seemed to be ringing in the corridors of my brain. I held my breath and listened. It came again, a horrid scream of panic and torture.

I was out of bed in a second, and only stopped to get my feet into slippers. The cry must have come from the temple. I tore downstairs expecting to hear the noise of an alarmed household. But there was no sound, and the awful cry was not repeated.

The door in the corridor was shut, as I expected. Behind it pandemonium seemed to be loose, for there was a howling like a tempest—and something more, a crackling like fire. I made for the front door, slipped off the chain, and found myself in the still, moonless night. Still, except for the rending gale that seemed to be raging in the house I had left.

From what I had seen on my evening's walk I knew that my one chance to get to the temple was by way of the quickset hedge. I thought I might manage to force a way between the end of it and the wall. I did it, at the cost of much of my raiment and my skin. Beyond was another rough lawn set with tangled shrubberies, and then a precipitous slope to the level of the lake. I scrambled along the sedgy margin, not daring to lift my eyes till I was on the temple steps.

The place was brighter than day with a roaring blast of fire. The very air seemed to be incandescent and to have become a flaming ether. And yet there were no flames—only a burning brightness. I could not enter, for the waft from it struck my face like a scorching hand and I felt my hair singe. . . .

I am short-sighted, as you know, and I may have been mistaken, but this is what I think I saw. From the altar a great tongue of flame seemed to shoot upwards and lick the roof, and from its pediment ran flaming streams. In front of it lay a body—Dubellay's—a naked body, already charred and black. There was nothing else, except that the Gorgon's head in the wall seemed to glow like a sun in hell.

I suppose I must have tried to enter. All I know is that I found myself staggering back, rather badly burned. I covered my eyes, and as I looked through my fingers I seemed to see the flames flowing under the wall, where there may have been lockers, or possibly another entrance. Then the great oak door suddenly shrivelled like gauze, and with a roar the fiery river poured into the house.

I ducked myself in the lake to ease the pain, and then ran back as hard as I could by the way I had come. Dubellay, poor devil, was beyond my aid. After that I am not very clear what happened. I know that the house burned like a haystack. I found one of the menservants on the lawn, and I think I helped to get the other down from his room by one of the rain-pipes. By the time the neighbours arrived the house was ashes, and I was pretty well mother-naked. They took me to the inn and put me to bed, and I remained there till after the inquest. The coroner's jury were puzzled, but they found it simply death by misadventure; a lot of country houses were burned that summer. There was nothing found of Dubellay; nothing remained of the house except a few blackened pillars; the altar and the sculptures were so cracked and scarred that no museum wanted them. The place has not been rebuilt, and for all I know they are there today. I am not going back to look for them.

Nightingale finished his story and looked round his audience.

'Don't ask me for an explanation,' he said, 'for I haven't any. You may believe if you like that the god Vaunus inhabited the temple which Dubellay built for him, and, when his votary grew scared and tried

Sidonius's receipt for shifting the dedication, became angry and slew him with his flaming wind. That wind seems to have been a perquisite of Vaunus. We know more about him now, for last year they dug up a temple of his in Wales.'

'Lightning,' some one suggested.

'It was a quiet night, with no thunderstorm,' said Nightingale.

'Isn't the countryside volcanic?' Peckwether asked. 'What about pockets of natural gas or something of the kind?'

'Possibly. You may please yourself in your explanation. I'm afraid I can't help you. All I know is that I don't propose to visit that valley again!'

'What became of your Theocritus?'

'Burned, like everything else. However, that didn't worry me much. Six weeks later came the war, and I had other things to think about.'

ROBERT E. HOWARD

THE TOWER OF THE ELEPHANT

Torches flared murkily on the revels in the Maul, where the thieves of the East held carnival by night. In the Maul they could carouse and roar as they liked, for honest people shunned the quarters, and watchmen, well paid with stained coins, did not interfere with their sport. Along the crooked, unpaved streets with their heaps of refuse and sloppy puddles, drunken roisterers staggered, roaring. Steel glinted in the shadows where rose the shrill laughter of women, and the sounds of scufflings and strugglings. Torchlight licked luridly from broken windows and wide-thrown doors, and out of those doors, stale smells of wine and rank sweaty bodies, clamour of drinking jacks and fists hammered on rough tables, snatches of obscene songs, rushed like a blow in the face.

In one of those dens merriment thundered to the low smoke-stained roof, where rascals gathered in every stage of rags and tatters—furtive cutpurses, leering kidnappers, quick-fingered thieves, swaggering bravoes with their wenches, strident-voiced women clad in tawdry finery. Native rogues were the dominant element—dark-skinned, dark-eyed Zamorians, with daggers at their girdles and guile in their hearts. But there were wolves of half a dozen outland nations there as well. There was a giant Hyperborean renegade, taciturn, dangerous, with a broadsword strapped to his great gaunt frame—for men wore steel openly in the Maul. There was a Shemitish counterfeiter, with his hook nose and curled blue-black beard. There was a bold-eyed Brythunian wench, sitting on the knee of a tawny-haired Gunderman—a wandering mercenary soldier, a deserter from some defeated army. And the fat gross rogue whose bawdy jests were causing all the shouts of mirth was a professional kidnapper come up from distant Koth to teach woman-stealing to Zamorians who were born with more knowledge of the art than he could ever attain. This man halted in his description of an intended victim's charms and thrust his muzzle into a huge tankard of frothing ale. Then blowing the foam from his fat lips, he said, 'By Bel, god of all thieves, I'll show them how to steal wenches; I'll have her over the Zamorian border before dawn, and there'll be a caravan waiting to receive her. Three hundred pieces of silver, a count of Ophir promised me for a sleek young Brythunian of the better class. It took me weeks, wandering among the border cities as a beggar, to find one I knew would suit. And is she a pretty baggage!'

He blew a slobbery kiss in the air.

'I know lords in Shem who would trade the secret of the Elephant Tower for her,' he said, returning to his ale.

A touch on his tunic sleeve made him turn his head, scowling at the interruption. He saw a tall, strongly made youth standing beside him. This person was as much out of place in that den as a grey wolf among mangy rats of the gutters. His cheap tunic could not conceal the hard, rangy lines of his powerful frame, the broad heavy shoulders, the massive chest, lean waist, and heavy arms. His skin was brown from outland suns, his eyes blue and smouldering; a shock of tousled black hair crowned his broad forehead. From his girdle hung a sword in a worn leather scabbard.

The Kothian involuntarily drew back; for the man was not one of any civilized race he knew.

'You spoke of the Elephant Tower,' said the stranger, speaking Zamorian with an alien accent. 'I've heard much of this tower; what is its secret?'

The fellow's attitude did not seem threatening, and the Kothian's courage was bolstered up by the ale and the evident approval of his audience. He swelled with self-importance.

'The secret of the Elephant Tower?' he exclaimed. 'Why, any fool knows that Yara the priest dwells there with the great jewel men call the Elephant's Heart, that is the secret of his magic.'

The barbarian digested this for a space.

'I have seen this tower,' he said. 'It is set in a great garden above the level of the city, surrounded by high walls. I have seen no guards. The walls would be easy to climb. Why has not somebody stolen this secret gem?'

The Kothian stared wide-mouthed at the other's simplicity, then burst into a roar of derisive mirth, in which the others joined.

'Harken to this heathen!' he bellowed. 'He would steal the jewel of

Yara!—Harken, fellow,' he said, turning portentously to the other, 'I suppose you are some sort of a northern barbarian—'

'I am a Cimmerian,' the outlander answered, in no friendly tone. The reply and the manner of it meant little to the Kothian; of a kingdom that lay far to the south, on the borders of Shem, he knew only vaguely of the northern races.

'Then give ear and learn wisdom, fellow,' said he, pointing his drinking jack at the discomfited youth. 'Know that in Zamora, and more especially in this city, there are more bold thieves than anywhere else in the world, even Koth. If mortal man could have stolen the gem, be sure it would have been filched long ago. You speak of climbing the walls, but once having climbed, you would quickly wish yourself back again. There are no guards in the gardens at night for a very good reason—that is, no human guards. But in the watch chamber, in the lower part of the tower, are armed men, and even if you passed those who roam the gardens by night, you must still pass through the soldiers, for the gem is kept somewhere in the tower above.'

'But if a man *could* pass through the gardens,' argued the Cimmerian, 'why could he not come at the gem through the upper part of the tower and thus avoid the soldiers?'

Again the Kothian gaped at him.

'Listen to him!' he shouted jeeringly. 'The barbarian is an eagle who would fly to the jewelled rim of the tower, which is only a hundred and fifty feet above the earth, with rounded sides slicker than polished glass!'

The Cimmerian glared about, embarrassed at the roar of mocking laughter that greeted this remark. He saw no particular humour in it and was too new to civilization to understand its discourtesies. Civilized men are more discourteous than savages because they know they can be impolite without having their skulls split, as a general thing. He was bewildered and chagrined and doubtless would have slunk away, abashed, but the Kothian chose to goad him further.

'Come, come!' he shouted. 'Tell these poor fellows, who have only been thieves since before you were spawned, tell them how you would steal the gem!'

'There is always a way, if the desire be coupled with courage,' answered the Cimmerian shortly, nettled.

The Kothian chose to take this as a personal slur. His face grew purple with anger.

'What!' he roared. 'You dare tell us our business, and intimate that we are cowards? Get along; get out of my sight!' And he pushed the Cimmerian violently.

'Will you mock me and then lay hands on me?' grated the barbarian, his quick rage leaping up; and he returned the push with an open-handed blow that knocked his tormentor back against the rude-hewn table. Ale splashed over the jack's lip, and the Kothian roared in fury, dragging at his sword.

'Heathen dog!' he bellowed. 'I'll have your heart for that!'

Steel flashed and the throng surged wildly back out of the way. In their flight they knocked over the single candle and the den was plunged in darkness, broken by the crash of upset benches, drum of flying feet, shouts, oaths of people tumbling over one another, and a single strident yell of agony that cut the din like a knife. When a candle was relighted, most of the guests had gone out by doors and broken windows, and the rest huddled behind stacks of wine kegs and under tables. The barbarian was gone; the centre of the room was deserted except for the gashed body of the Kothian. The Cimmerian, with the unerring instinct of the barbarian, had killed his man in the darkness and confusion.

II

The lurid lights and drunken revelry fell away behind the Cimmerian. He had discarded his torn tunic and walked through the night naked except for a loincloth and his high-strapped sandals. He moved with the supple ease of a great tiger, his steely muscles rippling under his brown skin.

He had entered the part of the city reserved for the temples. On all sides of him they glittered white in the starlight—snowy marble pillars and golden domes and silver arches, shrines of Zamora's myriad strange gods. He did not trouble his head about them; he knew that Zamora's religion, like all things of a civilized, long-settled people, was intricate and complex and had lost most of the pristine essence in a maze of formulas and rituals. He had squatted for hours in the courtyards of the philosophers, listening to the arguments of theologians and teachers, and come away in a haze of bewilderment, sure of only one thing, and that, that they were all touched in the head.

His gods were simple and understandable; Crom was their chief, and he lived on a great mountain, whence he sent forth dooms and death. It was useless to call on Crom, because he was a gloomy, savage god, and he hated weaklings. But he gave a man courage at birth, and the will and might to kill his enemies, which, in the Cimmerian's mind, was all any god should be expected to do.

His sandalled feet made no sound on the gleaming pave. No watchmen passed, for even the thieves of the Maul shunned the temples, where strange dooms had been known to fall on violators. Ahead of him he saw, looming against the sky, the Tower of the Elephant. He mused, wondering why it was so named. No one seemed to know. He had never seen an elephant, but he vaguely understood that it was a monstrous animal, with a tail in front as well as behind. This a wandering Shemite had told him, swearing that he had seen such beasts by the thousands in the country of the Hyrkanians; but all men knew what liars were the men of Shem. At any rate, there were no elephants in Zamora.

The shimmering shaft of the tower rose frostily in the stars. In the sunlight it shone so dazzlingly that few could bear its glare, and men said it was built of silver. It was round, a slim, perfect cylinder, a hundred and fifty feet in height, and its rim glittered in the starlight with the great jewels which crusted it. The tower stood among the waving, exotic trees of a garden raised high above the general level of the city. A high wall enclosed this garden, and outside the wall was a lower level, likewise enclosed by a wall. No lights shone forth; there seemed to be no windows in the tower—at least not above the level of the inner wall. Only the gems high above sparkled frostily in the starlight.

Shrubbery grew thick outside the lower, or outer wall. The Cimmerian crept close and stood beside the barrier, measuring it with his eye. It was high, but he could leap and catch the coping with his fingers. Then it would be child's play to swing himself up and over, and he did not doubt that he could pass the inner wall in the same manner. But he hesitated at the thought of the strange perils which were said to await within. These people were strange and mysterious to him; they were not of his kind—not even of the same blood as the more westerly Brythunians, Nemedians, Kothians, and Aquilonians, of whose civilized mysteries he had heard in times past. The people of Zamora were very ancient and, from what he had seen of them, very evil.

He thought of Yara, the high priest, who worked strange dooms from this jewelled tower, and the Cimmerian's hair prickled as he remembered a tale told by a drunken page of the court—how Yara had laughed in the face of a hostile prince, and held up a glowing, evil gem before him, and how rays shot blindingly from that unholy jewel, to envelop the prince, who screamed and fell down, and shrank to a withered blackened lump that changed to a black spider which scampered wildly about the chamber until Yara set his heel upon it.

Yara came not often from his tower of magic, and always to work evil on some man or some nation. The king of Zamora feared him more than he feared death, and kept himself drunk all the time because that fear was more that he could endure sober. Yara was very old—centuries old, men said, and added that he would live for ever because of the magic of his gem, which men called the Heart of the Elephant; for no better reason than this they named his hold the Elephant's Tower.

The Cimmerian, engrossed in these thoughts, shrank quickly against the wall. Within the garden someone was passing, who walked with a measured stride. The listener heard the clink of steel. So, after all, a guard did pace those gardens. The Cimmerian waited, expecting to hear him pass again on the next round; but silence rested over the mysterious gardens.

At last curiosity overcame him. Leaping lightly, he grasped the wall and swung himself up to the top with one arm. Lying flat on the broad coping, he looked down into the wide space between the walls. No shrubbery grew near him, though he saw some carefully trimmed bushes near the inner wall. The starlight fell on the even sward, and somewhere a fountain tinkled.

The Cimmerian cautiously lowered himself down on the inside and drew his sword, staring about him. He was shaken by the nervousness of the wild at standing thus unprotected in the naked starlight, and he moved lightly around the curve of the wall, hugging its shadow, until he was even with the shrubbery he had noticed. Then he ran quickly towards it, crouching low, and almost tripped over a form that lay crumpled near the edges of the bushes.

A quick look to right and left showed him no enemy, in sight at least, and he bent close to investigate. His keen eyes, even in the dim starlight, showed him a strongly built man in the silvered armour and crested helmet of the Zamorian royal guard. A shield and a spear lay near him, and it took but an instant's examination to show that he had been strangled. The barbarian glanced about uneasily. He knew that this man must be the guard he had heard pass his hiding place by the wall. Only a short time had passed, yet in that interval nameless hands had reached out of the dark and choked out the soldier's life.

Straining his eyes in the gloom, he saw a hint of motion through the shrubs near the wall. Thither he glided, gripping his sword. He made no more noise than a panther stealing through the night, yet the man he was stalking heard. The Cimmerian had a dim glimpse of a huge bulk close to the wall, felt relief that it was at least human; then the fellow wheeled quickly with a gasp that sounded like panic, made the first motion of a forward plunge, hands clutching, then recoiled as the Cimmerian's blade caught the starlight. For a tense instant neither spoke, standing ready for anything.

'You are no soldier,' hissed the stranger at last. 'You are a thief like myself.'

'And who are you?' asked the Cimmerian in a suspicious whisper.

'Taurus of Nemedia.'

The Cimmerian lowered his sword.

'I've heard of you. Men call you a prince of thieves.'

A low laugh answered him. Taurus was tall as the Cimmerian, and heavier; he was big-bellied and fat, but his every movement betokened a subtle dynamic magnetism, which was reflected in the keen eyes that glinted vitally, even in the starlight. He was barefooted and carried a coil of what looked like a thin, strong rope, knotted at regular intervals.

'Who are you?' he whispered.

'Conan, a Cimmerian,' answered the other. 'I came seeking a way to steal Yara's jewel, that men call the Elephant's Heart.'

Conan sensed the man's great belly shaking in laughter, but it was not derisive.

'By Bel, god of thieves!' hissed Taurus. 'I had thought only myself had courage to attempt *that* poaching. These Zamorians call themselves thieves—bah! Conan, I like your grit. I never shared an adventure with anyone; but, by Bel, we'll attempt this together if you're willing.'

'Then you are after the gem, too?'

'What else? I've had my plans laid for months; but you, I think, have acted on a sudden impulse, my friend.'

'You killed the soldier?'

'Of course. I slid over the wall when he was on the other side of the garden. I hid in the bushes; he heard me, or thought he heard something. When he came blundering over, it was no trick at all to get behind him and suddenly grip his neck and choke out his fool's life. He was like most men, half blind in the dark. A good thief should have eyes like a cat.'

'You made one mistake,' said Conan.

Taurus's eyes flashed angrily.

'I? I, a mistake? Impossible!'

'You should have dragged the body into the bushes.'

'Said the novice to the master of the art. They will not change the guard until past midnight. Should any come searching for him now and find his body, they would flee at once to Yara, bellowing the news, and give us time to escape. Were they not to find it, they'd go beating up the bushes and catch us like rats in a trap.'

'You are right,' agreed Conan.

'So. Now attend. We waste time in this cursed discussion. There are no guards in the inner garden—human guards, I mean, though there are sentinels even more deadly. It was their presence which baffled me for so long, but I finally discovered a way to circumvent them.'

'What of the soldiers in the lower part of the tower?'

'Old Yara dwells in the chambers above. By that route we will come—and go, I hope. Never mind asking me how. I have arranged a way. We'll steal down through the top of the tower and strangle old Yara before he can cast any of his accursed spells on us. At least we'll try; it's the chance of being turned into a spider or a toad, against the wealth and power of the world. All good thieves must know how to take risks.'

'I'll go as far as any man,' said Conan, slipping off his sandals.

'Then follow me.' And turning, Taurus leaped up, caught the wall and drew himself up. The man's suppleness was amazing, considering his bulk; he seemed almost to glide up over the edge of the coping. Conan followed him, and lying flat on the broad top, they spoke in wary whispers.

'I see no light,' Conan muttered. The lower part of the tower seemed much like that portion visible from outside the garden—a perfect, gleaming cylinder, with no apparent openings.

'There are cleverly constructed doors and windows,' answered Taurus, 'but they are closed. The soldiers breathe air that comes from above.'

The garden was a vague pool of shadows, where feathery bushes and low, spreading trees waved darkly in the starlight. Conan's wary soul felt the aura of waiting menace that brooded over it. He felt the burning glare of unseen eyes, and he caught a subtle scent that made the short hairs on his neck instinctively bristle as a hunting dog bristles at the scent of an ancient enemy.

'Follow me,' whispered Taurus; 'keep behind me, as you value your life.'

Taking what looked like a copper tube from his girdle, the Nemedian dropped lightly to the sward inside the wall. Conan was close behind him, sword ready, but Taurus pushed him back, close to the wall, and showed no inclination to advance, himself. His whole attitude was of tense expectancy, and his gaze, like Conan's, was fixed on the shadowy mass of shrubbery a few yards away. This shrubbery was shaken, although the breeze had died down. Then two great eyes blazed from the waving shadows, and behind them other sparks of fire glinted in the darkness.

'Lions!' muttered Conan.

'Aye. By day they are kept in subterranean caverns below the tower. That's why there are no guards in this garden.'

Conan counted the eyes rapidly.

'Five in sight; maybe more back in the bushes. They'll charge in a moment—'

'Be silent!' hissed Taurus, and he moved out from the wall, cautiously as if treading on razors, lifting the slender tube. Low rumblings rose from the shadows, and the blazing eyes moved forward. Conan could sense the great slavering jaws, the tufted tails lashing tawny sides. The air grew tense—the Cimmerian gripped his sword, expecting the charge and the irresistible hurtling of giant bodies. Then Taurus brought the mouth of the tube to his lips and blew powerfully. A long jet of yellowish powder shot from the other end of the tube and billowed out instantly in a thick green-yellow cloud that settled over the shrubbery, blotting out the glaring eyes.

Taurus ran back hastily to the wall. Conan glared without understanding. The thick cloud hid the shrubbery, and from it no sound came.

'What is that mist?' the Cimmerian asked uneasily.

'Death!' hissed the Nemedian. 'If a wind springs up and blows it back upon us, we must flee over the wall. But no, the wind is still, and now it is dissipating. Wait until it vanishes entirely. To breathe it is death.'

Presently only yellowish threads hung ghostily in the air; then they were gone, and Taurus motioned his companion forward. They stole toward the bushes, and Conan gasped. Stretched out in the shadows lay five great tawny shapes, the fire of their grim eyes dimmed for ever. A sweetish, cloying scent lingered in the atmosphere.

'They died without a sound!' muttered the Cimmerian. 'Taurus, what was that powder?'

'It was made from the black lotus, whose blossoms wave in the lost jungles of Khitai, where only the yellow-skulled priests of Yun dwell. Those blossoms strike dead any who smell of them.'

Conan knelt beside the great forms, assuring himself that they were indeed beyond power of harm. He shook his head; the magic of the exotic lands was mysterious and terrible to the barbarians of the north.

'Why can you not slay the soldiers in the tower in the same way?' he asked.

'Because that was all the powder I possessed. The obtaining of it was a feat which in itself was enough to make me famous among the thieves of the world. I stole it out of a caravan bound for Stygia, and I lifted it, in its cloth-of-gold bag, out of the coils of the great serpent which guarded it, without waking him. But come, in Bel's name! Are we to waste the night in discussion?'

They glided through the shrubbery to the gleaming foot of the tower, and there, with a motion enjoining silence, Taurus unwound his knotted cord, on one end of which was a strong steel hook. Conan saw his plan and asked no questions, as the Nemedian gripped the line a short distance below the hook and began to swing it about his head. Conan laid his ear to the smooth wall and listened, but could hear nothing. Evidently the soldiers within did not suspect the presence of intruders, who had made no more sound than the night wind blowing through the trees. But a strange nervousness was on the barbarian; perhaps it was the lion smell which was over everything.

Taurus threw the line with a smooth, rippling motion of his mighty arm. The hook curved upward and inward in a peculiar manner, hard to describe, and vanished over the jewelled rim. It apparently caught firmly, for cautious jerking and then hard pulling did not result in any slipping or giving.

'Luck the first cast,' murmured Taurus. 'I—'

It was Conan's savage instinct which made him wheel suddenly; for the death that was upon them made no sound. A fleeting glimpse showed the Cimmerian the giant tawny shape, rearing upright against the stars, towering over him for the death stroke. No civilized man could have moved half so quickly as the barbarian moved. His sword flashed frostily in the starlight with every ounce of desperate nerve and thew behind it, and man and beast went down together.

Cursing incoherently beneath his breath, Taurus bent above the mass and saw his companion's limbs move as he strove to drag himself from under the great weight that lay limply upon him. A glance showed the startled Nemedian that the lion was dead, its slanting skull split in half. He laid hold of the carcass and, by his aid, Conan thrust it aside and clambered up, still gripping his dripping sword.

'Are you hurt, man?' gasped Taurus, still bewildered by the stunning swiftness of that touch-and-go episode.

'No, by Crom!' answered the barbarian. 'But that was as close a call as

I've had in a life nowadays tame. Why did not the cursed beast roar as it charged?'

'All things are strange in this garden,' said Taurus. 'The lions strike silently—and so do other deaths. But come—little sound was made in that slaying, but the soldiers might have heard, if they are not asleep or drunk. That beast was in some other part of the garden and escaped the death of the flowers, but surely there are no more. We must climb this cord—little need to ask a Cimmerian if he can.'

'If it will bear my weight,' grunted Conan, cleansing his sword on the grass.

'It will bear thrice my own,' answered Taurus. 'It was woven from the tresses of dead women, which I took from their tombs at midnight, and steeped in the deadly wine of the upas tree, to give it strength. I will go first—then follow me closely.'

The Nemedian gripped the rope and, crooking a knee about it, began the ascent; he went up like a cat, belying the apparent clumsiness of his bulk. The Cimmerian followed. The cord swayed and turned on itself, but the climbers were not hindered; both had made more difficult climbs before. The jewelled rim glittered high above them, jutting out from the perpendicular of the wall, so that the cord hung perhaps a foot from the side of the tower—a fact which added greatly to the ease of the ascent.

Up and up they went, silently, the lights of the city spreading out further and further to their sight as they climbed, the stars above them more and more dimmed by the glitter of the jewels along the rim. Now Taurus reached up a hand and gripped the rim itself, pulling himself up and over. Conan paused a monment on the very edge, fascinated by the great frosty jewels whose gleams dazzled his eyes—diamonds, rubies, emeralds, sapphires, turquoises, moonstones, set thick as stars in the shimmering silver. At a distance their different gleams had seemed to merge into a pulsing white glare; but now, at close range, they shimmered with a million rainbow tints and lights, hypnotizing him with their scintillations.

'There is a fabulous fortune here, Taurus,' he whispered; but the Nemedian answered impatiently, 'Come on! If we secure the Heart, these and all other things shall be ours.'

Conan climbed over the sparkling rim. The level of the tower's top was some feet below the gemmed ledge. It was flat, composed of some dark blue substance, set with gold that caught the starlight, so that the whole looked like a wide sapphire flecked with shining gold dust. Across from the point where they had entered there seemed to be a sort of chamber, built upon the roof. It was of the same silvery material as the walls of the tower, adorned with designs worked in smaller gems; its single door was of gold, its surface cut in scales and crusted with jewels that gleamed like ice.

Conan cast a glance at the pulsing ocean of lights which spread far below them, then glanced at Taurus. The Nemedian was drawing up his cord and coiling it. He showed Conan where the hook had caught—a fraction of an inch of the point had sunk under a great blazing jewel on the inner side of the rim.

'Luck was with us again,' he muttered. 'One would think that our combined weight would have torn that stone out. Follow me; the real risks of the venture begin now. We are in the serpent's lair, and we know not where he lies hidden.'

Like stalking tigers they crept across the darkly gleaming floor and halted outside the sparkling door. With a deft and cautious hand Taurus tried it. It gave without resistance, and the companions looked in, tensed for anything. Over the Nemedian's shoulder Conan had a glimpse of a glittering chamber, the walls, ceiling, and floor of which were crusted with great, white jewels, which lighted it brightly and which seemed its only illumination. It seemed empty of life.

'Before we cut off our last retreat,' hissed Taurus, 'go you to the rim and look over on all sides; if you see any soldiers moving in the gardens, or anything suspicious, return and tell me. I will await you within this chamber.'

Conan saw scant reason in this, and a faint suspicion of his companion touched his wary soul, but he did as Taurus requested. As he turned away, the Nemedian slipped inside the door and drew it shut behind him. Conan crept about the rim of the tower, returning to his starting point without having seen any suspicious movement in the vaguely waving sea of leaves below. He turned toward the door—suddenly from within the chamber there sounded a strangled cry.

The Cimmerian leaped forward, electrified—the gleaming door swung open, and Taurus stood framed in the cold blaze behind him. He swayed and his lips parted, but only a dry rattle burst from his throat. Catching at the golden door for support, he lurched out upon the roof, then fell headlong, clutching at his throat. The door swung to behind him.

Conan, crouching like a panther at bay, saw nothing in the room behind the stricken Nemedian, in the brief instant the door was partly open—unless it was not a trick of the light which made it seem as if a shadow darted across the gleaming floor. Nothing followed Taurus out on the roof, and Conan bent above the man.

The Nemedian stared up with dilated, glazing eyes, that somehow held a terrible bewilderment. His hands clawed at his throat, his lips slobbered and gurgled; then suddenly he stiffened, and the astounded Cimmerian knew that he was dead. And he felt that Taurus had died without knowing what manner of death had stricken him. Conan glared bewilderedly at the cryptic golden door. In that empty room, with its glittering jewelled walls, death had come to the prince of thieves as swiftly and mysteriously as he had dealt doom to the lions in the gardens below.

Gingerly the barbarian ran his hands over the man's half-naked body, seeking a wound. But the only marks of violence were between his shoulders, high up near the base of his bull neck—three small wounds, which looked as if three nails had been driven deep in the flesh and withdrawn. The edges of these wounds were black, and a faint smell of putrefaction was evident. Poisoned darts? thought Conan—but in that case the missiles should be still in the wounds.

Cautiously he stole towards the golden door, pushed it open, and looked inside. The chamber lay empty, bathed in the cold, pulsing glow of the myriad jewels. In the very centre of the ceiling he idly noted a curious design—a black eight-sided pattern, in the centre of which four gems glittered with a red flame unlike the white blaze of the other jewels. Across the room there was another door, like the one in which he stood, except that it was not carved in the scale pattern. Was it from that door that death had come?—and having struck down its victim, had it retreated by the same way?

Closing the door behind him, the Cimmerian advanced into the chamber. His bare feet made no sound on the crystal floor. There were no chairs or tables in the chamber, only three or four silken couches, embroidered with gold and worked in strange serpentine designs, and several silver-bound mahogany chests. Some were sealed with heavy golden locks; others lay open, their carven lids thrown back, revealing heaps of jewels in a careless riot of splendour to the Cimmerian's astounded eyes. Conan swore beneath his breath; already he had looked upon more wealth that night than he had ever dreamed existed in all the world, and he grew dizzy thinking of what must be the value of the jewel he sought.

He was in the centre of the room now, going stooped forward, head thrust out warily, sword advanced, when again death struck at him soundlessly. A flying shadow that swept across the gleaming floor was his only warning, and his instinctive sidelong leap all that saved his life. He had a flashing glimpse of a hairy black horror that swung past him with a clashing of frothing fangs, and something splashed on his bare shoulder that burned like drops of liquid hell-fire. Springing back, sword high, he saw the horror strike the floor, wheel, and scuttle towards him with appalling speed—a gigantic black spider, such as men see only in nightmare dreams.

It was as large as a pig, and its eight thick hairy legs drove its ogreish body over the floor at headlong pace; its four evilly gleaming eyes shone with a horrible intelligence, and its fangs dripped venom that Conan knew, from the burning of his shoulder where only a few drops had splashed as the thing struck and missed, was laden with swift death. This was the killer that had dropped from its perch in the middle of the ceiling on a strand of web, on the neck of the Nemedian. Fools that they were, not to have suspected that the upper chambers would be guarded as well as the lower!

These thoughts flashed briefly through Conan's mind as the monster rushed. He leaped high, and it passed beneath him, wheeled, and charged back. This time he evaded its rush with a sidewise leap and struck back like a cat. His sword severed one of the hairy legs, and again he barely saved himself as the monstrosity swerved at him, fangs clicking fiendishly. But the creature did not press the pursuit; turning, it scuttled across the crystal floor and ran up the wall to the ceiling, where it crouched for an instant, glaring down at him with its fiendish red eyes. Then without warning it launched itself through space, trailing a strand of slimy greyish stuff.

Conan stepped back to avoid the hurtling body—then ducked frantically, just in time to escape being snared by the flying web-rope. He saw the monster's intent and sprang towards the door, but it was quicker, and a sticky strand cast across the door made him a prisoner. He dared not try to cut it with his sword; he knew the stuff would cling to the blade; and, before he could shake it loose, the fiend would be sinking its fangs into his back.

Then began a desperate game, the wits and quickness of the man matched against the fiendish craft and speed of the giant spider. It no longer scuttled across the floor in a direct charge, or swung its body through the air at him. It raced about the ceiling and the walls, seeking to snare him in the long loops of sticky grey web-strands, which it flung with a devilish accuracy. These strands were thick as ropes, and Conan knew that once they were coiled about him, his desperate strength would not be enough to tear him free before the monster struck.

All over the chamber went on that devil's dance, in utter silence except for the quick breathing of the man, the low scuff of his bare feet on the shining floor, the castanet rattle of the monstrosity's fangs. The grey strands lay in coils on the floor; they were looped along the walls; they overlaid the jewel-chests and silken couches, and hung in dusky festoons from the jewelled ceiling. Conan's steel-trap quickness of eye and muscle had kept him untouched, though the sticky loops had passed him so close they rasped his naked hide. He knew he could not always avoid them; he not only had to watch the strands swinging from the ceiling, but to keep his eye on the floor, lest he trip in the coils that lay there. Sooner or later a gummy loop would writhe about him, pythonlike, and then, wrapped like a cocoon, he would lie at the monster's mercy.

The spider raced across the chamber floor, the grey rope waving out behind it. Conan leaped high, clearing a couch—with a quick wheel the fiend ran up the wall, and the strand, leaping off the floor like a live thing, whipped about the Cimmerian's ankle. He caught himself on his hands as he fell, jerking frantically at the web which held him like a pliant vice, or the coil of a python. The hairy devil was racing down the wall to complete its capture. Stung to frenzy, Conan caught up a jewel chest and hurled it with all his strength. Full in the midst of the branching black legs the massive missile struck, smashing against the wall with a muffled sickening crunch. Blood and greenish slime spattered, and the shattered mass fell with the burst gem-chest to the floor. The crushed black body lay among the flaming riot of jewels that spilled over it; the hairy legs moved aimlessly, the dying eyes glittered redly among the twinkling gems.

Conan glared about, but no other horror appeared, and he set himself to working free of the web. The substance clung tenaciously to his ankle and his hands, but at last he was free, and taking up his sword, he picked his way among the grey coils and loops to the inner door. What horrors lay within he did not know. The Cimmerian's blood was up and, since he had come so far and overcome so much peril, he was determined to go through to the grim finish of the adventure, whatever that might be. And he felt that the jewel he sought was not among the many so carelessly strewn about the gleaming chamber.

Stripping off the loops that fouled the inner door, he found that it, like the other, was not locked. He wondered if the soldiers below were still unaware of his presence. Well, he was high above their heads, and if tales were to be believed, they were used to strange noises in the tower above them—sinister sounds, and screams of agony and horror.

Yara was on his mind, and he was not altogether comfortable as he opened the golden door. But he saw only a flight of silver steps leading down, dimly lighted by what means he could not ascertain. Down these he went silently, gripping his sword. He heard no sound and came presently to an ivory door, set with bloodstones. He listened, but no sound came from within; only thin wisps of smoke drifted lazily from beneath the door, bearing a curious exotic odour unfamiliar to the Cimmerian. Below him the silver stair wound down to vanish in the dimness, and up that shadowy well no sound floated; he had an eerie feeling that he was alone in a tower occupied only by ghosts and phantoms.

III

Cautiously he pressed against the ivory door, and it swung silently inward. On the shimmering threshold Conan stared like a wolf in strange surroundings, ready to fight or flee on the instant. He was looking into a large chamber with a domed golden ceiling; the walls were of green jade, the floor of ivory, partly covered with thick rugs. Smoke and exotic scent of incense floated up from a brazier on a golden tripod, and behind it sat an idol on a sort of marble couch. Conan stared aghast; the image had the body of a man, naked, and green in colour; but the head was one of nightmare and madness. Too large for the human body, it had no attributes of humanity. Conan stared at the wide flaring ears, the curling proboscis, on either side of which stood white tusks tipped with round golden balls. The eyes were closed, as if in sleep.

This then, was the reason for the name, the Tower of the Elephant, for the head of the thing was much like that of the beasts described by the Shemitish wanderer. This was Yara's god; where then should the gem be, but concealed in the idol, since the stone was called the Elephant's Heart?

As Conan came forward, his eyes fixed on the motionless idol, the eyes of the thing opened suddenly! The Cimmerian froze in his tracks. It was no image—it was a living thing, and he was trapped in its chamber!

That he did not instantly explode in a burst of murderous frenzy is a fact that measured his horror, which paralysed him where he stood. A civilized man in his position would have sought doubtful refuge in the conclusion that he was insane; it did not occur to the Cimmerian to doubt his senses. He knew he was face to face with a demon of the Elder World, and the realization robbed him of all his faculties except sight.

The trunk of the horror was lifted and quested about, the topaz eyes stared unseeingly, and Conan knew the monster was blind. With the thought came a thawing of his frozen nerves, and he began to back silently towards the door. But the creature heard. The sensitive trunk stretched towards him, and Conan's horror froze him again when the being spoke, in a strange, stammering voice that never changed its key or timbre. The Cimmerian knew that those jaws were never built or intended for human speech.

'Who is here? Have you come to torture me again, Yara? Will you never be done? Oh, Yag-kosha, is there no end to agony?'

Tears rolled from the sightless eyes, and Conan's gaze strayed to the limbs stretched on the marble couch. And he knew the monster would not rise to attack him. He knew the marks of the rack, and the searing brand of the flame, and tough-souled as he was, he stood aghast at the ruined deformities which his reason told him had once been limbs as comely as his own. And suddenly all fear and repulsion went from him to be replaced by a great pity. What this monster was, Conan could not know, but the evidences of its sufferings were so terrible and pathetic that a strange aching sadness came over the Cimmerian, he knew not why. He only felt that he was looking upon a cosmic tragedy, and he shrank with shame, as if the guilt of a whole race were laid upon him.

'I am not Yara,' he said. 'I am only a thief. I will not harm you.'

'Come near that I may touch you,' the creature faltered, and Conan came near unfearingly, his sword hanging forgotten in his hand. The sensitive trunk came out and groped over his face and shoulders, as a blind man gropes, and its touch was light as a girl's hand.

'You are not of Yara's race of devils,' sighed the creature. 'The clean, lean fierceness of the wastelands marks you. I know your people from of old, whom I knew by another name in the long, long ago when another world lifted its jewelled spires to the stars. There is blood on your fingers.'

'A spider in the chamber above and a lion in the garden,' muttered Conan.

'You have slain a man too, this night,' answered the other. 'And there is death in the tower above. I feel; I know.'

'Aye,' muttered Conan. 'The prince of all thieves lies there dead from the bite of a vermin.'

'So—and so!' the strange inhuman voice rose in a sort of low chant. 'A slaying in the tavern and a slaying on the roof—I know; I feel. And the third will make the magic of which not even Yara dreams—oh, magic of deliverance, green gods of Yag!'

Again tears fell as the tortured body was rocked to and fro in the grip of varied emotions. Conan looked on, bewildered.

Then the convulsions ceased; the soft, sightless eyes were turned towards the Cimmerian, the trunk beckoned.

'O man, listen,' said the strange being. 'I am foul and monstrous to you, am I not? Nay, do not answer; I know. But you would seem as strange to me, could I see you. There are many worlds besides this earth, and life takes many shapes. I am neither god nor demon, but flesh and blood like yourself, though the substance differ in part, and the form be cast in different mould.

'I am very old, O man of the waste countries; long and long ago I came to this planet with others of my world, from the green planet Yag, which circles for ever in the outer fringe of this universe. We swept through space on mighty wings that drove us through the cosmos quicker than light, because we had warred with the kings of Yag and were defeated and outcast. But we could never return, for on earth our wings withered from our shoulders. Here we abode apart from earthly life. We fought the strange and terrible forms of life which then walked the earth, so that we became feared and were not molested in the dim jungles of the East where we had our abode.

'We saw men grow from the ape and build the shining cities of Valusia, Kamelia, Commoria, and their sisters. We saw them reel before the thrusts of the heathen Atlanteans and Picts and Lemurians. We saw the oceans rise and engulf Atlantis and Lemuria, and the isles of the Picts, and the shining cities of civilization. We saw the survivors of Pictdom and Atlantis build their stone-age empire and go down to ruin, locked in bloody wars. We saw the Picts sink into abysmal savagery, the Atlanteans into apedom again. We saw new savages drift southward in conquering waves from the Arctic Circle to build a new civilization, with new kingdoms called Nemedia, and Koth, and Aquilonia, and their sisters. We saw your people rise under a new name from the jungles of the apes that had been Atlanteans. We saw the descendants of the Lemurians, who had survived the cataclysm, rise again through savagery and ride westward, as Hyrkanians. And we saw this race of devils, survivors of the ancient civilization that was before Atlantis sank, come once more into culture and power—this accursed kingdom of Zamora.

'All this we saw, neither aiding nor hindering the immutable cosmic law, and one by one we died; for we of Yag are not immortal, though our lives are as the lives of planets and constellations. At last I alone was left, dreaming of old times among the ruined temples of jungle-lost Khitai, worshipped as a god by an ancient yellow-skinned race. Then came Yara, versed in dark knowledge handed down through the days of barbarism, since before Atlantis sank.

'First he sat at my feet and learned wisdom. But he was not satisfied with what I taught him, for it was white magic, and he wished evil lore, to enslave kings and glut a fiendish ambition. I would teach him none of the black secrets I had gained, through no wish of mine, through the eons.

'But his wisdom was deeper than I had guessed; with guile gotten among the dusky tombs of dark Stygia, he trapped me into divulging a secret I had not intended to bare; and turning my own power upon me, he enslaved me. Ah, gods of Yag, my cup has been bitter since that hour!

'He brought me up from the lost jungles of Khitai where the grey apes danced to the pipes of the yellow priests, and offerings of fruit and wine heaped my broken altars. No more was I a god to kindly junglefolk—I was slave to a devil in human form.'

Again tears stole from the unseeing eyes.

'He pent me in this tower, which at his command I built for him in a single night. By fire and rack he mastered me, and by strange unearthly tortures you would not understand. In agony I would long ago have taken my own life, if I could. But he kept me alive—mangled, blinded, and broken—to do his foul bidding. And for three hundred years I have done his bidding, from this marble couch, blackening my soul with cosmic sins, and staining my wisdom with crimes, because I had no other choice. Yet not all my ancient secrets has he wrested from me, and my last gift shall be the sorcery of the Blood and the Jewel.

'For I feel the end of time draw near. You are the hand of Fate. I beg of you, take the gem you will find on yonder altar.'

Conan turned to the gold and ivory altar indicated, and took up a great round jewel, clear as crimson crystal; and he knew that this was the Heart of the Elephant.

'Now for the great magic, the mighty magic, such as earth has not seen before, and shall not see again, through a million million of millenniums. By my life-blood I conjure it, by blood born on the green breast of Yag, dreaming far-poised in the great, blue vastness of Space.

'Take your sword, man, and cut out my heart; then squeeze it so that the blood will flow over the red stone. Then go you down these stairs and enter the ebony chamber where Yara sits wrapped in lotus dreams of evil.

Speak his name and he will awaken. Then lay this gem before him, and say, "Yag-kosha gives you a last gift and a last enchantment." Then get you from the tower quickly; fear not, your way shall be made clear. The life of man is not the life of Yag, nor is human death the death of Yag. Let me be free of this cage of broken, blind flesh, and I will once more be Yogah of Yag, morning-crowned and shining, with wings to fly, and feet to dance, and eyes to see, and hands to break.'

Uncertainly Conan approached, and Yag-kosha, or Yogah, as if sensing his uncertainty, indicated where he should strike. Conan set his teeth and drove the sword deep. Blood streamed over the blade and his hand, and the monster started convulsively, then lay back quite still. Sure that life had fled, at least life as he understood it, Conan set to work on his grisly task and quickly brought forth something that he felt must be the strange being's heart, though it differed curiously from any he had ever seen. Holding the still pulsing organ over the blazing jewel, he pressed it with both hands, and a rain of blood fell on the stone. To his surprise, it did not run off, but soaked into the gem, as water is absorbed by a sponge.

Holding the jewel gingerly, he went out of the fantastic chamber and came upon the silver steps. He did not look back; he instinctively felt that some form of transmutation was taking place in the body on the marble couch, and he further felt that it was a sort not to be witnessed by human eyes.

He closed the ivory door behind him and without hesitation descended the silver steps. It did not occur to him to ignore the instructions given him. He halted at an ebony door, in the centre of which was a grinning silver skull, and pushed it open. He looked into a chamber of ebony and jet and saw, on a black silken couch, a tall, spare form reclining. Yara the priest and sorcerer lay before him, his eyes open and dilated with the fumes of the yellow lotus, far-staring, as if fixed on gulfs and nighted abysses beyond human ken.

'Yara!' said Conan, like a judge pronouncing doom. 'Awaken!'

The eyes cleared instantly and became cold and cruel as a vulture's. The tall, silken-clad form lifted erect and towered gauntly above the Cimmerian.

'Dog!' His hiss was like the voice of a cobra. 'What do you here?'

Conan laid the jewel on the great ebony table.

'He who sent this gem bade me say, "Yag-kosha gives a last gift and a last enchantment."'

Yara recoiled, his dark face ashy. The jewel was no longer crystal-clear; its murky depths pulsed and throbbed, and curious smoky waves of changing colour passed over its smooth surface. As if drawn hypnotically, Yara bent over the table and gripped the gem in his hands, staring into its shadowed depths, as if it were a magnet to draw the shuddering soul from his body. And as Conan looked, he thought that his eyes must be playing him tricks. For when Yara had risen up from his couch, the priest had seemed gigantically tall; yet now he saw that Yara's head would scarcely come to his shoulder. He blinked, puzzled, and for the first time that night doubted his own senses. Then with a shock he realized that the priest was shrinking in stature—was growing smaller before his very gaze.

With a detached feeling he watched, as a man might watch a play; immersed in a feeling of overpowering unreality, the Cimmerian was no longer sure of his own identity; he only knew that he was looking upon the external evidences of the unseen play of vast Outer forces, beyond his understanding.

Now Yara was no bigger than a child; now like an infant he sprawled on the table, still grasping the jewel. And now the sorcerer suddenly realized his fate, and he sprang up, releasing the gem. But still he dwindled, and Conan saw a tiny, pigmy figure rushing wildly about the ebony tabletop, waving tiny arms and shrieking in a voice that was like the squeak of an insect.

Now he had shrunk until the great jewel towered above him like a hill, and Conan saw him cover his eyes with his hands, as if to shield them from the glare, as he staggered about like a madman. Conan sensed that some unseen magnetic force was pulling Yara to the gem. Thrice he raced wildly about it in a narrowing circle, thrice he strove to turn and run out across the table; then with a scream that echoed faintly in the ears of the watcher, the priest threw up his arms and ran straight towards the blazing globe.

Bending close, Conan saw Yara clamber up the smooth, curving surface, impossibly, like a man climbing a glass mountain. Now the priest stood on the top, still with tossing arms, invoking what grisly names only the gods know. And suddenly he sank into the very heart of the jewel, as a man sinks into a sea, and Conan saw the smoky waves close over his head. Now he saw him in the crimson heart of the jewel, once more crystal-clear, as a man sees a scene far away, tiny with great distance. And into the heart came a green shining winged figure with the body of a man and the head of an elephant—no longer blind or crippled. Yara threw up his arms and fled as a madman flees, and on his heels came the avenger. Then, like the bursting of a bubble, the great jewel vanished in a rainbow burst of iridescent gleams, and the ebony tabletop lay bare and deserted—

as bare, Conan somehow knew, as the marble couch in the chamber above, where the body of that strange transcosmic being called Yag-kosha and Yogah had lain.

The Cimmerian turned and fled from the chamber, down the silver stairs. So mazed was he that it did not occur to him to escape from the tower by the way he had entered it. Down that winding, shadowy silver well he ran, and came into a larger chamber at the foot of the gleaming stairs. There he halted for an instant; he had come into the room of the soldiers. He saw the glitter of their silver corselets, the sheen of their jewelled sword-hilts. They sat slumped at the banquet board, their dusky plumes waving sombrely above their drooping helmeted heads; they lay among their dice and fallen goblets on the wine-stained, lapis-lazuli floor. And he knew that they were dead. The promise had been made, the word kept; whether sorcery or magic or the falling shadow of great green wings had stilled the revelry, Conan could not know, but his way had been made clear. And a silver door stood open, framed in the whiteness of dawn.

Into the waving green gardens came the Cimmerian and, as the dawn wind blew upon him with the cool fragrance of luxuriant growths, he started like a man waking from a dream. He turned back uncertainly, to stare at the cryptic tower he had just left. Was he bewitched and enchanted? Had he dreamed all that had seemed to have passed? As he looked he saw the gleaming tower sway against the crimson dawn, its jewel-crusted rim sparkling in the growing light, and crash into shining shards.

XEETHRA

Subtle and manifold are the nets of the Demon, who followeth his chosen from birth to death and from death to birth, throughout many lives.

THE TESTAMENTS OF CARNAMAGOS

Long had the wasting summer pastured its suns, like fiery red stallions, on the dun hills that crouched before the Mykrasian Mountains in wild easternmost Cincor. The peak-fed torrents were become tenuous threads or far-sundered, fallen pools; the granite boulders were shaled by the heat; the bare earth was cracked and creviced; and the low, meagre grasses were seared even to the roots.

So it occurred that the boy Xeethra, tending the black and piebald goats of his uncle Pornos, was obliged to follow his charges further each day on the combes and hilltops. In an afternoon of late summer he came to a deep, craggy valley which he had never before visited. Here a cool and shadowy tarn was watered by hidden well-springs; and the ledgy slopes about the tarn were mantled with herbage and bushes that had not wholly lost their vernal greenness.

Surprised and enchanted, the young goatherd followed his capering flock into this sheltered paradise. There was small likelihood that the goats of Pornos would stray afield from such goodly pasturage; so Xeethra did not trouble himself to watch them any longer. Entranced by his surroundings, he began to explore the valley, after quenching his thirst at the clear waters that sparkled like golden wine.

To him, the place seemed a veritable garden-pleasance. Forgetting the distance he had already come, and the wrath of Pornos if the flock should return late for the milking, he wandered deeper among the winding crags that protected the valley. On every hand the rocks grew sterner and wilder; the valley straitened; and he stood presently at its end, where a rugged wall forbade further progress.

Feeling a vague disappointment, he was about to turn and retrace his wanderings. Then, in the base of the sheer wall, he perceived the mysterious yawning of a cavern. It seemed that the rock must have opened only a little while before his coming: for the lines of cleavage were clearly marked, and the cracks made in the surrounding surface were unclaimed by the moss that grew plentifully elsewhere. From the cavern's creviced lip there sprang a stunted tree, with its newly broken roots hanging in air; and the stubborn taproot was in the rock at Xeethra's feet, where, it was plain, the tree had formerly stood.

Wondering and curious, the boy peered into the inviting gloom of the cavern, from which, unaccountably, a soft balmy air now began to blow. There were strange odours on the air, suggesting the pungency of temple incense, the languor and luxury of opiate blossoms. They disturbed the senses of Xeethra; and, at the same time, they seduced him with their promise of unbeholden marvellous things. Hesitating, he tried to remember certain legends that Pornos had told him: legends that concerned such hidden caverns as the one on which he had stumbled. But it seemed that the tales had faded now from his mind, leaving only a dim sense of things that were perilous, forbidden, and magical. He thought that the cavern was the portal of some undiscovered world—and the portal had opened expressly to permit his entrance. Being of a nature both venturesome and visionary, he was undeterred by the fears that others might have felt in his place. Overpowered by a great curiosity, he soon entered the cave, carrying for a torch a dry, resinous bough that had fallen from the tree in the cliff.

Beyond the mouth he was swallowed by a rough-arched passage that pitched downward like the gorge of some monstrous dragon. The torch's flame blew back, flaring and smoking in the warm aromatic wind that strengthened from unknown depths. The cave steepened perilously; but Xeethra continued his exploration, climbing down by the stair-like coigns and projections of the stone.

Like a dreamer in a dream, he was wholly absorbed by the mystery on which he had happened; and at no time did he recall his abandoned duty. He lost all reckoning of the time consumed in his descent. Then, suddenly, his torch was extinguished by a hot gust that blew upon him like the expelled breath of some prankish demon.

Feeling the assailment of a black panic, he tottered in darkness and sought to secure his footing on the dangerous incline. But, ere he could relume the blown-out torch, he saw that the night around him was not complete, but was tempered by a wan, golden glimmering from the depths below. Forgetting his alarm in a new wonder, he descended towards the mysterious light.

At the bottom of the long incline, Xeethra passed through a low cavern-mouth and emerged into sun-bright radiance. Dazzled and bewildered, he thought for a moment that his subterranean wanderings had brought him back to the outer air in some unsuspected land lying among the Mykrasian hills. Yet surely the region before him was no part of summer-stricken Cincor: for he saw neither hills nor mountains nor the black sapphire heaven from which the aging but despotic sun glared down with implacable drouth on the kingdoms of Zothique.

He stood on the threshold of a fertile plain that lapsed illimitably into golden distance under the measureless arch of a golden vault. Far off, through the misty radiance, there was a dim towering of unidentifiable masses that might have been spires and domes and ramparts. A level meadow lay at his feet, covered with close-grown curling sward that had the greenness of verdigris; and the sward, at intervals, was studded with strange blossoms appearing to turn and move like living eyes. Near at hand, beyond the meadow, was an orchard-like grove of tall, amply spreading trees amid whose lush leafage he descried the burning of numberless dark-red fruits. The plain, to all seeming, was empty of human life; and no birds flew in the fiery air or perched on the laden boughs. There was no sound other than the sighing of leaves: a sound like the hissing of many small hidden serpents.

To the boy from the parched hill-country, this realm was an Eden of untasted delights. But, for a little, he was stayed by the strangeness of it all, and by the sense of weird and preternatural vitality which informed the whole landscape. Flakes of fire appeared to descend and melt in the rippling air; the grasses coiled with verminous writhings; the flowery eyes returned his regard intently; the trees palpitated as if a sanguine ichor flowed within them in lieu of sap; and the undernote of adder-like hissings amid the foliage grew louder and sharper.

Xeethra, however, was deterred only by the thought that a region so fair and fertile must belong to some jealous owner who would resent his intrusion. He scanned the unpeopled plain with much circumspection. Then, deeming himself secure from observation, he yielded to the craving that had been roused within him by the red, luxuriant fruit.

The turf was elastic beneath him, like a living substance, as he ran forward to the nearest trees. Bowed with their shining globes, the branches drooped around him. He plucked several of the largest fruits and stored them thriftily in the bosom of his threadbare tunic. Then, unable to resist his appetence any longer, he began to devour one of the fruits. The rind broke easily under his teeth, and it seemed that a royal wine, sweet and puissant, was poured into his mouth from an overbrimming cup. He felt in his throat and bosom a swift warmth that almost suffocated him; and a strange fever sang in his ears and wildered his senses. It passed quickly, and he was startled from his bemusement by the sound of voices falling as if from an airy height.

He knew instantly that the voices were not those of men. They filled his ears with a rolling as of baleful drums, heavy with ominous echoes; yet it seemed that they spoke in articulate words, albeit of a strange language. Looking up between the thick boughs, he beheld a sight that inspired him with terror. Two beings of colossean stature, tall as the watchtowers of the mountain people, stood waist-high above the near treetops. It was as if they had appeared by sorcery from the green ground or the gold heavens: for surely the clumps of vegetation, dwarfed into bushes by their bulk, could never have concealed them from Xeethra's discernment.

The figures were clad in black armour, lustreless and gloomy, such as demons might wear in the service of Thasaidon, lord of the bottomless underworlds. Xeethra felt sure that they had seen him; and perhaps their unintelligible converse concerned his presence. He trembled, thinking now that he had trespassed on the gardens of genii. Peering fearfully from his covert, he could discern no features beneath the frontlets of the dark helms that were bowed towards him: but eye-like spots of yellowish-red fire, restless as marsh-lights, shifted to and fro in void shadow where the faces should have been.

It seemed to Xeethra that the rich foliage could afford no shelter from the scrutiny of these beings, the guardians of the land on which he had so rashly intruded. He was overwhelmed by a consciousness of guilt: the sibilant leaves, the drum-like voices of the giants, the eye-shaped flowers—all appeared to accuse him of trespass and thievery. At the same time he was perplexed by a queer and unwonted vagueness in regard to his own identity: somehow it was not Xeethra the goatherd . . . but another . . . who had found the bright garden-realm and had eaten the blood-dark fruit. This alien self was without name or formulable memory; but there was a flickering of confused lights, a murmur of indistinguishable voices, amid the stirred shadows of his mind. Again he felt the weird warmth, the swift-mounting fever, that had followed the devouring of the fruit.

From all this, he was aroused by a livid flash of light that clove downward towards him across the branches. Whether a bolt of levin had issued from the clear vault, or whether one of the armoured beings had brandished a great sword, he was never quite sure afterwards. The light seared his vision, he recoiled in uncontrollable fright, and found himself running, half-blind, across the open turf. Through whirling bolts of colour he saw before him, in a sheer, topless cliff, the cavern-mouth through which he had come. Behind him he heard a long rumbling as of summer thunder . . . or the laughter of colossi.

Without pausing to retrieve the still-burning brand he had left at the entrance, Xeethra plunged incontinently into the dark cave. Through Stygian murk he managed to grope his way upward on the perilous incline. Reeling, stumbling, bruising himself at every turn, he came at last to the outer exit, in the hidden valley behind the hills of Cincor.

To his consternation, twilight had fallen during his absence in the world beyond the cave. Stars crowded above the grim crags that walled the valley; and the skies of burnt-out purple were gored by the sharp horn of an ivory moon. Still fearing the pursuit of the giant guardians, and apprehending also the wrath of his uncle Pornos, Xeethra hastened back to the little tarn, collected his flock, and drove it homeward through the long, gloomy miles.

During that journey, it seemed that a fever burned and died within him at intervals, bringing strange fancies. He forgot his fear of Pornos, forgot, indeed, that he was Xeethra, the humble and disregarded goatherd. He was returning to another abode than the squalid hut of Pornos, built of clay and brushwood. In a high-domed city, gates of burnished metal would open for him, and fiery-coloured banners would stream on the perfumed air; and silver trumpets and the voices of blonde odalisques and black chamberlains would greet him as king in a thousand-columned hall. The ancient pomp of royalty, familiar as air and light, would surround him, and he, the King Amero, who had newly come to the throne, would rule as his fathers had ruled over all the kingdom of Calyz by the orient sea. Into his capital, on shaggy camels, the fierce southern tribesmen would bring a levy of date-wine and desert sapphires; and galleys from isles beyond the morning would burden his wharves with their semi-annual tribute of spices and strange-dyed fabrics. . . .

Surging and fading like pictures of delirium but lucid as daily memories, the madness came and went; and once again he was the nephew of Pornos, returning belated with the flock.

Like a downward-thrusting blade, the red moon had fixed itself in the sombre hills when Xeethra reached the rough wooden pen in which Pornos kept his goats. Even as Xeethra had expected, the old man was waiting at the gate, bearing in one hand a clay lantern and in the other a staff of briarwood. He began to curse the boy with half-senile vehemence, waving the staff, and threatening to beat him for his tardiness.

Xeethra did not flinch before the staff. Again, in his fancy, he was Amero, the young king of Calyz. Bewildered and astonished, he saw before him by the light of the shaken lantern a foul and rancid-smelling ancient whom he could not remember. Hardly could he understand the speech of Pornos; the man's anger puzzled but did not frighten him; and his nostrils, as if accustomed only to delicate perfumes, were offended by the goatish stench. As if for the first time, he heard the bleating of the tired flock, and gazed in wild surprise at the wattled pen and the hut beyond.

'Is it for this,' cried Pornos, 'that I have reared my sister's orphan at great expense? Accursed moon-calf! thankless whelp! If you have lost a milch-goat or a single kid, I shall flay you from thigh to shoulder.'

Deeming that the silence of the youth was due to mere obstinacy, Pornos began to beat him with the staff. At the first blow, the bright cloud lifted from Xeethra's mind. Dodging the briarwood with agility, he tried to tell Pornos of the new pasture he had found among the hills. At this the old man suspended his blows, and Xeethra went on to tell of the strange cave that had conducted him to an unguessed garden-land. To support his story, he reached within his tunic for the blood-red apples he had stolen; but, to his confoundment, the fruits were gone, and he knew not whether he had lost them in the dark or whether, perhaps, they had vanished by virtue of some indwelling necromancy.

Pornos, interrupting the youth with frequent scoldings, heard him at first with open unbelief. But he grew silent as the youth went on; and when the story was done, he cried out in a trembling voice:

'Ill was this day, for you have wandered among enchantments. Verily, there is no tarn such as you have described amid the hills; nor, at this season, has any herder found such pasturage. These things were illusion, designed to lead you astray; and the cave, I wot, was no honest cave but an entrance into hell. I have heard my fathers tell that the gardens of Thasaidon, king of the seven underworlds, lie near to the earth's surface in this region; and caves have opened ere this, like a portal, and the sons of men, trespassing unaware on the gardens, have been tempted by the fruit and eaten it. But madness comes thereof and much sorrow and long damnation: for the Demon, they say, forgetting not one stolen apple, will exact his price in the end. Woe! woe! the goat-milk will be soured for a whole moon by the grass of such wizard pasture; and, after all the food and care you have cost me, I must find another stripling to ward the flocks.'

Once more, as he listened, the burning cloud returned upon Xeethra.

'Old man, I know you not,' he said perplexedly. Then, using soft words of a courtly speech but half-intelligible to Pornos: 'It would seem that I have gone astray. Prithee, where lies the kingdom of Calyz? I am king thereof, being newly crowned in the high city of Shathair, over which my fathers have ruled for a thousand years.'

'Ai! Ai!' wailed Pornos. 'The boy is daft. These notions have come through the eating of the Demon's apple. Cease your maundering, and help me to milk the goats. You are none other than the child of my sister Askli, who was delivered these nineteen years agone after her husband, Outhoth, had died of a dysentery. Askli lived not long, and I, Pornos, have reared you as a son, and the goats have mothered you.'

'I must find my kingdom,' persisted Xeethra. 'I am lost in darkness, amid uncouth things, and how I have wandered here I cannot remember. Old man, I would have you give me food and lodging for the night. In the dawn I shall journey towards Shathair, by the orient main.'

Pornos, shaking and muttering, lifted his clay lantern to the boy's face. It seemed that a stranger stood before him, in whose wide and wondering eyes the flame of golden lamps was somehow reflected. There was no wildness in Xeethra's demeanour, but merely a sort of gentle pride and remoteness; and he wore his threadbare tunic with a strange grace. Surely, however, he was demented; for his manner and speech were past understanding. Pornos, mumbling under his breath, but no longer urging the boy to assist him, turned to the milking. . . .

Xeethra woke betimes in the white dawn, and peered with amazement at the mud-plastered walls of the hovel in which he had dwelt since birth. All was alien and baffling to him; and especially was he troubled by his rough garments and by the sun-swart tawniness of his skin: for such were hardly proper to the young King Amero, whom he believed himself to be. His circumstances were wholly inexplicable; and he felt an urgency to depart at once on his homeward journey.

He rose quietly from the litter of dry grasses that had served him for a bed. Pornos, lying in a far corner, still slept the sleep of age and senescence; and Xeethra was careful not to awaken him. He was both puzzled and repelled by this unsavoury ancient, who had fed him on the previous evening with coarse millet-bread and the strong milk and cheese of goats, and had given him the hospitality of a fetid hut. He had paid little heed to the mumblings and objurgations of Pornos; but it was plain that the old man doubted his claims to royal rank, and, moreover, was possessed of peculiar delusions regarding his identity.

Leaving the hovel, Xeethra followed an eastward-winding footpath amid the stony hills. He knew not whither the path would lead: but reasoned that Calyz, being the easternmost realm of the continent Zothique, was situated somewhere below the rising sun. Before him, in vision, the verdant vales of his kingdom hovered like a fair mirage, and the swelling domes of Shathair were as morning cumuli piled in the orient. These things, he deemed, were memories of yesterday. He could not recall the circumstances of his departure and his absence; but surely the land over which he ruled was not remote.

The path turned among lessening ridges, and Xeethra came to the small village of Cith, to whose inhabitants he was known. The place was alien to him now, seeming no more than a cirque of filthy hovels that reeked and festered under the sun. The people gathered about him, calling him by name, and staring and laughing oafishly when he enquired the road to Calyz. No one, it appeared, had ever heard of this kingdom or of the city of Shathair. Noting a strangeness in Xeethra's demeanour, and deeming that his queries were those of a madman, the people began to mock him. Children pelted him with dry clods and pebbles; and thus he was driven from Cith, following an eastern road that ran from Cincor into the neighbouring lowlands of the country of Zhel.

Sustained only by the vision of his lost kingdom, the youth wandered for many moons throughout Zothique. People derided him when he spoke of his kingship and made enquiry concerning Calyz; but many, thinking madness a sacred thing, offered him shelter and sustenance. Amid the far-stretching fruitful vineyards of Zhel, and into Istanam of the myriad cities; over the high passes of Ymorth, where snow tarried at the autumn's beginning; and across the salt-pale desert of Dhir, Xeethra followed that bright imperial dream which had now become his only memory. Always eastward he went, travelling sometimes with caravans whose members hoped that a madman's company would bring them good fortune; but oftener he went as a solitary wayfarer.

At whiles, for a brief space, his dream deserted him, and he was only the simple goatherd, lost in foreign realms, and homesick for the barren hills of Cincor. Then, once more, he remembered his kingship, and the opulent gardens of Shathair and the proud palaces, and the names and faces of them that had served him following the death of his father, King Eldamaque, and his own succession to the throne.

At midwinter, in the far city of Sha-Karag, Xeethra met certain sellers of amulets from Ustaim, who smiled oddly when he asked if they could direct him to Calyz. Winking among themselves when he spoke of his royal rank, the merchants told him that Calyz was situated several hundred leagues beyond Sha-Karag, below the orient sun.

'Hail, O King,' they said with mock ceremony. 'Long and merrily may you reign in Shathair.'

Very joyful was Xeethra, hearing word of his lost kingdom for the first time, and knowing now that it was more than a dream or a figment of madness. Tarrying no longer in Sha-Karag, he journeyed on with all possible haste. . . .

When the first moon of spring was a frail crescent at eve, he knew that he neared his destination. For Canopus burned high in the eastern heavens, mounting gloriously amid the smaller stars even as he had once seen it from his palace-terrace in Shathair.

His heart leapt with the gladness of homecoming; but much he marvelled at the wildness and sterility of the region through which he passed. It seemed that there were no travellers coming and going from Calyz; and he met only a few nomads, who fled at his approach like the creatures of the waste. The highway was overgrown with grasses and cacti, and was rutted only by the winter rains. Beside it, anon, he came to a stone terminus carved in the form of a rampant lion, that had marked the western boundary of Calyz. The lion's features had crumbled away, and his paws and body were lichened, and it seemed that long ages of desolation had gone over him. A chill dismay was born in Xeethra's heart: for only yesteryear, if his memory served him rightly, he had ridden past the lion with his father Eldamaque, hunting hyenas, and had remarked then the newness of the carving.

Now, from the high ridge of the border, he gazed down upon Calyz, which had lain like a long verdant scroll beside the sea. To his wonderment and consternation, the wide fields were sere as if with autumn; the rivers were thin threads that wasted themselves in sand; the hills were gaunt as the ribs of unceremented mummies; and there was no greenery other than the scant herbage which a desert bears in spring. Far off, by the purple main, he thought that he beheld the shining of the marble domes of Shathair; and, fearing that some blight of hostile sorcery had fallen upon his kingdom, he hastened towards the city.

Everywhere, as he wandered heartsick through the vernal day, he found that the desert had established its empire. Void were the fields, unpeopled the villages. The cots had tumbled into midden-like heaps of ruin; and it seemed that a thousand seasons of drouth had withered the fruitful orchards, leaving only a few black and decaying stumps.

In the late afternoon he entered Shathair, which had been the white mistress of the orient sea. The streets and the harbour were alike empty, and silence sat on the broken housetops and the ruining walls. The great bronze obelisks were greened with antiquity; the massy marmorean temples of the gods of Calyz leaned and slanted to their fall.

Tardily, as one who fears to confirm an expected thing, Xeethra came to the palace of the monarchs. Not as he recalled it, a glory of soaring marble half veiled by flowering almonds and trees of spice and high-pulsing fountains, but in stark dilapidation amid blasted gardens, the palace awaited him, while the brief, illusory rose of sunset faded upon its domes, leaving them wan as mausoleums.

How long the place had lain desolate, he could not know. Confusion filled him, and he was whelmed by utter loss and despair. It seemed that none remained to greet him amid the ruins; but, nearing the portals of the west wing, he saw, as it were, a fluttering of shadows that appeared to detach themselves from the gloom beneath the portico; and certain dubious beings, clothed in rotten tatters, came sidling and crawling before him on the cracked pavement. Pieces of their raiment dropped from them as they moved; and about them was an unnamed horror of filth, of squalor and disease. When they neared him, Xeethra saw that most of them were lacking in some member or feature, and that all were marked by the gnawing of leprosy.

His gorge rose within him, and he could not speak. But the lepers hailed him with hoarse cries and hollow croakings, as if deeming him another outcast who had come to join them in their abode amid the ruins.

'Who are ye that dwell in my palace of Shathair?' he enquired at length. 'Behold! I am King Amero, the son of Eldamaque, and I have returned from a far land to resume the throne of Calyz.'

At this, a loathsome cackling and tittering arose among the lepers. 'We alone are the kings of Calyz,' one of them told the youth. 'The land has been a desert for centuries, and the city of Shathair has long lain unpeopled save by such as we, who were driven out from other places. Young man, you are welcome to share the realm with us: for another king, more or less, is a small matter here.'

Thus, with obscene cachinnations, the lepers jeered at Xeethra and derided him; and he, standing amid the dark fragments of his dream, could find no words to answer them. However, one of the oldest lepers, well-nigh limbless and faceless, shared not in the mirth of his fellows, but seemed to ponder and reflect; and he said at last to Xeethra, in a voice issuing thickly from the black pit of his gaping mouth:

'I have heard something of the history of Calyz, and the names of Amero and Eldamaque are familiar to me. In bygone ages certain of the rulers were named thus; but I know not which of them was the son and which the father. Haply both are now entombed, with the rest of their dynasty, in the deep-lying vaults beneath the palace.'

Now, in the greying twilight, other lepers emerged from the shadowy ruin and gathered about Xeethra. Hearing that he laid claim to the kingship of the desert realm, certain of their number went away and returned presently, bearing vessels filled with rank water and mouldy victuals, which they proffered to Xeethra, bowing low with a mummery as of chamberlains serving a monarch.

Xeethra turned from them in loathing, though he was famished and athirst. He fled through the ashen gardens, among the dry fountain-mouths and dusty plots. Behind him he heard the hideous mirth of the lepers; but the sound grew fainter, and it seemed that they did not follow him. Rounding the vast palace in his flight, he met no more of these creatures. The portals of the south wing and the east wing were dark and empty, but he did not care to enter them, knowing that desolation and things worse than desolation were the sole tenants.

Wholly distraught and despairing, he came to the eastern wing and paused in the gloom. Dully, and with a sense of dream-like estrangement, he became aware that he stood on that very terrace above the sea, which he had remembered so often during his journey. Bare were the ancient flower-beds; the trees had rotted away in their sunken basins; and the great flags of the pavement were runnelled and broken. But the veils of twilight were tender upon the ruin; and the sea sighed as of yore under a purple shrouding; and the mighty star Canopus climbed in the east, with the lesser stars still faint around him.

Bitter was the heart of Xeethra, thinking himself a dreamer beguiled by some idle dream. He shrank from the high splendour of Canopus, as if from a flame too bright to bear; but, ere he could turn away, it seemed that a column of shadow, darker than the night and thicker than any cloud, rose upward before him from the terrace and blotted out the effulgent star. Out of the solid stone the shadow grew, towering tall and colossal; and it took on the outlines of a mailed warrior; and it seemed that the warrior looked down upon Xeethra from a great height with eyes that shone and shifted like fireballs in the darkness of his face under the lowering helmet.

Confusedly, as one who recalls an old dream, Xeethra remembered a boy who had herded goats upon summer-stricken hills; and who, one day, had found a cavern that opened portal-like on a garden-land of strangeness and marvel. Wandering there, the boy had eaten a blood-dark fruit and had fled in terror before the black-armoured giants who warded the garden. Again he was that boy; and still he was the King Amero, who had sought for his lost realm through many regions; and, finding it in the end, had found only the abomination of desolation.

Now, as the trepidation of the goatherd, guilty of theft and trespass, warred in his soul with the pride of the king, he heard a voice that rolled through the heavens like thunder from a high cloud in the spring night:

'I am the emissary of Thasaidon, who sends me in due season to all who have passed the nether portals and have tasted the fruit of his garden. No man, having eaten the fruit, shall remain thereafter as he was before; but to some the fruit brings oblivion, and to others, memory. Know, then, that in another birth, ages agone, you were indeed the young King Amero. The memory, being strong upon you, has effaced the remembrance of your present life, and has driven you forth to seek your ancient kingdom.'

'If this be true, then doubly am I bereft,' said Xeethra, bowing sorrowfully before the shadow. 'For, being Amero, I am throneless and realmless; and, being Xeethra, I cannot forget my former royalty and regain the content which I knew as a simple goatherd.'

'Harken, for there is another way,' said the shadow, its voice muted like the murmur of a far ocean. 'Thasaidon is the master of all sorceries, and a giver of magic gifts to those who serve him and acknowledge him as their lord. Pledge your allegiance, promise your soul to him; and in fee thereof the Demon will surely reward you. If it be your wish, he can wake again the buried past with his necromancy. Again, as King Amero, you shall reign over Calyz; and all things shall be as they were in the perished years; and the dead faces and the fields now desert shall bloom again.'

'I accept the bond,' said Xeethra. 'I plight my fealty to Thasaidon, and I promise my soul to him if he, in return, will give me back my kingdom.'

'There is more to be said,' resumed the shadow. 'Not wholly have you remembered your other life, but merely those years that correspond to your present youth. Living again as Amero, perhaps you will regret your royalty in time; and if such regret should overcome you, leading you to forget a monarch's duty, then the whole necromancy shall end and vanish like vapour.'

'So be it,' said Xeethra. 'This, too, I accept as part of the bargain.'

When the words ended, he beheld no longer the shadow towering against Canopus. The star flamed with a pristine splendour, as if no cloud had ever dimmed it; and, without sense of change or transition, he who watched the star was none other than King Amero; and the goatherd Xeethra, and the emissary, and the pledge given to Thasaidon, were as things that had never been. The ruin that had come upon Shathair was no more than the dream of some mad prophet; for in the nostrils of Amero the perfume of languorous flowers mingled with salt sea-balsams; and in his ears the grave murmur of ocean was pierced by the amorous plaint of lyres and a shrill laughter of slave-girls from the palace behind him. He heard the myriad noises of the nocturnal city, where his people feasted and made jubilee; and, turning from the star with a mystic pain and an obscure joy in his heart, Amero beheld the effulgent portals and windows of his father's house, and the far-mounting light from a thousand flambeaux that paled the stars as they passed over Shathair.

It is written in the old chronicles that King Amero reigned for many prosperous years. Peace and abundance were upon all the realm of Calyz; the drouth came not from the desert, nor violent gales from the main; and tribute was sent at the ordained seasons to Amero from the subject isles and outlying lands. And Amero was well content, dwelling superbly in rich-arrased halls, feasting and drinking royally, and hearing the praise of his lute-players and his chamberlains and his lemans.

When his life was a little past the meridian years, there came at whiles to Amero something of that satiety which lies in wait for the minions of fortune. At such times he turned from the cloying pleasures of the court and found delight in blossoms and leaves and the verses of olden poets. Thus was satiety held at bay; and, since the duties of the realm rested lightly upon him, Amero still found his kingship a goodly thing.

Then, in a latter autumn, it seemed that the stars looked disastrously upon Calyz. Murrain and blight and pestilence rode abroad as if on the wings of unseen dragons. The coast of the kingdom was beset and sorely harried by pirate galleys. Upon the west, the caravans coming and going through Calyz were assailed by redoubtable bands of robbers; and certain fierce desert peoples made war on the villages lying near to the southern border. The land was filled with turmoil and death, with lamentations and many miseries.

Deep was Amero's concern, hearing the distressful complaints that were brought before him daily. Being but little skilled in kingcraft, and wholly untried by the ordeals of dominion, he sought counsel of his courtlings but was ill advised by them. The troubles of the realm multiplied upon him; uncurbed by authority, the wild peoples of the waste grew bolder, and the pirates gathered like vultures of the sea. Famine and drouth divided his realm with the plague; and it seemed to Amero, in his sore perplexity, that such matters were beyond all medication; and his crown was become a too onerous burden.

Striving to forget his own impotence and the woeful plight of his kingdom, he gave himself to long nights of debauch. But the wine refused its oblivion, and love had now forfeited its rapture. He sought other divertissements, calling before him strange maskers and mummers and buffoons, and assembling outlandish singers and the players of uncouth instruments. Daily he made proclamation of a high reward to any that could bemuse him from his cares.

Wild songs and sorcerous ballads of yore were sung to him by immortal minstrels; the black girls of the north, with amber-dappled limbs, danced before him their weird lascivious measures; the blowers of the horns of chimeras played a mad and secret tune; and savage drummers pounded a troublous music on drums made from the skin of cannibals; while men clothed with the scales and pelts of half-mythic monsters ramped or crawled grotesquely through the halls of the palace. But all these were vain to beguile the king from his grievous musings.

One afternoon, as he sat heavily in his hall of audience, there came to him a player of pipes who was clad in tattered homespun. The eyes of the man were bright as newly stirred embers, and his face was burned to a cindery blackness, as if by the ardour of outland suns. Hailing Amero with small servility, he announced himself as a goatherd who had come to Shathair from a region of valleys and mountains lying sequestered beyond the bourn of sunset.

'O King, I know the melodies of oblivion,' he said, 'and I would play for you, though I desire not the reward you have offered. If haply I succeed in diverting you, I shall take my own guerdon in due time.'

'Play, then,' said Amero, feeling a faint interest rise within him at the bold speech of the piper.

Forthwith, on his pipes of reed, the black goatherd began a music that was like the falling and rippling of water in quiet vales, and the passing of wind over lonely hilltops. Subtly the pipes told of freedom and peace and forgetfulness lying beyond the sevenfold purple of outland horizons.

Dulcetly they sang of a place where the years came not with an iron trampling, but were soft of tread as a zephyr shod with flower petals. There the world's turmoil and troubling were lost upon measureless leagues of silence, and the burdens of empire were blown away like thistledown. There the goatherd, tending his flock on solitary fells, was possessed of tranquillity sweeter than the power of monarchs.

As he listened to the piper, a sorcery crept upon the mind of Amero. The weariness of kingship, the cares and perplexities, were as dream-bubbles lapsing in some Lethean tide. He beheld before him, in sun-bright verdure and stillness, the enchanted vales evoked by the music; and he himself was the goatherd, following grassy paths, or lying oblivious of the vulture hours by the margin of lulled waters.

Hardly he knew that the low piping had ceased. But the vision darkened, and he who had dreamt of a goatherd's peace was again a troubled king.

'Play on!' he cried to the black piper. 'Name your own guerdon—and play.'

The eyes of the goatherd burned like embers in a dark place at evening. 'Not till the passing of ages and the falling of kingdoms shall I require of you my reward,' he said enigmatically. 'Howbeit, I shall play for you once more.'

So, through the afternoon, King Amero was beguiled by that sorcerous piping which told ever of a far land of ease and forgetfulness. With each playing it seemed that the spell grew stronger upon him; and more and more was his royalty a hateful thing; and the very grandeur of his palace oppressed and stifled him. No longer could he endure the heavily jewelled yoke of duty; and madly he envied the carefree lot of the goatherd.

At twilight he dismissed the ministrants who attended him, and held speech alone with the piper.

'Lead me to this land of yours,' he said, 'where I too may dwell as a simple herder.'

Clad in mufti, so that his people might not recognize him, the king stole from the palace through an unguarded postern, accompanied by the piper. Night, like a formless monster with the crescent moon for its lowered horn, was crouching beyond the town; but in the streets the shadows were thrust back by a flaming of myriad cressets. Amero and his guide were unchallenged as they went towards the outer darkness. And the king repented not his forsaken throne: though he saw in the city a continual passing of biers laden with the victims of the plague; and faces gaunt with famine rose up from the shadows as if to accuse him of recreancy. These he heeded not: for his eyes were filled with the dream of a green silent valley, in a land lost beyond the turbid flowing of time with its wreckage and tumult.

Now, as Amero followed the black piper, there descended upon him a sudden dimness; and he faltered in weird doubt and bewilderment. The street-lights flickered before him, and swiftly they expired in the gloom. The loud murmuring of the city fell away in a vast silence; and, like the shifting of some disordered dream, it seemed that the tall houses crumbled stilly and were gone even as shadows, and the stars shone over broken walls. Confusion filled the thoughts and the senses of Amero; and in his heart was a black chill of desolation; and he seemed to himself as one who had known the lapse of long empty years, and the loss of high splendour; and who stood now amid the extremity of age and decay. In his nostrils was a dry mustiness such as the night draws from olden ruin; and it came to him, as a thing foreknown and now remembered obscurely, that the desert was lord in his proud capital of Shathair.

'Where have you led me?' cried Amero to the piper.

For all reply, he heard a laughter that was like the peal of derisive thunder. The muffled shape of the goatherd towered colossally in the gloom, changing, growing, till its outlines were transformed to those of a giant warrior in sable armour. Strange memories thronged the mind of Amero, and he seemed to recall darkly something of another life. . . . Somehow, somewhere, for a time, he had been the goatherd of his dreams, content and forgetful . . . somehow, somewhere, he had entered a strange bright garden and had eaten a blood-dark fruit. . . .

Then, in a flaring as of infernal levin, he remembered all, and knew the mighty shadow that towered above him like a Terminus reared in hell. Beneath his feet was the cracked pavement of the seaward terrace; and the stars above the emissary were those that precede Canopus; but Canopus himself was blotted out by the Demon's shoulder. Somewhere in the dusty darkness, a leper laughed and coughed thickly, prowling about the ruined palace in which had once dwelt the kings of Calyz. All things were even as they had been before the making of that bargain through which a perished kingdom had been raised up by the powers of hell.

Anguish choked the heart of Xeethra as if with the ashes of burnt-out pyres and the shards of heaped ruin. Subtly and manifoldly had the Demon tempted him to his loss. Whether these things had been dream or necromancy or verity he knew not with sureness; nor whether they had happened once or had happened often. In the end there was only dust and dearth; and he, the doubly accurst, must remember and repent forevermore all that he had forfeited.

He cried out to the emissary: 'I have lost the bargain that I made with Thasaidon. Take now my soul and bear it before him where he sits aloft on his throne of ever-burning brass; for I would fulfil my bond to the uttermost.'

'There is no need to take your soul,' said the emissary, with an ominous rumble as of departing storm in the desolate night. 'Remain here with the lepers, or return to Pornos and his goats, as you will: it matters little. At all times and in all places your soul shall be part of the dark empire of Thasaidon.'

CATHERINE L. MOORE

JIREL MEETS MAGIC

Over Guischard's fallen drawbridge thundered Joiry's warrior lady, sword swinging, voice shouting hoarsely inside her helmet. The scarlet plume of her crest rippled in the wind. Straight into the massed defenders at the gate she plunged, careering through them by the very impetuosity of the charge, the weight of her mighty warhorse opening up a gap for the men at her heels to widen. For a while there was tumult unspeakable there under the archway, the yells of fighters and the clang of mail on mail and the screams of stricken men. Jirel of Joiry was a shouting battle-machine from which Guischard's men reeled in bloody confusion as she whirled and slashed and slew in the narrow confines of the gateway, her great stallion's iron hoofs weapons as potent as her own whistling blade.

In her full armour she was impregnable to the men on foot, and the horse's armour protected him from their vengeful blades, so that alone, almost, she might have won the gateway. By sheer weight and impetuosity she carried the battle through the defenders under the arch. They gave way before the mighty warhorse and his screaming rider. Jirel's swinging sword and the stallion's trampling feet cleared a path for Joiry's men to follow, and at last into Guischard's court poured the steel-clad hordes of Guischard's conquerors.

Jirel's eyes were yellow with blood-lust behind the helmet bars, and her voice echoed savagely from the steel cage that confined it, 'Giraud! Bring me Giraud! A gold piece to the man who brings me the wizard Giraud!'

She waited impatiently in the courtyard, reining her excited charger in mincing circles over the flags, unable to dismount alone in her heavy armour and disdainful of the threats of possible arbalesters in the arrow-slits that looked down upon her from Guischard's frowning grey walls. A crossbow shaft was the only thing she had to fear in her impregnable mail.

She waited in mounting impatience, a formidable figure in her bloody armour, the great sword lying across her saddlebow and her eager, angry voice echoing hoarsely from the helmet, 'Giraud! Make haste, you varlets! Bring me Giraud!'

There was such bloodthirsty impatience in that hollowly booming voice that the men who were returning from searching the castle hung back as they crossed the court towards their lady in reluctant twos and threes, failure eloquent upon their faces.

'What!' screamed Jirel furiously. 'You, Giles! Have you brought me Giraud? Watkin! Where is that wizard Giraud? Answer me, I say!'

'We've scoured the castle, my lady,' said one of the men fearfully as the angry voice paused. 'The wizard is gone.'

'Now God defend me!' groaned Joiry's lady. 'God help a poor woman served by fools! Did you search among the slain?'

'We searched everywhere, Lady Jirel. Giraud has escaped us.'

Jirel called again upon her Maker in a voice that was blasphemy in itself.

'Help me down, then, you hell-spawned knaves,' she grated. 'I'll find him myself. He must be here!'

With difficulty they got her off the sidling horse. It took two men to handle her, and a third to steady the charger. All the while they struggled with straps and buckles she cursed them hollowly, emerging limb by limb from the casing of steel and swearing with a soldier's fluency as the armour came away. Presently she stood free on the bloody flagstones, a slim, straight lady, keen as a blade, her red hair a flame to match the flame of her yellow eyes. Under the armour she wore a tunic of link-mail from the Holy Land, supple as silk and almost as light, and a doeskin shirt to protect the milky whiteness of her skin.

She was a creature of the wildest paradox, this warrior lady of Joiry, hot as a red coal, chill as steel, satiny of body and iron of soul. The set of her chin was firm, but her mouth betrayed a tenderness she would have died before admitting. But she was raging now.

'Follow me, then, fools!' she shouted. 'I'll find that God-cursed wizard and split his head with this sword if it takes me until the day I die. I swear it. I'll teach him what it costs to ambush Joiry men. By heaven, he'll pay with his life for my ten who fell at Massy Ford last week. The foul spell-brewer! He'll learn what it means to defy Joiry!'

Breathing threats and curses, she strode across the court, her men following reluctantly at her heels and casting nervous glances upward at the grey towers of Guischard. It had always borne a bad name, this ominous castle of the wizard Giraud, a place where queer things happened, which no man entered uninvited and whence no prisoner had ever escaped, though the screams of torture echoed often from its walls. Jirel's men would have followed her straight through the gates of hell, but they stormed Guischard at her heels with terror in their hearts and no hope of conquest.

She alone seemed not to know fear of the dark sorcerer. Perhaps it was because she had known things so dreadful that mortal perils held no terror for her—there were whispers at Joiry of their lady, and of things that had happened there which no man dared think on. But when Guischard fell, and the wizard's defenders fled before Jirel's mighty steed and the onrush of Joiry's men, they had plucked up heart, thinking that perhaps the ominous tales of Giraud had been gossip only, since the castle fell as any ordinary lord's castle might fall. But now—there were whispers again, and nervous glances over the shoulder, and men huddled together as they re-entered Guischard at their lady's hurrying heels. A castle from which a wizard might vanish into thin air, with all the exits watched, must be a haunted place, better burned and forgotten. They followed Jirel reluctantly, half ashamed but fearful.

In Jirel's stormy heart there was no room for terror as she plunged into the gloom of the archway that opened upon Guischard's great central hall. Anger that the man might have escaped her was a torch to light the way, and she paused in the door with eager anticipation, sweeping the corpse-strewn hall at a glance, searching for some clue to explain how her quarry had disappeared.

'He can't have escaped,' she told herself confidently. 'There's no way out. He *must* be here somewhere.' And she stepped into the hall, turning over the bodies she passed with a careless foot to make sure that death had not robbed her of vengeance.

An hour later, as they searched the last tower, she was still telling herself that the wizard could not have gone without her knowledge. She had taken special pains about that. There was a secret passage to the river, but she had had that watched. And an underwater door opened into the moat, but he could not have gone that way without meeting her men. Secret paths and open, she had found them all and posted a guard at each, and Giraud had not left the castle by any door that led out. She climbed the stairs of the last tower wearily, her confidence shaken.

An iron-barred oaken door closed the top of the steps, and Jirel drew back as her men lifted the heavy crosspieces and opened it for her. It had not been barred from within. She stepped into the little round room inside, hope fading completely as she saw that it too was empty, save for the body of a page-boy lying on the uncarpeted floor. Blood had made a congealing pool about him, and as Jirel looked she saw something which roused her flagging hopes. Feet had trodden in that blood, not the mailed feet of armed men, but the tread of shapeless cloth shoes such as surely none but Giraud would have worn when the castle was besieged and falling, and every man's help needed. Those bloody tracks led straight across the room towards the wall, and in that wall—a window.

Jirel stared. To her a window was a narrow slit deep in stone, made for the shooting of arrows, and never covered save in the coldest weather. But this window was broad and low, and instead of the usual animal pelt for hangings a curtain of purple velvet had been drawn back to disclose shutters carved out of something that might have been ivory had any beast alive been huge enough to yield such great unbroken sheets of whiteness. The shutters were unlatched, swinging slightly ajar, and upon them Jirel saw the smear of bloody fingers.

With a little triumphant cry she sprang forward. Here, then, was the secret way Giraud had gone. What lay beyond the window she could not guess. Perhaps an unsuspected passage, or a hidden room. Laughing exultantly, she swung open the ivory shutters.

There was a gasp from the men behind her. She did not hear it. She stood quite still, staring with incredulous eyes. For those ivory gates had opened upon no dark stone hiding-place or secret tunnel. They did not even reveal the afternoon sky outside, nor did they admit the shouts of her men still subduing the last of the defenders in the court below. Instead she was looking out upon a green woodland over which brooded a violet day like no day she had ever seen before. In paralysed amazement she looked down, seeing not the bloody flags of the courtyard far below, but a mossy carpet at a level with the floor. And on that moss she saw the mark of blood-stained feet. This window might be a magic one, opening into strange lands, but through it had gone the man she swore to kill, and where he fled she must follow.

She lifted her eyes from the tracked moss and stared out again through the dimness under the trees. It was a lovelier land than anything seen even in dreams; so lovely that it made her heart ache with its strange, unearthly enchantment—green woodland hushed and brooding in the hushed violet day. There was a promise of peace there, and forgetfulness and rest. Suddenly the harsh, shouting, noisy world behind her seemed very far away and chill. She moved forward and laid her hand upon the ivory shutters, staring out.

The shuffle of the scared men behind her awakened Jirel from the enchantment that had gripped her. She turned. The dreamy magic of the woodland loosed its hold as she faced the men again, but its memory lingered. She shook her red head a little, meeting their fearful eyes. She nodded towards the open window.

'Giraud has gone out there,' she said. 'Give me your dagger, Giles. This sword is too heavy to carry far.'

'But lady—Lady Jirel—dear lady—you can't go out there—Saint Guilda save us! Lady Jirel!'

Jirel's crisp voice cut short the babble of protest.

'Your dagger, Giles. I've sworn to slay Giraud, and slay him I shall, in whatever land he hides. Giles!'

A man-at-arms shuffled forward with averted face, handing her his dagger. She gave him the sword she carried and thrust the long-bladed knife into her belt. She turned again to the window. Green and cool and lovely, the woodland lay waiting. She thought as she set her knee upon the sill that she must have explored this violet calm even had her oath not driven her; for there was an enchantment about the place that drew her irresistibly. She pulled up her other knee and jumped lightly. The mossy ground received her without a jar.

For a few moments Jirel stood very still, watching, listening. Bird songs trilled intermittently about her, and breezes stirred the leaves. From very far away she thought she caught the echoes of a song when the wind blew, and there was something subtly irritating about its simple melody that seemed to seesaw endlessly up and down on two notes. She was glad when the wind died and the song no longer shrilled in her ears.

It occurred to her that before she ventured far she must mark the window she had entered by, and she turned curiously, wondering how it looked from this side. What she saw sent an inexplicable little chill down her back. Behind her lay a heap of mouldering ruins, moss-grown, crumbling into decay. Fire had blackened the stones in ages past. She could see that it must have been a castle, for the original lines of it were not yet quite lost. Only one low wall remained standing now, and in it opened the window through which she had come. There was something hauntingly familiar about the lines of those mouldering stones, and she turned away with a vague unease, not quite understanding why. A little path wound away under the low-hanging trees, and she followed it slowly, eyes alert for signs that Giraud had passed this way. Birds trilled drowsily in the leaves overhead, queer, unrecognizable songs like the music of no birds she knew. The violet light was calm and sweet about her.

She had gone on in the bird-haunted quiet for many minutes before she caught the first hint of anything at odds with the perfect peace about her. A whiff of wood-smoke drifted to her nostrils on a vagrant breeze. When she rounded the next bend of the path she saw what had caused it. A tree lay across the way in a smother of shaking leaves and branches. She knew that she must skirt it, for the branches were too tangled to penetrate, and she turned out of the path, following the trunk towards its broken base.

She had gone only a few steps before the sound of a curious sobbing came to her ears. It was the gasp of choked breathing, and she had heard sounds like that too often before not to know that she approached death in some form or another. She laid her hand on her knife-hilt and crept forward softly.

The tree trunk had been severed as if by a blast of heat, for the stump was charred black and still smoking. Beyond the stump a queer tableau was being enacted, and she stopped quite still, staring through the leaves.

Upon the moss a naked girl was lying, gasping her life out behind the hands in which her face was buried. There was no mistaking the death-sound in that failing breath, although her body was unmarked. Hair of a strange green-gold pallor streamed over her bare white body, and by the fragility and tenuosity of that body Jirel knew that she could not be wholly human.

Above the dying girl a tall woman stood. And that woman was a magnet for Jirel's fascinated eyes. She was generously curved, sleepy-eyed. Black hair bound her head sleekly, and her skin was like rich, dark, creamy velvet. A violet robe wrapped her carelessly, leaving arms and one curved shoulder bare, and her girdle was a snake of something like purple glass. It might have been carved from some vast jewel, save for its size and unbroken clarity. Her feet were thrust bare into silver sandals. But it was her face that held Jirel's yellow gaze.

The sleepy eyes under heavily drooping lids were purple as gems, and the darkly crimson mouth curled in a smile so hateful that fury rushed up in Jirel's heart as she watched. That lazy purple gaze dwelt aloofly upon the gasping girl on the moss. The woman was saying in a voice as rich and deep as thick-piled velvet,

'—nor will any other of the dryad folk presume to work forbidden magic in my woodlands for a long, long while to come. Your fate shall be a deadly example to them, Irsla. You dared too greatly. None who defy Jarisme live. Hear me, Irsla!'

The sobbing breath had slowed as the woman spoke, as if life were slipping fast from the dryad-girl on the moss; and as she realized it the speaker's arm lifted and a finger of white fire leaped from her outstretched hand, stabbing the white body at her feet. And the girl Irsla started like one shocked back into life.

'Hear me out, dryad! Let your end be a warning to—'

The girl's quickened breath slowed again as the white brilliance left her, and again the woman's hand rose, again the light-blade stabbed. From behind her shielding hands the dryad gasped.

'Oh, mercy, mercy, Jarisme! Let me die!'

'When I have finished. Not before. Life and death are mine to command here, and I am not yet done with you. Your stolen magic—'

She paused, for Irsla had slumped once more upon the moss, breath scarcely stirring her. As Jarisme's light-dealing hand rose for the third time Jirel leapt forward. Partly it was intuitive hatred of the lazy-eyed woman, partly revolt at this cat-and-mouse play with a dying girl for victim. She swung her arm in an arc that cleared the branches from her path, and called out in her clear, strong voice,

'Have done, woman! Let her die in peace.'

Slowly Jarisme's purple eyes rose. They met Jirel's hot yellow glare. Almost physical impact was in that first meeting of their eyes, and hatred flashed between them instantly, like the flash of blades—the instinctive hatred of total opposites, born enemies. Each stiffened subtly, as cats do in the instant before combat. But Jirel thought she saw in the purple gaze, behind all its kindling anger, a faint disquiet, a nameless uncertainty.

'Who are you?' asked Jarisme, very softly, very dangerously.

Something in that unsureness behind her angry eyes prompted Jirel to answer boldly.

'Jirel of Joiry. I seek the wizard Giraud, who fled me here. Stop tormenting that wretched girl and tell me where to find him. I can make it worth your while.'

Her tone was imperiously mandatory, and behind Jarisme's drooping lips an answering flare of anger lighted, almost drowning out that faint unease.

'You do not know me,' she observed, her voice very gentle. 'I am the sorceress Jarisme, and high ruler over all this land. Did you think to buy me, then, earth-woman?'

Jirel smiled her sweetest, most poisonous smile.

'You will forgive me,' she purred. 'At the first glance at you I did not think your price could be high. . . .'

A petty malice had inspired the speech, and Jirel was sorry as it left her lips, for she knew that the scorn which blazed up in Jarisme's eyes was justified. The sorceress made a contemptuous gesture of dismissal.

'I shall waste no more of my time here,' she said. 'Get back to your little lands, Jirel of Joiry, and tempt me no further.'

The purple gaze rested briefly on the motionless dryad at her feet, flicked Jirel's hot eyes with a glance of scorn which yet did not wholly hide that curious uncertainty in its depths. One hand slid behind her, oddly as if she were seeking a door-latch in empty air. Then like a heat-shimmer the air danced about her, and in an instant she was gone.

Jirel blinked. Her ears had deceived her as well as her eyes, she thought, for as the sorceress vanished a door closed softly somewhere. Yet look though she would, the green glade was empty, the violet air untroubled. No Jarisme anywhere—no door. Jirel shrugged after a moment's bewilderment. She had met magic before.

A sound from the scarcely breathing girl upon the moss distracted her, and she dropped to her knees beside the dying dryad. There was no mark or wound upon her, yet Jirel knew that death could be only a matter of moments. And dimly she recalled that, so legend said, a tree-sprite never survived the death of its tree. Gently she turned the girl over, wondering if she were beyond help.

At the feel of those gentle hands the dryad's lids quivered and rose. Brook-brown eyes looked up at Jirel, with green swimming in their deeps like leaf-reflections in a woodland pool.

'My thanks to you,' faltered the girl in a ghostly murmur. 'But get you back to your home now—before Jarisme's anger slays you.'

Jirel shook her red head stubbornly.

'I must find Giraud first, and kill him, as I have sworn to do. But I will wait. Is there anything I can do?'

The green-reflecting eyes searched hers for a moment. The dryad must have read resolution there, for she shook her head a little.

'I must die—with my tree. But if you are determined—hear me. I owe you—a debt. There is a talisman—braided in my hair. When I—am dead—take it. It is Jarisme's sign. All her subjects wear them. It will guide you to her—and to Giraud. He is ever beside her. I know. I think it was her anger at you—that made her forget to take it from me, after she had dealt me my death. But why she did not slay you—I do not know.

Jarisme is quick—to kill. No matter—listen now. If you must have Giraud—you must take a risk that no one here—has ever taken—before. Break this talisman—at Jarisme's feet. I do not know—what will happen then. Something—very terrible. It releases powers—even she can not control. It may—destroy you too. But—it is—a chance. May you—have—all good—'

The faltering voice failed. Jirel, bending her head, caught only meaningless murmurs that trailed away to nothing. The green-gold head dropped suddenly forward on her sustaining arm. Through the forest all about her went one long, quivering sigh, as if an intangible breeze ruffled the trees. Yet no leaves stirred.

Jirel bent and kissed the dryad's forehead, then laid her very gently back on the moss. And as she did so her hand in the masses of strangely coloured hair came upon something sharp and hard. She remembered the talisman. It tingled in her fingers as she drew it out—an odd little jagged crystal sparkling with curious aliveness from the fire burning in its heart.

When she had risen to her feet, leaving the dead dryad lying upon the moss which seemed so perfectly her couch, she saw that the inner brilliance streaming in its wedge-shaped pattern through the crystal was pointing a quivering apex forward and to the right. Irsla had said it would guide her. Experimentally she twisted her hand to the left. Yes, the shaking light shifted within the crystal, pointing always towards the right, and Jarisme.

One last long glance she gave to the dryad on the moss. Then she set off again down the path, the little magical thing stinging her hand as she walked. And as she went she wondered. This strong hatred which had flared so instinctively between her and the sorceress was hot enough to burn any trace of fear from her mind, and she remembered that look of uncertainty in the purple gaze that had shot such hatred at her. Why? Why had she not been slain as Irsla was slain, for defiance of this queer land's ruler?

For a while she paced unheedingly along under the trees. Then abruptly the foliage ceased and a broad meadow lay before her, green in the clear, violet day. Beyond the meadow the slim shaft of a tower rose dazzlingly white, and towards it in steady radiance that magical talisman pointed.

From very far away she thought she still caught the echoes of that song when the wind blew, an irritating monotony that made her ears ache. She was glad when the wind died and the song no longer shrilled in her ears.

Out across the meadow she went. Far ahead she could make out purple mountains like low clouds on the horizon, and here and there in the distances clumps of woodland dotted the meadows. She walked on more rapidly now, for she was sure that the white tower housed Jarisme, and with her Giraud. And she must have gone more swiftly than she knew, for with almost magical speed the shining shaft drew nearer.

She could see the arch of its doorway, bluely violet within. The top of the shaft was battlemented, and she caught splashes of colour between the teeth of the stone scarps, as if flowers were massed there and spilling blossoms against the whiteness of the tower. The singsong music was louder than ever, and much nearer. Jirel's heart beat a bit heavily as she advanced, wondering what sort of a sorceress this Jarisme might be, what dangers lay before her in the path of her vow's fulfilment. Now the white tower rose up over her, and she was crossing the little space before the door, peering in dubiously. All she could see was dimness and violet mist.

She laid her hand upon the dagger, took a deep breath and stepped boldly in under the arch. In the instant her feet left the solid earth she saw that this violet mist filled the whole shaft of the tower, that there was no floor. Emptiness engulfed her, and all reality ceased.

She was falling through clouds of violet blankness, but in no recognizable direction. It might have been up, down, or sidewise through space. Everything had vanished in the violet nothing. She knew an endless moment of vertigo and rushing motion; then the dizzy emptiness vanished in a breath and she was standing in a gasping surprise upon the roof of Jarisme's tower.

She knew where she was by the white battlements ringing her round, banked with strange blossoms in muted colours. In the centre of the circular, marble-paved place a low couch, cushioned in glowing yellow, stood in the midst of a heap of furs. Two people sat side by side on the couch. One was Giraud. Black-robed, dark-visaged, he stared at Jirel with a flicker of disquiet in his small, dull eyes. He said nothing.

Jirel dismissed him with a glance, scarcely realizing his presence. For Jarisme had lowered from her lips a long, silver flute. Jirel realized that the queer, maddening music must have come from that gleaming length, for it no longer echoed in her ears. Jarisme was holding the instrument now in midair, regarding Jirel over it with a purple-eyed gaze that was somehow thoughtful and a little apprehensive, though anger glowed in it, too.

'So,' she said richly, in her slow, deep voice. 'For the second time you defy me.'

At these words Giraud turned his head sharply and stared at the sorceress' impassive profile. She did not return his gaze, but after a moment he looked quickly back at Jirel, and in his eyes too she saw that flicker of alarm, and with it a sort of scared respect. It puzzled her, and she did not like being puzzled. She said a little breathlessly.

'If you like, yes. Give me that skulking potion-brewer beside you and set me down again outside this damned tower of trickery. I came to kill your pet spellmonger here for treachery done me in my own world by this creature who dared not stay to face me.'

Her peremptory words hung in the air like the echoes of a gong. For a while no one spoke. Jarisme smiled more subtly than before, an insolent, slow smile that made Jirel's pulses hammer with the desire to smash it down the woman's lush, creamy throat. At last Jarisme said, in a voice as rich and deep as thick-piled velvet,

'Hot words, hot words, soldier-woman! Do you really imagine that your earthly squabbles matter to Jarisme?'

'What matters to Jarisme is of little moment to me,' Jirel said contemptuously. 'All I want is this skulker here, whom I have sworn to kill.'

Jarisme's slow smile was maddening. 'You demand it of me—Jarisme?' she asked with soft incredulity. 'Only fools offend me, woman, and they but once. None commands me. You will have to learn that.'

Jirel smiled thinly. 'At what price, then, do you value your pet cur?'

Giraud half rose from the couch at that last insult, his dark face darker with a surge of anger. Jarisme pushed him back with a lazy hand.

'This is between your—friend—and me,' she said. 'I do not think, soldier'—the appellation was the deadliest of insults in the tone she used— 'that any price you could offer would interest me.'

'And yet your interest is very easily caught.' Jirel flashed a contemptuous glance at Giraud, restive under the woman's restraining hand.

Jarisme's rich pallor flushed a little. Her voice was sharper as she said, 'Do not tempt me too far, earthling.'

Jirel's yellow eyes defied her. 'I am not afraid.'

The sorceress' purple gaze surveyed her slowly. When Jarisme spoke again a tinge of reluctant admiration lightened the slow scorn of her voice.

'No—you are not afraid. And a fool not to be. Fools annoy me, Jirel of Joiry.'

She laid the flute down on her knee and lazily lifted a ringless hand. Anger was glowing in her eyes now, blotting out all trace of that little haunting fear. But Giraud caught the rising hand, bending, whispering urgently in her ear. Jirel caught a part of what he said, '—what happens to those who tamper with their own destiny—' And she saw the anger fade from the sorceress's face as apprehension brightened there again. Jarisme looked at Jirel with a long, hard look and shrugged her ample shoulders.

'Yes,' she murmured. 'Yes, Giraud. It is wisest so.' And to Jirel, 'Live, then, earthling. Find your way back to your own land if you can, but I warn you, do not trouble me again. I shall not stay my hand if our paths ever cross in the future.'

She struck her soft, white palms together sharply. And at the sound the roof-top and the violet sky and the banked flowers at the parapets whirled around Jirel in dizzy confusion. From very far away she heard that clap of peremptory hands still echoing, but it seemed to her that the great, smokily coloured blossoms were undergoing an inexplicable transformation. They quivered and spread and thrust upward from the edges of the tower to arch over her head. Her feet were pressing a mossy ground, and the sweet, earthy odours of a garden rose about her. Blinking, she stared around as the world slowly steadied.

She was no longer on the roof-top. As far as she could see through the tangled stems, great flowering plants sprang up in the gloaming of a strange, enchanted forest. She was completely submerged in greenery, and the illusion of under-water filled her eyes, for the violet light that filtered through the leaves was diffused and broken into a submarine dimness. Uncertainly she began to grope her way forward, staring about to see what sort of a miracle had enfolded her.

It was a bower in fairyland. She had come into a tropical garden of great, muted blooms and jungle silences. In the diffused light the flowers nodded sleepily among the leaves, hypnotically lovely, hypnotically soporific with their soft colours and drowsy, never-ending motion. The fragrance was overpowering. She went on slowly, treading moss that gave back no sound. Here under the canopy of leaves was a little separate world of colour and silence and perfume. Dreamily she made her way among the flowers.

Their fragrance was so strongly sweet that it went to her head, and she walked in a waking dream. Because of this curious, scented trance in which she went she was never quite sure if she had actually seen that motion among the leaves, and looked closer, and made out a huge, incredible serpent of violet transparency, a giant replica of the snake that girdled Jarisme's waist, but miraculously alive, miraculously supple and gliding, miraculously twisting its soundless way among the blossoms and staring at her with impassive, purple eyes.

While it glided along beside her she had other strange visions too, and could never remember just what they were, or why she caught familiar traces in the tiny, laughing faces that peered at her from among the flowers, or half believed the wild, impossible things they whispered to her, their laughing mouths brushing her ears as they leaned down among the blossoms.

The branches began to thin at last, as she neared the edge of the enchanted place. She walked slowly, half conscious of the great transparent snake like a living jewel writhing along soundlessly at her side, her mind vaguely troubled in its dream by the fading remembrance of what those little, merry voices had told her. When she came to the very edge of the bowery jungle and broke out into clear daylight again she stopped in a daze, staring round in the brightening light as the perfumes slowly cleared from her head.

Sanity and realization returned to her at last. She shook her red head dizzily and looked round, half expecting, despite her returning clarity, to see the great serpent gliding across the grass. But there was nothing. Of course she had dreamed. Of course those little laughing voices had not told her that—that—she clutched after the vanishing tags of remembrance, and caught nothing. Ruefully she laughed and brushed away the clinging memories, looking round to see where she was.

She stood at the crest of a little hill. Below her the flower-fragrant jungle nodded, a little patch of enchanted greenery clothing the slopes of the hill. Beyond and below green meadows stretched away to a far-off line of forest which she thought she recognized as that in which she had first met Jarisme. But the white tower which had risen in the midst of the meadows was magically gone. Where it had stood, unbroken greenery lay under the violet clarity of the sky.

As she stared round in bewilderment a faint prickling stung her palm, and she glanced down, remembering the talisman clutched in her hand. The quivering light was streaming in a long wedge towards some point behind her. She turned. She was in the foothills of those purple mountains she had glimpsed from the edge of the woods. High and shimmering, they rose above her. And, hazily in the heat-waves that danced among their heights, she saw the tower.

Jirel groaned to herself. Those peaks were steep and rocky. Well, no help for it. She must climb. She growled a soldier's oath in her throat and turned wearily towards the rising slopes. They were rough and deeply slashed with ravines. Violet heat beat up from the reflecting rocks, and tiny, brilliantly coloured things scuttled from her path—orange lizards and coral red scorpions and little snakes like bright blue jewels.

It seemed to her as she stumbled upward among the broken stones that the tower was climbing too. Time after time she gained upon it, and time after time when she lifted her eyes after a gruelling struggle up steep ravines, that mocking flicker of whiteness shimmered still high and unattainable on some distant peak. It had the mistiness of unreality, and if her talisman's guide had not pointed steadily upward she would have thought it an illusion to lead her astray.

But after what seemed hours of struggle, there came the time when, glancing up, she saw the shaft rising on the topmost peak of all, white as snow against the clear violet sky. And after that it shifted no more. She took heart now, for at last she seemed to be gaining. Every laborious step carried her nearer that lofty shining upon the mountain's highest peak.

She paused after a while, looking up and wiping the moisture from her forehead where the red curls clung. As she stood there something among the rocks moved, and out from behind a boulder a long, slinking feline creature came. It was not like any beast she had ever seen before. Its shining pelt was fabulously golden, brocaded with queer patterns of darker gold, and down against its heavy jaws curved two fangs whiter than ivory. With a grace as gliding as water it paced down the ravine towards her.

Jirel's heart contracted. Somehow she found the knife-hilt in her hand, though she had no recollection of having drawn it. She was staring hard at the lovely and terrible cat, trying to understand the haunting familiarity about its eyes. They were purple, like jewels. Slowly recognition dawned. She had met that purple gaze before, insolent under sleepy lids. Jarisme's eyes. Yes, and the snake in her dream had watched her with a purple stare too. Jarisme?

She closed her hand tightly about the crystal, knowing that she must conceal from the sorceress her one potent weapon, waiting until the time came to turn it against its maker. She shifted her knife so that light glinted down the blade. They stood quite still for a moment, yellow-eyed woman and fabulous, purple-eyed cat, staring at each other with hostility eloquent in every line of each. Jirel clenched her knife tight, warily eyeing the steel-clawed paws on which the golden beast went so softly. They could have ripped her to ribbons before the blade struck home.

She saw a queer expression flicker across the sombre purple gaze that met hers, and the beautiful cat crouched a little, tail jerking, lip twitched back to expose shining fangs. It was about to spring. For an interminable moment she waited for that hurtling golden death to launch itself upon her, tense, rigid, knife steady in her hand. . . .

It sprang. She dropped to one knee in the split second of its leaping, instinctively hiding the crystal, but thrusting up her dagger in defence. The great beast sailed easily over her head. As it hurtled past, a peal of derisive laughter rang in her ears, and she heard quite clearly the sound of a slamming door. She scrambled up and whirled in one motion, knife ready. The defile was quite empty in the violet day. There was no door anywhere. Jarisme had vanished.

A little shaken, Jirel sheathed her blade. She was not afraid. Anger burned out all trace of fear as she remembered the scorn in that ringing laugh. She took up her course again towards the tower, white and resolute, not looking back.

The tower was drawing near again. She toiled upward. Jarisme showed no further sign of her presence, but Jirel felt eyes upon her, purple eyes, scornful and sleepy. She could see the tower clearly, just above her at the crest of the highest peak, up to which a long arc of steps curved steeply. They were very old, these steps, so worn that many were little more than irregularities on the stone. Jirel wondered what feet had worn them so, to what door they had originally led.

She was panting when she reached the top and peered in under the arch of the door. To her surprise she found herself staring into a broad, semicircular hallway, whose walls were lined with innumerable doors. She remembered the violet nothingness into which she had stepped the last time she crossed the sill, and wondered as she thrust a tentative foot over it if the hall were an illusion and she were really about to plunge once more into that cloudy abyss of falling. But the floor was firm.

She stepped inside and paused, looking round in some bewilderment and wondering where to turn now. She could smell peril in the air. Almost she could taste the magic that hovered like a mist over the whole enchanted place. Little warning prickles ran down her back as she went forward very softly and pushed open one of those innumerable doors. Behind it a gallery stretched down miles of haze-shrouded extent. Arrow-straight it ran, the arches of the ceiling making an endless parade that melted into violet distance. And as she stood looking down the cloudy vista, something like a puff of smoke obscured her vision for an instant—smoke that eddied and billowed and rolled away from the shape of that golden cat which had vanished in the mountain ravine.

It paced slowly down the hall towards her, graceful and lovely, muscles rippling under the brocaded golden coat and purple eyes fixed upon her in a scornful stare. Jirel's hand went to the knife in her belt, hatred choking up in her throat as she met the purple eyes. But in the corridor a voice was echoing softly, Jarisme's voice, saying,

'Then it is war between us, Jirel of Joiry. For you have defied my mercy, and you must be punished. Your punishment I have chosen—the simplest, and the subtlest, and the most terrible of all punishments, the worst that could befall a human creature. Can you guess it? No? Then wonder a while, for I am not prepared yet to administer it fully . . . or shall I kill you now? Eh-h-h? . . .'

The curious, long-drawn query melted into a purring snarl, and the great cat's lip lifted, a flare of murderous light flaming up in the purple eyes. It had been pacing nearer all the while that light voice had echoed in the air. Now its roar crescendoed into a crashing thunder that rang from the walls, and the steel springs of its golden body tightened for a leap straight at Jirel's throat. Scarcely a dozen paces away, she saw the brocaded beauty of it crouching, taut and poised, saw the powerful body quiver and tighten—and spring. In instinctive panic she leaped back and slammed the door in its face.

Derisive laughter belled through the air. A cloud of thin smoke eddied through the crack around the door and puffed in her face with all the insolence of a blow. Then the air was clear again. The red mist of murder swam before Jirel's eyes. Blind with anger, breath beating thickly in her throat, she snatched at the door again, ripping the dagger from her belt. Through that furious haze she glared down the corridor. It was empty. She closed the door a second time and leaned against it, trembling with anger, until the mist had cleared from her head and she could control her shaking hand well enough to replace the dagger.

When she had calmed a little she turned to scan the hall, wondering what to do next. And she saw that there was no escape now, even had she wished, for the door she had entered by was gone. All about her now closed the door-studded walls, enigmatic, imprisoning. And the very fact of their presence was an insult, suggesting that Jarisme had feared she would flee if the entrance were left open. Jirel forced herself into calmness again. She was not afraid, but she knew herself in deadly peril.

She was revolving the sorceress's threat as she cast about for some indication to guide her next step. The simplest and subtlest and most terrible of punishments—what could it be? Jirel knew much of the ways of torture—her dungeons were as blood-stained as any of her neighbours'—but she knew too that Jarisme had not meant only the pain of the flesh. There was a subtler menace in her words. It would be a feminine vengeance, and more terrible than anything iron and fire could inflict. She knew that. She knew also that no door she could open now would lead to freedom, but she could not stay quiet, waiting. She glanced along the rows of dark, identical panels. Anything that magic could contrive might lie behind them. In the face of peril more deadly than death she could not resist the temptation to pull open the nearest one and peer within.

A gust of wind blew in her face and rattled the door. Dust was in that wind, and bitter cold. Through an inner grille of iron, locked across the opening, she saw a dazzle of whiteness like sun on snow in the instant before she slammed the door shut on the piercing gust. But the incident had whetted her curiosity. She moved along the wall and opened another.

This time she was looking through another locked grille into a dimness of grey smoke shot through with flame. The smell of burning rose in her nostrils, and she could hear faintly, as from vast distances, the sound of groans and the shivering echo of screams. Shuddering, she closed the door.

When she opened the next one she caught her breath and stared. Before her a thick crystal door separated her from bottomless space. She pressed her face to the cold glass and stared out and down. Nothingness met her gaze. Dark and silence and the blaze of unwinking stars. It was day outside the tower, but she looked into fathomless night. And as she stared, a long streak of light flashed across the blackness and faded. It was not a shooting star. By straining her eyes she could make out something like a thin sliver of silver flashing across the dark, its flaming tail fading behind it in the sky. And the sight made her ill with sudden vertigo. Bottomless void reeled around her, and she fell back into the hallway, slamming the door upon that terrifying glimpse of starry nothingness.

It was several minutes before she could bring herself to try the next door. When she did, swinging it open timorously, a familiar sweetness of flower perfume floated out and she found herself gazing through a grille of iron bars deep into that drowsy jungle of blossoms and scent and silence which she had crossed at the mountain's foot. A wave of remembrance washed over her. For an instant she could hear those tiny, laughing voices again, and she felt the presence of the great snake at her side, and the wild, mirth-ridden secrets of the little grey voices rang in her ears. Then she was awake again, and the memory vanished as dreams do, leaving nothing but tantalizing fragments of forgotten secrets drifting through her mind. She knew as she stared that she could step straight into that flowery fairyland again if the bars would open. But there was no escape from this magical place, though she might look through any number of opening doors into far lands and near.

She was beginning to understand the significance of the hall. It must be from here that Jarisme by her magical knowledge journeyed into other lands and times and worlds through the doors that opened between her domain and those strange, outland places. Perhaps she had sorcerer friends there, and paid them visits and brought back greater knowledge, stepping from world to world, from century to century, through her enchanted doorways. Jirel felt certain that one of these enigmatic openings would give upon that mountain pass where the golden cat with its scornful purple eyes had sprung at her, and vanished, and laughed backward as the door slammed upon it, and upon the woodland glade where the dryad died. But she knew that bars would close these places away even if she could find them.

She went on with her explorations. One door opened upon a steamy fern-forest of gigantic growths, out of whose deeps floated musky, reptilian odours, and the distant sound of beasts bellowing hollowly. And another upon a grey desert stretching flat and lifeless to the horizon, wan under the light of a dim red sun.

But at last she came to one that opened not into alien lands but upon a stairway winding down into solid rock whose walls showed the mark of the tools that had hollowed them. No sound came up the shaft of the stairs, and a grey light darkened down their silent reaches. Jirel peered in vain for some hint of what lay below. But at last, because inactivity had palled upon her and she knew that all ways were hopeless for escape, she entered the doorway and went slowly down the steps. It occurred to her that possibly she might find Jarisme below, engaged in some obscure magic in the lower regions, and she was eager to come to grips with her enemy.

The light darkened as she descended, until she was groping her way through obscurity round and round the curving stairs. When the steps ended at a depth she could not guess, she could tell that she had emerged into a low-roofed corridor only by feeling the walls and ceiling that met her exploring hands, for the thickest dark hid everything. She made her slow way along the stone hall, which wound and twisted and dipped at unexpected angles until she lost all sense of direction. But she knew she had gone a long way when she began to see the faint gleam of light ahead.

Presently she began to catch the faraway sound of a familiar song—Jarisme's monotonous little flute melody on two notes, and she was sure then that her intuition had been true, that the sorceress was down here somewhere. She drew her dagger in the gloom and went on more warily.

An arched opening ended the passage. Through the arch poured a blaze of dancing white luminance. Jirel paused, blinking and trying to make out what strange place she was entering. The room before her was filled with the baffling glitter and shimmer and mirage of reflecting surfaces so bewilderingly that she could not tell which was real and which mirror, and which dancing light. The brilliance dazzled in her face and dimmed into twilight and blazed again as the mirrors shifted. Little currents of dark shivered through the chaos and brightened into white sparkle once more. That monotonous music came to her through the quivering lights and reflections, now strongly, now faintly in the distance.

The whole place was a chaos of blaze and confusion. She could not know if the room were small or large, a cavern or a palace hall. Queer reflections danced through the dazzle of it. She could see her own image looking back at her from a dozen, a score, a hundred moving planes that grotesquely distorted her and then flickered out again, casting a blaze of light in her blinded eyes. Dizzily she blinked into the reeling wilderness of planes.

Then she saw Jarisme in her violet robe watching her from a hundred identical golden couches reflected upon a hundred surfaces. The figure held a flute to its lips, and the music pulsed from it in perfect time with the pulsing of the sorceress's swelling white throat. Jirel stared round in confusion at the myriad Jarismes all piping the interminable monotones. A hundred sensual, dreamy faces turned to her, a hundred white arms dropped as the flute left a hundred red mouths that Jarisme might smile ironic welcome a hundredfold more scornful for its multiplicity.

When the music ceased, all the flashing dazzle suddenly stilled. Jirel blinked as the chaos resolved itself into shining order, the hundred Jarismes merging into one sleepy-eyed woman lounging upon her golden couch in a vast crystal-walled chamber shaped like the semicircular half of a great, round, domed room. Behind the couch a veil of violet mist hung like a curtain shutting off what would have formed the other half of the circular room.

'Enter,' said the sorceress with the graciousness of one who knows herself in full command of the situation. 'I thought you might find the way here. I am preparing a ceremony which will concern you intimately. Perhaps you would like to watch? This is to be an experiment, and for that reason a greater honour is to be yours than you can ever have known before; for the company I am assembling to watch your punishment is a more distinguished one than you could understand. Come here, inside the circle.'

Jirel advanced, dagger still clenched in one hand, the other closed about her bit of broken crystal. She saw now that the couch stood in the centre of a ring engraved in the floor with curious, cabalistic symbols. Beyond it the cloudy violet curtain swayed and eddied within itself, a vast, billowing wall of mist. Dubiously she stepped over the circle and stood eyeing Jarisme, her yellow gaze hot with rigidly curbed emotion. Jarisme smiled and lifted the flute to her lips again.

As the irritating two notes began their seesawing tune Jirel saw something amazing happen. She knew then that the flute was a magic one, and the song magical too. The notes took on a form that overstepped the boundaries of the aural and partook in some inexplicable way of all the other senses too. She could feel them, taste them, smell them, see them. In a queer way they were visible, pouring in twos from the flute and dashing outward like little needles of light. The walls reflected them, and those reflections became swifter and brighter and more numerous until the air was full of flying slivers of silvery brilliance, until shimmers began to dance among them and over them, and that bewildering shift of mirrored planes started up once more. Again reflections crossed and dazzled and multiplied in the shining air as the flute poured out its flashing double notes.

Jirel forgot the sorceress beside her, the music that grated on her ears, even her own peril, in watching the pictures that shimmered and vanished in the mirrored surfaces. She saw flashes of scenes she had glimpsed through the doors of Jarisme's hallway. She saw stranger places than that, passing in instant-brief snatches over the silvery planes. She saw jagged black mountains with purple dawns rising behind them and stars in unknown figures across the dark skies; she saw grey seas flat and motionless beneath grey clouds; she saw smooth meadows rolling horizon-ward under the glare of double suns. All these and many more awoke to the magic of Jarisme's flute, and melted again to give way to others.

Jirel had the strange fancy, as the music went on, that it was audible in those lands whose brief pictures were flickering across the background of its visible notes. It seemed to be piercing immeasurable distances, ringing across the cloudy seas, echoing under the double suns, calling insistently in strange lands and far, unknown places, over deserts and mountains that man's feet had never trod, reaching other worlds and other times and crying its two-toned monotony through the darkness of interstellar space. All of this, to Jirel, was no more than a vague realization that it must be so. It meant nothing to her, whose world was a flat plane arched by the heaven-pierced bowl of the sky. Magic, she told herself, and gave up trying to understand.

Presently the tempo of the fluting changed. The same two notes still shrilled endlessly up and down, but it was no longer a clarion call ringing across borderlands into strange worlds. Now it was slower, statelier. And the notes of visible silver that had darted crazily against the crystal walls and reflected back again took on an order that ranked them into one shining plane. Upon that plane Jirel saw the outlines of a familiar scene gradually take shape. The great door-lined hall above mirrored itself in faithful replica before her eyes. The music went on changelessly.

Then, as she watched, one of those innumerable doors quivered. She held her breath. Slowly it swung open upon that grey desert under the red sun which she had seen before she closed it quickly away behind concealing panels. Again as she looked, that sense of utter desolation and weariness and despair came over her, so uncannily dreary was the scene. Now the door stood wide, its locked grille no longer closing it, and as the music went on she could see a dazzle like a jagged twist of lightning begin to shimmer in its aperture. The gleam strengthened. She saw it quiver once, twice, then sweep forward with blinding speed through the open doorway. And as she tried to follow it with her eyes another moving door distracted her.

This time the steamy fern-forest was revealed as the panels swung back. But upon the threshold sprawled something so frightful that Jirel's free hand flew to her lips and a scream beat up in her throat. It was black—shapeless and black and slimy. And it was alive. Like a heap of putrescently shining jelly it heaved itself over the door-sill and began to flow across the floor, inching its way along like a vast blind amoeba. But she knew without being told that it was horribly wise, horribly old. Behind it a black trail of slime smeared the floor.

Jirel shuddered and turned her eyes away. Another door was swinging open. Through it she saw a place she had not chanced upon before, a country of bare red rock strewn jaggedly under a sky so darkly blue that it might have been black, with stars glimmering in it more clearly than stars of earth. Across this red, broken desert a figure came striding that she knew could be only a figment of magic, so tall it was, so spidery-thin, so grotesquely human despite its bulbous head and vast chest. She could not see it clearly, for about it like a robe it clutched a veil of blinding light. On those incredibly long, thin legs it stepped across the door-sill, drew its dazzling garment closer about it, and strode forward. As it neared, the light was so blinding that she could not look upon it. Her averted eyes caught the motion of a fourth door.

This time she saw that flowery ravine again, dim in its underwater illusion of diffused light. And out from among the flowers writhed a great serpent-creature, not of the transparent crystal she had seen in her dream, but iridescently scaled. Nor was it entirely serpent, for from the thickened neck sprang a head which could not be called wholly unhuman. The thing carried itself as proudly as a cobra, and as it glided across the threshold its single, many-faceted eye caught Jirel's in the reflection. The eye flashed once, dizzyingly, and she reeled back in sick shock, the violence of that glance burning through her veins like fire. When she regained control of herself many other doors were standing open upon scenes both familiar and strange. During her daze other denizens of those strange worlds must have entered at the call of the magic flute.

She was just in time to see an utterly indescribable thing flutter into the hall from a world which so violated her eyes that she got no more than a glimpse of it as she flung up outraged hands to shut it out. She did not lower that shield until Jarisme's amused voice said in an undertone, 'Behold your audience, Jirel of Joiry,' and she realized that the music had ceased and a vast silence was pressing against her ears. Then she looked out, and drew a long breath. She was beyond surprise and shock now, and she stared with the dazed incredulity of one who knows herself in a nightmare.

Ranged outside the circle that enclosed the two women sat what was surely the strangest company ever assembled. They were grouped with a queer irregularity which, though meaningless to Jirel, yet gave the impression of definite purpose and design. It had a symmetry so strongly marked that even though it fell outside her range of comprehension she could not but feel the rightness of it.

The light-robed dweller in the red barrens sat there, and the great black blob of shapeless jelly heaved gently on the crystal floor. She saw others she had watched enter, and many more. One was a female creature whose robe of peacock iridescence sprang from her shoulders in great drooping wings and folded round her like a bat's leathery cloak. And her neighbour was a fat grey slug of monster size, palpitating endlessly. One of the crowd looked exactly like a tall white lily swaying on a stalk of silver pallor, but from its chalice poured a light so ominously tinted that she shuddered and turned her eyes away.

Jarisme had risen from her couch. Very tall and regal in her violet robe, she rose against the backdrop of mist which veiled the other half of the room. As she lifted her arms, the incredible company turned to her with an eager expectancy. Jirel shuddered. Then Jarisme's flute spoke softly. It was a different sort of music from the clarion that called them together, from the stately melody which welcomed them through the opening doors. But it harped still on the two seesawing notes, with low, rippling sounds so different from the other two that Jirel marvelled at the range of the sorceress's ability on the two notes.

For a few moments as the song went on, nothing happened. Then a motion behind Jarisme caught Jirel's eye. The curtain of violet mist was swaying. The music beat at it and it quivered to the tune. It shook within itself, and paled and thinned, and from behind it a light began to glow. Then on a last low monotone it dissipated wholly and Jirel was staring at a vast globe of quivering light which loomed up under the stupendous arch that soared outward to form the second half of the chamber.

As the last clouds faded she saw that the thing was a huge crystal sphere, rising upon the coils of a translucent purple base in the shape of a serpent. And in the heart of the globe burned a still flame, living, animate, instinct with a life so alien that Jirel stared in utter bewilderment. It was a thing she knew to be alive—yet she knew it could *not* be alive. But she recognized even in her daze of incomprehension its relation to the tiny fragment of crystal she clutched in her hand. In that too the still flame burned. It stung her hand faintly in reminder that she possessed a weapon which could destroy Jarisme, though it might destroy its wielder in the process. The thought gave her a sort of desperate courage.

Jarisme was ignoring her now. She had turned to face the great globe with lifted arms and shining head thrown back. And from her lips a piercingly sweet sound fluted, midway between hum and whistle. Jirel had the wild fancy that she could see that sound arrowing straight into the heart of the vast sphere bulking so high over them all. And in the heart of that still, living flame a little glow of red began to quiver.

Through the trembling air shrilled a second sound. From the corner of her eye Jirel could see that a dark figure had moved forward into the circle and fallen to its knees at the sorceress's side. She knew it for Giraud. Like two blades the notes quivered in the utter hush that lay upon the assembly, and in the globe that red glow deepened.

One by one, other voices joined the chorus, queer, uncanny sounds some of them, from throats not shaped for speech. No two voices blended. The chorus was one of single, unrelated notes. And as each voice struck the globe, the fire burned more crimson, until its still pallor had flushed wholly into red. High above the rest soared Jarisme's knife-keen fluting. She lifted her arms higher, and the voices rose in answer. She lowered them, and the blade-like music swooped down an almost visible arc to a lower key. Jirel felt that she could all but see the notes spearing straight from each singer into the vast sphere that dwarfed them all. There was no melody in it, but a sharply definite pattern as alien and unmistakable as the symmetry of their grouping in the room. And as Jarisme's arms rose, lifting the voices higher, the flame burned more deeply red, and paled again as the voices fell.

Three times that stately, violet-robed figure gestured with lifted arms, and three times the living flame deepened and paled. Then Jarisme's voice soared in a high, triumphant cry and she whirled with spread arms, facing the company. In one caught breath, all voices ceased. Silence fell upon them like a blow. Jarisme was no longer priestess, but goddess, as she fronted them in that dead stillness with exultant face and blazing eyes. And in one motion they bowed before her as corn bows under wind. Alien things, shapeless monsters, faceless, eyeless, unrecognizable creatures from unknowable dimensions, abased themselves to the crystal floor before the splendour of light in Jarisme's eyes. For a moment of utter silence the tableau held. Then the sorceress's arms fell.

Ripplingly the company rose. Beyond Jarisme the vast globe had paled again into that living, quiet flame of golden pallor. Immense, brooding, alive, it loomed up above them. Into the strained stillness Jarisme's low voice broke. She was speaking in Jirel's native tongue, but the air, as she went on, quivered thickly with something like waves of sound that were pitched for other organs than human ears. Every word that left her lips made another wave through the thickened air. The assembly shimmered before Jirel's eyes in that broken clarity as a meadow quivers under heat waves.

'Worshippers of the Light,' said Jarisme sweetly, 'be welcomed from your far dwellings into the presence of the Flame. We who serve it have called you to the worship, but before you return, another sort of ceremony is to be held, which we have felt will interest you all. For we have called it truly the simplest and subtlest and most terrible of all punishments for a human creature.

'It is our purpose to attempt a reversal of this woman's physical and mental self in such a way as to cause her body to become rigidly motionless while her mind—her soul—looks eternally backward along the path it has travelled. You who are human, or have known humanity, will understand what deadly torture that can be. For no human creature, by the laws that govern it, can have led a life whose intimate review is anything but pain. To be frozen into eternal reflections, reviewing all the futility and pain of life, all the pain that thoughtless or intentional acts have caused others, all the spreading consequences of every act—that, to a human being, would be the most dreadful of all torments.'

In the silence that fell as her voice ceased, Giraud laid a hand on Jarisme's arm. Jirel saw terror in his eyes.

'Remember,' he uttered, 'remember, for those who tamper with their known destiny a more fearful thing may come than—'

Jarisme shrugged off the restraining hand impatiently. She turned to Jirel.

'Know, earthling,' she said in a queerly strained voice, 'that in the books of the future it is written that Jarisme the Sorceress must die at the hands of the one human creature who defies her thrice—and that human creature a woman. Twice I have been weak, and spared you. Once in the forest, once on the rooftop, you cast your puny defiance in my face, and I stayed my hand for fear of what is written. But the third time shall not come. Though you are my appointed slayer, you shall not slay. With my own magic I break Fate's sequence, now, and we shall see!'

In the blaze of her purple eyes Jirel saw that the moment had come. She braced herself, fingers closing about the fragment of crystal in her hand uncertainly as she hesitated, wondering if the time had come for the breaking of her talisman at the sorceress's feet. She hesitated too long, though her waiting was only a split second in duration. For Jarisme's magic was more supremely simple than Jirel could have guessed. The sorceress turned a blazing purple gaze upon her and sharply snapped her plump fingers in the earthwoman's face.

At the sound Jirel's whole world turned inside out about her. It was the sheerest physical agony. Everything vanished as that terrible shift took place. She felt her own body being jerked inexplicably around in a reversal like nothing that any living creature could ever have experienced before. It was a backward-facing in a direction which could have had no existence until that instant. She felt the newness in the second before sight came to her—a breathless, soundless, new-born *now* in which she was the first dweller, created simultaneously with the new plane of being. Then sight broke upon her consciousness.

The thing spread out before her was so stupendous that she would have screamed if she had possessed an animate body. All life was open to her gaze. The sight was too immeasurable for her to grasp it fully—too vast for her human consciousness to look upon at all save in flashing shutter-glimpses without relation or significance. Motion and immobility existed simultaneously in the thing before her. Endless activity shuttling to and fro—yet the whole vast panorama was frozen in a timeless calm through which a mighty pattern ran whose very immensity was enough to strike terror into her soul. Threaded through it the backward trail of her own life stretched. As she gazed upon it such floods of conflicting emotion washed over her that she could not see anything clearly, but she was fiercely insisting to her inner consciousness that she would not—*would not*—look back, dared not, could not—and all the while her sight was running past days and weeks along the path which led inexorably towards the one scene she could not bear to think of.

Very remotely, as her conscious sight retraced the backward way, she was aware of overlapping planes of existence in the stretch of limitless activity before her. Shapes other than human, scenes that had no meaning to her, quivered and shifted and boiled with changing lives—yet lay motionless in the mighty pattern. She scarcely heeded them. For her, of all that panoramic impossibility one scene alone had meaning—the one scene towards which her sight was racing now, do what she would to stop it—the one scene that she knew she could never bear to see again.

Yet when her sight reached that place the pain did not begin at once. She gazed almost calmly upon that little interval of darkness and flaring light, the glare of torches shining upon a girl's bent red head and on a man's long body sprawled motionless upon flagstones. In the deepest stillness she stared. She felt no urge to look further, on beyond the scene into the past. This was the climax, the centre of all her life—this torch-lit moment on the flagstones. Vividly she was back again in the past, felt the hardness of the cold flags against her knees, and the numbness of her heart as she stared down into a dead man's face. Timelessly she dwelt upon that long-ago heartbreak, and within her something swelled unbearably.

That something was a mounting emotion too great to have name, too complexly blending agony and grief and hatred and love—and rebellion; so strong that all the rest of the stupendous thing before her was blotted out in the gathering storm of what seethed in her innermost consciousness. She was aware of nothing but that overwhelming emotion. And it was boiling into one great unbearable explosion of violence in which rage took precedence over all. Rage at life for permitting such pain to be. Rage at Jarisme for forcing her into memory. Such rage that everything shook before it, and melted and ran together in a heat of rebellion, and—something snapped. The panorama reeled and shivered and collapsed into the dark of semi-oblivion.

Through the clouds of her half-consciousness the agony of change stabbed at her. Half understanding, she welcomed it, though the piercing anguish of that reversal was so strong it dragged her out of her daze again and wrung her anew in the grinding pain of that change which defied all natural laws. In heedless impatience she waited for the torture to pass. Exultation was welling up in her, for she knew that her own violence had melted the spell by which Jarisme held her. She knew what she must do when she stood free again, and conscious power flowed intoxicatingly through her.

She opened her eyes. She was standing rigidly before the great fire-quickened globe. The amazing company was grouped around her intently, and Jarisme, facing her, had taken one angry, incredulous step forward as she saw her own spell break. Upon that tableau Jirel's hot yellow eyes opened, and she laughed in grim exultation and swung up her arm. Violet light glinted upon crystal.

In the instant Jarisme saw what she intended, convulsive terror wiped all other expression from her face. A cry of mingled inarticulateness thundered up from the transfixed crowd. Giraud started forward from among them, frantic hands clawing out toward her.

'No, no!' shrieked Jarisme. 'Wait!'

It was too late. The crystal dashed itself from Jirel's down-swinging arm, the light in it blazing. With a splintering crash it struck the floor at the sorceress's sandalled feet and flew into shining fragments.

For an instant nothing happened. Jirel held her breath, waiting. Giraud had flung himself flat on the shining floor, reaching out for her in a last desperate effort. His hands had flown out to seize her, and found only her ankles. He clung to them now with a paralysed grip, his face hidden between his arms. Jarisme cowered motionless, arms clasped about her head as if she were trying to hide. The motley throng of watchers was rigid in fatalistic quiet. In tense silence they waited.

Then in the great globe above them the pale flame flickered. Jarisme's gaspingly caught breath sounded loud in the utter quiet. Again the flame shook. And again. Then abruptly it went out. Darkness stunned them for a moment; then a low muttering roar rumbled up out of the stillness, louder and deeper and stronger until it pressed unbearably upon Jirel's ears and her head was one great aching surge of sound. Above that roar a sharply crackling noise broke, and the crystal walls of the room trembled, reeled dizzily—split open in long jagged rents through which the violet day poured in thin fingers of light. Overhead the shattering sound of falling walls roared loud. Jarisme's magic tower was crumbling all around them. Through the long, shivering cracks in the walls the pale violet day poured more strongly, serene in the chaos.

In that clear light Jirel saw a motion among the throng. Jarisme had risen to her full height. She saw the sleek black head go up in an odd, defiant, desperate poise, and above the soul-shaking tumult she heard the sorceress's voice scream,

'Urda! Urda-sla!'

In the midst of the roar of the falling walls for the briefest instant a deathly silence dropped. And out of that silence, like an answer to the sorceress's cry, came a Noise, an indescribable, intolerable loudness like the crack of cyclopean thunder. And suddenly in the sky above them, visible through the crumbling crystal walls, a long black wedge opened. It was like a strip of darkest midnight splitting the violet day, a midnight through which stars shone unbearably near, unbearably bright.

Jirel stared up in dumb surprise at that streak of starry night cleaving the daylit sky. Jarisme stood rigid, arms outstretched, defiantly fronting the thunderous dark whose apex was drawing nearer and nearer, driving downward like a vast celestial spear. She did not flinch as it reached towards the tower. Jirel saw the darkness sweep forward like a racing shadow. Then it was upon them, and the earth shuddered under her feet, and from very far away she heard Jarisme scream.

When consciousness returned to her, she sat up painfully and stared around. She lay upon green grass, bruised and aching, but unharmed. The violet day was serene and unbroken once more. The purple peaks had vanished. No longer was she high among mountains. Instead, the green meadow where she had first seen Jarisme's tower stretched about her. In its dissolution it must have returned to its original site, flashing back along the magical ways it had travelled as the sorceress's magic was broken. For the tower too was gone. A little distance away she saw a heap of marble blocks outlining a rough circle, where that white shaft had risen. But the stones were weathered and cracked like the old, old stones of an ancient ruin.

She had been staring at this for many minutes, trying to focus her bewildered mind upon its significance, before the sound of groaning which had been going on for some time impressed itself on her brain. She turned. A little way off, Giraud lay in a tangle of torn black robes. Of Jarisme and the rest she saw no sign. Painfully she got to her feet and staggered to the wizard, turning him over with a disdainful toe. He opened his eyes and stared at her with a cloudy gaze into which recognition and realization slowly crept.

'Are you hurt?' she demanded.

He pulled himself to a sitting position and flexed his limbs experimentally. Finally he shook his head, more in answer to his own investigation than to her query, and got slowly to his feet. Jirel's eyes sought the weapon at his hip.

'I am going to kill you now,' she said calmly. 'Draw your sword, wizard.'

The little dull eyes flashed up to her face. He stared. Whatever he saw in the yellow gaze must have satisfied him that she meant what she said, but he did not draw, nor did he fall back. A tight little smile drew his mouth askew, and he lifted his black-robed arms. Jirel saw them rise, and her gaze followed the gesture automatically. Up they went, up. And then in the queerest fashion she lost all control of her own eyes, so that they followed some invisible upward line which drew her on and on skyward until she was rigidly staring at a fixed point of invisibility at the spot where the lines of Giraud's arms would have crossed, where they extended to a measureless distance. Somehow she actually saw that point, and could not look away. Gripped in the magic of those lifted arms, she stood rigid, not even realizing what had happened, unable even to think in the moveless magic of Giraud.

His little mocking chuckle reached her from immeasurably far away.

'Kill me?' he was laughing thickly. 'Kill me, Giraud? Why, it was you who saved me, Joiry! Why else should I have clung to your ankles so tightly? For I knew that when the Light died, the only one who could hope to live would be the one who slew it—nor was that a certainty, either. But I took the risk, and well I did, or I would be with Jarisme now in the outer dark whence she called up her no-god of the void to save her from oblivion. I warned her what would happen if she tampered with Fate. And I would rather—yes, much rather—be here, in this pleasant violet land which I shall rule alone now. Thanks to you, Joiry! Kill me, eh? I think not!'

That thick, mocking chuckle reached her remotely, penetrated her magic-stilled mind. It echoed round and round there, for a long while, before she realized what it meant. But at last she remembered, and her mind woke a little from its inertia, and such anger swept over her that its heat was an actual pain. Giraud, the runaway sorcerer, laughing at Joiry! Holding Jirel of Joiry in his spell! Mocking her! Blindly she wrenched at the bonds of magic, blindly urged her body forward. She could see nothing but that non-existent point where the lifted arms would have crossed, in measureless distances, but she felt the dagger-hilt in her hand, and she lunged forward through invisibility, and did not even know when the blade sank home.

Sight returned to her then in a stunning flood. She rubbed dazed eyes and shook herself and stared round the green meadow in the violet day uncomprehendingly, for her mind was not yet fully awake. Not until she looked down did she remember.

Giraud lay there. The black robes were furled like wings over his quiet body, but red in a thick flood was spreading on the grass, and from the tangled garments her dagger-hilt stood up. Jirel stared down at him, emotionless, her whole body still almost asleep from the power of the dead man's magic. She could not even feel triumph. She pulled the blade free automatically and wiped it on his robes. Then she sat down beside the body and rested her head in her hands, forcing herself to awaken.

After a long while she looked up again, the old hot light rising in her eyes, life flushing back into her face once more. Shaking off the last shreds of the spell, she got to her feet, sheathing the dagger. About her the violet-misted meadows were very still. No living creature moved anywhere in sight. The trees were motionless in the unstirring air. And beyond the ruins of the marble tower she saw the opening in the woods out of which her path had come, very long ago.

Jirel squared her shoulders and turned her back upon her vow fulfilled, and without a backward glance set off across the grass towards the tree-hid ruins which held the gate to home.

FRITZ LEIBER

THE BLEAK SHORE

'So you think a man can cheat death and outwit doom?' said the small, pale man, whose bulging forehead was shadowed by a black cowl.

The Gray Mouser, holding the dice box ready for a throw, paused and quickly looked sideways at the questioner.

'I said that a cunning man can cheat death for a long time.'

The Silver Eel bustled with pleasantly raucous excitement. Fighting men predominated and the clank of swordsmen's harness mingled with the thump of tankards, providing a deep obbligato to the shrill laughter of the women. Swaggering guardsmen elbowed the insolent bravos of the young lords. Grinning slaves bearing open wine jars dodged nimbly between. In one corner a slave girl was dancing, the jingle of her silver anklet bells inaudible in the din. Outside the small, tight-shuttered windows a dry, whistling wind from the south filled the air with dust that eddied between the cobblestones and hazed the stars. But here all was jovial confusion.

The Gray Mouser was one of a dozen at the gaming table. He was dressed all in grey—jerkin, silken shirt, and mouseskin cap—but his dark, flashing eyes and cryptic smile made him seem more alive than any of the others, save for the huge copper-haired barbarian next to him, who laughed immoderately and drank tankards of the sour wine of Lankhmar as if it were beer.

'They say you're a skilled swordsman and have come close to death many times,' continued the small, pale man in the black robe, his thin lips barely parting as he spoke the words.

But the Mouser had made his throw, and the odd dice of Lankhmar had stopped with the matching symbols of the eel and serpent uppermost, and he was raking in triangular golden coins. The barbarian answered for him.

'Yes, the grey one handles a sword daintily enough—almost as well as myself. He's also a great cheat at dice.'

'Are you, then, Fafhrd?' asked the other. 'And do you, too, truly think a man can cheat death, be he ever so cunning a cheat at dice?'

The barbarian showed his white teeth in a grin and peered puzzledly at the small, pale man whose sombre appearance and manner contrasted so strangely with the revellers thronging the low-ceilinged tavern fumy with wine.

'You guess right again,' he said in a bantering tone. 'I am Fafhrd, a Northerner, ready to pit my wits against any doom.' He nudged his companion. 'Look, Mouser, what do you think of this little black-coated mouse who's sneaked in through a crack in the floor and wants to talk with you and me about death?'

The man in black did not seem to notice the jesting insult. Again his bloodless lips hardly moved, yet his words were unaffected by the surrounding clamour, and impinged on the ears of Fafhrd and the Gray Mouser with peculiar clarity.

'It is said you two came close to death in the Forbidden City of the Black Idols, and in the stone trap of Angarngi, and on the misty island in the Sea of Monsters. It is also said that you have walked with doom on the Cold Waste and through the Mazes of Klesh. But who may be sure of these things, and whether death and doom were truly near? Who knows but what you are both braggarts who have boasted once too often? Now I have heard tell that death sometimes calls to a man in a voice only he can hear. Then he must rise and leave his friends and go to whatever place death shall bid him, and there meet his doom. Has death ever called to you in such a fashion?'

Fafhrd might have laughed, but did not. The Mouser had a witty rejoinder on the tip of his tongue, but instead he heard himself saying: 'In what words might death call?'

'That would depend,' said the small man. 'He might look at two such as you and say the Bleak Shore. Nothing more than that. The Bleak Shore. And when he said it three times you would have to go.'

This time Fafhrd tried to laugh, but the laugh never came. Both of them could only meet the gaze of the small man with the white, bulging forehead, stare stupidly into his cold, cavernous eyes. Around them the tavern roared with mirth at some jest. A drunken guardsman was bellowing a song. The gamblers called impatiently to the Mouser to stake his next wager. A giggling woman in red and gold stumbled past the small, pale man, almost brushing away the black cowl that covered his pate. But he did not move. And Fafhrd and the Gray Mouser continued to stare—fascinatedly, helplessly—into his chill, black eyes, which now seemed to them twin tunnels leading into a far and evil distance. Something deeper than fear gripped them in iron paralysis. The tavern became faint and soundless, as if viewed through many thicknesses of glass. They saw only the eyes and what lay beyond the eyes, something desolate, dreary, and deadly.

'The Bleak Shore,' he repeated.

Then those in the tavern saw Fafhrd and the Gray Mouser rise and without sign or word of leave-taking walk together to the low oaken door. A guardsman cursed as the huge Northerner blindly shoved him out of the way. There were a few shouted questions and mocking comments—the Mouser had been winning—but these were quickly hushed, for all perceived something strange and alien in the manner of the two. Of the small, pale, black-robed man none took notice. They saw the door open. They heard the dry moaning of the wind and a hollow flapping that probably came from the awnings. They saw an eddy of dust swirl up from the threshold. Then the door was closed and Fafhrd and the Mouser were gone.

No one saw them on their way to the great stone docks that bank the east side of the River Hlal from one end of Lankhmar to the other. No one saw Fafhrd's north-rigged, red-sailed sloop cast off and slip out into the current that slides down to the squally Inner Sea. The night was dark and the dust kept men indoors. But the next day they were gone, and the ship with them, and its Mingol crew of four—these being slave prisoners, sworn to life service, whom Fafhrd and the Mouser had brought back from an otherwise unsuccessful foray against the Forbidden City of the Black Idols.

About a fortnight later a tale came back to Lankhmar from Earth's End, the little harbour town that lies furthest of all towns to the west, on the very margin of the shipless Outer Sea; a tale of how a north-rigged sloop had come into port to take on an unusually large amount of food and water—unusually large because there were only six in the crew; a sullen, white-skinned Northern barbarian; an unsmiling little man in grey; and four squat, stolid, black-haired Mingols. Afterward the sloop had sailed straight into the sunset. The people of Earth's End had watched the red sail until nightfall, shaking their heads at its audacious progress. When this tale was repeated in Lankhmar, there were others who shook their heads, and some who spoke significantly of the peculiar behaviour of the two companions on the night of their departure. And as the weeks dragged on into months and the months slowly succeeded one another, there were many who talked of Fafhrd and the Gray Mouser as two dead men.

Then Ourph the Mingol appeared and told his curious story to the dockmen of Lankhmar. There was some difference of opinion as to the validity of the story, for although Ourph spoke the soft language of Lankhmar moderately well, he was an outsider, and, after he was gone, no one could prove that he was or was not one of the four Mingols who had sailed with the north-rigged sloop. Moreover, his story did not answer several puzzling questions, which is one of the reasons that many thought it untrue.

'They were mad,' said Ourph, 'or else under a curse, those two men, the great one and the small one. I suspected it when they spared our lives under the very walls of the Forbidden City. I knew it for certain when they sailed west and west and west, never reefing, never changing course, always keeping the star of the ice fields on our right hand. They talked little, they slept little, they laughed not at all. Ola, they were cursed! As for us four—Teevs, Larlt, Ouwenyis, and I—we were ignored but not abused. We had our amulets to keep off evil magics. We were sworn slaves to the death. We were men of the Forbidden City. We made no mutiny.

'For many days we sailed. The sea was stormless and empty around us, and small, very small; it looked as if it bent down out of sight to the north and the south and the awful west, as if the sea ended an hour's sail from where we were. And then it began to look that way to the east, too. But the great Northerner's hand rested on the steering oar like a curse, and the small grey one's hand was as firm. We four sat mostly in the bow, for there was little enough sail-tending, and diced our destinies at night and morning, and gambled for our amulets and clothes—we would have played for our hides and bones, were we not slaves.

'To keep track of the days, I tied a cord around my right thumb and moved it over a finger each day until it passed from right little finger to left little finger and came to my left thumb. Then I put it on Teev's right thumb. When it came to his left thumb he gave it to Larlt. So we numbered the days and knew them. And each day the sky became emptier and the sea smaller, until it seemed that the end of the sea was but a bow-shot away from our stem and sides and stern. Teevs said that we were upon an enchanted patch of water that was being drawn through the air towards the red star that is Hell. Surely Teevs may have been right. There cannot be so much water to the west. I have crossed the Inner Sea and the Sea of Monsters—and I say so.

'It was when the cord was around Larlt's left ring finger that the great storm came at us from the south-west. For three days it blew stronger and stronger, smiting the water into great seething waves; crags and gullies piled mast-high with foam. No other men have seen such waves nor should see them; they are not churned for us or for our oceans. Then I had further proof that our masters were under a curse. They took no notice of the storm; they let it reef their sails for them. They took no notice when Teevs was washed overboard, when we were half swamped and filled to the gunwales with spume, our bailing buckets foaming like tankards of beer. They stood in the stern, both braced against the steering oar, both drenched by the following waves, staring straight ahead, seeming to hold converse with creatures that only the bewitched can hear. Ola! They were accursed! Some evil demon was preserving their lives for a dark reason of his own. How else came we safe through the storm?

'For when the cord was on Larlt's left thumb, the towering waves and briny foam gave way to a great black sea swell that the whistling wind rippled but did not whiten. When the dawn came and we first saw it, Ouwenyis cried out that we were riding by magics upon a sea of black sand; and Larlt averred that we were fallen during the storm into the ocean of sulphurous oil that some say lies under the earth—for Larlt has seen the black, bubbling lakes of the Far East; and I remembered what Teevs had said and wondered if our patch of water had not been carried through the thin air and plunged into a wholly different sea on a wholly different world. But the small grey one heard our talk and dipped a bucket over the side and soused us with it, so that we knew our hull was still in water and that the water was salt—wherever that water might be.

'And then he bid us patch the sails and make the sloop shipshape. By midday we were flying west at a speed even greater than we had made during the storm, but so long were the swells and so swift did they move with us that we could only climb five or six in a whole day. By the Black Idols, but they were long!

'And so the cord moved across Ouwenyis' fingers. But the clouds were as leaden dark above us as was the strange sea heavy around our hull, and we knew not if the light that came through them was that of the sun or of some wizard moon, and when we caught sight of the stars they seemed strange. And still the white hand of the Northerner lay heavy on the steering oar, and still he and the grey one stared straight ahead. But on the third day of our flight across that black expanse the Northerner broke silence. A mirthless, terrible smile twisted his lips, and I heard him mutter 'the Bleak Shore'. Nothing more than that. The grey one nodded, as if there were some portentous magic in the words. Four times I heard the words pass the Northerner's lips, so that they were imprinted on my memory.

'The days grew darker and colder, and the clouds slid lower and lower, threatening, like the roof of a great cavern. And when the cord was on Ouwenyis's pointing finger we saw a leaden and motionless extent ahead of us, looking like the swells, but rising above them, and we knew that we were come to the Bleak Shore.

'Higher and higher that shore rose, until we could distinguish the towering basalt crags, rounded like the sea swell, studded here and there with grey boulders, whitened in spots as if by the droppings of gigantic birds—yet we saw no birds, large or small. Above the cliffs were dark clouds, and below them was a strip of pale sand, nothing more. Then the Northerner bent the steering oar and sent us straight in, as though he intended our destruction; but at the last moment he passed us at mast length by a rounded reef that hardly rose above the crest of the swell and found us harbour room. We sent the anchor over and rode safe.

'Then the Northerner and the grey one, moving like men in a dream, accoutred themselves, a shirt of light chain mail and a rounded, uncrested helmet for each—both helmets and shirts white with salt from the foam and spray of the storm. And they bound their swords to their sides, and wrapped great cloaks about them, and took a little food and a little water, and bade us unship the small boat. And I rowed them ashore and they stepped out onto the beach and walked towards the cliffs. Then, although I was much frightened, I cried out after them, "Where are you going? Shall we follow? What shall we do?" For several moments there was no reply. Then, without turning his head, the grey one answered, his voice a low, hoarse, yet far-carrying whisper. And he said. "Do not follow. We are dead men. Go back if you can."

'And I shuddered and bowed my head to his words and rowed out to the ship. Ouwenyis and Larlt and I watched them climb the high, rounded crags. The two figures grew smaller and smaller, until the Northerner was no more than a tiny, slim beetle and his grey companion almost invisible, save when they crossed a whitened space. Then a wind came down from the crags and blew the swell away from the shore, and we knew we could make sail. But we stayed—for were we not sworn slaves? And am I not a Mingol?

'As evening darkened, the wind blew stronger, and our desire to depart—if only to drown in the unknown sea—became greater. For we did not like the strangely rounded basalt crags of the Bleak Shore; we did not like it that we saw no gulls or hawks or birds of any kind in the leaden air, no seaweed on the beach. And we all three began to catch glimpses of something shimmering at the summit of the cliffs. Yet it was not until the third hour of night that we upped anchor and left the Bleak Shore behind.

'There was another great storm after we were out several days, and perhaps it hurled us back into the seas we know. Ouwenyis was washed overboard and Larlt went mad from thirst, and towards the end I knew not myself what was happening. Only I was cast up on the southern coast near Quarmall and, after many difficulties, am come here to Lankhmar. But my dreams are haunted by those black cliffs and by visions of the whitening bones of my masters, and their grinning skulls staring empty-eyed at something strange and deadly.'

Unconscious of the fatigue that stiffened his muscles, the Gray Mouser wormed his way past the last boulder, finding shallow handholds and footholds at the juncture of the granite and black basalt, and finally stood erect on the top of the rounded crags that walled the Bleak Shore. He was aware that Fafhrd stood at his side, a vague, hulking figure in whitened chain-mail vest and helmet. But he saw Fafhrd vaguely, as if through many thicknesses of glass. The only things he saw clearly—and it seemed he had been looking at them for an eternity—were two cavernous, tunnel-like black eyes, and beyond them something desolate and deadly, which had once been across the Outer Sea but was now close at hand. So it had been, ever since he had risen from the gaming table in the low-ceilinged tavern in Lankhmar. Vaguely he remembered the staring people of Earth's End, the foam, and fury of the storm, the curve of the black sea swell, and the look of terror on the face of Ourph the Mingol; these memories, too, came to him as if through many thicknesses of glass. Dimly he realized that he and his companion were under a curse, and that they were now come to the source of that curse.

For the flat landscape that spread out before them was without sign of life. In front of them the basalt dipped down to form a large hollow of black sand—tiny particles of iron ore. In the sand were half embedded more than two score of what seemed to the Gray Mouser to be inky-black, oval boulders of various sizes. But they were too perfectly rounded, too regular in form to be boulders, and slowly it was borne in on the Mouser's consciousness that they were not boulders, but monstrous black eggs, a few small, some so large that a man could not have clasped his arms around them, one big as a tent.

Scattered over the sand were bones, large and small. The Mouser recognized the tusked skull of a boar, and two smaller ones—wolves. There was the skeleton of some great predatory cat. Beside it lay the bones of a horse, and beyond them the rib cage of a man or ape. The bones lay all around the huge black eggs—a whitely gleaming circle.

From somewhere a toneless voice sounded, thin but clear, in accents of command, saying: 'For warriors, a warrior's doom.'

The Mouser knew the voice, for it had been echoing in his ears for weeks, ever since it had first come from the lips of a pale little man with a bulging forehead, wearing a black robe and sitting near him in a tavern in Lankhmar. And a more whispering voice came to him from within, saying, *He seeks always to repeat past experience, which has always been in his favour.*

Then he saw that what lay before him was not utterly lifeless. Movement of a sort had come to the Bleak Shore. A crack had appeared in one of the great black eggs, and then in another, and the cracks were branching, widening as bits of shell fell to the black, sandy floor.

The Mouser knew that this was happening in answer to the first voice, the thin one. He knew this was the end to which the thin voice had called him across the Outer Sea. Powerless to move further, he dully watched the slow progress of this monstrous birth. Under the darkening leaden sky he watched twin deaths hatching out for him and his companion.

The first hint of their nature came in the form of a long, swordlike claw which struck out through a crack, widening it further. Fragments of shell fell more swiftly.

The two creatures which emerged in the gathering dusk held enormity even for the Mouser's drugged mind. Shambling things, erect like men but taller, with reptilian heads boned and crested like helmets, feet clawed like a lizard's, shoulders topped with bony spikes, forelimbs each terminating in a single yard-long claw. In the semidarkness they seemed like hideous caricatures of fighting men, armoured and bearing swords. Dusk did not hide the yellow of their blinking eyes.

Then the voice called again: 'For warriors, a warrior's doom.'

At those words the bonds of paralysis dropped from the Mouser. For an instant he thought he was waking from a dream. But then he saw the new-hatched creatures racing towards them, a shrill, eager screeching issuing from their long muzzles. From beside him he heard a quick, rasping sound as Fafhrd's sword whipped from its scabbard. Then the Mouser drew his own blade, and a moment later it crashed against a steel-like claw which thrust at his throat. Simultaneously, Fafhrd parried a like blow from the other monster.

What followed was nightmare. Claws that were swords slashed and stabbed. Not so swiftly that they could not be parried, though there were four against two. Counter-thrusts glanced off impenetrable bony armour. Both creatures suddenly wheeled, striking at the Mouser. Fafhrd drove in from the side, saving him. Slowly the two companions were driven back towards where the crag sheered off. The beasts seemed tireless, creatures of bone and metal rather than flesh. The Mouser foresaw the end. He and Fafhrd might hold them off for a while longer, but eventually fatigue would supervene; their parries would become slower, weaker; the beasts would have them.

As if in anticipation of this, the Mouser felt a claw nick his wrist. It was then that he remembered the dark, cavernous eyes that had drawn them across the Outer Sea, the voice that had loosed doom upon them. He was gripped with a strange, mad rage—not against the beasts but their master. He seemed to see the black, dead eyes staring at him from the black sand. Then he lost control of his actions. When the two monsters next attempted a double attack on Fafhrd he did not turn to help, but instead dodged past and dashed down into the hollow, towards the embedded eggs.

Left to face the monsters alone, Fafhrd fought like a madman himself, his great sword whistling as his last resources of energy jolted his muscles. He hardly noticed when one of the beasts turned back to pursue his comrade.

The Mouser stood among the eggs, facing one of a glossier hue and smaller than most. Vindictively he brought his sword crashing down upon it. The blow numbed his hand, but the egg shattered.

Then the Mouser knew the source of the evil of the Bleak Shore, lying here and sending its spirit abroad, lying here and calling men to doom. Behind him he heard the scrabbling steps and eager screeching of the monster chosen for his destruction. But he did not turn. Instead, he raised his sword and brought it down whirring on the half-embryonic creature gloating in secret over the creatures he had called to death, down on the bulging forehead of the small pale man with the thin lips.

Then he waited for the finishing blow of the claw. It did not come. Turning, he saw the monster sprawled on the black sand. Around him, the deadly eggs were crumbling to dust. Silhouetted against the lesser darkness of the sky, he saw Fafhrd stumbling towards him, sobbing out vague words of relief and wonder in a deep, throaty voice. Death was gone from the Bleak Shore, the curse cut off at the root. From out of the night sounded the exultant cry of a sea bird, and Fafhrd and the Mouser thought of the long, trackless road leading back to Lankhmar.

RAY BRADBURY

HOMECOMING

'Here they come,' said Cecy, lying there flat in her bed.

'Where are they?' cried Timothy from the doorway.

'Some of them are over Europe, some over Asia, some of them over the Islands, some over South America!' said Cecy, her eyes closed, the lashes long, brown, and quivering.

Timothy came forward upon the bare plankings of the upstairs room. 'Who are they?'

'Uncle Einar and Uncle Fry, and there's Cousin William, and I see Frulda and Helgar and Aunt Morgiana and Cousin Vivian, and I see Uncle Johann! They're all coming fast!'

'Are they up in the sky?' cried Timothy, his little grey eyes flashing. Standing by the bed, he looked no more than his fourteen years. The wind blew outside, the house was dark and lit only by starlight.

'They're coming through the air and travelling along the ground, in many forms,' said Cecy, in her sleeping. She did not move on the bed; she thought inward on herself and told what she saw. 'I see a wolflike thing coming over a dark river—at the shallows—just above a waterfall, the starlight shining up his pelt. I see a brown oak leaf blowing far up in the sky. I see a small bat flying. I see many other things, running through the forest trees and slipping through the highest branches; and they're *all* coming this way!'

'Will they be here by tomorrow night?' Timothy clutched the bed-clothes. The spider on his lapel swung like a black pendulum, excitedly dancing. He leaned over his sister. 'Will they all be here in time for the Homecoming?'

'Yes, yes, Timothy, yes,' sighed Cecy. She stiffened. 'Ask no more of me. Go away now. Let me travel in the places I like best.'

'Thanks, Cecy,' he said. Out in the hall, he ran to his room. He hurriedly made his bed. He had just awakened a few minutes ago, at sunset, and as the first stars had risen, he had gone to let his excitement about the party run with Cecy. Now she slept so quietly there was not a sound. The spider hung on a silvery lasso about Timothy's slender neck as he washed his face. 'Just think, Spid, tomorrow night is Allhallows Eve!'

He lifted his face and looked into the mirror. His was the only mirror allowed in the house. It was his mother's concession to his illness. Oh, if only he were not so afflicted! He opened his mouth, surveyed the poor, inadequate teeth nature had given him. No more than so many corn kernels—round, soft and pale in his jaws. Some of the high spirit died in him.

It was now totally dark and he lit a candle to see by. He felt exhausted. This past week the whole family had lived in the fashion of the old country. Sleeping by day, rousing at sunset to move about. There were blue hollows under his eyes. 'Spid, I'm no good,' he said, quietly, to the little creature. 'I can't even get used to sleeping days like the others.'

He took up the candleholder. Oh, to have strong teeth, with incisors like steel spikes. Or strong hands, even, or a strong mind. Even to have the power to send one's mind out, free, as Cecy did. But, no, he was the imperfect one, the sick one. He was even—he shivered and drew the candle flame closer—afraid of the dark. His brothers snorted at him. Bion and Leonard and Sam. They laughed at him because he slept in a bed. With Cecy it was different; her bed was part of her comfort for the composure necessary to send her mind abroad to hunt. But Timothy, did he sleep in the wonderful polished boxes like the others? He did not! Mother let him have his own bed, his own room, his own mirror. No wonder the family skirted him like a holy man's crucifix. If only the wings would sprout from his shoulder blades. He bared his back, stared at it. And sighed again. No chance. Never.

Downstairs were exciting and mysterious sounds, the slithering black crape going up in all the halls and on the ceilings and doors. The sputter of burning black tapers in the banistered stair well. Mother's voice, high and firm. Father's voice, echoing from the damp cellar. Bion walking from outside the old country house lugging vast two-gallon jugs.

'I've just got to go to the party, Spid,' said Timothy. The spider whirled at the end of its silk, and Timothy felt alone. He would polish cases, fetch toadstools and spiders, hang crape, but when the party started he'd be ignored. The less seen or said of the imperfect son the better.

All through the house below, Laura ran.

'The Homecoming!' she shouted gaily. 'The Homecoming!' Her footsteps everywhere at once.

Timothy passed Cecy's room again, and she was sleeping quietly. Once a month she went belowstairs. Always she stayed in bed. Lovely Cecy. He felt like asking her, 'Where are you now, Cecy? And in who? And what's happening? Are you beyond the hills? And what goes on there?' But he went on to Ellen's room instead.

Ellen sat at her desk, sorting out many kinds of blond, red, and black hair and little scimitars of fingernail gathered from her manicurist job at the Mellin Village beauty parlour fifteen miles over. A sturdy mahogany case lay in one corner with her name on it.

'Go away,' she said, not even looking at him. 'I can't work with you gawking.'

'Allhallows Eve, Ellen; just think!' he said, trying to be friendly.

'Hunh!' She put some fingernail clippings in a small white sack, labelled them. 'What can it mean to you? What do you know of it? It'll scare the hell out of you. Go back to bed.'

His cheeks burned. 'I'm needed to polish and work and help serve.'

'If you don't go, you'll find a dozen raw oysters in your bed tomorrow,' said Ellen, matter-of-factly. 'Goodbye, Timothy.'

In his anger, rushing downstairs, he bumped into Laura.

'Watch where you're going!' she shrieked from clenched teeth.

She swept away. He ran to the open cellar door, smelled the channel of moist earthy air rising from below. 'Father?'

'It's about time,' Father shouted up the steps. 'Hurry down, or they'll be here before we're ready!'

Timothy hesitated only long enough to hear the million other sounds in the house. Brothers came and went like trains in a station, talking and arguing. If you stood in one spot long enough the entire household passed with their pale hands full of things. Leonard with his little black medical case, Samuel with his large dusty ebony-bound book under his arm, bearing more black crape, and Bion excursioning to the car outside and bringing in many more gallons of liquid.

Father stopped polishing to give Timothy a rag and a scowl. He thumped the huge mahogany box. 'Come on, shine this up, so we can start on another. Sleep your life away.'

While waxing the surface, Timothy looked inside.

'Uncle Einar's a big man, isn't he, Papa?'

'Unh.'

'How big is he?'

'The size of the box'll tell you.'

'I was only asking. Seven feet tall?'

'You talk a lot.'

About nine o'clock Timothy went out into the October weather. For two hours in the now-warm, now-cold wind he walked the meadows collecting toadstools and spiders. His heart began to beat with anticipation again. How many relatives had Mother said would come? Seventy? One hundred? He passed a farmhouse. If only you knew what was happening at our house, he said to the glowing windows. He climbed a hill and looked at the town, miles away, settling into sleep, the town hall clock high and round white in the distance. The town did not know, either. He brought home many jars of toadstools and spiders.

In the little chapel belowstairs a brief ceremony was celebrated. It was like all the other rituals over the years, with Father chanting the dark lines, mother's beautiful white ivory hands moving in the reverse blessings, and all the children gathered except Cecy, who lay upstairs in bed. But Cecy was present. You saw her peering, now from Bion's eyes, now Samuel's, now Mother's, and you felt a movement and now she was in you, fleetingly and gone.

Timothy prayed to the Dark One with a tightened stomach. 'Please, please, help me grow up, help me be like my sisters and brothers. Don't let me be different. If only I could put the hair in the plastic images as Ellen does, or make people fall in love with me as Laura does with people, or read strange books as Sam does, or work in a respected job like Leonard and Bion do. Or even raise a family one day, as mother and father have done. . . .'

At midnight a storm hammered the house. Lightning struck outside in amazing, snow-white bolts. There was a sound of an approaching, probing, sucking tornado, funnelling and nuzzling the moist night earth. Then the front door, blasted half off its hinges, hung stiff and discarded, and in trooped Grandmama and Grandpapa, all the way from the old country!

From then on people arrived each hour. There was a flutter at the side window, a rap on the front porch, a knock at the back. There were fey noises from the cellar; autumn wind piped down the chimney throat, chanting. Mother filled the large crystal punch bowl with a scarlet fluid poured from the jugs Bion had carried home, Father swept from room to room lighting more tapers. Laura and Ellen hammered up more wolfsbane. And Timothy stood amidst this wild excitement, no expression to his face, his hands trembling at his sides, gazing now here, now there. Banging of doors, laughter, the sound of liquid pouring, darkness, sound of wind, the webbed thunder of wings, the padding of feet, the welcoming bursts of talk at the entrances, the transparent rattlings of casements, the shadows passing, coming, going, wavering.

'Well, well, and *this* must be Timothy!'

'What?'

A chilly hand took his hand. A long hairy face leaned down over him. 'A good lad, a fine lad,' said the stranger.

'Timothy,' said his mother. 'This is Uncle Jason.'

'Hello, Uncle Jason.'

'And over here——' Mother drifted Uncle Jason away. Uncle Jason peered back at Timothy over his caped shoulder, and winked.

Timothy stood alone.

From off a thousand miles in the candled darkness, he heard a high fluting voice; that was Ellen. 'And my brothers, they *are* clever. Can you guess their occupations, Aunt Morgiana?'

'I have no idea.'

'They operate the undertaking establishment in town.'

'What!' A gasp.

'Yes!' Shrill laughter. 'Isn't that priceless!'

Timothy stood very still.

A pause in the laughter. 'They bring home sustenance for Mama, Papa and all of us,' said Laura. 'Except, of course, Timothy. . . .'

An uneasy silence. Uncle Jason's voice demanded. 'Well? come now. What about Timothy?'

'Oh, Laura, your tongue,' said mother.

Laura went on with it. Timothy shut his eyes. 'Timothy doesn't—well—doesn't *like* blood. He's delicate.'

'He'll learn,' said mother. 'He'll learn,' she said very firmly. 'He's my son, and he'll learn. He's only fourteen.'

'But I was raised on the stuff,' said Uncle Jason, his voice passing from one room on into another. The wind played the trees outside like harps. A little rain spatted on the windows— 'raised on the stuff,' passing away into faintness.

Timothy bit his lips and opened his eyes.

'Well, it was all my fault.' Mother was showing them into the kitchen now. 'I tried forcing him. You can't force children, you only make them sick, and then they never get a taste for things. Look at Bion, now, he was thirteen before he . . .'

'I understand,' murmured Uncle Jason. 'Timothy will come around.'

'I'm sure he will,' said mother, defiantly.

Candle flames quivered as shadows crossed and recrossed the dozen musty rooms. Timothy was cold. He smelled the hot tallow in his nostrils and instinctively he grabbed at a candle and walked with it around and about the house, pretending to straighten the crape.

'*Timothy*,' someone whispered behind a patterned wall, hissing and sizzling and sighing the words, '*Timothy is afraid of the dark*.'

Leonard's voice. Hateful Leonard!

'I like the candle, that's all,' said Timothy in a reproachful whisper.

More lightning, more thunder. Cascades of roaring laughter. Bangings and clickings and shouts and rustles of clothing. Clammy fog swept through the front door. Out of the fog, settling his wings, stalked a tall man.

'Uncle Einar!'

Timothy propelled himself on his thin legs, straight through the fog, under the green webbing shadows. He threw himself across Einar's arms. Einar lifted him.

'You've wings, Timothy!' He tossed the boy light as thistles. 'Wings, Timothy: fly!' Faces wheeled under. Darkness rotated. The house blew away. Timothy felt breezelike. He flapped his arms. Einar's fingers caught and threw him once more to the ceiling. The ceiling rushed down like a charred wall. 'Fly, Timothy!' shouted Einar, loud and deep. 'Fly with wings! Wings!'

He felt an exquisite ecstasy in his shoulder blades, as if roots grew, burst to explode and blossom into new, moist membrane. He babbled wild stuff; again Einar hurled him high.

The autumn wind broke in a tide on the house, rain crashed down, shaking the beams, causing chandeliers to tilt their enraged candle lights. And the one hundred relatives peered out from every black, enchanted room, circling inward, all shapes and sizes, to where Einar balanced the child like a baton in the roaring spaces.

'Enough!' shouted Einar, at last.

Timothy, deposited on the floor timbers, exaltedly, exhaustedly fell against Uncle Einar, sobbing happily. 'Uncle, uncle, uncle!'

'Was it good, flying? Eh, Timothy?' said Uncle Einar, bending down, patting Timothy's head. 'Good, good.'

It was coming towards dawn. Most had arrived and were ready to bed down for the daylight, sleep motionlessly with no sound until the following sunset, when they would shout out of their mahogany boxes for the revelry.

Uncle Einar, followed by dozens of others, moved towards the cellar.

Mother directed them downward to the crowded row on row of highly polished boxes. Einar, his wings like sea-green tarpaulins tented behind him, moved with a curious whistling down the passageway; where his wings touched they made a sound of drumheads gently beaten.

Upstairs, Timothy lay wearily thinking, trying to like the darkness. There was so much you could do in darkness that people couldn't criticize you for, because they never saw you. He *did* like the night, but it was a qualified liking: sometimes there was so much night he cried out in rebellion.

In the cellar, mahogany doors sealed downward, drawn in by pale hands. In corners, certain relatives circled three times to lie, heads on paws, eyelids shut. The sun rose. There was a sleeping.

Sunset. The revel exploded like a bat nest struck full, shrieking out, fluttering, spreading. Box doors banged wide. Steps rushed up from cellar damp. More late guests, kicking on front and back portals, were admitted.

It rained, and sodden visitors laid their capes, their water-pelleted hats, their sprinkled veils upon Timothy who bore them to a closet. The rooms were crowd-packed. The laughter of one cousin, shot from one room, angled off the wall of another, ricocheted, banked and returned to Timothy's ears from a fourth room, accurate and cynical.

A mouse ran across the floor.

'I know you, Niece Leibersrouter!' exclaimed father around him but not to him. The dozens of towering people pressed in against him, elbowed him, ignored him.

Finally, he turned and slipped away up the stairs.

He called softly. 'Cecy. Where are you now, Cecy?'

She waited a long while before answering. 'In the Imperial Valleys,' she murmured faintly. 'Beside the Salton Sea, near the mud pots and the steam and the quiet. I'm inside a farmer's wife. I'm sitting on a front porch. I can make her move if I want, or do anything or think anything. The sun's going down.'

'What's it like, Cecy?'

'You can hear the mud pots hissing,' she said, slowly, as if speaking in a church. 'Little grey heads of steam push up the mud like bald men rising in the thick syrup, head first, out in the broiling channels. The grey heads rip like rubber fabric, collapse with noises like wet lips moving. And feathery plumes of steam escape from the ripped tissue. And there is a smell of deep sulphurous burning and old time. The dinosaur has been abroiling here ten million years.'

'Is he done yet, Cecy?'

The mouse spiralled three women's feet and vanished into a corner. Moments later a beautiful woman rose up out of nothing and stood in the corner, smiling her white smile at them all.

Something huddled against the flooded pane of the kitchen window. It sighed and wept and tapped continually, pressed against the glass, but Timothy could make nothing of it, he saw nothing. In imagination he was outside staring in. The rain was on him, the wind at him, and the taper-dotted darkness inside was inviting. Waltzes were being danced; tall thin figures pirouetted to outlandish music. Stars of light flickered off lifted bottles; small clods of earth crumbled from casques, and a spider fell and went silently legging over the floor.

Timothy shivered. He was inside the house again. Mother was calling him to run here, run there, help, serve, out to the kitchen now, fetch this, fetch that, bring the plates, heap the food—on and on—the party happened.

'Yes, he's done. Quite done.' Cecy's calm sleeper's lips turned up. The languid words fell slowly from her shaping mouth. 'Inside this woman's skull I am, looking out, watching the sea that does not move, and is so quiet it makes you afraid. I sit on the porch and wait for my husband to come home. Occasionally, a fish leaps, falls back, starlight edging it. The valley, the sea, the few cars, the wooden porch, my rocking chair, myself, the silence.'

'What now, Cecy?'

'I'm getting up from my rocking chair,' she said.

'Yes?'

'I'm walking off the porch, towards the mud pots. Planes fly over, like primordial birds. Then it is quiet, so quiet.'

'How long will you stay inside her, Cecy?'

'Until I've listened and looked and felt enough: until I've changed her life some way. I'm walking off the porch and along the wooden boards. My feet knock on the planks, tiredly, slowly.'

'And now?'

'Now the sulphur fumes are all around me. I stare at the bubbles as they break and smooth. A bird darts by my temple, shrieking. Suddenly I am in the bird and fly away! And as I fly, inside my new small glass-bead eyes I see a woman below me, on a boardwalk, take one, two, three steps forward into the mud pots. I hear a sound as of a boulder plunged into molten depths. I keep flying, circle back. I see a white hand, like a spider, wriggle and disappear into the grey lava pool. The lava seals over. Now I'm flying home, swift, swift, swift!'

Something clapped hard against the window, Timothy started.

Cecy flicked her eyes wide, bright, full, happy, exhilarated.

'Now I'm *home*!' she said.

After a pause, Timothy ventured, 'The Homecoming's on. And everybody's here.'

'Then why are you upstairs?' She took his hand. 'Well, ask me.' She smiled slyly. 'Ask me what you came to ask.'

'I didn't come to ask anything,' he said. 'Well, almost nothing. Well—oh, Cecy!' It came from him in one long rapid flow. 'I want to do something at the party to make them look at me, something to make me good as them, something to make me belong, but there's nothing I can do and I feel funny and, well, I thought you might . . .'

'I might,' she said, closing her eyes, smiling inwardly. 'Stand up straight. Stand very still.' He obeyed. 'Now, shut your eyes and blank out your thought.'

He stood very straight and thought of nothing, or at least thought of thinking nothing.

She sighed. 'Shall we go downstairs now, Timothy?' Like a hand into a glove, Cecy was within him.

'Look everybody!' Timothy held the glass of warm red liquid. He held up the glass so that the whole house turned to watch him. Aunts, uncles, cousins, brothers, sisters!

He drank it straight down.

He jerked a hand at his sister Laura. He held her gaze, whispering to her in a subtle voice that kept her silent, frozen. He felt tall as the trees as he walked to her. The party now slowed. It waited on all sides of him, watching. From all the room doors the faces peered. They were not laughing. Mother's face was astonished. Dad looked bewildered, but pleased and getting prouder every instant.

He nipped Laura, gently, over the neck vein. The candle flames swayed drunkenly. The wind climbed around on the roof outside. The relatives stared from all the doors. He popped toadstools into his mouth, swallowed, then beat his arms against his flanks and circled. 'Look, Uncle Einar! I can fly, at last!' Beat went his hands. Up and down pumped his feet. The faces flashed past him.

At the top of the stairs flapping, he heard his mother cry, 'Stop, Timothy!' far below. 'Hey!' shouted Timothy, and leaped off the top of the well, thrashing.

Halfway down, the wings he thought he owned dissolved. He screamed. Uncle Einar caught him.

Timothy flailed whitely in the receiving arms. A voice burst out of his lips, unbidden. 'This is Cecy! This is Cecy! Come see me, all of you, upstairs, first room on the left!' Followed by a long trill of high laughter. Timothy tried to cut it off with his tongue.

Everybody was laughing. Einar set him down. Running through the crowding blackness as the relatives flowed upstairs towards Cecy's room to congratulate her, Timothy banged the front door open.

'Cecy, I hate you, I hate you!'

By the sycamore tree, in deep shadow. Timothy spewed out his dinner, sobbed bitterly and threshed in a pile of autumn leaves. Then he lay still. From his blouse pocket, from the protection of the matchbox he used for his retreat, the spider crawled forth. Spid walked along Timothy's arm. Spid explored up his neck to his ear and climbed in the ear to tickle it. Timothy shook his head. 'Don't, Spid. Don't.'

The feathery touch of a tentative feeler probing his eardrum set Timothy shivering. 'Don't, Spid!' He sobbed somewhat less.

The spider travelled down his cheek, took a station under the boy's nose, looked up into the nostrils as if to seek the brain, and then clambered softly up over the rim of the nose to sit, to squat there peering at Timothy with green gem eyes until Timothy filled with ridiculous laughter. 'Go away, Spid!'

Timothy sat up, rustling the leaves. The land was very bright with the moon. In the house he could hear the faint ribaldry as Mirror, Mirror was played. Celebrants shouted dimly muffled, as they tried to identify those of themselves whose reflections did not, had not ever appeared in a glass.

'Timothy.' Uncle Einar's wings spread and twitched and came in with a sound like kettledrums. Timothy felt himself plucked up like a thimble and set upon Einar's shoulder. 'Don't feel badly, Nephew Timothy. Each to his own, each in his own way. How much better things are for you. How rich. The world's dead for us. We've seen so much of it, believe me. Life's best to those who live the least of it. It's worth more per ounce, Timothy, remember that.'

The rest of the black morning, from midnight on, Uncle Einar led him about the house, from room to room, weaving and singing. A horde of late arrivals set the entire hilarity off afresh. Great-great-great-great and a thousand more great-greats Grandmother was there, wrapped in Egyptian cerements. She said not a word, but lay straight as a burnt ironing board against the wall, her eye hollows cupping a distant, wise, silent glimmering. At the breakfast, at four in the morning, one-thousand-odd-greats Grandmama was stiffly seated at the head of the longest table.

The numerous young cousins caroused at the crystal punch bowl. Their shiny olive-pit eyes, their conical, devilish faces and curly bronze hair hovered over the drinking table, their hard-soft, half-girl half-boy bodies wrestling against each other as they got unpleasantly, sullenly drunk. The wind got higher, the stars burned with fiery intensity, the noises redoubled, the dances quickened, the drinking became more positive. To Timothy there were thousands of things to hear and watch. The many darknesses roiled, bubbled, the many faces passed and repassed. . . .

'Listen!'

The party held its breath. Far away the town clock struck its chimes, saying six o'clock. The party was ending. In time to the rhythm of the striking clock, their one hundred voices began to sing songs that were four hundred years old, songs Timothy could not know. Arms twined, circling slowly, they sang, and somewhere in the cold distance of morning the town clock finished out its chimes and quieted.

Timothy sang. He knew no words, no tune, yet the words and tune came round and high and good. And he gazed at the closed door at the top of the stairs.

'Thanks Cecy,' he whispered. 'You're forgiven. Thanks.'

Then he just relaxed and let the words move, with Cecy's voice, free from his lips.

Goodbyes were said, there was a great rustling. Mother and Father stood at the door to shake hands and kiss each departing relative in turn. The sky beyond the open door coloured in the east. A cold wind entered. And Timothy felt himself seized and settled in one body after another, felt Cecy press him into Uncle Fry's head so he stared from the wrinkled leather face, then leaped in a flurry of leaves up over the house and awakening hills. . . .

Then, loping down a dirt path, he felt his red eyes burning, his fur pelt rimed with morning, as inside Cousin William he panted through a hollow and dissolved away. . . .

Like a pebble in Uncle Einar's mouth, Timothy flew in a webbed thunder, filling the sky. And then he was back, for all time, in his own body.

In the growing dawn, the last few were embracing and crying and thinking how the world was becoming less a place for them. There had been a time when they had met every year, but now decades passed with no reconciliation. 'Don't forget,' someone cried, 'we meet in Salem in 1970!'

Salem. Timothy's numbed mind turned the words over. Salem, 1970. And there would be Uncle Fry and a thousand-times-great Grandmother in her withered cerements, and Mother and Father and Ellen and Laura and Cecy and all the rest. But would he be there? Could he be certain of staying alive until then?

With one last withering blast, away they all went, so many scarves, so many fluttery mammals, so many sere leaves, so many whining and clustering noises, so many midnights and insanities and dreams.

Mother shut the door. Laura picked up a broom. 'No,' said Mother. 'We'll clean tonight. Now we need sleep.' And the Family vanished down cellar and upstairs. And Timothy moved in the crape-littered hall, his head down. Passing a party mirror, he saw the pale mortality of his face all cold and trembling.

'Timothy,' said Mother.

She came to touch her hand on his face. 'Son,' she said, 'We love you. Remember that. We all love you. No matter how different you are, no matter if you leave us one day.' She kissed his cheek. 'And if and when you die, your bones will lie undisturbed, we'll see to that. You'll lie at ease forever, and I'll come visit every Allhallows Eve and tuck you in the more secure.'

The house was silent. Far away the wind went over a hill with its last cargo of dark bats, echoing, chittering.

Timothy walked up the steps, one by one, crying to himself all the way.

HENRY KUTTNER

SEE YOU LATER

Old Yancey was just about the meanest man in the world. I never seen a feller so downright, sot-in-his ways, short-sighted, plain, ornery mean. What happened to him reminded me of what another feller told me oncet, quite a spell ago. Fergit exactly who it was—name of Louis, maybe, or could be Tamerlane—but one time he said he wished the whole world had only one haid, so's he could chop it off.

Trouble with Yancey, he got to the point where he figgered everybody in the world was again' him, and blamed if he warn't right. That was a real spell of trouble, even for us Hogbens.

Oh, Yancey was a regular stinker, all right. The whole Tarbell family was bad-eyed, but Yancey made even them plumb disgusted. He lived up in a little one-room shanty back of the Tarbell place, and wouldn't let nobody near, except to push vittles through the cut-out moon in the door.

Seems like some ten years back there was a new survey or something and the way it worked out, through some funny legal business, Yancey had to prove he'd got squatter's rights on his land. He had to prove it by living there for a year or something. 'Bout then he had an argument with his wife and moved out to the property line, and said he was a-gonna let the land go right back to the government, for all he cared, and that'd show the whole family. He knew his wife sot store by her turnip patch and was afraid the government would take it away.

The way it turned out, nobody wanted the land anyhow. It was all up and down and had too many rocks in it, but Yancey's wife kept on worriting and begging Yancey to come back, which he was just too mean to do.

Yancey Tarbell couldn't have been oncommon comfortable up in that little shack, but he was short-sighted as he was mean. After a spell Mrs Tarbell died of being hit on the haid with a stone she was throwing up the slope at the shack, and it bounced back at her. So that left only the eight Tarbell boys and Yancey. He stayed right where he was, though.

He might have stayed there till he shrivelled up and went to glory, except the Tarbells started feuding with us. We stood it as long as we could, on account of they couldn't hurt us. Uncle Les, who was visiting us, got skittery, though, and said he was tired of flying up like a quail, two or three miles in the air, every time a gun went off behind a bush. The holes in his hide closed up easy enough, but he said it made him dizzy, on account of the air being thinned out that high up.

This went on for a while, leastwise, and nobody got hurt, which seemed to rile the eight Tarbell boys. So one night they all come over in a bunch with their shooting irons and busted their way in. We didn't want no trouble.

Uncle Lem—who's Uncle Les's twin except they was born quite a spell apart—he was asleep for the winter, off in a holler tree somewheres, so he was out of it. But the baby, bless his heart, is gitting kind of awkward to shift around, being as how he's four hunnerd years old and big for his age—'bout three hunnerd pounds, I guess.

We could of all hid out or gone down to Piperville in the valley for a mite, but then there was Grandpaw in the attic, and I'd got sort of fond of the little Perfesser feller we keep in a bottle. Didn't want to leave him on account of the bottle might of got smashed in the ruckus, if the eight Tarbell boys were likkered up enough.

The Perfesser's cute—even though he never did have much sense. Used to say we was mutants, whatever they are, and kept shooting off his mouth about some people he knowed called chromosomes. Seems like they got mixed up with what the Perfesser call hard radiations and had some young 'uns which was either dominant mutations or Hogbens, but I allus got it mixed up with the Roundhead plot, back when we was living in the old country. 'Course I don't mean the *real* old country. That got sunk.

So, seeing as how Grandpaw told us to lay low, we waited till the eight Tarbell boys busted down the door, and then we all went invisible, including the baby. Then we waited for the thing to blow over, only it didn't.

After stomping around and ripping up things a lot, the eight Tarbell boys come down in the cellar. Now, that was kind of bad, because we was caught by surprise. The baby had gone invisible, like I say, and so had the tank we keep him in, but the tank couldn't move around fast like we could.

One of the eight Tarbell boys went and banked into it and hit hisself a smart crack on the shank bone. How he cussed! It was shameful for a growing boy to hear, except Grandpaw kin outcuss anybody I ever heard, so I didn't larn nothing.

Well—he cussed a lot, jumped around, and all of a sudden his squirrel rifle went off. Must have had a hair trigger. That woke up the baby, who got scared and let out a yell. It was the blamedest yell I'd ever heard out of the baby yet, and I've seen men go all white and shaky when he bellers. Our Perfesser feller told us oncet the baby emitted a subsonic. Imagine!

Anyhow, seven of the eight Tarbell boys dropped daid, all in a heap, without even time to squeal. The eighth one was up at the haid of the cellar steps, and he got all quivery and turned around and ran. I guess he was so dizzy he didn't know where he was heading. 'Fore he knowed it, he was up in the attic, where he stepped right square on Grandpaw.

Now, the fool thing was this: Grandpaw was so busy telling us what to do he'd entirely fergot to go invisible hisself. And I guess one look at Grandpaw just plumb finished the eighth Tarbell boy. He fell right down, daid as a skun coon. Cain't imagine why, though I got to admit Grandpaw wasn't looking his best that week. He'd been sick.

'You all right, Grandpaw?' I asked, sort of shaking him out. He cussed me.

''Twarn't my fault,' I told him.

''Sblood!' he said, mad-like. 'What rabble of canting jolt-heads have I sired? Put me down, you young scoundrel.' So I put him back on the gunny sack and he turned around a couple of times and shut his eyes. After that, he said he was going to take a nap and not to wake him up for nothing, bar Judgment Day. He meant it, too.

So we had to figger out for ourselves what was best to do. Maw said it warn't our fault, and all we could do was pile the eight Tarbell boys in a wheelbarrow and take 'em back home, which I done. Only I got to feeling kind of shy on the way, on account of I couldn't figger out no real polite way to mention what had happened. Besides, Maw had told me to break the news gentle. 'Even a polecat's got feelings,' she said.

So I left the wheelbarrow with the eight Tarbell boys in it behind some scrub brush, and I went on up the slope to where I could see Yancey sitting, airing hisself out in the sun and reading a book. I still hadn't studied out what to say. I just traipsed along slow-like, whistling 'Yankee Doodle.' Yancey didn't pay me no mind for a while.

He's a little, mean, dirty man with chin whiskers. Couldn't be much more'n five feet high. There was tobacco juice on his whiskers, but I might have done old Yancey wrong in figgering he was only sloppy. I heard he used to spit in his beard to draw flies, so's he could ketch 'em and pull off their wings.

Without looking, he picked up a stone, and flang it past my head. 'Shet up an' go way,' he said.

'Just as you say, Mr Yancey,' I told him, mighty relieved, and started to. But then I remembered Maw would probably whup me if I didn't mind her orders, so I sort of moved around quiet till I was in back of Yancey and looking over his shoulder at what he was reading. It looked like a book. Then I moved around a mite more till I was upwind of him.

He started cackling in his whiskers.

'That's a real purty picture, Mr Yancey,' I said.

He was giggling so hard it must of cheered him up.

'Ain't it, though!' he said, banging his fist on his skinny old rump. 'My, my! Makes me feel full o' ginger just to look at it.'

It wasn't a book, though. It was a magazine, the kind they sell down at the village, and it was opened at a picture. The feller that made it could draw real good. Not so good as an artist I knowed once, over in England. He went by the name of Crookshank or Crookback or something like that, unless I'm mistook.

Anyway, this here that Yancey was looking at was quite a picture. It showed a lot of fellers, all exactly alike, coming out of a big machine which I could tell right off wouldn't work. But all these fellers was as like as peas in a pod. Then there was a red critter with bugged-out eyes grabbing a girl, I dunno why. It was sure purty.

'Wisht something like that could really happen,' Yancey said.

'It ain't so hard,' I told him. 'Only that gadget's all wrong. All you need is a washbasin and some old scrap iron.'

'Hey?'

'That thing there,' I said. 'The jigger that looks like it's making one feller into a whole lot of fellers. It ain't built right.'

'I s'pose you could do it better?' he snapped, sort of mad.

'We did, once,' I said. 'I forget what Paw had on his mind, but he owed a man name of Cadmus a little favour. Cadmus wanted a lot of fighting men in a real hurry, so Paw fixed it so's Cadmus could split hisself up into a passel of soldiers. Shucks. I could do it myself.'

'What are you blabbering about?' Yancey asked. 'You ain't looking at the right thing. This here red critter's what I mean. See what he's a-gonna do? Gonna chaw that there purty gal's haid off, looks like. See the tusks on him? Heh, heh, heh. I wisht I was a critter like that. I'd chaw up plenty of people.'

'You wouldn't chaw up your own kin, though, I bet,' I said, seeing a way to break the news gentle.

'Tain't right to bet,' he told me. 'Allus pay your debts, fear no man, and don't lay no wagers. Gambling's a sin. I never made no bets and I allus paid my debts.' He stopped, scratched his whiskers, and sort of sighed. 'All except one,' he added, frowning.

'What was that?'

'Oh, I owed a feller something. Only I never could locate him afterward. Must be nigh on thutty years ago. Seems like I got likkered up and got on a train. Guess I robbed somebody, too, 'cause I had a roll big enough to choke a hoss. Never tried that, come to think of it. You keep hosses?'

'No, sir,' I said. 'We was talking about your kin.'

'Shet up,' old Yancey said. 'Well, now, I had myself quite a time.' He licked his whiskers. 'Ever heard tell of a place called New York? In some furrin country, I guess. Can't understand a word nobody says. Anyway, that's where I met up with this feller. I often wisht I could find him again. An honest man like me hates to think of dying without paying his lawful debts.'

'Did your eight boys owe any debts?' I asked.

He squinted at me, slapped his skinny leg, and nodded.

'Now I know,' he said. 'Ain't you the Hogben boy?'

'That's me. Saunk Hogben.'

'I heard tell 'bout you Hogbens. All witches, ain't you?'

'No, sir.'

'I heard what I heard. Whole neighbourhood's buzzing. Hexers, that's what. You get outa here, go on, git!'

'I'm a-going,' I said. 'I just come by to say it's real unfortunate you couldn't chaw up your own kin if'n you was a critter like in that there picture.'

'Ain't nobody big enough to stop me!'

'Maybe not,' I said, 'but they've all gone to glory.'

When he heard this, old Yancey started to cackle. Finally, when he got his breath back, he said, 'Not them! Them varmints have gone plumb smack to perdition, right where they belong. How'd it happen?'

'It was sort of an accident,' I said. 'The baby done kilt seven of them and Grandpaw kilt the other, in a way of speaking. No harm intended.'

'No harm done,' Yancey said, cackling again.

'Maw sent her apologies, and what do you want done with the remains? I got to take the wheelbarrow back home.'

'Take 'em away. I don't want 'em. Good riddance to bad rubbish,' old Yancey said, so I said all right and started off. But then he yelled out and told me he'd changed his mind. Told me to dump 'em where they was. From what I could make out, which wasn't much because he was laughing so hard, he wanted to come down and kick 'em.

So I done like he said and then went back home and told Maw, over a mess of catfish and beans and pot-likker. She made some hush puppies, too. They was good. I sat back, figgering I'd earned a rest, and thunk a mite, feeling warm and nice around the middle. I was trying to figger what a bean would feel like, down in my tummy. But it didn't seem to have no feelings.

It couldn't of been more than a half hour later when the pig yelled outside like he was getting kicked, and then somebody knocked on the door. It was Yancey. Minute he come in, he pulled a bandana out of his britches and started sniffing. I looked at Maw, wide-eyed. I couldn't tell her nothing.

Paw and Uncle Les was drinking corn in a corner, and giggling a mite. I could tell they was feeling good because of the way the table kept rocking, the one between them. It wasn't touching neither one, but it kept jiggling, trying to step fust on Paw's toes and then on Uncle Les's. They was doing it inside their haids, trying to ketch the other one off guard.

It was up to Maw, and she invited old Yancey to set down a spell and have some beans. He just sobbed.

'Something wrong, neighbour?' Maw asked, polite.

'It sure is,' Yancey said, sniffling. 'I'm a real old man.'

'You surely are,' Maw told him. 'Mebbe not as old as Saunk here, but you look awful old.'

'Hey?' Yancey said, staring at her. 'Saunk? Saunk ain't more'n seventeen, big as he is.'

Maw near looked embarrassed. 'Did I say Saunk?' she covered up, quick-like. 'I meant this Saunk's grandpaw. His name's Saunk too.' It wasn't; even Grandpaw don't remember what his name was first, it's been so long. But in his time he's used a lot of names like Elijah and so forth. I ain't even sure they had names in Atlantis, where Grandpaw come from in the first place. Numbers or something. It don't signify, anyhow.

Well, seems like old Yancey kept snuffling and groaning and moaning, and made out like we'd kilt his eight boys and he was all alone in the world. He hadn't cared a mite half an hour ago, though, and I said so. But he pointed out he hadn't rightly understood what I was talking about then, and for me to shet up.

'Ought to had a bigger family,' he said. 'They used to be two more boys, Zeb and Robbie, but I shot 'em one time. Didn't like the way they was looking ory-eyed at me. The point is, you Hogbens ain't got no right to kill my boys.'

'We didn't go for to do it,' Maw said. 'It was more or less an accident. We'd be right happy to make it up to you, one way or another.'

'That's what I was counting on,' old Yancey said. 'It seems like the least you could do, after acting up like you done. It don't matter whether the baby kilt my boys, like Saunk says and he's a liar. The idea is that I figger all you Hogbens are responsible. But I guess we could call it square if'n you did me a little favour. It ain't really right for neighbours to hold bad feelings.'

'Any favour you name,' Maw said, 'if it ain't out of line.'

''Tain't much,' old Yancey said. 'I just want you to split me up into a rabble, sort of temporary.'

'Hey, you been listening to Medea?' Paw said, being drunk enough not to know no better. 'Don't you believe her. That was purely a prank she played on Pelias. After he got chopped up he stayed daid; he didn't git young like she said he would.'

'Hey?' Yancey said. He pulled that old magazine out of his pocket and it fell open right to that purty picture. 'This here,' he said. 'Saunk tells me you kin do it. And everybody round here knows you Hogbens are witches. Saunk said you done it once with a feller name of Messy.'

'Guess he means Cadmus,' I said.

Yancey waved the magazine. I saw he had a queer kind of gleam in his eye.

'It shows right here,' he said, wild-like. 'A feller steps inside this here gimmick and then he keeps coming out of it, dozens of him, over and over. Witchcraft. Well, I know about you Hogbens. You may fool the city folk, but you don't fool me none. You're all witches.'

'We ain't,' Paw said from the corner. 'Not no more.'

'You are so,' Yancey said. 'I heard stories. I even seen him'—he pointed right at Uncle Les—'I seen him flying around in the air. And if that ain't witchcraft I don't know what is.'

'Don't you, honest?' I asked. 'That's easy. It's when you get some——'

But Maw told me to shet up.

'Saunk told me you kin do it,' he said. 'An' I been sitting and studying and looking over this here magazine. I got me a fine idea. Now, it stands to reason, everybody knows a witch kin be in two places at the same time. Couldn't a witch mebbe git to be in three places at the same time?'

'Three's as good as two,' Maw said. 'Only there ain't no witches. It's like this here science you hear tell about. People make it up out of their haids. It ain't natcheral.'

'Well, then,' Yancey said, putting the magazine down. 'Two or three or a whole passel. How many people are there in the world, anyway?'

'Two billion, two hunnerd fifty million, nine hunnerd and fifty-nine thousand, nine hunnerd and nineteen,' I said.

'Then——'

'Hold on a minute,' I said. 'Now it's two billion, two hunnerd fifty million, nine hunnerd and fifty-nine thousand, nine hunnerd and twenty. Cute little tyke, too.'

'Boy or girl?' Maw asked.

'Boy,' I told her.

'Then why can't you make me be in two billion whatever it was places at the same time? Mebbe for just a half a minute or so. I ain't greedy. That'd be long enough, anyhow.'

'Long enough for what?' Maw asked.

Yancey give me a sly look. 'I got me a problem,' he said. 'I want to find a feller. Trouble is, I dunno if I kin find him now. It's been a awful long time. But I got to, somehow or other. I ain't a-gonna rest easy in my grave unless I done paid all my debts, and for thutty years I been owing this feller something. It lays heavy on my conscience.'

'That's right honourable of you, neighbour,' Maw said.

Yancey snuffled and wiped his nose on his sleeve.

'It's a-gonna be a hard job,' he said. 'I put it off mebbe a mite too long. The thing is, I was figgering on sending my eight boys out to look for this feller sometime, so you kin see why it's busted me all up, the way them no-good varmints up and got kilt without no warning. How am I gonna find that feller I want now?'

Maw looked troubled and passed Yancey the jug.

'Whoosh!' he said, after a snort. 'Tastes like real hell-fire for certain. Whoosh!' Then he took another swig, sucked in some air, and scowled at Maw.

'If'n a man plans on sawing down a tree and his neighbour busts the saw, seems to me that neighbour ought to lend his own saw. Ain't that right?'

'Sure is,' Maw said. 'Only we ain't got eight boys to lend you.'

'You got something better,' Yancey said. 'Black, wicked magic, that's what. I ain't saying yea or nay 'bout that. It's your own affair. But seeing as how you kilt off them wuthless young 'uns of mine, so's I can't do like I was intending—why, then it looks like you ought to be willing to help me in some other way. Long as I kin locate that feller and pay him what I owe him, I'm satisfied. Now, ain't it the gospel truth that you kin split me up into a passel of me-critters?'

'Why, I guess we kin do that, I s'pose,' Maw said.

'An' ain't it gospel that you kin fix it so's every dang one of them me-critters will travel real fast and see everybody in the whole, entire world?'

'That's easy,' I said.

'If'n I kin git to do that,' Yancey said, 'it'd be easy for me to spot that feller and give him what he's got coming to him.' He snuffled. 'I allus been honest. I'm skeered of dying unless I pay all my debts fust. Danged if'n I want to burn through all eternity like you sinful Hogbens are a-gonna.'

'Shucks,' Maw said, 'I guess we kin help out, neighbour, being as how you feel so het up about it. Yes, sir, we'll do like you want.'

Yancey brightened up considerable.

'Promise?' he asked. 'Swear it, on your word an' honour?'

Maw looked kind of funny, but Yancey pulled out his bandanna again, so she busted down and made her solemn promise. Right away Yancey cheered up.

'How long will the spell take?' he asked.

'There ain't no spell,' I said. 'Like I told you, all I need is some scrap iron and a washbasin. 'Twon't take long.'

'I'll be back real soon,' Yancey said, sort of cackling, and run out, laughing his haid off. Going through the yard, he kicked out at a chicken, missed, and laughed some more. Guess he was feeling purty good.

'You better go on and make that gadget so's it'll be ready,' Maw told me. 'Git going.'

'Yes, Maw,' I said, but I sat there for a second or two, studying. She picked up the broomstick.

'You know, Maw——'

'Well?'

'Nothing,' I said, and dodged the broomstick. I went on out, trying to git clear what was troubling me. Something was, only I couldn't tell what. I felt kind of unwilling to make that there gadget, which didn't make right good sense, since there didn't seem to be nothing really wrong.

I went out behind the woodshed, though, and got busy. Took me 'bout ten minutes, but I didn't hurry much. Then I come back to the house with the gadget and said I was done. Paw told me to shet up.

Well, I sat there and looked at the gimmick and still felt trouble on my mind. Had to do with Yancey, somehow or other. Finally I noticed he'd left his old magazine behind, so I picked it up and started reading the story right under that picture, trying to make sense out of it. Durned if I could.

It was all about some crazy hillbillies who could fly. Well, that ain't no trick but what I couldn't figger out was whether the feller that writ it was trying to be funny or not. Seems to me people are funny enough anyhow, without trying to make 'em funnier.

Besides, serious things ought to be treated serious, and from what our Perfesser feller told me once, there's an awful lot of people what really believe in science and take it tremendous serious. He allus got a holy light in his eye when he talked about it. The only good thing about that story, it didn't have no girls in it. Girls make me feel funny.

I didn't seem to be gitting nowheres, so I went down to the cellar and played with the baby. He's kind of big for his tank these days. He was glad to see me. Winked all four of his eyes at me, one after the other. Real cute.

But all the time there was something about that magazine that kept nagging at me. I felt itchy inside, like when before they had that big fire in London, some while ago. Quite a spell of sickness they had then, too.

It reminded me of something Grandpaw had told me once, that he'd got the same sort of skitters just before Atlantis foundered. 'Course, Grandpaw kin sort of look into the future—which ain't much good, really, on account of it keeps changing around. I cain't do that myself yet. I ain't growed up enough. But I had a kind of hunch that something real bad was around, only it hadn't happened quite yet.

I almost decided to wake up Grandpaw, I felt so troubled. But around then I heard tromping upstairs, so I clomb up to the kitchen, and there was Yancey, swigging down some corn Maw'd give him. Minute I looked at the old coot, I got that feeling agin.

Yancey said, 'Whoosh,' put down the jug, and wanted to know if we was ready. So I pointed at the gadget I'd fixed up and said that was it, all right, and what did he think about it?

'That little thing?' Yancey asked. 'Ain't you a-gonna call up Old Scratch?'

'Ain't no need,' Uncle Les said. 'Not with you here, you little water moccasin, you.'

Yancey looked right pleased. 'That's me,' he said. 'Mean as a moccasin, and fulla pizen. How does it work?'

'Well,' I said, 'it sort of splits you up into a lot of Yanceys, is all.'

Paw had been setting quiet, but he must of tuned in inside the haid of some perfesser somewheres, on account of he started talking foolish. He don't know any four-bit words hisself.

I wouldn't care to know 'em myself, being as how they only mix up what's simple as cleaning a trout.

'Each human organism,' Paw said, showing off like crazy, 'is an electromagnetic machine, emitting a pattern of radiations, both from brain and body. By reversing polarity, each unit of you, Yancey, will be automatically attracted to each already existent human unit, since unlikes attract. But first you will step on Saunk's device and your body will be broken down——'

'Hey!' Yancey yelped.

Paw went right on, proud as a peacock.

'—into a basic electronic matrix, which can then be duplicated to the point of infinity, just as a type face may print millions of identical copies of itself in reverse—negative instead of positive.

'Since space is no factor where electronic wave-patterns are concerned, each copy will be instantly attracted to the space occupied by every other person in the world,' Paw was going on, till I like to bust. 'But since two objects cannot occupy the same space-time, there will be an automatic spacial displacement, and each Yancey-copy will be repelled to approximately two feet away from each human being.'

'You forgot to draw a pentagram,' Yancey said, looking around nervous-like. 'That's the awfullest durn spell I ever heard in all my born days. I thought you said you wasn't gonna call up Old Scratch?'

Maybe it was on account of Yancey was looking oncommon like Old Scratch hisself just then, but I just couldn't stand it no longer—having this funny feeling inside me. So I woke up Grandpaw. I did it inside my haid, the baby helping, so's nobody noticed. Right away there was a stirring in the attic, and Grandpaw heaved hisself around a little and woke up. Next thing I knew he was cussing a blue streak.

Well, the whole family heard that, even though Yancey couldn't. Paw stopped showing off and shet up.

'Dullards!' Grandpaw said, real mad. 'Rapscallions! Certes, y-wist it was no wonder I was having bad dreams. Saunk, you've put your foot in it now. Have you no sense of process? Didn't you realize what this caitiff schmo was planning, the stinkard? Get in the groove, Saunk, ere manhood's state shall find thee unprepared.' Then he added something in Sanskrit. Living as long as Grandpaw has, he gits mixed up in his talk sometimes.

'Now, Grandpaw,' Maw thunk, 'what's Saunk been and done?'

'You've all done it!' Grandpaw yelled. 'Couldn't you add cause and effect? Saunk, what of the picture y-wrought in Yancey's pulp mag? Wherefore hys sodien change of herte, when obviously the stinkard hath no more honour than a lounge lizard? Do you want the world depopulated before its time? Ask Yancey what he's got in his britches pocket, dang you!'

'Mr Yancey,' I said, 'what have you got in your britches pocket?'

'Hey?' he said, reaching down and hauling out a big, rusty monkey wrench. 'You mean this? I picked it up back of the shed.' He was looking real sly.

'What you aiming to do with that?' Maw asked, quick.

Yancey give us all a mean look. 'Ain't no harm telling you,' he said. 'I aim to hit everybody, every durn soul in the whole, entire world, right smack on top of the haid, and you promised to help me do it.'

'Lawks a-mercy,' Maw said.

'Yes, siree,' Yancey giggled. 'When you hex me, I'm a-gonna be in every place everybody else is, standing right behind 'em. I'll whang 'em good. Thataway, I kin be sure I'll get even. One of them people is just bound to be the feller I want, and he'll git what I been owing him for thutty years.'

'What feller?' I said. 'You mean the one you met up with in New York you was telling me about? I figgered you just owed him some money.'

'Never said no sech thing,' Yancey snapped. 'A debt's a debt, be it money or a bust in the haid. Ain't nobody a-gonna step on my corn and git away with it, thutty years or no thutty years.'

'He stepped on your corn?' Paw asked. 'That's all he done?'

'Yup. I was likkered up at the time, but I recollect I went down some stairs to where a lot of trains was rushing around under the ground.'

'You was drunk.'

'I sure was,' Yancey said. 'Couldn't be no sech thing—trains running underground! But I sure as shooting wasn't dreaming 'bout the feller what stepped on my corn. Why, I kin still feel it. I got mad. It was so crowded I couldn't even move for a mite, and I never even got a good look at the feller what stepped on me.

'By the time I hit out with my stick, he must of got away. Never knew what he looked like. Might have been a female, but that don't signify. I just ain't a-gonna die till I pay my debts and git even with everybody what ever done me dirt. I allus got even with every dang soul what done me wrong, and most everybody I ever met did.'

Riled up a whole lot was Yancey Tarbell. He went right on from there:

'So I figgered, since I never found out just who this feller was what stepped on my corn, I better make downright sure and take a lick at everybody, man, woman, and child.'

'Now you hold your hosses,' I said. 'Ain't no children could have been alive thutty years ago, an' you know it.'

'Makes no difference,' Yancey snapped. 'I was a-thinking, and I got an awful idea: suppose that feller went and died. Thutty years is a long time. But then I figgered, even if he did up and die, chances are he got married and had kids fust. If'n I can't git even with him, I kin get even with his children. The sins of the father—that's Scripture. If'n I hit everybody in the world, I can't go fur wrong.'

'You ain't hitting no Hogbens,' Maw said. 'None of us been in New York since afore you was born. I mean, we ain't never been there. So you kin just leave us out of it. How'd you like to git young again or something like that? We kin fix that for you instead, if you'll give up this here wicked idea.'

'I ain't a-gonna,' Yancey said, stubborn. 'You give your gospel word to help me.'

'Well, we ain't bound to keep a promise like that,' Maw said, but then Grandpaw chimed in from the attic.

'The Hogben word is sacred,' he told us. 'It's our bond. We must keep our promise to this booby. But, having kept it, we are not bound further.'

'Oh?' I said, sort of gitting a thought. 'That being the case—Mr Yancey, just what did we promise, exact?'

He waved the monkey wrench at me.

'I'm a-gonna git split up into as many people as there are people in the world, and I'm a-gonna be standing right beside all of 'em. You give your word to help me do that. Don't you try to wiggle out of it.'

'I ain't wiggling,' I said. 'Only we better git it clear, so's you'll be satisfied and won't have no kick coming. One thing, though. You got to be the same size as everybody you visit.'

'Hey?'

'I kin fix it easy. When you step on this here gadget, there'll be two billion, two hunnerd fifty million, nine hunnerd and fifty-nine thousand, nine hunnerd and twenty Yanceys all over the world. S'posin', now, one of these here Yanceys find himself standing next to a big feller seven feet tall. That wouldn't be so good, would it?'

'I want to be eight feet high,' Yancey said.

'No, sir. The Yancey who goes to visit a feller that high is a-gonna be just that high hisself, exactly. And the one who visits a baby only two feet high is a-gonna be only two feet high hisself. What's fair's fair. You agree to that, or it's all off. Only other thing, you'll be just exactly as strong as the feller you're up again'.'

I guess he seen I was firm. He hefted the monkey wrench.

'How'll I git back?' he asked.

'We'll take care of that,' I said. 'I'll give you five seconds. That's long enough to swing a monkey wrench, ain't it?'

'It ain't very long.'

'If'n you stay longer, somebody might hit back.'

'So they might,' he said, turning pale under the dirt. 'Five seconds is plenty.'

'Then if'n we do just that, you'll be satisfied? You won't have no kick coming?'

He swung the monkey wrench and laughed.

'Suits me fine and dandy,' he said. 'I'll bust their haids good. Heh, heh, heh.'

'Then you step right on here,' I said, showing him. 'Wait a mite, though. I better try it fust, to make sure it works right.'

I picked up a stick of firewood from the box by the stove and winked at Yancey. 'You git set,' I said. 'The minute I git back, you step right on here.'

Maw started to say something, but all of a sudden Grandpaw started laughing in the attic. I guess he was looking into the future again.

I stepped on the gadget, and it worked slick as anything. Afore I could blink, I was split up into two billion, two hunnerd and fifty million, nine hunnerd and fifty-nine thousand, nine hunnerd and nineteen Saunk Hogbens.

There was one short, o'course, on account of I left out Yancey, and o'course the Hogbens ain't listed in no census. But here I was, standing right in front of everybody in the whole, entire world except the Hogben fam'ly and Yancey hisself. It was plumb onreasonable.

Never did I know there was so many faces in this world! They was all colours, some with whiskers, some without, some with clothes on, some naked as needles, some awful big and some real short, and half of them was in daylight and half was in the night-time. I got downright dizzy.

For just a flash, I thought I could make out some of the people I knowed down in Piperville, including the Sheriff, but he got mixed up with a lady in a string of beads who was casing a kangaroo-critter, and she turned into a man dressed up fit to kill who was speechifyin' in a big room somewheres.

My, I was dizzy.

I got ahold of myself and it was about time, too, for just about then near everybody in the whole world noticed me. 'Course, it must have looked like I'd popped out of thin air, right in front of them, real sudden, and—well, you ever had near two billion, two hunnerd and fifty million, nine hunnerd and fifty-nine thousand, nine hunnerd and nineteen people looking you right square in the eye? It's just awful. I forgot what I'd been intending. Only I sort of heard Grandpaw's voice telling me to hurry up.

So I pushed that stick of firewood I was holding, only now it was two billion, two hunnerd and fifty million, nine hunnerd and fifty-nine thousand, nine hunnerd and nineteen sticks, into just about the same number of hands and let go. Some of the people let go too, but most of 'em held on to it. Then I tried to remember the speech I was a-gonna make, telling 'em to git in the fust lick at Yancey afore he could swing that monkey wrench.

But I was too confounded. It was funny. Having all them people looking right at me made me so downright shy, I couldn't even open my mouth. What made it worse was that Grandpaw yelled I had only one second left, so there wasn't even time to make a speech. In just one second, I was a-gonna flash back to our kitchen, and then old Yancey was all ready to jump in the gadget and swing that monkey wrench. And I hadn't warned nobody. All I'd done was give everybody a little old stick of firewood.

My, how they stared! I felt plumb naked. Their eyes bugged right out. And just as I started to thin out around the edges like a biscuit, I—well, I don't know what come over me. I guess it was feeling so oncommon shy. Maybe I shouldn't of done, but—

I done it!

Then I was back in the kitchen. Grandpaw was laughing fit to kill in the attic. The old gentleman's got a funny kind of sense of humour, I guess. I didn't have no time for him then, though, for Yancey jumped past me and into the gadget. And he disappeared into thin air, the way I had. Split up, like I'd been, into as many people as there was in the world, and standing right in front of 'em.

Maw and Paw and Uncle Les was looking at me real hard. I sort of shuffled.

'I fixed it,' I said. 'Seems like a man who's mean enough to hit little babies over the haid deserves what he's'—I stopped and looked at the gadget—'what he's been and got,' I finished, on account of Yancey had tumbled out of thin air, and a more whupped-up old rattlesnake I never seen. My!

Well, I guess purty near everybody in the whole world had took a whang at Mr Yancey. He never even had a chance to swing that monkey wrench. The whole world had got in the fust lick.

Yes, siree. Mr Yancey looked plumb ruined.

But he could still yell. You could of heard him a mile off. He kept screaming that he'd been cheated. He wanted another chance, and this time he was taking his shooting iron and a bowie knife. Finally Maw got disgusted, took him by the collar, and shook him up till his teeth rattled.

'Quoting Scripture!' she said, madlike. 'You little dried-up scraggle of downright pizen! The Good Book says an eye for an eye, don't it? We kept our word, and there ain't nobody kin say different.'

'That's the truth, certes,' Grandpaw chimed in from the attic.

'You better go home and git some arnicy,' Maw said, shaking Yancey some more. 'And don't you come round here no more, never again, or we'll set the baby on you.'

'But I didn't git even!' Yancey squalled.

'I guess you ain't a-gonna, ever,' I said. 'You just cain't live long enough to git even with everybody in the whole world, Mr Yancey.'

By and by, that seemed to strike Yancey all in a heap. He turned a rich colour like beet soup, made a quacking noise, and started cussing. Uncle Les reached for the poker, but there wasn't no need.

'The whole dang world done me wrong!' Yancey squealed, and clapped his hands to his haid. 'I been flummoxed! Why in tarnation did they hit me fust? There's something funny about——'

'Hush up,' I said, all of a sudden realizing the trouble wasn't over, like I'd thought. 'Listen, anybody hear anything from the village?'

Even Yancey shet up whilst we listened. 'Don't hear a thing,' Maw said.

'Saunk's right,' Grandpaw put in. 'That's what's wrong.'

Then everybody got it—that is, everybody except Yancey. Because about now there ought to of been quite a rumpus down at Piperville. Don't fergit me and Yancey went visiting the whole world, which includes Piperville, and people don't take a thing like that quiet. There ought to of been some yelling going on, at least.

'What are you all standing round dumb as mutes for?' Yancey busted out. 'You got to help me git even!'

I didn't pay him no mind. I sat down and studied the gadget. After a minute I seen what it was I'd done wrong. I guess Grandpaw seen it about as quick as I did. You oughta heard him laugh. I hope it done the old gentleman good. He has a right peculiar sense of humour sometimes.

'I sort of made a mistake in this gadget, Maw,' I said. 'That's why it's so quiet down in Piperville.'

'Aye, by my troth,' Grandpaw said, still laughing. 'Saunk had best seek cover. Twenty-three skiddoo, kid.'

'You done something you shouldn't, Saunk,' Maw said.

'Blabber, blabber, blabber!' Yancey yelled. 'I want my rights! I want to know what it was Saunk done that made everybody in the world hit me over the haid! He must of done something. I never had no time to—'

'Now you leave the boy alone, Mr Yancey,' Maw said. 'We done what we promised, and that's enough. You git outa here and simmer down afore you say something you regret.'

Paw winked at Uncle Les, and before Yancey could yell back at Maw the table sort of bent its legs down like they had knees in 'em and snuck up behind Yancey real quiet. Then Paw said to Uncle Les, 'All together now, let 'er go,' and the table straightened up its legs and give Yancey a terrible bunt that sent him flying out the door.

The last we heard of Yancey was the whoops he kept letting out whenever he hit the ground all the way down the hill. He rolled half the way to Piperville, I found out later. And when he got there he started hitting people over the haid with his monkey wrench.

I guess he figgered he might as well make a start the hard way.

They put him in jail for a spell to cool off, and I guess he did, 'cause afterward he went back to that little shack of his'n. I hear he don't do nothing but set around with his lips moving, trying to figger a way to git even with the hull world. I don't calc'late he'll ever hit on it, though.

At that time I wasn't paying him much mind. I had my own troubles. As soon as Paw and Uncle Les got the table back in place, Maw lit into me again.

'Tell me what happened, Saunk,' she said. 'I'm a-feared you done something wrong when you was in that gadget. Remember you're a Hogben, son. You got to behave right when the whole world's looking at you. You didn't go and disgrace us in front of the entire human race, did you, Saunk.'

Grandpaw laughed agin. 'Not yet, he hasn't,' he said. Then down in the basement I heard the baby give a kind of gurgle and I knowed he could see it too. That's surprising, kinda. We never know for sure about the baby. I guess he really kin see a little bit into the future too.

'I just made a little mistake, Maw,' I said. 'Could happen to anybody. It seems the way I fixed that gadget up, it split me into a lot of Saunks, all right, but it sent me ahead into next week too. That's why there ain't no ruckus yet down in Piperville.'

'My land!' Maw said. 'Child, you do things so careless!'

'I'm sorry, Maw,' I said. 'Trouble is, too many people in Piperville know me. I'd better light out for the woods and pick me a nice holler tree. I'll be needing it, come next week.'

'Saunk,' Maw said, 'you been up to something. Sooner or later I'll find out, so you might as well tell me now.'

Well, shucks, I knowed she was right. So I told her, and I might as well tell you, too. You'll find out anyhow, come next week. It just shows you can't be too careful. This day next week, everybody in the whole world is a-gonna be mighty surprised when I show up out of thin air, hand 'em all a stick of firewood, and then r'ar back and spit right smack in their eye.

I s'pose that there two billion, two hunnerd and fifty million, nine hunnerd and fifty-nine thousand, nine hunnerd and nineteen includes everybody on earth.

Everybody!

Sometime next week, I figger.

See you later.

LIANE THE WAYFARER

Through the dim forest came Liane the Wayfarer, passing along the shadowed glades with a prancing lightfooted gait. He whistled, he carolled, he was plainly in high spirits. Around his finger he twirled a bit of wrought bronze—a circlet graved with angular crabbed characters, now stained black.

By excellent chance he had found it, banded around the root of an ancient yew. Hacking it free, he had seen the characters on the inner surface—rude forceful symbols, doubtless the cast of a powerful antique rune . . . Best take it to a magician and have it tested for sorcery.

Liane made a wry mouth. There were objections to the course. Sometimes it seemed as if all living creatures conspired to exasperate him. Only this morning, the spice merchant—what a tumult he had made dying! How carelessly he had spewed blood on Liane's cock comb sandals! Still, thought Liane, every unpleasantness carried with it compensation. While digging the grave he had found the bronze ring.

And Liane's spirits soared; he laughed in pure joy. He bounded, he leapt. His green cape flapped behind him, the red feather in his cap winked and blinked . . . But still—Liane slowed his step—he was no whit closer to the mystery of the magic, if magic the ring possessed.

Experiment, that was the word!

He stopped where the ruby sunlight slanted down without hindrance from the high foliage, examined the ring, traced the glyphs with his fingernail. He peered through. A faint film, a flicker? He held it at arm's length. It was clearly a coronet. He whipped off his cap, set the band on his brow, rolled his great golden eyes, preened himself . . . Odd. It slipped down on his ears. It tipped across his eyes. Darkness. Frantically Liane clawed it off . . . A bronze ring, a hand's-breadth in diameter. Queer.

He tried again. It slipped down over his head, his shoulders. His head was in the darkness of a strange separate space. Looking down, he saw the level of the outside light dropping as he dropped the ring.

Slowly down . . . Now it was around his ankles—and in sudden panic, Liane snatched the ring up over his body, emerged blinking into the maroon light of the forest.

He saw a blue-white, green-white flicker against the foliage. It was a Twk-man, mounted on a dragon-fly, and light glinted from the dragon-fly's wings.

Liane called sharply, 'Here, sir! Here, sir!'

The Twk-man perched his mount on a twig. 'Well, Liane, what do you wish?'

'Watch now, and remember what you see.' Liane pulled the ring over his head, dropped it to his feet, lifted it back. He looked up to the Twkman, who was chewing a leaf. 'And what did you see?'

'I saw Liane vanish from mortal sight—except for the red curled toes of his sandals. All else was as air.'

'Ha!' cried Liane. 'Think of it! Have you ever seen the like?'

The Twk-man asked carelessly, 'Do you have salt? I would have salt.'

Liane cut his exultations short, eyed the Twk-man closely.

'What news do you bring me?'

'Three erbs killed Florejin the Dream-builder, and burst all his bubbles. The air above the manse was coloured for many minutes with the flitting fragments.'

'A gram.'

'Lord Kandive the Golden has built a barge of carven mo-wood ten lengths high, and it floats on the River Scaum for the Regatta, full of treasure.'

'Two grams.'

'A golden witch named Lith has come to live on Thamber Meadow. She is quiet and very beautiful.'

'Three grams.'

'Enough,' said the Twk-man, and leaned forward to watch while Liane weighed out the salt in a tiny balance. He packed it in small panniers hanging on each side of the ribbed thorax, then twitched the insect into the air and flicked off through the forest vaults.

Once more Liane tried his bronze ring, and this time brought it entirely past his feet, stepped out of it and brought the ring up into the darkness beside him. What a wonderful sanctuary! A hole whose opening could be hidden inside the hole itself! Down with the ring to his feet, step through, bring it up his slender frame and over his shoulders, out into the forest with a small bronze ring in his hand.

Ho! and off to Thamber Meadow to see the beautiful golden witch.

Her hut was a simple affair of woven reeds—a low dome with two round windows and a low door. He saw Lith at the pond bare-legged among the water shoots, catching frogs for her supper. A white kirtle was gathered up tight around her thighs; stock-still she stood and the dark water rippled rings away from her slender knees.

She was more beautiful than Liane could have imagined, as if one of Florejin's wasted bubbles had burst here on the water. Her skin was pale creamed stirred gold, her hair a denser, wetter gold. Her eyes were like Liane's own, great golden orbs, and hers were wide apart, tilted slightly.

Liane strode forward and planted himself on the bank. She looked up startled, her ripe mouth half-open.

'Behold, golden witch, here is Liane. He has come to welcome you to Thamber; and he offers you his friendship, his love . . .'

Lith bent, scooped a handful of slime from the bank and flung it into his face.

Shouting the most violent curses, Liane wiped his eyes free, but the door to the hut had slammed shut.

Liane strode to the door and pounded it with his fist.

'Open and show your witch's face, or I burn the hut!'

The door opened, and the girl looked forth, smiling. 'What now?'

Liane entered the hut and lunged for the girl, but twenty thin shafts darted out, twenty points pricking his chest. He halted, eyebrows raised, mouth twitching.

'Down, steel,' said Lith. The blades snapped from view. 'So easily could I seek your vitality,' said Lith, 'had I willed.'

Liane frowned and rubbed his chin as if pondering. 'You understand,' he said earnestly, 'what a witless thing you do. Liane is feared by those who fear fear, loved by those who love love. And you—' his eyes swam the golden glory of her body—'you are ripe as a sweet fruit, you are eager, you glisten and tremble with love. You please Liane, and he will spend much warmness on you.'

'No, no,' said Lith, with a slow smile. 'You are too hasty.'

Liane looked at her in surprise. 'Indeed?'

'I am Lith,' said she. 'I am what you say I am. I ferment, I burn, I seethe. Yet I may have no lover but him who has served me. He must be brave, swift, cunning.'

'I am he,' said Liane. He chewed at his lip. 'It is not usually thus. I detest this indecision.' He took a step forward. 'Come, let us—'

She backed away. 'No, no. You forget. How have you served me, how have you gained the right to my love?'

'Absurdity!' stormed Liane. 'Look at me! Note my perfect grace, the beauty of my form and feature, my great eyes, as golden as your own, my manifest will and power . . . It is you who should serve me. That is how I will have it.' He sank upon a low divan. 'Woman, give me wine.'

She shook her head. 'In my small domed hut I cannot be forced. Perhaps outside on Thamber Meadow—but in here, among my blue and red tassels, with twenty blades of steel at my call, you must obey me. . . . So choose. Either arise and go, never to return, or else agree to serve me on one small mission, and then have me and all my ardour.'

Liane sat straight and stiff. An odd creature, the golden witch. But, indeed, she was worth some exertion, and he would make her pay for her impudence.

'Very well, then,' he said blandly. 'I will serve you. What do you wish? Jewels? I can suffocate you in pearls, blind you with diamonds. I have two emeralds the size of your fist, and they are green oceans, where the gaze is trapped and wanders forever among vertical green prisms . . .'

'No, no jewels—'

'An enemy, perhaps. Ah, so simple. Liane will kill you ten men. Two steps forward, thrust—*thus!*' He lunged. 'And souls go thrilling up like bubbles in a beaker of mead.'

'No. I want no killing.'

He sat back, frowning. 'What, then?'

She stepped to the back of the room and pulled at a drape. It swung aside, displaying a golden tapestry. The scene was a valley bounded by two steep mountains, a broad valley where a placid river ran, past a quiet village and so into a grove of trees. Golden was the river, golden the mountains, golden the trees—golds so various, so rich, so subtle that the effect was like a many-coloured landscape. But the tapestry had been rudely hacked in half.

Liane was entranced. 'Exquisite, exquisite . . .'

Lith said, 'It is the Magic Valley of Ariventa so depicted. The other half has been stolen from me, and its recovery is the service I wish of you.'

'Where is the other half?' demanded Liane. 'Who is the dastard?'

Now she watched him closely. 'Have you ever heard of Chun? Chun the Unavoidable?'

Liane considered. 'No.'

'He stole the half to my tapestry, and hung it in a marble hall, and this hall is in the ruins to the north of Kaiin.'

'Ha!' muttered Liane.

'The hall is by the Place of Whispers, and is marked by a leaning column with a black medallion of a phoenix and a two-headed lizard.'

'I go,' said Liane. He rose. 'One day to Kaiin, one day to steal, one day to return. Three days.'

Lith followed him to the door. 'Beware of Chun the Unavoidable,' she whispered.

And Liane strode away whistling, the red feather bobbing in his green cap. Lith watched him, then turned and slowly approached the golden tapestry. 'Golden Ariventa,' she whispered, 'my heart cries and hurts with longing for you . . .'

The Derna is a swifter, thinner river than the Scaum, its bosomy sister to the south. And where the Scaum wallows through a broad dale, purple with horse-blossom, pocked white and grey with crumbling castles, the Derna has sheared a steep canyon, overhung by forested bluffs.

An ancient flint road long ago followed the course of the Derna, but now the exaggeration of the meandering has cut into the pavement, so that Liane, treading the road to Kaiin, was occasionally forced to leave the road and make a detour through banks of thorn and the tubegrass which whistled in the breeze.

The red sun, drifting across the universe like an old man creeping to his death-bed, hung low to the horizon when Liane breasted Porphiron Scar, looked across white-walled Kaiin and the blue bay of Sanreale beyond.

Directly below was the market-place, a medley of stalls selling fruit, slabs of pale meat, molluscs from the slime banks, dull flagons of wine. And the quiet people of Kaiin moved among the stalls, buying their sustenance, carrying it loosely to their stone chambers.

Beyond the market-place rose a bank of ruined columns, like broken teeth—legs to the arena built two hundred feet from the ground by Mad King Shin; beyond, in a grove of bay trees, the glossy dome of the palace was visible, where Kandive the Golden ruled Kaiin and as much of Ascolais as one could see from a vantage on Porphiron Scar.

The Derna, no longer a flow of clear water, poured through a network of dank canals and subterranean tubes, and finally seeped past rotting wharves into the Bay of Sanreale.

A bed for the night, thought Liane; then to his business in the morning.

He leapt down the zigzag steps—back, forth, back, forth—and came out into the market-place. And now he put on a grave demeanour. Liane the Wayfarer was not unknown in Kaiin, and many were ill-minded enough to work him harm.

He moved sedately in the shade of the Pannone Wall, turned through a narrow cobbled street, bordered by old wooden houses glowing the rich brown of old stump-water in the rays of the setting sun, and so came to a small square and the high stone face of the Magician's Inn.

The host, a small fat man, sad of eye, with a small fat nose the identical shape of his body, was scraping ashes from the hearth. He straightened his back and hurried behind the counter of his little alcove.

Liane said, 'A chamber, well-aired, and a supper of mushrooms, wine and oysters.'

The innkeeper bowed humbly.

'Indeed, sir—and how will you pay?'

Liane flung down a leather sack, taken this very morning. The innkeeper raised his eyebrows in pleasure at the fragrance.

'The ground buds of the spase-bush, brought from a far land,' said Liane.

'Excellent, excellent . . . Your chamber, sir, and your supper at once.'

As Liane ate, several other guests of the house appeared and sat before the fire with wine, and the talk grew large, and dwelt on wizards of the past and the great days of magic.

'Great Phandaal knew a lore now forgot,' said one old man with hair dyed orange. 'He tied white and black strings to the legs of sparrows and sent them veering to his direction. And where they wove their magic woof, great trees appeared, laden with flowers, fruit, nuts, or bulbs of rare liqueurs. It is said that thus he wove Great Da Forest on the shores of Sanra Water.'

'Ha,' said a dour man in a garment of dark blue, brown and black, 'this I can do.' He brought forth a bit of string, flicked it, whirled it, spoke a quiet word, and the vitality of the pattern fused the string into a tongue of red and yellow fire, which danced, curled, darted back and forth along the table till the dour man killed it with a gesture.

'And this I can do,' said a hooded figure in a black cape sprinkled with silver circles. He brought forth a small tray, laid it on the table and sprinkled therein a pinch of ashes from the hearth. He brought forth a whistle and blew a clear tone, and up from the tray came glittering motes, flashing the prismatic colours red, blue, green, yellow. They floated up a foot and burst in coruscations of brilliant colours, each a beautiful star-shaped pattern, and each burst sounded a tiny repetition of the original tone—the clearest, purest sound in the world. The motes became fewer, the magician blew a different tone, and again the motes floated up to burst in glorious ornamental spangles. Another time—another swarm of motes. At last the magician replaced his whistle, wiped off the tray, tucked it inside his cloak and lapsed back to silence.

Now the other wizards surged forward, and soon the air above the table swarmed with visions, quivered with spells. One showed the group nine new colours of ineffable charm and radiance; another caused a mouth to form on the land-lord's forehead and revile the crowd, much to the landlord's discomfiture, since it was his own voice. Another displayed a green glass bottle from which the face of a demon peered and grimaced; another a ball of pure crystal which rolled back and forward to the command of the sorcerer who owned it, and who claimed it to be an earring of the fabled master Sankaferrin.

Liane had attentively watched all, crowing in delight at the bottled imp, and trying to cozen the obedient crystal from its owner, without success.

And Liane became pettish, complaining that the world was full of rock-hearted men, but the sorcerer with the crystal earring remained indifferent, and even when Liane spread out twelve packets of rare spice he refused to part with his toy.

Liane pleaded, 'I wish only to please the witch Lith.'

'Please her with the spice, then.'

Liane said ingenuously, 'Indeed, she has but one wish, a bit of tapestry which I must steal from Chun the Unavoidable.'

And he looked from face to suddenly silent face.

'What causes such immediate sobriety? Ho, Landlord, more wine!'

The sorcerer with the earring said, 'If the floor swam ankle-deep with wine—the rich red wine of Tanvilkat—the leaden print of that name would still ride the air.'

'Ha,' laughed Liane, 'let only a taste of that wine pass your lips, and the fumes would erase all memory.'

'See his eyes,' came a whisper. 'Great and golden.'

'And quick to see,' spoke Liane. 'And these legs—quick to run, fleet as starlight on the waves. And this arm—quick to stab with steel. And my magic—which will set me to a refuge that is out of all cognizance.' He gulped wine from a beaker. 'Now behold. This is magic from antique days.' He set the bronze band over his head, stepped through, brought it up inside the darkness. When he deemed that sufficient time had elapsed, he stepped through once more.

The fire glowed, the landlord stood in his alcove, Liane's wine was at hand. But of the assembled magicians, there was no trace.

Liane looked about in puzzlement. 'And where are my wizardly friends?'

The landlord turned his head. 'They took to their chambers; the name you spoke weighed on their souls.'

And Liane drank his wine in frowning silence.

Next morning he left the inn and picked a roundabout way to the Old Town—a grey wilderness of tumbled pillars, weathered blocks of sandstone, slumped pediments with crumbled inscriptions, flagged terraces overgrown with rusty moss. Lizards, snakes, insects crawled the ruins; no other life did he see.

Threading a way through the rubble, he almost stumbled on a corpse—the body of a youth, one who stared at the sky with empty eye-sockets.

Liane felt a presence. He leapt back, rapier half-bared. A stooped old man stood watching him. He spoke in a feeble, quavering voice: 'And what will you have in the Old Town?'

Liane replaced his rapier. 'I seek the Place of Whispers. Perhaps you will direct me.'

The old man made a croaking sound at the back of his throat. 'Another? Another? When will it cease . . .' He motioned to the corpse. 'This one came yesterday seeking the Place of Whispers. He would steal from Chun the Unavoidable. See him now.' He turned away. 'Come with me.' He disappeared over a tumble of rock.

Liane followed. The old man stood by another corpse with eye-sockets bereft and bloody. 'This one came four days ago, and he met Chun the Unavoidable . . . And over there behind the arch is still, a great warrior in cloison armor. And there—and there—' he pointed, pointed. 'And there—and there—like crushed flies.'

He turned his watery blue gaze back to Liane. 'Return, young man, return—lest your body lie here in its green cloak to rot on the flagstones.'

Liane drew his rapier and flourished it. 'I am Liane the Wayfarer; let them who offend me have fear. And where is the Place of Whispers?'

'If you must know,' said the old man, 'it is beyond that broken obelisk. But you go to your peril.'

'I am Liane the Wayfarer. Peril goes with me.'

The old man stood like a piece of weathered statuary as Liane strode off.

And Liane asked himself, suppose this old man were an agent of Chun, and at this minute were on his way to warn him? . . . Best to take all precautions. He leapt up on a high entablature and ran crouching back to where he had left the ancient.

Here he came, muttering to himself, leaning on his staff. Liane dropped a block of granite as large as his head. A thud, a croak, a gasp—and Liane went his way.

He strode past the broken obelisk, into a wide court—the Place of Whispers. Directly opposite was a long wide hall, marked by a leaning column with a big black medallion, the sign of a phoenix and a two-headed lizard.

Liane merged himself with the shadow of a wall, and stood watching like a wolf, alert for any flicker of motion.

All was quiet. The sunlight invested the ruins with dreary splendour. To all sides, as far as the eye could reach, was broken stone, a wasteland leached by a thousand rains, until now the sense of man had departed and the stone was one with the ratural earth.

The sun moved across the dark-blue sky. Liane presently stole from his vantage-point and circled the hall. No sight nor sign did he see.

He approached the building from the rear and pressed his ear to the stone. It was dead, without vibration. Around the side—watching up, down, to all sides; a breach in the wall. Liane peered inside. At the back hung half a golden tapestry. Otherwise the hall was empty.

Liane looked up, down, this side, that. There was nothing in sight. He continued around the hall.

He came to another broken place. He looked within. To the rear hung the golden tapestry. Nothing else, to right or left, no sight or sound.

Liane continued to the front of the hall and sought into the eaves; dead as dust.

He had a clear view of the room. Bare, barren, except for the bit of golden tapestry.

Liane entered, striding with long soft steps. He halted in the middle of the floor. Light came to him from all sides except the rear wall. There were a dozen openings from which to flee and no sound except the dull thudding of his heart.

He took two steps forward. The tapestry was almost at his fingertips.

He stepped forward and swiftly jerked the tapestry down from the wall.

And behind was Chun the Unavoidable.

Liane screamed. He turned on paralysed legs and they were leaden, like legs in a dream which refused to run.

Chun dropped out of the wall and advanced. Over his shiny black back he wore a robe of eyeballs threaded on silk.

Liane was running, fleetly now. He sprang, he soared. The tips of his toes scarcely touched the ground. Out the hall, across the square, into the wilderness of broken statues and fallen columns. And behind came Chun, running like a dog.

Liane sped along the crest of a wall and sprang a great gap to a shattered fountain. Behind came Chun.

Liane darted up a narrow alley, climbed over a pile of refuse, over a roof, down into a court. Behind came Chun.

Liane sped down a wide avenue lined with a few stunted old cypress trees, and he heard Chun close at his heels. He turned into an archway, pulled his bronze ring over his head, down to his feet. He stepped through, brought the ring up inside the darkness. Sanctuary. He was alone in a dark magic space, vanished from earthly gaze and knowledge. Brooding silence, dead space . . .

He felt a stir behind him, a breath of air. At his elbow a voice said, 'I am Chun the Unavoidable.'

Lith sat on her couch near the candles, weaving a cap from frogskins. The door to her hut was barred, the windows shuttered. Outside, Thamber Meadow dwelled in darkness.

A scrape at her door, a creak as the lock was tested. Lith became rigid and stared at the door.

A voice said, 'Tonight, O Lith, tonight it is two long bright threads for you. Two because the eyes were so great, so large, so golden . . .'

Lith sat quiet. She waited an hour; then, creeping to the door, she listened. The sense of presence was absent. A frog croaked nearby.

She eased the door ajar, found the threads and closed the door. She ran to her golden tapestry and fitted the threads into the ravelled warp.

And she stared at the golden valley, sick with longing for Ariventa, and tears blurred out the peaceful river, the quiet golden forest. 'The cloth slowly grows wider . . . One day it will be done, and I will come home. . . .'

MANLY WADE WELLMAN

THE DESRICK ON YANDRO

The folks at the party clapped me such an encore, I sang that song.

The lady had stopped her car at the roadside when she saw my thumb out and my silver-strung guitar under my arm. Asked me my name, I told her John. Asked where I was headed, I told her nowhere special. Asked could I play that guitar, I played it as we rolled along. Then she invited me most kindly to her country house, to sing to her friends, and they'd be obliged, she said. And I went.

The people there were fired up with what they'd drunk, lots of ladies and men in costly clothes, and I had my bothers not getting drunk, too. But, shoo, they liked what I played and sang. Staying off worn out songs, I smote out what they'd never heard before—*Witch in the Wilderness* and *Rebel Soldier* and *Vandy, Vandy, I've Come to Court You*. When they clapped and hollered for more, I sang the Yandro song, like this:

> I'll build me a desrick on Yandro's high hill,
> Where the wild beasts can't reach me or hear my sad cry.
> For he's gone, he's gone away, to stay a little while,
> But he'll come back if he comes ten thousand miles.

Then they strung around and made me more welcome than any stranger could call for, and the hostess lady said I must stay to supper, and sleep there that night. But at that second, everybody sort of pulled away, and one man came up and sat down by me.

I'd been aware that, when first he came in, things stilled down, like with little boys when a big bully shows himself. He was built short and broad, his clothes were sporty, cut handsome and costly. His buckskin hair was combed across his head to baffle folks he wasn't getting bald. His round, pink face wasn't soft, and his big, smiling teeth reminded you there was a bony skull under that meat. His pale eyes, like two gravel bits, prodded me and made me remember I needed a haircut and a shine.

'You said Yandro, young man,' said this fellow. He said it almost like a charge in court, with me the prisoner.

'Yes, sir. The song's mountainy, not too far from the Smokies. I heard it in a valley, and the highest peak over that valley's called Yandro. Now,' I said, 'I've had scholar-men argue me it really means yonder—yonder high hill. But the peak's called Yandro. Not a usual name.'

'No, John.' He smiled toothy and fierce. 'Not a usual name. I'm like the peak. I'm called Yandro, too.'

'How you, Mr Yandro?' I said.

'I never heard of that peak or valley, nor, I imagine, did my father before me. But my grandfather—Joris Yandro—came from the Southern mountains. He was young, with small education, but lots of energy and ambition.' Mr Yandro swelled up inside his fancy clothes. 'He went to New York, then Chicago. His fortunes prospered. His son—my father—and then I, we contrived to make them prosper still more.'

'You're to be honoured,' I said, my politest; but I judged, with no reason to be sure, that he might not be too honourable about how he made his money, or used it. The way the others drew from him made me reckon he scared them, and that kind of folks scares worst where their money pocket's located.

'I've done all right,' he said, not caring who heard the brag. 'I don't think anybody for a hundred miles around here can turn a deal or make a promise without clearing it with me. John, I own this part of the world.'

Again he showed his teeth. 'You're the first one ever to tell me about where my grandfather might have come from. Yandro's high hill, eh? How do we get there, John?'

I tried to think of the way from highway to side way, side way to trail, and so in and around and over. 'I fear,' I said, 'I could show you better than I could tell you.'

'All right, you'll show me,' he said, with no notion I might want something different. 'I can afford to make up my mind on a moment's notice like that. I'll call the airport and charter a plane. We leave now.'

'I asked John to stay tonight,' said my hostess lady.

'We leave now,' said Mr Yandro, and she shut right up, and I saw how it was. Everybody was scared of him. Maybe they'd be pleasured if I took him out of there for a while.

'Get your plane,' I said. 'We leave now.'

He meant that thing. Not many hours had died before the hired plane set us down at the airport between Asheville and Hendersonville. A taxi rode us into Hendersonville. Mr Yandro woke up a used car man and bought a fair car from him. Then, on my guiding, Mr Yandro took out in the dark for that part of the mountains I pointed out to him.

The sky stretched over us with no moon at all, only a many stars, like little stitches of blazing thread in a black quilt. For real light, only our headlamps—first on a paved road twining around one slope and over another and behind a third, then a gravel road and pretty good, then a dirt road and pretty bad.

'What a stinking country!' said Mr Yandro as we chugged along a ridge as lean as a butcher knife.

I didn't say how I resented that word about a country that stoops to none for prettiness. 'Maybe we ought to have waited till day,' I said.

'I never wait,' he sniffed. 'Where's the town?'

'No town. Just the valley. Three–four hours away. We'll be there by midnight.'

'Oh, God. Let's have some of that whiskey I brought.' He reached for the glove compartment, but I shoved his hand away.

'Not if you're going to drive these mountain roads, Mr Yandro.'

'Then you drive a while, and I'll take a drink.'

'I don't know how to drive a car, Mr Yandro.'

'Oh, God,' he said again, and couldn't have scorned me more if I'd said I didn't know how to wash my face. 'What is a desrick, exactly?'

'Only old-aged folks use the word any more. It's the kind of cabin they used to make, strong logs and a door you can bar, and loophole windows. So you could stand off Indians, maybe.'

'Or the wild beasts can't reach you,' he quoted, and snickered. 'What wild beasts do they have up here in the Forgotten Latitudes?'

'Can't rightly say. A few bears, a wildcat or two. Used to be wolves, and a bounty for killing them. I'm not sure what else.'

True enough, I wasn't sure about the tales I'd heard. Not anyway when Mr Yandro was ready to sneer at them for foolishment.

Our narrow road climbed a great slant of rock one way, then doubled back to climb opposite, and became a double rut, with an empty, hell-scary drop of thousands of feet beside the car. Finally, Mr Yandro edged us into a sort of nick beside the road and shut off the power. He shook. Fear must have been a new feel in his bones.

'Want some of the whiskey, John?' he asked, and drank.

'Thank you, no. We walk from here, anyway. Beyond's the valley.'

He grumped and mad-whispered, but out he got. I took a flashlight and my silver-strung guitar and led out. It was a downways walk, on a narrow trail where even mules would be nervous. And not quiet enough to be easy.

There were mountain night noises, like you never get used to, not even if you're born and raised there, and live and die there. Noises too soft and sneaky to be real murmuring voices. Noises like big flapping wings far off and then near. And, above and below the trail, noises like heavy soft paws keeping pace with you, sometimes two paws, sometimes four, sometimes many. They stay with you, noises like that, all the hours you grope along the night trail, all the way down to the valley so low, till you bless God for the little crumb of light that means a human home, and you ache and pray to get to that home, be it ever so humble, so you'll be safe in the light.

I've wondered since if Mr Yandro's constant blubber and chatter was a string of curses or a string of prayers.

The light we saw was a pine-knot fire inside a little coop above the stream that giggled in the valley bottom. The door was open, and someone sat on the threshold.

'Is that a desrick?' panted and puffed Mr Yandro.

'No, it's newer made. There's Miss Tully at the door, sitting up to think.'

Miss Tully remembered me and welcomed us. She was eighty or ninety, without a tooth in her mouth to clamp her stone-bowl pipe, but she stood straight as a pine on the split-slab floor, and the firelight showed no grey in her tight-combed black hair. 'Rest your hats,' said Miss Tully. 'So this stranger man's name is Mr Yandro. Funny, you coming just now. You're looking for the desrick on Yandro, it's still right where it's been,' and she pointed with her pipe stem off into the empty dark across the valley and up.

Inside, she gave us two chairs bottomed with juniper bark and sat on a stool next to the shelf with herbs in pots, and one or two old paper books, *The Long Lost Friend* and *Egyptian Secrets*, and *Big Albert* the one they say can't be thrown away or given away, only got rid of by burying with a funeral prayer, like a human corpse. 'Funny,' she said again, 'you coming along as the seventy-five years are up.'

We questioned, and she told us what we'd come to hear. 'I was just a little pigtail girl back then,' she said, 'when Joris Yandro courted Polly Wiltse, the witch girl. Mr Yandro, you favour your grandsire a right much. He wasn't as stout-built as you, and younger by years, when he left.'

Even the second time hearing it, I listened hard. It was like a many such tale at the start. Polly Wiltse was sure enough a witch, not just a study-witch like Miss Tully, and Polly Wiltse's beauty would melt the heart of nature and make a dumb man cry out, 'Praise God Who made her!' But none dared court her save only Joris Yandro, who was handsome for a man like she was lovely for a girl. For it was his wish to get her to show him the gold on top of the mountain named for his folks, that only Polly Wiltse and her witching could find.

'Certain sure there's gold in these mountains,' I answered Mr Yandro's interrupting question. 'Before ever the California rush started, folks mined and minted gold in these parts, the history-men say.'

'Gold,' he repeated, both respectful and greedy. 'I was right to come.'

Miss Tully said that Joris Yandro coaxed Polly Wiltse to bring down gold to him, and he carried it away and never came back. And Polly Wiltse pined and mourned like a sick bird, and on Yandro's top she built her desrick. She sang the song, the one I'd sung, it was part of a long spell and charm. Three quarters of a century would pass, seventy-five years, and her lover would come back.

'But he didn't,' said Mr Yandro. 'My grandfather died up north.'

'He sent his grandson, who favours him,' said Miss Tully. 'The song you heard brought you back at the right time.' She thumbed tobacco into her pipe. 'All the Yandro kin moved away, pure down scared of Polly Wiltse's singing.'

'In her desrick, where the wild beasts can't reach her,' quoted Mr Yandro, and chuckled. 'John says they have bears and wildcats up here.' He expected her to say I was stretching it.

'Oh, there's other creatures, too. Scarce animals, like the Toller.'

'The Toller?' he said.

'It's the hugest flying thing there is, I guess,' said Miss Tully. 'Its voice tolls like a bell, to tell other creatures their feed's near. And there's the Flat. It lies level with the ground, and not much higher. It can wrap you like a blanket.' She lighted the pipe. 'And the Bammat. Big, the Bammat is.'

'The Behemoth, you mean,' he suggested.

'No, the Behemoth's in the Bible. The Bammat's something hairy-like, with big ears and a long wiggly nose and twisty white teeth sticking out of its mouth—'

'Oh!' And Mr Yandro trumpeted his laughter. 'You've got some story about the Mammoth. Why, they've been extinct—dead and forgotten—for thousands of years.'

'Not for so long, I've heard tell,' she said, puffing.

'Anyway,' he went on arguing, 'the Mammoth—the Bammat, as you call it—is of the elephant family. How would anything like that get up in the mountains?'

'Maybe folks hunted it there,' said Miss Tully, 'and maybe it stays there so folks will think it's dead and gone a thousand years. And there's the Behinder.'

'And what,' said Mr Yandro, 'might the Behinder look like?'

'Can't rightly say, Mr Yandro. For it's always behind the man or woman it wants to grab. And there's the Skim—it kites through the air—and the Culverin, that can shoot pebbles with its mouth.'

'And you believe all that?' sneered Mr Yandro, the way he always sneered at everything, everywhere.

'Why else should I tell it?' she replied. 'Well, sir, you're back where your kin used to live, in the valley where they named the mountain for them. I can let you two sleep on my front stoop tonight.'

'I came to climb the mountain and see the desrick,' said Mr Yandro with that anxious hurry to him that I kept wondering about.

'You can't climb up there until it's light,' she told him, and she made us two quilt pallets on the split-slab stoop.

I was tired and glad to stretch out, but Mr Yandro grumbled, as if we were wasting time. At sunup next morning, Miss Tully fried us some side meat and slices of hominy grit porridge, and she fixed us a snack to carry, and a gourd to put water in. Mr Yandro held out a ten dollar bill.

'No, I thank you,' said Miss Tully. 'I bade you stay, and I won't take money for that.'

'Oh, everybody takes money from me,' he snickered, and threw it on the door-sill at her feet. 'Go on, it's yours.'

Quick as a weasel, Miss Tully's hand grabbed a stick of stove wood.

'Lean down and take back that money-bill, Mister,' she said.

He did as she told him. With the stick she pointed out across the stream that ran through the thickets below us, and up the height beyond. She acted as if there wasn't any trouble a second before.

'That's the Yandro Mountain,' she said. 'There, on the highest point, where it looks like the crown of a hat, thick with trees all the way up, stands the desrick built by Polly Wiltse. You look close with the sun rising, and you can maybe make it out.'

I looked hard. There for sure it was, far off and high up, and tiny, but I could see it. It looked a lean sort of a building.

'How about trails going up?' I asked her.

'There's trails up there, John, but nobody walks them.'

'Now, now,' said Mr Yandro, 'if there's a trail, somebody must walk it.'

'May be a lot in what you say, but I know nobody in this valley would set foot to such a trail. Not with what they say's up there.'

He laughed at her, as I wouldn't have dared. 'You mean the Bammat,' he said. 'And the Flat, and the Skim, and the Culverin.'

'And the Toller,' she added for him. 'And the Behinder. Only a gone gump would go up there.'

We headed away down to the waterside, and crossed on logs laid on top of rocks. On the far side a trail led along, and when the sun was an hour higher we were at the foot of Yandro's high hill, and a trail went up there, too.

We rested. Mr Yandro needed rest worse than I did. Moving most of the night before, unused to walking and climbing, he had a gaunted look to his heavy face, and his clothes were sweated, and dust dulled out his shoes. But he grinned at me.

'So she's waited seventy-five years,' he said, 'and so I look like the man she's waiting for. And so there's gold up there. More gold than my grandfather could have carried off.'

'You believe what you've been hearing,' I said, and it was a mystery.

'John, a wise man knows when to believe the unusual, and how it will profit him. She's up there, waiting, and so is the gold.'

'What when you find it?' I asked.

'My grandfather was able to go off and leave her. It sounds like a good example to me.' He grinned wider and toothier. 'I'll give you part of the gold.'

'No thanks, Mr Yandro.'

'You don't want your pay? Why did you come here with me?'

'Just made up my mind on a moment's notice, like you.'

He scowled then, but he looked up at the height. 'How long will it take to climb, John?'

'Depends on how fast we climb, how well we keep up the pace.'

'Then let's go,' and he started up the trail.

It wasn't folks' feet had worn that trail. I saw a hoofmark.

'Deer,' grunted Mr Yandro; and I said, 'Maybe.'

We scrambled up on a rightward slant, then leftward. The trees marched in close around us, with branches above that filtered only soft green light. Something rustled, and we saw a brown, furry shape, big as a big cat, scuttling out of sight.

'Woodchuck,' wheezed Mr Yandro; again I said, 'Maybe.'

After an hour's working upward we rested, and after two hours more we rested again. Around 11 o'clock we reached an open space where clear light touched the middle, and there we sat on a log and ate the corn bread and smoked meat Miss Tully had fixed. Mr Yandro mopped his face with a fancy handkerchief, and gobbled food for strength to glitter his eye at me. 'What are you glooming about?' he said. 'You look as if you'd call me a name if you weren't afraid.'

'I've held my tongue,' I said, 'by way of manners, not fear. I'm just thinking about how and why we came so far and sudden to this place.'

'You sang me a song, and I heard, and thought I'd come to where my people originated. Now I have a hunch about profit. That's enough for you.'

'It's not just that gold story,' I said. 'You're more than rich enough.'

'I'm going up there,' said Mr Yandro, 'because, by God, that old hag down there said everybody's afraid to do it. And you said you'd go with me.'

'I'll go right to the top with you,' I said.

I forebore to say that something had come close and looked from among the trees behind him. It was big and broad-headed, with elephant ears to right and left, and white tusks like bannisters on a spiral staircase. But it was woolly-shaggy, like a buffalo bull. The Bammat. How could such a thing move so quiet-like?

He drank from his whiskey bottle, and on we climbed. We could hear those noises in the woods and brush, behind rocks and down little gulleys, as if the mountain side thronged with living things as thick as fleas on a possum dog and another sight sneakier. I didn't let on I was nervous.

'Why are you singing under your breath?' he grunted after a while.

'I'm not singing,' I said. 'I need my breath for climbing.'

'I hear you!' he charged me, like a lawyer in court.

We'd stopped dead on the trail, and I heard it, too.

It was soft, almost like some half-remembered song in your mind. It was the Yandro song, all right:

Look away, look away, look away over Yandro,
Where them wild things are flyin'
From bough to bough, and a-mating with their mates,
So why not me with mine?

'That singing comes from up above us,' I told Mr Yandro.

'Then,' he said, 'we must be nearly at the top.'

As we started climbing again, I could hear the noises to right and left in the woods, and then I realized they'd quieted down when we stopped. They moved when we moved, they waited when we waited. There were lots of them. Soft noises, but lots of them.

Which is why I myself, and probably Mr Yandro too, didn't pause any more on the way up, even on a rocky stretch where we had to climb on all fours. It may have been an hour after noon when we came to the top.

Right there was a circle-shaped clearing, with the trees thronged around it all the way except an open space towards the slope. Those trees had mist among and between them, quiet and fluffy, like spider webbing. And at the open space, on the lip of the way down, perched the desrick.

Old-aged was what it looked. It stood high and looked the higher, because it was built so narrow of unnotched logs, set four above four, hogpen fashion, as tall as a tall tobacco barn. The spaces between the logs were clinked shut with great masses and wads of clay. The steep-pitched roof was of shingles, cut long and narrow, so that they looked almost like thatch. There was one big door, made of an axe-chopped plank, and the hinges must have been inside, for I could see none. And one window, covered with what must have been rawhide scraped thin, with a glow of soft light coming through.

'That's it,' puffed Mr Yandro. 'The desrick.'

I looked at him then, and knew what most he wanted on this earth. He wanted to be boss. Money was just something to greaten him. His idea of greatness was bigness. He wanted to do all the talking, and have everybody else do the listening. He had his eyes hung on that desrick, and he licked his lips, like a cat over a dish of cream.

'Let's go in,' he said.

'Not where I'm not invited,' I told him, as flatly as anybody could ever tell him. 'I said I'd come to the top. This is the top.'

'Come with me,' he said. 'My name's Yandro. This mountain's name is Yandro. I can buy and sell every man, woman and child in this part of the country. If I say it's all right to go into a house, it's all right to go into a house.'

He meant that thing. The world and everybody in it was just there to let him walk on. He took a step towards the desrick. Somebody hummed inside, not the words of the song, but the tune. Mr Yandro snorted at me, to show how small he reckoned me because I held back, and he headed towards the big door.

'If she's there, she'll show me the gold,' he said.

But I couldn't have moved from where I stood at the edge of the clearing. I was aware of a sort of closing in all around the edge, among the trees and brushy clumps. Not that the closing in could be seen, but there was a *gong-gong* further off, the voice of the Toller norating to the other creatures their feed was near. And above the treetops sailed a round, flat thing, like a big plate being pitched high. A Skim. Then another Skim. And the blood inside my body was cold and solid as ice, and my voice turned to a handful of sand in my throat.

I knew, plain as paint, that if I tried to back up, to turn around even, my legs would fail and I'd fall down. With fingers like twigs with sleet stuck to them, I dragged around my guitar, to pluck at the silver strings, because silver is protection against evil.

But I didn't. For out of the bushes near me the Bammat stuck its broad woolly head, and it shook that head at me once, for silence. It looked me between the eyes, steadier than a beast should look at a man, and shook its head. I wasn't to make any noise. And I didn't. When the Bammat saw that I'd be quiet, it paid me no more mind, and I knew I wasn't to be included in what would happen then.

Mr Yandro was knocking at the axe-chopped door. He waited, and knocked again. I heard him growl, something about how he wasn't used to waiting for people to answer his knock.

Inside, the humming had died out. After a moment, Mr Yandro moved around to where the window was, and picked at the rawhide.

I could see, but he couldn't, as around from behind the corner of the desrick flowed something. It lay out on the ground like a broad, black, short-furred carpet rug. But it moved, humping and then flattening out, the way a measuring worm moves. It moved pretty fast, right towards Mr Yandro from behind and to one side. The Toller said *gong-gong-gong*, from closer in.

'Anybody in there?' bawled Mr Yandro. 'Let me in!'

The crawling carpet brushed its edge against his foot. He looked down at it, and his eyes stuck out all of a sudden, like two door knobs. He knew what it was, and named it at the top of his voice.

'The Flat!'

Humping against him, it tried to wrap around his foot and leg. He gasped out something I'd never want written down for my last words, and pulled loose and ran, fast and straight, towards the edge of the clearing.

Gong-gong, said the Toller, and Mr Yandro tried to slip along next to the trees. But, just ahead of him, the Culverin hoved itself half into sight on its many legs. It pointed its needle-shaped mouth and spit a pebble. I heard the pebble ring on Mr Yandro's head. He staggered against a tree. And I saw what nobody's ever supposed to see.

The Behinder flung itself on his shoulders. Then I knew why nobody's supposed to see one. I wish I hadn't. To this day I can see it, as plain as a fence at noon, and forever I will be able to see it. But talking about it's another matter. Thank you, I won't try.

Then everything else was out—the Bammat, the Culverin, and all the others. They were hustling him across towards the desrick, and the door moved slowly and quietly open for him to come in.

As for me, I was out of their minds, and I hoped and prayed they wouldn't care if I just went on down the trail as fast as I could set one foot below the other.

Scrambling and scrambling down, without a noise to keep me company, I figured that I'd probably had my unguessed part in the whole thing. Seventy-five years had to pass, and then Mr Yandro come there to the desrick. And it needed me, or somebody like me, to meet him and sing the song that would put it in his head and heart to come to where his granddaddy had courted Polly Wiltse, just as though it was his own whim.

No. No, of course, he wasn't the man who had made Polly Wiltse love him and then had left her. But he was the man's grandson, of the same blood and the same common, low-down, sorry nature that wanted money and power, and didn't care who he hurt so he could have both. And he looked like Joris Yandro. Polly Wiltse would recognize him.

I haven't studied much about what Polly Wiltse was like, welcoming him into the desrick on Yandro, after waiting inside for three quarters of a century. Anyway, I never heard of him following me down. Maybe he's been missed. But I'll lay you anything you name he's not been mourned.

THEODORE STURGEON

THE SILKEN-SWIFT . . .

There's a village by the Bogs, and in the village is a Great House. In the Great House lived a squire who had land and treasures and, for a daughter, Rita.

In the village lived Del, whose voice was a thunder in the inn when he drank there; whose corded, cabled body was golden-skinned, and whose hair flung challenges back to the sun.

Deep in the Bogs, which were brackish, there was a pool of purest water, shaded by willows and wide-wondering aspen, cupped by banks of a moss most marvellously blue. Here grew mandrake, and there were strange pipings in mid-summer. No one ever heard them but a quiet girl whose beauty was so very contained that none of it showed. Her name was Barbara.

There was a green evening, breathless with growth, when Del took his usual way down the lane beside the manor and saw a white shadow adrift inside the tall iron pickets. He stopped and the shadow approached, and became Rita. 'Slip around to the gate,' she said, 'and I'll open it for you.'

She wore a gown like a cloud and a silver circlet round her head. Night was caught in her hair, moonlight in her face, and in her great eyes, secrets swam.

Del said, 'I have no business with the squire.'

'He's gone,' she said. 'I've sent the servants away. Come to the gate.'

'I need no gate.' He leaped and caught the top bar of the fence, and in a continuous fluid motion went high and across and down beside her. She looked at his arms, one, the other; then up at his hair. She pressed her small hands tight together and made a little laugh, and then she was gone through the tailored trees, lightly, swiftly, not looking back. He followed, one step for three of hers, keeping pace with a new pounding in the sides of his neck. They crossed a flower bed and a wide marble terrace. There was an open door, and when he passed through it he stopped, for she was nowhere in sight. Then the door clicked shut behind him and he whirled. She was there, her back to the panel, laughing up at him in the dimness. He thought she would come to him then, but instead she twisted by, close, her eyes on his. She smelt of violets and sandalwood. He followed her into a great hall, quite dark but full of the subdued lights of polished wood, cloisonné, tooled leather and gold-threaded tapestry. She flung open another door, and they were in a small room with a carpet made of rosy silences, and a candle-lit table. Two places were set, each with five different crystal glasses and old silver as prodigally used as the iron pickets outside. Six teak-wood steps rose to a great oval window. 'The moon,' she said, 'will rise for us there.'

She motioned him to a chair and crossed to a sideboard, where there was a rack of decanters—ruby wine and white; one with a strange brown bead; pink, and amber. She took down the first and poured. Then she lifted the silver domes from the salvers on the table, and a magic of fragrance filled the air. There were smoking sweets and savouries, rare seafood and slivers of fowl, and morsels of strange meat wrapped in flower petals, spitted with foreign fruits and tiny soft sea-shells. All about were spices, each like a separate voice in the distant murmur of a crowd: saffron and sesame, cumin and marjoram and mace.

And all the while Del watched her in wonder, seeing how the candles left the moonlight in her face, and how completely she trusted her hands, which did such deftnesses without supervision—so composed she was, for all the silent secret laughter that tugged at her lips, for all the bright dark mysteries that swirled and swam within her.

They ate, and the oval window yellowed and darkened while the candlelight grew bright. She poured another wine, and another, and with the courses of the meal they were as May to the crocus and as frost to the apple.

Del knew it was alchemy and he yielded to it without question. That which was purposely over-sweet would be piquantly cut; this induced thirst would, with exquisite timing, be quenched. He knew she was watching him; he knew she was aware of the heat in his cheeks and the tingle at his fingertips. His wonder grew, but he was not afraid.

In all this time she spoke hardly a word; but at last the feast was over and they rose. She touched a silken rope on the wall, and panelling slid aside. The table rolled silently into some ingenious recess and the panel returned. She waved him to an L-shaped couch in one corner, and as he sat close to her, she turned and took down the lute which hung on the wall behind her. He had his moment of confusion; his arms were ready for her, but not for the instrument as well. Her eyes sparkled, but her composure was unshaken.

Now she spoke, while her fingers strolled and danced on the lute, and her words marched and wandered in and about the music. She had a thousand voices, so that he wondered which of them was truly hers. Sometimes she sang; sometimes it was a wordless crooning. She seemed at times remote from him, puzzled at the turn the music was taking, and at other times she seemed to hear the pulsing roar in his ear-drums, and she played laughing syncopations to it. She sang words which he almost understood:

Bee to blossom, honey dew,
Claw to mouse, and rain to tree,
Moon to midnight, I to you;
Sun to starlight, you to me . . .

and she sang something wordless:

Ake ya rundefle, rundefle fye,
Orel ya rundefle kown,
En yea, en yea, ya bunderbee bye
En sor, en see, en sown.

which he also almost understood.

In still another voice she told him the story of a great hairy spider and a little pink girl who found it between the leaves of a half-open book; and at first he was all fright and pity for the girl, but then she went on to tell of what the spider suffered, with his home disrupted by this yawping giant, and so vividly did she tell of it that at the end he was laughing at himself and all but crying for the poor spider.

So the hours slipped by, and suddenly, between songs, she was in his arms; and in the instant she had twisted up and away from him, leaving him gasping. She said, in still a new voice, sober and low, 'No, Del. We must wait for the moon.'

His thighs ached and he realized that he had half-risen, arms out, hands clutching and feeling the extraordinary fabric of her gown though it was gone from them; and he sank back to the couch with an odd, faint sound that was wrong for the room. He flexed his fingers and, reluctantly, the sensation of white gossamer left them. At last he looked across at her and she laughed and leapt high lightly, and it was as if she stopped in midair to stretch for a moment before she alighted beside him, bent and kissed his mouth, and leapt away.

The roaring in his ears was greater, and at this it seemed to acquire a tangible weight. His head bowed; he tucked his knuckles into the upper curve of his eye sockets and rested his elbows on his knees. He could hear the sweet sussurrus of Rita's gown as she moved about the room; he could sense the violets and sandalwood. She was dancing, immersed in the joy of movement and of his nearness. She made her own music, humming, sometimes whispering to the melodies in her mind.

And at length he became aware that she had stopped; he could hear nothing, though he knew she was still near. Heavily he raised his head. She was in the centre of the room, balanced like a huge white moth, her eyes quite dark now with their secrets quiet. She was staring at the window, poised, waiting.

He followed her gaze. The big oval was black no longer, but dusted over with silver light. Del rose slowly. The dust was a mist, a loom, and then, at one edge, there was a shard of the moon itself creeping and growing.

Because Del stopped breathing, he could hear her breath; it was rapid and so deep it faintly strummed her versatile vocal cords.

'Rita . . .'

Without answering she ran to the sideboard and filled two small glasses. She gave him one, then, 'Wait,' she breathed, 'oh, wait!'

Spellbound, he waited while the white stain crept across the window. He understood suddenly that he must be still until the great oval was completely filled with direct moonlight, and this helped him, because it set a foreseeable limit to his waiting; and it hurt him, because nothing in life, he thought, had ever moved so slowly. He had a moment of rebellion, in which he damned himself for falling in with her complex pacing; but with it he realized that now the darker silver was wasting away, now it was a finger's breadth, and now a thread, and now, and *now*—.

She made a brittle feline cry and sprang up the dark steps to the window. So bright was the light that her body was a jet cameo against it. So delicately wrought was her gown that he could see the epaulettes of silver light the moon gave her. She was so beautiful his eyes stung.

'Drink,' she whispered. 'Drink with me, darling, darling . . .'

For an instant he did not understand her at all, and only gradually did he become aware of the little glass he held. He raised it towards her and drank. And of all the twists and titillations of taste he had had this night, this was the most startling; for it had no taste at all, almost no substance, and a temperature almost exactly that of blood. He looked stupidly down at the glass and back up at the girl. He thought that she had turned about and was watching him, though he could not be sure, since her silhouette was the same.

And then he had his second of unbearable shock, for the light went out.

The moon was gone, the window, the room, Rita was gone.

For a stunned instant he stood tautly, stretching his eyes wide. He made a sound that was not a word. He dropped the glass and pressed his palms to his eyes, feeling them blink, feeling the stiff silk of his lashes against them. Then he snatched the hands away, and it was still dark, and more than dark; this was not a blackness. This was like trying to see with an elbow or with a tongue; it was not black, it was *Nothingness.*

He fell to his knees.

Rita laughed.

An odd, alert part of his mind seized on the laugh and understood it, and horror and fury spread through his whole being; for this was the laugh which had been tugging at her lips all evening, and it was a hard, cruel, self-assured laugh. And at the same time, because of the anger or in spite of it, desire exploded whitely within him. He moved towards the sound, groping, mouthing. There was a quick, faint series of rustling sounds from the steps, and then a light, strong web fell around him. He struck out at it, and recognized it for the unforgettable thing it was—her robe. He caught at it, ripped it, stamped upon it. He heard her bare feet run lightly down and past him, and lunged, and caught nothing. He stood, gasping painfully.

She laughed again.

'I'm blind,' he said hoarsely. 'Rita, I'm blind!'

'I know,' she said coolly, close beside him. And again she laughed.

'What have you done to me?'

'I've watched you be a dirty animal of a man,' she said.

He grunted and lunged again. His knees struck something—a chair, a cabinet—and he fell heavily. He thought he touched her foot.

'Here, lover, here!' she taunted.

He fumbled about for the thing which had tripped him, found it, used it to help him upright again. He peered uselessly about.

'Here, lover!'

He leaped, and crashed into the door jamb: cheekbone, collarbone, hip-bone, ankle were one straight blaze of pain. He clung to the polished wood.

After a time he said, in agony, 'Why?'

'No man has ever touched me and none ever will,' she sang. Her breath was on his cheek. He reached and touched nothing, and then he heard her leap from her perch on a statue's pedestal by the door, where she had stood high and leaned over to speak.

No pain, no blindness, not even the understanding that it was her witch's brew working in him could quell the wild desire he felt at her nearness. Nothing could tame the fury that shook him as she laughed. He staggered after her, bellowing.

She danced around him, laughing. Once she pushed him into a clattering rack of fire-irons. Once she caught his elbow from behind and spun him. And once, incredibly, she sprang past him and, in midair, kissed him again on the mouth.

He descended into Hell, surrounded by the small, sure patter of bare feet and sweet cool laughter. He rushed and crashed, he crouched and bled and whimpered like a hound. His roaring and blundering took an echo, and that must have been the great hall. Then there were walls that seemed more than unyielding; they struck back. And there were panels to lean against, gasping, which became opening doors as he leaned. And always the black nothingness, the writhing temptation of the pat-pat of firm flesh on smooth stones, and the ravening fury.

It was cooler, and there was no echo. He became aware of the whisper of the wind through trees. The balcony, he thought; and then, right in his ear, so that he felt her warm breath, 'Come, lover . . .' and he sprang. He sprang and missed, and instead of sprawling on the terrace, there was nothing, and nothing, and then, when he least expected it, a shower of cruel thumps as he rolled down the marble steps.

He must have had a shred of consciousness left, for he was vaguely aware of the approach of her bare feet, and of the small, cautious hand that touched his shoulder and moved to his mouth, and then his chest. Then it was withdrawn, and either she laughed or the sound was still in his mind.

Deep in the Bogs, which were brackish, there was a pool of purest water, shaded by willows and wide-wondering aspens, cupped by banks of a moss most marvellously blue. Here grew mandrake, and there were strange pipings in midsummer. No one ever heard them but a quiet girl whose beauty was so very contained that none of it showed. Her name was Barbara.

No one noticed Barbara, no one lived with her, no one cared. And Barbara's life was very full, for she was born to receive. Others are born wishing to receive, so they wear bright masks and make attractive sounds like cicadas and operettas, so others will be forced, one way or another, to give to them. But Barbara's receptors were wide open, and always had been, so that she needed no substitute for sunlight through a tulip petal, or the sound of morning-glories climbing, or the tangy sweet smell of formic acid which is the only death cry possible to an ant, or any other of the thousand things overlooked by folk who can only wish to receive. Barbara had a garden and an orchard, and took things in to market when she cared to, and the rest of the time she spent in taking what was given. Weeds grew in her garden, but since they were welcomed, they grew only where they could keep the watermelons from being sunburned. The rabbits were welcome, so they kept to the two rows of carrots, the one of lettuce, and the one of tomato vines which were planted for them, and they left the rest alone. Goldenrod shot up beside the bean hills to lend a hand upward, and the birds ate only the figs and peaches from the waviest top branches, and in return patrolled the lower ones for caterpillars and egg-laying flies. And if a fruit stayed green for two weeks longer until Barbara had time to go to market, or if a mole could channel moisture to the roots of the corn, why it was the least they could do.

For a brace of years Barbara had wandered more and more, impelled by a thing she could not name—if indeed she was aware of it at all. She knew only that over-the-rise was a strange and friendly place, and that it was a fine thing on arriving there to find another rise to go over. It may very well be that she now needed someone to love, for loving is a most receiving thing, as anyone can attest who has been loved without returning it. It is the one who is loved who must give and give. And she found her love, not in her wandering, but at the market. The shape of her love, his colours and sounds, were so much with her that when she saw him first it was without surprise; and thereafter, for a very long while, it was quite enough that he lived. He gave to her by being alive, by setting the air athrum with his mighty voice, by his stride, which was, for a man afoot, the exact analogue of what the horseman calls a 'perfect seat'.

After seeing him, of course, she received twice and twice again as much as ever before. A tree was straight and tall for the magnificent sake of being straight and tall, but wasn't straightness a part of him, and being tall? The oriole gave more now than song, and the hawk more than walking the wind, for had they not hearts like his, warm blood and his same striving to keep it so for tomorrow? And more and more, over-the-rise was the place for her, for only there could there be more and still more things like him.

But when she found the pure pool in the brackish Bogs, there was no more over-the-rise for her. It was a place without hardness or hate, where the aspens trembled only for wonder, and where all contentment was rewarded. Every single rabbit there was *the* champion nose-twinkler, and every waterbird could stand on one leg the longest, and proud of it. Shelf-fungi hung to the willow-trunks, making that certain, single purple of which the sunset is incapable, and a tanager and a cardinal gravely granted one another his definition of 'red'.

Here Barbara brought a heart light with happiness, large with love, and set it down on the blue moss. And since the loving heart can receive more than anything else, so it is most needed, and Barbara took the best bird songs, and the richest colours, and the deepest peace, and all the other things which are most worth giving. The chipmunks brought her nuts when she was hungry and the prettiest stones when she was not. A green snake explained to her, in pantomime, how a river of jewels may flow uphill, and three mad otters described how a bundle of joy may slip and slide down and down and be all the more joyful for it. And there was the magic moment when a midge hovered, and then a honeybee, and then a humblebee, and at last a humming-bird; and there they hung, playing a chord in A sharp minor.

Then one day the pool fell silent, and Barbara learned why the water was pure.

The aspens stopped trembling.

The rabbits all came out of the thicket and clustered on the blue bank, backs straight, ears up, and all their noses as still as coral.

The waterbirds stepped backwards, like courtiers, and stopped on the brink with their heads turned sidewise, one eye closed, the better to see with the other.

The chipmunks respectfully emptied their cheek pouches, scrubbed their paws together and tucked them out of sight; then stood still as tent pegs.

The pressure of growth around the pool ceased: the very grass waited.

The last sound of all to be heard—and by then it was very quiet—was the soft *whick!* of an owl's eyelids as it awoke to watch.

He came like a cloud, the earth cupping itself to take each of his golden hooves. He stopped on the bank and lowered his head, and for a brief moment his eyes met Barbara's, she looked into a second universe of wisdom and compassion. Then there was the arch of the magnificent neck, the blinding flash of his golden horn.

And he drank, and he was gone. Everyone knows the water is pure, where the unicorn drinks.

How long had he been there? How long gone? Did time wait too, like the grass?

'And couldn't he stay?' she wept. 'Couldn't he stay?'

To have seen the unicorn is a sad thing; one might never see him more. But then—to have seen the unicorn!

She began to make a song.

It was late when Barbara came in from the Bogs, so late the moon was bleached with cold and fleeing to the horizon. She struck the highroad just below the Great House and turned to pass it and go out to her garden house.

Near the locked main gate an animal was barking. A sick animal, a big animal. . . .

Barbara could see in the dark better than most, and soon saw the creature clinging to the gate, climbing, uttering that coughing moan as it went. At the top it slipped, fell outward, dangled; then there was a ripping sound, and it fell heavily to the ground and lay still and quiet.

She ran to it, and it began to make the sound again. It was a man, and he was weeping.

It was her love, her love, who was tall and straight and so very alive—her love, battered and bleeding, puffy, broken, his clothes torn, crying.

Now of all times was the time for a lover to receive, to take from the loved one his pain, his trouble, his fear. 'Oh, hush, hush,' she whispered, her hands touching his bruised face like swift feathers. 'It's all over now. It's all over.'

She turned him over on his back and knelt to bring him up sitting. She lifted one of his thick arms around her shoulder. He was very heavy, but she was very strong. When he was upright, gasping weakly, she looked up and down the road in the waning moonlight. Nothing, no one. The Great House was dark. Across the road, though, was a meadow with high hedgerows which might break the wind a little.

'Come, my love, my dear love,' she whispered. He trembled violently.

All but carrying him, she got him across the road, over the shallow ditch, and through a gap in the hedge. She almost fell with him there. She gritted her teeth and set him down gently. She let him lean against the hedge, and then ran and swept up great armfuls of sweet broom. She made a tight springy bundle of it and set it on the ground beside him, and put a corner of her cloak over it, and gently lowered his head until it was pillowed. She folded the rest of the cloak about him. He was very cold.

There was no water near, and she dared not leave him. With her kerchief she cleaned some of the blood from his face. He was still very cold. He said, 'You devil. You rotten little devil.'

'Shh.' She crept in beside him and cradled his head. 'You'll be warm in a minute.'

'Stand still,' he growled. 'Keep running away.'

'I won't run away,' she whispered. 'Oh, my darling, you've been hurt, so hurt. I won't leave you. I promise I won't leave you.'

He lay very still. He made the growling sound again.

'I'll tell you a lovely thing,' she said softly. 'Listen to me, think about the lovely thing,' she crooned.

'There's a place in the bog, a pool of pure water, where the trees live beautifully, willow and aspen and birch, where everything is peaceful, my darling, and the flowers grow without tearing their petals. The moss is blue and the water is like diamonds.'

'You tell me stories in a thousand voices,' he muttered.

'Shh. Listen, my darling. This isn't a story, it's a real place. Four miles north and a little west, and you can see the trees from the ridge with the two dwarf oaks. And I know why the water is pure!' she cried gladly. 'I know why!'

He said nothing. He took a deep breath and it hurt him, for he shuddered painfully.

'The unicorn drinks there,' she whispered. 'I *saw* him!'

Still he said nothing. She said, 'I made a song about it. Listen, this is the song I made:

And He—suddenly gleamed! My dazzled eyes
Coming from outer sunshine to this green
And secret gloaming, met without surprise
The vision. Only after, when the sheen
And splendour of his going fled away,
I knew amazement, wonder and despair,
That he should come—and pass—and would not stay,
The Silken-swift—the gloriously Fair!
That he should come—and pass—and would not stay.
So that, forever after, I must go,
Take the long road that mounts against the day,
Travelling in the hope that I shall know

Again that lifted moment, high and sweet,
Somewhere—on purple moor or windy hill—
Remembering still his wild and delicate feet,
The magic and the dream—remembering still!

His breathing was more regular. She said, 'I truly *saw* him!'

'I'm blind,' he said. 'Blind, I'm blind.'

'Oh, my dear . . .'

He fumbled for her hand, found it. For a long moment he held it. Then, slowly, he brought up his other hand and with them both he felt her hand, turned it about, squeezed it. Suddenly he grunted, half sitting. 'You're here!'

'Of course, darling. Of course I'm here.'

'Why?' he shouted. 'Why? *Why?* Why all of this? Why blind me?' He sat up, mouthing, and put his great hand on her throat. 'Why do all that if . . .' The words ran together into an animal noise. Wine and witchery, anger and agony boiled in his veins.

Once she cried out.

Once she sobbed.

'Now,' he said, 'you'll catch no unicorns. Get away from me.' He cuffed her.

'You're mad. You're sick,' she cried.

'Get away,' he said ominously.

Terrified, she rose. He took the cloak and hurled it after her. It almost toppled her as she ran away, crying silently.

After a long time, from behind the hedge, the sick, coughing sobs began again.

Three weeks later Rita was in the market when a hard hand took her upper arm and pressed her into the angle of a cottage wall. She did not start. She flashed her eyes upward and recognized him, and then said composedly, 'Don't touch me.'

'I need you to tell me something,' he said. 'And tell me you *will!*' His voice was as hard as his hand.

'I'll tell you anything you like,' she said. 'But don't touch me.'

He hesitated, then released her. She turned to him casually. 'What is it?' Her gaze darted across his face and its almost-healed scars. The small smile tugged at one corner of her mouth.

His eyes were slits. 'I have to know this: why did you make up all that . . . prettiness, that food, that poison . . . just for me? You could have had me for less.'

She smiled. 'Just for you? It was your turn, that's all.'

He was genuinely surprised. 'It's happened before?'

She nodded. 'Whenever it's the full of the moon—and the squire's away.'

'You're lying!'

'You forget yourself!' she said sharply. Then, smiling, 'It is the truth, though.'

'I'd've heard talk—'

'Would you now? And tell me—how many of your friends know about your humiliating adventure?'

He hung his head.

She nodded. 'You see? They go away until they're healed, and they come back and say nothing. And they always will.'

'You're a devil . . . why do you do it? Why?'

'I told you,' she said openly. 'I'm a woman in my own way. No man will ever touch me, though. I am virgin and shall remain so.'

'You're *what*?' he roared.

She held up a restraining, ladylike glove. 'Please,' she said, pained.

'Listen,' he said, quietly now, but with such intensity that for once she stepped back a pace. He closed his eyes, thinking hard. 'You told me—the pool, the pool of the unicorn, and a song, wait. "The Silken-swift, the gloriously Fair . . ." Remember? And then I—I saw to it that *you*'d never catch a unicorn!'

She shook her head, complete candour in her face. 'I like that, "the Silken-swift". Pretty. But believe me—no! That isn't mine.'

He put his face close to hers, and though it was barely a whisper, it came out like bullets. 'Liar! Liar! I couldn't forget. I was sick, I was hurt, I was poisoned, but I know what I did!' He turned on his heel and strode away.

She put the thumb of her glove against her upper teeth for a second, then ran after him. 'Del!'

He stopped but, rudely, would not turn. She rounded him, faced him. 'I'll not have you believing that of me—it's the one thing I have left,' she said tremulously.

He made no attempt to conceal his surprise. She controlled her expression with a visible effort, and said, 'Please. Tell me a little more—just about the pool, the song, whatever it was.'

'You don't remember?'

'I don't *know*!' she flashed. She was deeply agitated.

He said, with mock patience, 'You told me of a unicorn pool out on the Bogs. You said you had seen *him* drink there. You made a song about it. And then I—'

'Where? Where was this?'

'You forget so soon?'

'Where? Where did it happen?'

'In the meadow, across the road from your gate, where you followed me,' he said. 'Where my sight came back to me, when the sun came up.'

She looked at him blankly, and slowly her face changed. First the imprisoned smile struggling to be free, and then—she was herself again, and she laughed. She laughed a great ringing peal of the laughter that had plagued him so, and she did not stop until he put one hand behind his back, then the other, and she saw his shoulders swell with the effort to keep from striking her dead.

'You animal!' she said, goodhumouredly. 'Do you know what you've done? Oh, you . . . you *animal*!' She glanced around to see that there were no ears to hear her. 'I left you at the foot of the terrace steps,' she told him. Her eyes sparkled. 'Inside the gates, you understand? And you . . .'

'Don't laugh,' he said quietly.

She did not laugh. 'That was someone else out there. Who, I can't imagine. But it wasn't I.'

He paled. 'You followed me out.'

'On my soul I did not,' she said soberly. Then she quelled another laugh.

'That can't be,' he said. 'I couldn't have . . .'

'But you were blind, blind and crazy, Del-my-lover!'

'Squire's daughter, take care,' he hissed. Then he pulled his big hand through his hair. 'It can't be. It's three weeks; I'd have been accused . . .'

'There are those who wouldn't,' she smiled. 'Or—perhaps she will, in time.'

'There has never been a woman so foul,' he said evenly, looking her straight in the eye. 'You're lying—you know you're lying.'

'What must I do to prove it—aside from that which I'll have no man do?'

His lip curled. 'Catch the unicorn,' he said.

'If I did, you'd believe I was virgin?'

'I must,' he admitted. He turned away, then said, over his shoulder, 'But—*you*?'

She watched him thoughtfully until he left the market-place. Her eyes sparkled; then she walked briskly to the goldsmith's, where she ordered a bridle of woven gold.

If the unicorn pool lay in the Bogs nearby, Rita reasoned, someone who was familiar with that brackish wasteland must know of it. And when she made a list in her mind of those few who travelled the Bogs, she knew whom to ask. With that, the other deduction came readily. Her laughter drew stares as she moved through the market-place.

By the vegetable stall she stopped. The girl looked up patiently.

Rita stood swinging one expensive glove against the other wrist, half-smiling. 'So you're the one.' She studied the plain, inward-turning, peaceful face until Barbara had to turn her eyes away. Rita said, without further preamble, 'I want you to show me the unicorn pool in two weeks.'

Barbara looked up again, and now it was Rita who dropped her eyes. Rita said, 'I can have someone else find it, of course. If you'd rather not.' She spoke very clearly, and people turned to listen. They looked from Barbara to Rita and back again, and they waited.

'I don't mind,' said Barbara faintly. As soon as Rita had left, smiling, she packed up her things and went silently back to her house.

The goldsmith, of course, made no secret of such an extraordinary commission; and that, plus the gossips who had overheard Rita talking to Barbara, made the expedition into a cavalcade. The whole village turned out to see; the boys kept firmly in check so that Rita might lead the way; the young bloods ranged behind her (some a little less carefree than they might be) and others snickering behind their hands. Behind them the girls, one or two a little pale, others eager as cats to see the squire's daughter fail, and perhaps even . . . but then, only she had the golden bridle.

She carried it casually, but casualness could not hide it, for it was not wrapped, and it swung and blazed in the sun. She wore a flowing white robe, trimmed a little short so that she might negotiate the rough bogland; she had on a golden girdle and little gold sandals, and a gold chain bound her head and hair like a coronet.

Barbara walked quietly a little behind Rita, closed in with her own thoughts. Not once did she look at Del, who strode somberly by himself.

Rita halted a moment and let Barbara catch up, then walked beside her. 'Tell me,' she said quietly, 'why did you come? It needn't have been you.'

'I'm his friend,' Barbara said. She quickly touched the bridle with her finger. 'The unicorn.'

'Oh,' said Rita. 'The unicorn.' She looked archly at the other girl. 'You wouldn't betray all your friends, would you?'

Barbara looked at her thoughtfully, without anger. 'If—when you catch the unicorn,' she said carefully, 'what will you do with him?'

'What an amazing question! I shall keep him, of course!'

'I thought I might persuade you to let him go.'

Rita smiled, and hung the bridle on her other arm. 'You could never do that.'

'I know,' said Barbara. 'But I thought I might, so that's why I came.' And before Rita could answer, she dropped behind again.

The last ridge, the one which overlooked the unicorn pool, saw a series of gasps as the ranks of villagers topped it, one after the other, and saw what lay below; and it was indeed beautiful.

Surprisingly, it was Del who took it upon himself to call out, in his great voice, 'Everyone wait here!' And everyone did; the top of the ridge filled slowly, from one side to the other, with craning, murmuring people. And then Del bounded after Rita and Barbara.

Barbara said, 'I'll stop here.'

'Wait,' said Rita imperiously. Of Del she demanded, 'What are you coming for?'

'To see fair play,' he growled. 'The little I know of witchcraft makes me like none of it.'

'Very well,' she said calmly. Then she smiled her very own smile. 'Since you insist, I'd rather enjoy Barbara's company too.'

Barbara hesitated. 'Come, he won't hurt you, girl,' said Rita. 'He doesn't know you exist.'

'Oh,' said Barbara, wonderingly.

Del said gruffly, 'I do so. She has the vegetable stall.'

Rita smiled at Barbara, the secrets bright in her eyes. Barbara said nothing, but came with them.

'You should go back, you know,' Rita said silkily to Del, when she could. 'Haven't you been humiliated enough yet?'

He did not answer.

She said, 'Stubborn animal! Do you think I'd have come this far if I weren't sure?'

'Yes,' said Del, 'I think perhaps you would.'

They reached the blue moss. Rita shuffled it about with her feet and then sank gracefully down to it. Barbara stood alone in the shadows of the willow grove. Del thumped gently at an aspen with his fist. Rita, smiling, arranged the bridle to cast, and laid it across her lap.

The rabbits stayed hid. There was an uneasiness about the grove. Barbara sank to her knees, and put out her hand. A chipmunk ran to nestle in it.

This time there was a difference. This time it was not the slow silencing of living things that warned of his approach, but a sudden babble from the people on the ridge.

Rita gathered her legs under her like a sprinter, and held the bridle poised. Her eyes were round and bright, and the tip of her tongue showed between her white teeth. Barbara was a statue. Del put his back against his tree, and became as still as Barbara.

Then from the ridge came a single, simultaneous intake of breath, and silence. One knew without looking that some stared speechless, that some buried their faces or threw an arm over their eyes.

He came.

He came slowly this time, his golden hooves choosing his paces like so many embroidery needles. He held his splendid head high. He regarded the three on the bank gravely, and then turned to look at the ridge for a moment. At last he turned, and came round the pond by the willow grove. Just on the blue moss, he stopped to look down into the pond. It seemed that he drew one deep clear breath. He bent his head then, and drank, and lifted his head to shake away the shining drops.

He turned towards the three spellbound humans and looked at them each in turn. And it was not Rita he went to, at last, nor Barbara. He came to Del, and he drank of Del's eyes with his own just as he had partaken of the pool—deeply and at leisure. The beauty and wisdom were there, and the compassion, and what looked like a bright white point of anger. Del knew that the creature had read everything then, and that he knew all three of them in ways unknown to human beings.

There was a majestic sadness in the way he turned then, and dropped his shining head, and stepped daintily to Rita. She sighed, and rose up a little, lifting the bridle. The unicorn lowered his horn to receive it—

—and tossed his head, tore the bridle out of her grasp, sent the golden thing high in the air. It turned there in the sun, and fell into the pond.

And the instant it touched the water, the pond was a bog and the birds rose mourning from the trees. The unicorn looked up at them, and shook himself. Then he trotted to Barbara and knelt, and put his smooth, stainless head in her lap.

Barbara's hands stayed on the ground by her sides. Her gaze roved over the warm white beauty, up to the tip of the golden horn and back.

The scream was frightening. Rita's hands were up like claws, and she had bitten her tongue; there was blood on her mouth. She screamed again. She threw herself off the now withered moss toward the unicorn and Barbara. 'She can't be!' Rita shrieked. She collided with Del's broad right hand. 'It's wrong, I tell you, she, you, I. . . .'

'I'm satisfied,' said Del, low in his throat. 'Keep away, squire's daughter.'

She recoiled from him, made as if to try to circle him. He stepped forward. She ground her chin into one shoulder, then the other, in a gesture of sheer frustration, turned suddenly and ran towards the ridge. 'It's mine, it's mine,' she screamed. 'I tell you, it can't be hers, don't you understand? I never once, I never did, but she, but she—'

She slowed and stopped, then, and fell silent at the sound that rose from the ridge. It began like the first patter of rain on oak leaves, and it gathered voice until it was a rumble and then a roar. She stood looking up, her face working, the sound washing over her. She shrank from it.

It was laughter.

She turned once, a pleading just beginning to form on her face. Del regarded her stonily. She faced the ridge then, and squared her shoulders, and walked up the hill, to go into the laughter, to go through it, to have it follow her all the way home and all the days of her life.

Del turned to Barbara just as she bent over the beautiful head. She said, 'Silken-swift . . . go free.'

The unicorn raised its head and looked up at Del. Del's mouth opened. He took a clumsy step forward, stopped again. 'You!'

Barbara's face was wet. 'You weren't to know,' she choked. 'You weren't ever to know . . . I was so glad you were blind, because I thought you'd never know.'

He fell on his knees beside her. And when he did, the unicorn touched her face with his satin nose, and all the girl's pent-up beauty flooded outward. The unicorn rose from his kneeling, and whickered softly. Del looked at her, and only the unicorn was more beautiful. He put out his hand to the shining neck, and for a moment felt the incredible silk of the mane flowing across his fingers. The unicorn reared then, and wheeled, and in a great leap was across the bog, and in two more was on the crest of the further ridge. He paused there briefly, with the sun on him, and then was gone.

Barbara said, 'For us, he lost his pool, his beautiful pool.'

And Del said, 'He will get another. He must.' With difficulty he added, 'He, couldn't be . . . punished . . . for being so gloriously Fair.'

POUL ANDERSON

OPERATION AFREET

It was sheer bad luck, or maybe their Intelligence was better than we knew, but the last raid, breaking past our air defences, had spattered the Weather Corps tent from here to hell. Supply problems being what they were, we couldn't get replacements for weeks, and meanwhile the enemy had control of the weather. Our only surviving Corpsman, Major Jackson, had to save what was left of his elementals to protect us against thunderbolts; so otherwise we took whatever they chose to throw at us. At the moment, it was rain.

There's nothing so discouraging as a steady week of cold rain. The ground turns liquid, and runs up into your boots, and they get so heavy you can barely pick them up. Your uniform is a drenched rag around your shivering skin, the rations are soggy, the rifles have to have extra care, and always the rain drums down on your helmet till you hear it in dreams. You'll never forget that endless grey washing and beating; ten years later a rainstorm will still make you feel depressed.

The one consolation, I thought, was that they couldn't very well attack us from the air while it went on. Doubtless they'd yank the cloud cover away when they were ready to strafe us, but our broomsticks could scramble as fast as their carpets could arrive. Meanwhile, we slogged ahead, a whole division of us with auxiliaries—the 45th, the Lightning Busters, pride of the United States Army, turned into a wet misery of men and dragons hunting through the Oregon hills for the invader.

I made a slow way through the camp. Water ran off the pup tents and gurgled in the slit trenches. Our sentries were, of course, wearing Tarnkappen, but I could see their footprints form in the mud and hear the boots squelch and the steady stream of tired monotonous cursing.

I passed by the Air Force strip—they were bivouacked with us, to give support as needed. A couple of men were on guard outside the knock-down hangar, not bothering with invisibility. Their blue uniforms were as mucked and bedraggled as my OD's, but they had shaved and their insignia—the winged broomstick and the anti-Evil Eye beads—were polished. They saluted me, and I returned the gesture idly. *Esprit de corps*, wild blue yonder, nuts.

Beyond was the armour. The boys had erected portable shelters for their beasts, so I only saw steam rising out of the cracks and caught the rank reptile smell. Dragons hate rain, and their drivers were having a hell of a time controlling them.

Nearby lay Petrological Warfare, with a pen full of hooded basilisks writhing and hissing and striking out with their crowned heads at the men feeding them. Personally, I doubted the practicality of that whole corps. You have to get a basilisk quite close to a man, and looking straight at him, for petrifaction; and the aluminium-foil suit and helmet you must wear to deflect the influence of your pets is an invitation to snipers. Then, too, when human carbon is turned to silicon, you have a radioactive isotope, and maybe get such a dose of radiation yourself that the medics have to give you St John's Wort plucked from a graveyard in the dark of the moon.

So, in case you didn't know, cremation hasn't simply died out as a custom; it's become illegal under the National Defense Act. We have to have plenty of old-fashioned cemeteries. Thus does the age of science pare down our liberties.

I went on past the engineers, who were directing a gang of zombies carving another drainage ditch, and on to General Vanbrugh's big tent. When the guard saw my Tetragrammaton insignia, Intelligence Corps, and the silver bar on my shoulder, he saluted and let me in. I came to a halt before the desk and brought my hand up.

'Captain Matuchek reporting, sir,' I said.

Vanbrugh looked at me from beneath shaggy grey brows. He was a large man with a face like weathered rock, very much Regular Army, but we all liked him as well as you can like a buck general. 'At ease,' he said. 'Sit down. This'll take a while.'

I found a folding chair and lowered myself into it. There were two others already seated whom I didn't know. One was a plump man with a round red face and a fluffy white beard, a major with the crystal-ball emblem of the Signal Corps. The other was a young woman. In spite of my weariness, I blinked and looked twice at her. She was worth it—a tall green-eyed redhead with straight high-cheeked features and a figure too good for the WAC clothes or any other. Captain's bar, Cavalry spider . . . or Sleipnir, if you want to be official about it.

'Major Harrigan,' grumfed the general. 'Captain Graylock. Captain Matuchek. We may as well get down to business.'

He spread a map out before us. I leaned over and looked at it. Positions were indicated, our own and the enemy's. They still held the Pacific seaboard from Alaska halfway down through Oregon, though that was considerable improvement from a year ago, when the Battle of the Mississippi had turned the tide.

'Now then,' said Vanbrugh, 'I'll tell you the overall situation. This is a dangerous mission, you don't have to volunteer, but I want you to know how important it is.'

What I knew, just then, was that I'd been told to volunteer or else. That was the Army, at least in a major war like this, and in principle I couldn't object. I'd been a contented Hollywood actor when the Saracen Caliphate attacked us. I wanted to go back to more of the same, but that meant finishing the war.

'You can see we're driving them back,' said the general, 'and the occupied countries are all set to revolt as soon as they get a fighting chance. The British have been organizing the underground and arming them while readying for a cross-Channel jump. But we have to give the enemy a decisive blow, break this whole front and roll 'em up. That'll be the signal. If we succeed, the war will be over this year. Otherwise it might drag on for another three.'

I knew it. The whole Army knew it. Official word hadn't been passed yet, but somehow you feel when a big push is impending. There is a tension in the air.

His stumpy finger traced along the map. 'The 9th Armoured Division is here, the 12th Broomborne here, the 14th Cavalry here, the Salamanders here where we know they've concentrated their firebreathers. The Marines are all set to establish a beachhead and retake Seattle, now that the Navy's bred enough Krakens. One good goose, and we'll have 'em running.'

Major Harrigan snuffled into his beard and stared gloomily at a crystal ball. It was clouded and vague; the enemy had been jamming all our crystals till they were no use whatsoever, though naturally we'd retaliated. Captain Graylock tapped impatiently on the desk with a perfectly manicured nail. She was so clean and crisp and efficient, I decided I didn't like her looks after all. Not while I had half an inch of beard bristling from my chin.

'But apparently there's something gone wrong, sir,' I ventured.

'There is, damn it,' said Vanbrugh. 'In Trollburg.'

I nodded. The Saracens held that town. It was a key position, sitting as it did on US Highway 20 and guarding the approach to Salem and Portland.

'I take it we're supposed to seize Trollburg, sir,' I murmured.

Vanbrugh scowled. 'That's the job for the 45th,' he grunted. 'If we muff it, they can sally out against the 9th, cut it off, and throw the whole operation akilter. But now Major Harrigan and Captain Graylock come from the 14th to tell me the Trollburg garrison has an afreet.'

I whistled, and a chill crawled along my spine. The Caliphate had exploited the Powers recklessly—that was one reason why the rest of the Moslem world regarded them as heretics and hated them as much as we did—but I never thought they'd go so far as breaking Solomon's Seal. An afreet getting out of hand could wreck a whole nation.

'I hope they haven't more than one,' I whispered.

'No, they don't,' said the Graylock woman. Her voice was low and could have been pleasant if it weren't so brisk. 'They've been dredging the Red Sea in hopes of finding another Solly bottle, but this seems to be the only one left.'

'Bad enough, though,' I said. The effort to keep my tone steady helped calm me down all around. 'How'd you find out?'

'We're with the 14th,' said Graylock unnecessarily. Her Cavalry badge had surprised me, though; normally, the only recruits the Army can dig up to ride unicorns are pickle-faced, dried-up schoolteachers and the like.

'I'm only a liaison officer,' said Major Harrigan quickly. 'I go by broomstick myself.' I grinned at that. No American male, unless he's in holy orders, likes to admit he's qualified to control a unicorn. He saw me and flushed angrily.

Graylock went on, as if dictating. She kept her voice flat, though little else. 'We had the luck to capture a bimbashi in a commando attack. I questioned him.'

'They're pretty close-mouthed, those noble sons of . . . um . . . the desert,' I said. I'd bent the Geneva Convention myself, occasionally, but didn't relish the idea of breaking it completely—even if the enemy had no such scruples.

'Oh, we practised no brutality,' said Graylock. 'We housed him and fed him very well. But the moment a bite of food was in his throat, I'd turn it into pork. He gave up pretty fast, and spilled all he knew.'

I had to laugh aloud, and even Vanbrugh chuckled; but she sat there perfectly dead-pan. Organic-organic transformation, which merely shuffles molecules around without transforming atoms, has no radiation hazards but naturally requires a good knowledge of chemistry. That's the real reason the average dogface hates the technical corps: pure envy of a man who can turn K rations into steak and French fries. The quartermasters have enough trouble conjuring up the rations themselves, without branching into fancy dishes.

'So you found out they have an afreet in Trollburg,' said the general. 'What about their strength?'

'A small division, sir. You can take the place readily enough, if that demon can only be immobilized,' said Harrigan.

'Yes. So I see.' Vanbrugh swivelled his eyes around to me. 'Well, captain, are you game? If you can carry it off, it'll mean a Silver Star at least—pardon me, a bronze one.'

'I . . .' I paused, fumbling after words. I was more interested in promotion and ultimate discharge, but that might follow too. However, quite apart from my own neck, there was a practical objection. 'Sir, I don't know a damn thing about the job. I nearly flunked Demonology I in college.'

'That'll be my part of it,' said Graylock.

'You!' I picked my jaw off the floor again, but couldn't find anything else to say.

'I was head witch of the Arcane Agency in New York before the war,' she said coldly. Now I knew where she got that personality: the typical big-city career girl. I can't stand them. 'I know as much about handling demons as anyone on this coast. Your task will be to escort me safely to the place and back.'

'Yeah,' I said weakly. 'Yeah, that's all.'

Vanbrugh cleared his throat. He didn't like sending a woman on such a mission, but with time as short as it was, he had no choice. 'Captain Matuchek is one of the best werewolves in the business,' he complimented me. He only compliments people he doesn't expect to see again. *Ave, Caesar, morituri te salutant. No, that isn't what I mean, but what the hell? I'll be dead long enough to figure out a better phrasing.*

'I think two adepts can get past their guards,' he went on. 'It's up to you. We attack at noon tomorrow, and that afreet had better be out of action by then. Now, here's a geodetic survey map of the town and the approaches . . .'

He didn't waste time asking me if I had really volunteered.

I guided Captain Graylock back to the tent I shared with two brother officers. Darkness was creeping through the long cold slant of rain; it would be night soon. We plodded through the muck in silence until we were under the canvas. My tent-mates were out on picket duty, so we had the place to ourselves. I lit the Glory Hand and sat down on the sodden floor.

'Have a chair,' I said, pointing to our one camp stool. It was an animated job we'd bought in San Francisco—not very bright, but it would carry our duffel and come when called. It shifted uneasily at the unfamiliar weight, then went back to sleep.

Graylock took out a pack of Wings and raised her brows. I nodded my thanks, and the cigarette flapped over to my mouth. Personally, I smoke Luckies in the field: self-striking tobacco is convenient when your matches may be wet. When I was a civilian and could afford it, my brand was Philip Morris, because the little red-coated smoke sprite can mix you a drink while you wait.

We puffed for a bit in silence, listening to the rain. 'Well,' I said at last, 'I suppose you have transportation.'

'My personal broomstick,' she said. 'I don't like this GI Willys. Give me a Cadillac anytime. I've souped it up, too.'

'And you have your grimoires and powders and whatnot?'

'Just some chalk. No material agency is much use against a powerful demon.'

'Yeh? What about the sealing wax on the Solly bottle?'

'It isn't the wax that holds an afreet in, but the seal. The spells are all symbolic; in fact, it's believed their effect is purely psychosomatic.' She hollowed the flat planes of her cheeks, sucking in smoke, and I saw what a good bone structure she had. 'We may have a chance to test that theory tonight.'

'Well, then, you'll want a light pistol loaded with silver slugs—they have weres of their own, you know. I'll take a carbine and a .45 and a few grenades.'

'How about a squirter?'

I frowned. The notion of using holy water as a weapon has always struck me as blasphemous, though the chaplain said it was all right against Low World critters. 'No good to us,' I said. 'The Moslems don't have that ritual, so of course they don't use any beings that can be controlled by it. Let's see, I'll want my polaroid flash too.'

Ike Abrams stuck his big nose in the tent flap. 'Would you and the lady captain like some eats, sir?' he asked.

'Why, sure,' I said. 'Hate to spend my last night on Midgard standing in a chow line.' When he had gone, I explained to the girl: 'Ike's only a private now, but we were good friends in Hollywood—he scripted *Call of the Wild* and *Silver Chief* for me—and he's kind of appointed himself my orderly. He'll bring us some food here.'

'You know,' she remarked, 'that's one good thing about the technological age. Did you know there used to be anti-Semitism in this country?'

'No. Was there?'

'Quite a bit. Especially a completely false belief that Jews were cowards and never found in the front lines. Now, when they live in an era where their religion forbids them to cast spells, they're all dogfaces and commandos, and everybody knows it.'

I myself had gotten tired of comic-strip supermen and pulp-magazine heroes having such monotonously Yiddish names—don't Anglo-Saxons belong in our culture too?—but it was a good point she made. And it showed she was a trifle more than just a money machine. A bare trifle.

'What'd you do in civilian life?' I asked, chiefly to drown out the incessant noise of the rain.

'I told you,' she snapped, once again the ice factory. 'I was with the Arcane Agency. Advertising, public relations, and so on.'

'Oh, well,' I said. 'Hollywood is as phoney, so I shouldn't sneer.'

I couldn't help it, though. Those Madison Avenue characters give me a pain in the rear end. Using the good Art to puff up some self-important nobody, or to sell a product whose main virtue is its complete similarity to other brands of the same. The SPCA has cracked down on training nixies to make fountains spell out words, or cramming young salamanders into glass tubes to light up Broadway, but I can still think of better uses for slick paper than trumpeting Ma Chère perfume. Which is actually a love potion anyway, though you know what postal regulations are.

'You don't understand,' she said. 'It's part of our economy—part of our whole society. Do you think the average backyard warlock is capable of repairing . . . oh, say a lawn sprinkler? Hell, no! He'd probably let loose the water elementals and flood half a township if it weren't for the inhibitory spells. And we, Arcane, undertook the campaign to convince the Hydros they had to respect our symbols. I told you it's psychosomatic when you're dealing with these really potent beings. For that job, I had to go down in an aqualung!'

I stared at her with more respect. Ever since it was found out how to degauss the ruinous effects of cold iron, and the age of science began, the world has needed some pretty bold people. Apparently she was one of them.

Just then Abrams brought in two plates of rations. He looked wistful, and I would have invited him to join us except that our mission was secret.

Captain Graylock 'chanted the coffee into martinis (not quite dry enough) and the dogfood into steaks (a turn too well done); but you can't expect the finer sensibilities in a woman, and it was the best chow I'd had in a month. She relaxed a bit over the brandy, and I learned that her repellent crispness was only defensive armour against the slick types she dealt with, and we found out that our first names were Stephen and Virginia. But then it was quite dark outside, and time to get going.

You may think it was sheer lunacy to send only two people, and one of them a woman, into an enemy division on a task like this. It would seem to call for a Ranger brigade, at least. But present-day science has transformed war as well as industry, medicine, and ordinary life. Our mission was forlorn in any event, and we wouldn't have gained enough by numbers to make reinforcements worthwhile. We simply had to do it ourselves.

You see, it takes so much study and practice to become adept that only a very small minority qualify. If you're a born were—and only about five per cent of the population is—you can transform, you know instinctively; and if you don't have those chromosomes, you can't by any means. Also, everyone can learn a few simple spells, enough to operate a broomstick or a vacuum cleaner or a turret lathe, but many years of work are required for ability to do more. Just imagine somebody trying alchemy without a thorough knowledge of nuclear physics. He'd either get a radioactive isotope that would kill him, or blow up half a county.

My scientific friends tell me that the Art involves regarding the universe as a set of Cantorian infinities. Within any given class, the part is equal to the whole and so on. One good witch could do all the running we were likely to need; a larger party would simply be more liable to detection, and would risk valuable personnel. So Vanbrugh had very rightly sent us two alone.

The trouble with sound military principles is that sometimes you personally get caught in them.

Virginia and I turned our backs on each other while we changed clothes. She got into an outfit of slacks and combat jacket, I into the rubberized garment which would fit me as well in wolf-shape. We put on our helmets, hung our equipment around us, and turned about. Even in the baggy green battle clothes she looked good.

'Well,' I said tonelessly, 'shall we go?'

I wasn't afraid, of course. Every recruit is immunized against panic when they put the geas on him. But I didn't like the prospect.

'The sooner the better, I suppose,' she answered. Stepping to the entrance, she whistled.

Her stick swooped down and landed just outside. It had been stripped of all the fancy chrome, but was still a neat job—the foamrubber seats had good shock absorbers and well-designed back rests, unlike Army transport. Her familiar was a gigantic tomcat, black as a furry midnight, with two malevolent yellow eyes. He arched his back and spat indignantly. The weatherproofing spell kept the rain off, of course, but he didn't like the damp in the air.

Virginia chucked him under the chin. 'Oh, so Svartalf,' she murmured. 'Good cat, rare sprite, prince of darkness, if we outlive this night you shall sleep on cloudy cushions and lap cream from a golden bowl.' He cocked his ears and raced his motor.

I climbed into the rear seat, put my feet in the stirrups, and leaned back. The girl mounted in front of me and crooned to the stick. It swished upward, the ground fell away and camp was hidden in gloom. Both of us had been given witch-sight—infra-red vision, actually—so we didn't need lights.

When we got above the clouds, there was a giant vault of stars overhead and a swirling dim whiteness below. I saw a couple of P-56s circling on patrol, fast jobs with six brooms to lift their weight of armour and machine-guns. We left them behind and streaked northward. I rested the carbine on my lap and sat listening to the split air whining past. Underneath us, in the harsh-edged murk of the hills, I spied occasional flashes; an artillery duel was going on. So far no one had been able to cast a spell fast enough to turn or implode a shell. I'd heard rumours that General Electric was developing a gadget which could recite the formula in a few microseconds, but meanwhile the big guns went on talking.

Trollburg was only a few miles from our position. I saw it as a vague sprawling mass, blacked out against our cannon and bombers. It would have been nice to have an atomic weapon just then, but as long as the Tibetans keep those anti-nuclear-warfare prayer wheels turning such thoughts must remain merely wistful. I felt my belly muscles tighten. The cat bottled out his tail and swore. Virginia sent the broomstick slanting down.

We landed in a clump of trees and she turned to me. 'Their outposts must be somewhere near,' she whispered. 'I didn't dare try landing on a rooftop, we could have been seen too easily. We'll have to go in from here.'

I nodded. 'All right. Wait till I come back.'

I turned the flash on myself. It was hard to believe that transforming had depended on a bright full moon till only ten years ago. Then Weiner showed that the process was simply one of polarized light of the right wavelengths, triggering the pineal gland, and the Polaroid Corporation made another million dollars or so from its Were-Wish Lens. It's not easy to keep up with this fearful and wonderful age we live in, but I wouldn't trade.

The usual rippling, twisting sensations, the brief drunken dizziness and half-ecstatic pain, went through me. Atoms reshuffled into whole new molecules, nerves grew some endings and lost others, bone was briefly fluid and muscles like stretched rubber. Then it was past, and I shook myself, stuck my tail out the flap of the skintight pants, and muzzled Virginia's hand.

She stroked my neck, behind the helmet. 'Good boy,' she whispered. 'Go get 'em.'

I turned and faded into the brush.

A lot of writers have tried to describe how it feels to be were, and all of them have failed because human language doesn't have the words. My vision was not so good now, the stars were blurred above me and I couldn't see to right or left and the world took on a colourless flatness. But I heard with a clarity that made the night almost a roar, way up into the supersonic; and a universe of smells roiled in my nostrils, wet grass and teeming dirt, the hot sweet little odour of a scampering fieldmouse, the clean tang of oil and guns, a faint harshness of smoke—poor stupefied humanity, half dead to the world's earthy glories!

The psychological part is the hardest to convey. I was a wolf now, with a wolf's nerves and glands and instincts, a wolf's sharp but limited intelligence. I had a man's memories and a man's purposes, but they were unreal, dreamlike, it took an effort of trained will to hold to them and not go hallooing off after the nearest jackrabbit. No wonder weres had a bad name in the old days, before they themselves understood the mental changes involved and got the right habits drilled into them from badyhood.

I weigh a hundred and eighty pounds, and the conservation of mass holds good like all laws of nature, so I was a pretty big wolf. But it was easy to flow through the bushes and meadows and gullies, another drifting shadow. I was quite close to the town when I caught the near smell of man.

I flattened, the grey fur bristling along my spine, and waited. The sentry came by. He was a tall bearded fellow with little gold earrings that glimmered wanly under the stars. The turban wrapped around his helmet bulked monstrous against the Milky Way.

I let him go and followed his path until I saw the next one. They were placed around Trollburg, each pacing a hundred-yard arc and meeting his opposite number at either end of it. No simple task to—

Something murmured in my ears and I crouched. One of their aircraft ghosted overhead; I could see two men and a couple of machine-guns squatting on top of the carpet. It circled low and lazily, above the ring of sentries. Trollburg was well guarded.

Somehow, Virginia and I had to get through that picket. I wished that the transformation had left me with full human reasoning powers. My wolf-impulse was simply to jump on the nearest man, but that would bring the whole garrison down on my hairy ears.

Wait—maybe that was what was needed!

I loped back to the thicket. The Svartalf cat scratched at me and zoomed up a tree. Virginia Graylock started, her pistol sprang into her hand, then she relaxed and laughed a bit nervously. I could work the flash hung about my neck, even as I was, but it went more quickly with her fingers.

'Well?' she asked when I was human again. 'What'd you find out?'

I described the situation, and saw her frown and bite her lip. It was really too shapely a lip for such purposes. 'Not so good,' she reflected. 'I was afraid of something like this.'

'Look,' I said, 'can you locate that afreet in a hurry?'

'Oh, yes. I've studied at Congo U. and did quite well at witch-smelling. What of it?'

'If I attack one of those guards, and make enough noise doing it, their attention will all be turned that way. You should have an even chance to fly across the line unobserved, and once you're in the town your Tarnkappe—'

She shook her red head. 'I didn't bring one. Their sniffers are just as good as ours. Invisibility is actually obsolete.'

'Mmm— . . . yeh, I suppose you're right. Well, anyhow, you can take advantage of the darkness to get to the afreet house. From there on, you'll have to play by ear.'

'I suspected we'd have to do something like this,' she replied. Then with a softness that astonished me: 'But Steve, it's a long chance for you to take.'

'Not unless they hit me with silver, and most of their cartridges are plain lead. They use a tracer principle like us; every tenth round is argent. I've got at least a ninety per cent probability of getting away with it.'

'You're a liar,' she said. 'But a brave liar.'

I wasn't brave at all. It's inspiring enough to think of Valley Forge, or the Alamo, or Bataan, or Casablanca where our outnumbered army stopped three Panzer divisions of von Ogerhaus's Afrika Korps—but only when you're safe and comfortable yourself. Down underneath the antipanic geas, there was a cold knot in my guts. But I couldn't see any other way to do the job, and failure to attempt it would mean court-martial.

'I'll run their legs off once they start chasing me,' I told her. 'When I've shaken 'em, I'll try to circle back and join you.'

'All right.' Suddenly she stood up on tiptoe and kissed me. The impact was explosive.

I stood for a moment, looking at her. 'What are you doing Saturday night?' I asked, a little shakily.

She laughed. 'Don't get ideas, Steve. I'm in the Cavalry.'

'Yeh, but the war won't last forever.' I grinned at her, a reckless fighting grin that made her eyes linger. It's helpful having acting experience.

We settled the details as well as we could. She herself had no soft touch: the afreet would be well guarded, and was dangerous enough in itself. The chances of our both seeing daylight were not good.

I turned back to wolf-shape and licked her hand. She rumpled my fur, and I slipped off into the darkness.

I had chosen a sentry well off the highway, across which there would surely be barriers. A man could be seen to either side of my victim, tramping slowly up and down. I glided behind a stump near the middle of his beat and waited for him.

When he came, I sprang. I had a dark brief vision of eyes and teeth white in the bearded face, I heard him yelp and smelled the upward spurt of his fear, then we shocked together. He went down on his back, threshing, and I snapped for the throat. My jaws closed on his arm, and blood was hot and salt on my tongue.

He screamed again, and I sensed the call going down the line. The nearest Saracens ran to help. I tore out the gullet of the first man and bunched myself for a leap at the next.

He fired. The bullet went through me in a jag of pain and the impact sent me staggering. But he didn't know how to deal with a were. He should have dropped on one knee and fired steadily till he got to the silver bullet; if necessary, he should have fended me off, even pinned me with his bayonet, while he shot. This one kept running towards me, calling on the Allah of his heretical sect.

My tissues knitted as I plunged to meet him. I got past the bayonet and gun muzzle, hitting him hard enough to knock the weapon loose but not to bowl him over. He braced his legs, grabbed my neck, and hung on.

I swung my left hind leg back of his ankle and shoved. He fell with me on top, the position an infighting werewolf always tries for. My head swivelled, I gashed open his arm and broke his grip.

Before I could settle the business, three others had piled on me. Their trench scimitars went up and down, in between my ribs and out again. Lousy training they'd had. I snapped my way free of the heap—there were half a dozen by then—and broke loose.

Through the sweat and blood I caught the faintest whiff of Chanel No 5, and something in me laughed. Virginia had sped past the confusion, riding her stick a foot above ground, and was inside Trollburg. Now it was up to me to lead a chase and not stop a silver slug while doing it.

I howled to taunt the men spilling from the outlying houses, let them have a good look at me, and then made off across the rolling fields. I was taking it easy, not to outpace them at once, relying on zigzags to keep me unpunctured. They took the bait and followed, stumbling and shouting.

As far as they knew, this had been just a commando raid. Their pickets would be re-formed by now and the garrison alerted, but they had no way of telling what we had really planned. Maybe we'd pull the operation off after all—

Something swooped overhead, one of their damned carpets. It rushed down on me like a hawk, the guns spitting. I made for the nearest patch of woods.

Into the trees! Given half a break, I could—

They didn't give it. There was a bounding behind me. I caught the acrid smell and whimpered. A weretiger could go as fast as I.

For a moment I remembered an old guide I'd had, up in Alaska, and wished to hell he were here. He was a were-Kodiak bear. Then I whirled and met the tiger before he could pounce.

He was a big one, five hundred pounds at least. His eyes smouldered above the great fangs, and he lifted a paw that could crack my spine like a dry twig. I rushed in, snapping, and danced back before he could strike.

I could hear the enemy, blundering around in the underbrush trying to find us. The tiger leaped. I evaded him and bolted for the nearest thicket. Maybe I could go where he couldn't. He romped through the woods behind me, roaring.

I saw a narrow space between two giant oaks, too small for him, and hurried that way. But it was too small for me too. In the instant that I was stuck, he caught up and the lights went out.

Consciousness returned a few minutes later. It had been a cuff on the neck. My head hammered, and I retched.

'Get up.' Someone stuck a boot in my ribs.

I lurched to my feet. They'd used my flash to make me human again, then removed it with the rest of my equipment. There were a score of them holding their guns on me. Tiger Boy was standing nearby; in man-shape he was almost seven feet tall and monstrously fat. Squinting through the headache, I saw he wore the insignia of an emir—which was a military rank these days rather than a title, but important.

'Come,' he said. He led the way, and I was hustled along behind.

I saw their carpets in the sky and heard the howling of their own weres looking for spoor of other Americans. I was still too groggy to care very much.

We entered the town, its pavement sounding hollow under the boots, and went towards the centre. Trollburg wasn't big, maybe 5,000 population. Most of the streets were empty. I saw a few Caliphate soldiers, anti-aircraft guns poking into the sky, a dragon lumbering past with flames flickering around its jaws and cannon projecting from the armoured howdah. No trace of the civilians, but I knew well enough what had happened to them. The attractive young women were in the officers' harems, the rest dead or locked away pending shipment to the slave markets.

By the time we got to the hotel which was now enemy headquarters, my aches had subsided and my brain was clear. That was a very mixed blessing under the circumstances. I was taken upstairs to a suite and told to stand before a table. The emir sat down behind it, half a dozen guards lined the walls, and a young pasha of Intelligence seated himself nearby.

The emir's big face turned to that one, and he spoke a few words—I suppose to the effect of 'I'll handle this, you take notes.' Then he looked at me. His eyes were the pale tiger-green.

'Now then,' he said in good English, 'we shall have some questions. Identify yourself, please.'

I told him mechanically that I was called Sherrinford Mycroft, captain AUS, and gave him my serial number.

'That is not your real name, is it?' he asked.

'Of course not!' I replied. 'I know the Geneva Convention, and you're not going to cast name-spells on me. Sherrinford Mycroft is my official johnsmith.'

'The Caliphate has not subscribed to the Geneva Convention,' said the emir quietly, 'and stringent measures are sometimes necessary in a jihad. What was the purpose of this raid?'

'I am not required to answer that,' I said. Silence would have served the same end, delay to gain time for Virginia, but not as well.

'You may be persuaded to do so,' he said.

If this had been a movie, I'd have told him I was picking daisies, then kept on wisecracking while they brought out the thumbscrews. In practice it would have fallen a little flat.

'All right,' I said. 'I was scouting.'

'Only one of you?'

'A few others. I hope they got away.' That might keep his boys busy hunting for a while.

'You lie,' he said dispassionately.

'I can't help it if you don't believe me,' I shrugged.

His eyes narrowed. 'I shall soon know if you speak truth,' he said. 'If not, then may Eblis have mercy on you.'

I couldn't help it, I jerked where I stood and sweat pearled out on my skin. The emir laughed. He had an unpleasant laugh, a sort of whining growl deep in his fat throat, like a tiger playing with its kill.

'Think it over,' he advised. Then he turned to some papers on the table.

It was very quiet in the room. The guards stood as if cast in bronze. The young shavetail dozed beneath his turban. Behind the emir's back, a window looked out on a blankness of night. The only sound was the loud ticking of a clock and the rustle of the papers. It only seemed to deepen the silence.

I was tired, my head ached, my mouth tasted foul and thirsty. The sheer physical weariness of having to stand was meant to help wear me down. It occurred to me that the emir must be getting scared of us, to take all this trouble himself with a lone prisoner. That was kudos for the triumphant American cause, but small consolation to me.

My eyes flickered, studying the tableau. There wasn't much to see, the usual dreary hotel furnishings. The emir had cluttered his desk with a number of objects: a crystal ball useless because of our own jamming, a fine cut-glass bowl looted from somebody's house, a set of nice crystal wineglasses, a cigar humidor of quartz glass, a decanter full of what looked like good scotch. I guess he just liked crystal.

He helped himself to a cigar—waving his hand to make the humidor spring open and a Havana fly into his mouth and light itself. As the minutes crawled by, an ashtray soared up from time to time to receive from him. I supposed that everything he had was 'chanted so it would rise and move easily; a man that fat, paying the price of being a really big werebeast, needed such conveniences.

It was very quiet. The light glared down on all of us. It was somehow hideously wrong to see a good ordinary GE St Elmo shining on those turbaned alien heads.

I began to get the forlorn glimmerings of an idea. How to put it into effect I didn't yet know, but just to pass the time I began to compose some spells.

It must have been after half an hour, though it seemed more like half a century, when the door opened and a fennec, the small fox of the African desert, trotted in. The emir looked up as it went into a closet, to find darkness in which to use its flash. The fellow who came out was, naturally, a dwarf barely one foot high. He prostrated himself and spoke rapidly in a thin thready voice.

'So.' The emir's chins turned slowly around to me. 'It is reported that no trace was found of other tracks than yours. You have lied.'

'Didn't I tell you?' I asked. My throat felt stiff and strange. 'We used owls and bats. I was the only wolf.'

'Be still,' he said tonelessly. 'I know very well that the only werebats are vampires, and that all vampires are—what you say—4F in all armies.'

That was true enough, of course. Every so often, some armchair general asks why we don't raise a force of Draculas. The answer is routine: they're too light and flimsy; they can't endure sunshine; if they don't get a steady blood ration they're apt to turn on their comrades; and you can't possibly use them with or against Italian troops. I swore at myself, but my mind had been too numb to think straight.

'I believe you are concealing something,' went on the emir. He gestured at his glasses and decanter, which supplied him with a shot of scotch, and sipped judiciously. The Caliphate sect is also heretical with respect to strong drink—they maintain that the Prophet forbade only wine, but said nothing about beer, gin, whisky, brandy, rum, or akvavit.

'We shall have to use stronger measures,' he said at last. 'I was hoping to avoid them.' He nodded at his guards.

Two of them held my arms, while the pasha began working me over. He was good at it. The werefennec watched avidly, the emir puffed his cigar and went on with his paperwork. After a very long few minutes, he gave an order. They let me go, and even set forth a chair for me, which I needed badly.

I sat breathing hard. The emir looked up with a certain gentleness. 'I regret this,' he said. 'It is not enjoyable.' Oddly enough, I believed him. 'Let us hope you will be reasonable before we have to inflict permanent injuries. Meanwhile, would you like a cigar?'

It was the old third degree procedure. Knock a man around for a while, then show him kindness. You'd be surprised how often that makes him blubber and break.

'We are in need of information about your troops and their plans,' said the emir. 'If you will co-operate and accept the true faith, you can have an honoured position with us. We like to have good men in the Caliphate.' He smiled. 'After the war, you could select your harem out of Hollywood if you desired.'

'And if I don't squeal—' I murmured.

He spread his hands. 'You will have no further wish for a harem. The choice is yours.'

'Let me think,' I begged. 'It isn't easy.'

'Please do,' he answered urbanely, and returned to his papers.

I sat as relaxed as possible, drawing the smoke into my throat and letting strength flow back. The Army geas could be broken by their technicians only if I gave my free consent, and I didn't want to. I considered the window behind the emir. It was a two-storey drop to the street.

Most likely, I'd just get myself killed. But even that was preferable to any other offer I'd had.

I went over the spells I'd rigged up. A real technician has to know at least one arcane language—Latin, Greek, classical Arabic, Sanskrit, Old Norse, or the like—for the usual reasons of sympathetic science. Extraordinary phenomena are not strongly influenced by ordinary speech. But except for the usual tag-ends of incantations, just enough to operate the gadgets of daily life, I was no scholar.

However, I knew one slightly esoteric dialect quite well. I didn't know if it would work, but I'd try.

My muscles tautened as I moved. It was a shuddersome effort to be casual. I knocked the end of ash off my cigar. As I lifted the thing again, it picked up some ash from the emir's.

I got the rhyme straight in my mind, put the cigar to my lips, and subvocalized the spell:

Ashes-way of the urningbay,
upward-way ownay eturningray,
as-way the arksspay do yflay,
ikestray imhay in the eye-way!

I closed my right eye and brought the glowing cigar end almost against the lid.

The emir's El Fumo leaped up and ground itself into *his* right eye.

Even as he screamed and fell backward, I was on my feet. I'd marked the werefennec, and one stride brought me over to him. I broke his vile little neck with a backhanded cuff and yanked off the flash that hung from it.

The guards howled and plunged for me. I went over the table and down on top of the emir, snatching up his decanter en route. He clawed at me, wild with pain, I saw the ghastly hollowness of his eyesocket, and all the time I was hanging on to the vessel and shouting:

Ingthay of ystalcray,
ebay a istralmay!
As-way I-way owthray,
yflay ouyay osay!

As I finished, I broke free of the emir and hurled the decanter at the guards. It was lousy poetics, and might not have worked if the fat man hadn't already sensitized all his stuff. As it was, the ball, the ashtray, the bowl, the glasses, the humidor, and the windowpanes all took off after the decanter. The air was full of flying glass.

I didn't stay to watch the results, but went out that window like an exorcized devil. I landed in a ball on the sidewalk, bounced up, and began running.

Soldiers were around, and bullets started sleeting after me. I set a record reaching the nearest alley. My witch-sight showed me a broken window, and I wriggled through that. Crouching beneath the sill, I heard the pursuit go by.

It was the back room of a looted grocery store, plenty dark enough for my purposes. I hung the flash around my neck, turned it on myself, and made the changeover. They'd return in a minute, and I didn't want to be vulnerable.

Wolf, I snuffled around after another exit. There was a rear door standing half open, and I slipped into a courtyard full of ancient packing cases. They made a good hideout. I lay there, striving to control my lupine nature which wanted to pant, while they swarmed through the area.

When they were gone again, I tried to consider my situation. The temptation was to hightail it out of this poor, damned place; I could probably make it, and had technically fulfilled my share of the mission. But the job wasn't really complete, and Virginia was all alone with the afreet—if she still lived—and . . .

When I tried to call up her image, it came as a she-wolf and a furry odour. I shook my head angrily. Weariness and desperation were submerging my reason and letting the animal instincts take over. I'd better do whatever had to be done fast.

I cast about. The town smells were confusing, but I picked up the faintest sulphurous whiff and trotted cautiously towards it. I kept to the shadows, and was seen once but not challenged—they must have supposed I was one of theirs. The brimstone reek grew stronger.

They kept the afreet in the court-house, a good solid building. I went through the little park in front of it, snuffed the wind carefully, and then dashed over the street and up the steps. Four enemy soldiers sprawled there, throats cut open, and the broomstick was parked by the door. It had a twelve-inch switchblade in the handle, and Virginia had used it like a flying lance.

The man side of me, which had been entertaining stray romantic thoughts, backed up in a cold sweat; but the wolf grinned. I poked at the door. She'd 'chanted the lock open and left it that way. I stuck my nose in, and almost had it clawed off before Svartalf recognized me. He jerked his tail curtly, and I went through and across the lobby. The stinging smell was coming from upstairs, and I followed it through a thick darkness.

There was light in one second-floor room. I thrust the door ajar and peered in. Virginia was there. She had drawn the curtains and lit the Elmos to see by. She was still busy with her precautions, started a little on spying me but went on with the chant. I parked my shaggy behind near the door and watched.

She'd chalked the usual figure, same as the Pentagon in Washington, and a Star of David inside that. The Solly bottle was at the centre. It didn't look like much, an old flask of hard-baked clay with its hollow handle bent over and returning inside—just a Klein bottle, with Solomon's Seal in red wax at the mouth. She'd loosened her hair, and it floated in a ruddy cloud about the pale beautiful face.

The wolf of me wondered why we didn't just make off with the crock—it was safely stoppered. The man reminded him that undoubtedly the emir had taken precautions and would have sympathetic means to uncork it from afar. We had to put the demon out of action—somehow . . . but nobody on our side knew a great deal about his race.

Virginia finished her spell, drew the bung, and sprang outside the pentacle as smoke boiled from the flask. She almost didn't make it, the afreet came out so fast. I stuck my tail between my legs and snarled. She was scared too, trying hard not to show it, but I caught the adrenalin odour.

The afreet had to bend almost double under the ceiling. He was a monstrous grey thing, more or less anthropoid, but with horns and long ears, a mouthful of fangs and eyes like hot embers. Turned loose, he could scatter any army on earth. Controlling him before he laid the country waste would be another problem.

Smoke swirled from his mouth as he roared something in Arabic. Virginia looked very small and helpless under his looming wings. Her voice wasn't quite as cool as it should have been: 'Speak English, Marid. Or are you too ignorant?'

The demon huffed indignantly. 'O spawn of a thousand baboons!' he answered. It was like conversing with a thunderstorm. 'O thou white and gutless infidel thing, which I could break with my littlest finger, come in to me if thou darest!'

I was worried, not only by the chance of his breaking loose but by the racket he was making. It could be heard for a quarter mile.

'Be still, accursed of God!' said Virginia. It shut him up for a moment—like all his hell-breed, he flinched at holy names, though they couldn't stop him for long. She stood back a little, hands on hips, head thrown back to meet the gaze that smouldered above her. 'Suleiman bin-Daoud, on whom be peace, didn't jug you for nothing, I see. Back to your prison and never come forth again, lest the anger of Heaven smite you!'

The afreet sneered. 'Know that Suleiman the Wise is dead these three thousand years,' he retorted. 'Long and long have I brooded in my narrow cell, I who once raged free through earth and sky and Jehannum below, and now am I released to work my vengeance on the puny sons of Adam.' He shoved against the invisible barrier, but one of that type has a rated strength of several million p.s.i. and it held firm enough for the moment. 'O thou shameless unveiled harlot with hair of hell, know that I am Rashid the mighty, the glorious in power, the smiter of dragons! Come in here and fight like a man!'

I moved close to the girl, my hackles lifted. The hand that touched my head was cold and wet. 'Paranoid type,' she whispered. 'All these harmful Low Worlders are psycho. It's our only chance. I don't know any spells to compel him directly, but—'Aloud she answered: 'Shut up, Rashid, and listen to me. I too am of your race, and to be respected as such.'

'Thou?' He howled with laughter. 'Thou of the Marid race? Why, thou fish-faced antling, if thou'dst come in here I'd show thee thou'rt not even fit to—'. The rest was pretty obscene.

'No, hear me,' said the girl. 'Look well.' She made the cross sign which forbids lying. 'The name is the being, and my name is Ginny.'

It seemed reckless to give her *nomen* to him, but he started in surprise. 'Art indeed?' he asked.

'Yes. Now will you listen to me? I came to give you good advice, as one jinni to another. I have powers too, you know, though I employ them in the service of Allah, the Omnipotent, the All-Knowing, the Compassionate.'

He glared malevolently, but supposing her to be one of his species he was ready to put on a crude show of courtesy—until he could lure her inside, or break the barrier. 'Well, then,' he rumbled. 'why camest thou to disturb my rest? Tomorrow I go forth to destroy the infidel host.' He got caught up again in his dreams of glory. 'Aye, well will I rip them, and trample them, and break and gut and flay them. Well will they learn the power of Rashid the bright-winged, the fiery, the merciless, the wise, the . . .'

Virginia waited out his adjectives, then said gently: 'But Rashid, why must you wreak harm? All you earn by it is hate.'

A whine crept into his bass. 'Aye, thou speakest sooth. All the world hates me. All conspire against me. Had he not had the aid of traitors, Suleiman had never locked me away. All I have sought to do has been thwarted by envious ill-wishers—aye, but tomorrow comes the day of reckoning!'

Virginia lit a cigarette with a very steady hand and blew smoke at him. 'But how you can trust the emir and his cohorts?' she asked. 'He too is your enemy. He only wants to make a cat's-paw of you, then back in the bottle!'

'Why—why—' The afreet swelled till the spacewarp barrier creaked. Lightning crackled from his nostrils. It hadn't occurred to him before, his race isn't very bright; but of course a trained psychologist would know how to follow out paranoid logic.

'Have you not known enmity all your long days?' continued Virginia quickly. 'Think back, Rashid. Was not the very first thing you remember the cruel act of a spitefully envious world?'

'Aye—it was.' The maned head nodded, and the voice dropped very low. 'On the day I was hatched . . . aye, my mother's wingtip smote me so I reeled.'

'Perhaps it was accidental,' said Virginia.

'Nay. Ever she favoured my older brother—the lout!'

Virginia sat down cross-legged. 'Tell me about it,' she advised. Her tone dripped sympathy.

I felt a lessening of the great forces that surged within the barrier. The afreet was squatting on his hams, eyes half shut, going back down a memory trail of millennia. Virginia guided him, a hint here and there. I didn't know what she was driving at, surely you couldn't psychoanalyse the monster in half a night, but—

'. . . aye, and I was scarce turned three centuries when I fell into a pit my foes must have dug for me.'

'Surely you could fly out of it,' murmured Virginia.

The afreet's eyes rolled, and something twisted his face into still more gruesome furrows. 'It was a pit, I say!'

'Not by any chance a lake?' she enquired.

'Nay!' His wings thundered. 'No such damnable thing . . . 'twas dark, and wet, but—nay, not wet either, a cold which burned—'

I saw dimly that the girl had a lead. She dropped long lashes to hide the sudden gleam in her gaze. Even as a wolf, I could realize what a shock it must have been to an aerial demon, nearly drowning, his fires hissing into steam, and how he must ever after deny to himself that it had happened. But what use could she make of—

Svartalf the cat streaked in and skidded to a halt. Every hair on him stood up, and his eyes blistered me. He spat something and went out again with me following.

Down in the lobby I heard voices, and looking through the door I saw a few soldiers milling about. They'd come by, perhaps to investigate the noise, seen the dead guards, and now they must have sent after reinforcements.

Whatever Ginny was trying to do, she had to have time for it. I went out that door in one grey leap and tangled with the Saracens. We boiled into a clamouring pile. I was almost pinned flat by their numbers, but kept my jaws free and used them. Then Svartalf rode that broomstick above the fight, stabbing.

We carried a few of their weapons back into the lobby in our jaws, and sat down to wait. I figured it was better to remain wolf and be immune to everything but silver than to have the convenience of hands. Svartalf regarded a tommy-gun thoughtfully, propped it up, and crouched over it.

I was in no hurry. Every minute we were left alone, or held off the coming attack, was a minute gained for Ginny. I laid my head on my forepaws and dozed off. It was much too soon that I heard boots ringing on pavement.

There must have been a good hundred of them. I saw their dark mass, and the gleam of starlight off their weapons. They hovered around the squad we'd liquidated, then whooped and charged up the steps.

Svartalf braced himself and worked the tommy-gun. The recoil sent him skating back across the lobby, swearing, but he got a couple. I met the rest in the doorway.

Slash, snap, leap in, leap out, rip them and gash them and howl in their faces! They were jammed together in the entrance, slow and clumsy, it was a brief whirl of teeth and then they retreated. They left half a dozen dead and wounded.

I peered through the glass in the door and saw my friend the emir. There was a bandage over his eye, but he was rushing around exhorting his men with more energy than I'd expected. Groups of them broke loose from the main bunch and ran to either side. They'd be coming in the windows and the other doors.

I whined as I realized we'd left the broomstick outside. There could be no escape now, not even for Ginny. The protest became a snarl as I heard glass breaking and rifles blowing off locks.

That Svartalf was a smart cat. He found the tommy-gun again and somehow, clumsy though paws are, managed to shoot out the lights. Then he and I retreated to the stairway.

They came at us in the dark, blind as most men are. I let them fumble around, and the first one who groped to the stairs was killed quietly. The second had time to yell. Then the whole gang of them crowded after him.

They couldn't shoot in the gloom and press without potting their own people. Excited to mindlessness, they attacked me with scimitars, which I didn't object to at all. Svartalf raked their legs and I tore them up—whick, snap, clash, Allah Akbar and teeth in the night!

The stair was narrow enough for me to hold, and their own casualties hampered them, but the sheer weight of a hundred brave men forced me back a tread at a time. Otherwise one could have tackled me and a dozen more have piled on top. As it was, we gave the houris a few fresh customers for every foot we lost.

I have no clear memory of the fight. You seldom do. But it must have been about twenty minutes before they fell back at an angry growl. Then the emir himself stood at the foot of the stairs, lashing his tail and rippling his gorgeously striped hide.

I shook myself wearily and braced my feet for the last round. The one-eyed tiger came slowly up the stairs. Svartalf spat, then suddenly zipped down the banister past him and disappeared in the gloom. Well, he had his own neck to think about. . . .

We were almost nose to nose when the emir lifted a paw full of swords and brought it down. I dodged it somehow and flew for his throat. All I got was a mouthful of baggy skin, but I hung on and tried to work my way inward.

He roared and shook his head till I swung like a bell clapper. I shut my eyes and clamped on tight. He raked my ribs with those long claws. I skipped away but kept my teeth where they were. Lunging, he fell on top of me, and his own jaws clashed shut. Pain jagged through my tail, and I let go to howl.

He pinned me down with one paw, raising the other to break my spine. Somehow, crazed with the hurt, I writhed free and struck upward. It was his remaining eye which was glaring down on me, and I bit it out of his head.

He screamed! A sweep of one paw sent me kiting up to slam against the banister. I lay there with the wind knocked from me while the blind tiger rolled over in his agony. The beast drowned the man, and he went down the stairs and wrought havoc among his own soldiers.

A broomstick whizzed above the melee. Good old Svartalf! He'd only gone to fetch our transportation. I saw him ride towards the door of the afreet, and got groggily up to meet the next wave of Saracens.

They were still trying to control their boss down there. I gulped for breath and stood watching and smelling and listening. My tail seemed ablaze. I looked and saw that half of it was gone.

A tommy-gun began stuttering. I heard blood rattle in the emir's lungs. He was hard to kill. *That's the end of you, Steve Matuchek*, thought the man of me. *They'll do what they should have done in the first place, stand beneath you and sweep you with their fire, every tenth round argent.*

The emir fell and lay gasping out his life. I waited for his men to collect their wits and remember me.

Ginny appeared on the landing, astride the broomstick. Her voice seemed to come from very far away. 'Steve! Quick—up here!'

I shook my head dazedly, trying to understand. I was too tired, too canine. She stuck her fingers in her mouth and whistled. That fetched me.

She slung me across her lap and hung on tight as Svartalf piloted the stick. A gun began firing blindly from below. We went out a second-storey window and into the sky.

A carpet swooped near. Svartalf arched his back and poured on the Power. That Cadillac had legs! We left the enemy sitting there, and I passed out. . . .

When I came to, I was lying prone on a cot in a hospital tent. There was bright daylight outside, the earth lay wet and steaming. A medic looked up as I groaned. 'Hello, hero,' he said. 'Better stay in that position for a while. How're you feeling?'

I waited till full consciousness returned, then accepted the bouillon he gave me. 'How am I?' I whispered—they'd humanized me, of course.

'Not too bad, considering. There was some infection of your wounds, but we cleaned that up with a new antibiotic. You should be quite well in a month or less.'

I lay thinking about that. A field hospital just doesn't have the equipment to stick micropins in bacteria. Often it doesn't even have the enlarged anatomical dummies on which the surgeon can do a sympathetic operation. 'What technique do you mean?' I asked.

'Oh, one of our boys has the Evil Eye,' he said. 'He looks at the germs through a microscope.'

I didn't enquire further, knowing very well that de Kruif would be writing it up in a few months. Something else nagged at me. 'The attack . . . has it begun?'

'The—Oh. That. That was two days ago, Rin-Tin-Tin. You've been kept under asphodel. We mopped 'em up all along the line. Last I heard, they were across the Washington border and still running.'

I sighed and went back to sleep. Even the noise as the medic dictated a report to his typewriter couldn't hold me awake.

Ginny came in the next day, with Svartalf riding her shoulder. Sunlight striking through the tent flap turned her hair to hot copper. 'Hello, Captain Matuchek,' she said. 'I came to see how you were, soon as I could get leave.'

I raised myself on my elbows, and whistled at the cigarette she offered.

When it was between my lips, I said slowly: 'Come off it, Ginny. We didn't exactly go on a date that night, but I think we're properly introduced.'

'Yes.' She sat down on the cot and stroked my hair. It felt good. Svartalf purred at me, and I wished I could respond.

'How about the afreet?' I asked after a while.

'Still in his bottle.' She grinned. 'I doubt if anybody'll ever be able to get him out again assuming anybody would want to.'

'But what did you *do*?'

'A simple application of Papa Freud's principles. If it's ever written up, I'll have every Jungian in the country on my neck, but it worked. I got him to spinning out his memories and illusions, and soon found he had a hydrophobic complex—which is fear of water, Rover, not rabies—'

'You can call me Rover,' I growled, 'but if you ever call me Fido, gives a paddling.'

She didn't ask why I assumed I'd be close enough in future for such laying on of hands. That encouraged me vastly. Indeed, she even blushed, but went on: 'Having gotten the key to his personality, I found it simple to play on his phobia. I pointed out how common a substance water is and how difficult total dehydration is. He got more and more scared. When I showed him that all animal tissue, including his own, is about eighty per cent water, that was it. He crept back into his bottle and went into a catatonic state.'

After a moment, she added thoughtfully: 'I'd like to have him for my mantelpiece, but I suppose he'll wind up in the Smithsonian. So I'll just write a little treatise on the military uses of psychiatry.'

'Aren't bombs and dragons and elfshot gruesome enough?' I demanded with a shudder.

Poor simple elementals! They think they're fiendish, but ought to take lessons from the human race.

As for me, I could imagine certain drawbacks to getting hitched with a witch, but still and all—'C'mere, youse.'

She did.

I haven't kept many souvenirs of the war. It was an ugly time and best forgotten. But there's one keepsake I'll always have, in spite of the plastic surgeons' best efforts. As a wolf, I have a stumpy tail, and as a man I don't like to sit down in wet weather.

It's a hell of a thing to get a Purple Heart for.

AVRAM DAVIDSON

THE SINGULAR EVENTS WHICH OCCURRED IN THE HOVEL ON THE ALLEY OFF OF EYE STREET

In 1961, the year when the dragons were so bad, a young man named George Laine, an industrial alchemist by profession, attended the coronation of the new president in Washington. The guilds were in high favour with the president-select, John V (the first of that name since John IV C. Coolidge), who sent to each and every of their delegation, as a mark of his esteem, garments of vertue worthy of the occasion, viz. a silken hat, a pair of galoshes with silvern buckles, a greatcoat with a collar of black samite, cuff-links enchased in gold, and a pen-and-pencil set of malachite and electrum which it were guaranteed to write under water and over butter: both, as it happened, essential to the practice of industrial alchemy.

The ceremonies proceeded without any untowardness. The Supreme Justice of the Chief Court placed on the President's head the sacred beaver with the star-spangled band and declared that 'Regardless of rape, crude, choler, or national ore or gin, any resemblance is purely coincidental.' The Chairman of the Board of Augurs of the Federal Reserve System pronounced a curse in weirdmane and in womrath on anyone who should presume to send gold o'er the white-waved seas. The new Veep, wearing the ritual ten-gallon hat, and mounted on a palomino, cantered up and down before the Selectoral College, and uttered the prescribed challenge: 'Whosoever doth deny that the Honourable John V Fitz-Kenneth is the rightful Chief Executive of Thiscountry lies, and is an SOB.' The out-going Jester raised the liturgical *hwyl* of *We want Wilkie*, and was smitten twice with a slap-stick and thrice with a bladder, both wielded by his successor. The Fall River Chamber of Commerce and

Horror presented the ceremonial breakfast of cold mutton soup, sliced bananas, and an axe: it was ceremoniously refused. A Boston Brahmin, clad in cutaway, *dhoti*, and sacred thread, offered a salver bearing two curried cod-fish balls; the new President ate both whilst the Brahmin intoned,

> Eat it up, wear it out,
> Make it do, or do without;

after which he, the BB, hurried to wash himself in sacred 6 per cent Charles water to remove the impurity of feeding with a lower caste.

George Laine and his fellows of the alchemists and other guilds were not forgotten even afterwards; for Prex Jax (as the newsguild had already termed him *in parvo*) sent them out great smoking helpings of buffalo hump, bear paws, caponized peacocks, pemmican, ptarmigan, succotash, and syllabub, from the high table where he was dining with his notables, including Surgeon-General Doctor Caligari, who had just been raised to Cabinet rank.

It was during these moments of revelry and mirth that George choked on a quartern of orange in an Old Fashioned Cocktail, all went black before his face, and, on awakening to find himself bound with silken cords in a hovel on an alley off of Eye Street, knew that he had been ensorceled.

There was a bim looking bemused at him with a bodkin in her bosom, and he wotted well it were for lack of wit anent her that he bode bound: for who was she but Yancey-Courtney Belleregarde, a Drum Majorette 1/c, who had been sitting in his lap that time he raised the dram-glass to his lips.

'I say, that bodkin must hurt something dreadful,' he said (not having attended the NY High School of Gallanterie Trades in vain); 'untie me and I'll have it out for you in a trice: there's a good gel, do.'

The bim smiled scornfully. Her lips were as red as the chassis of a new-model Jaguar of the first enamelling. 'Not on your tin-type, Cully,' she said. 'Rats. Nit.' She spoke in the Archaic tongue of the bim-folk, which is akin to elf-talk, and cognate with 23 Skiddoo (unlawful for a man to know until he has passed his finals in The Deep School, and been awarded the right to wear the Navel Plug, with two Pips).

'Nix on the soft-soap, Charlie,' she said; 'I only keep the bodkin there because these, now, sorcelsacquets don't have any pockets in them, as if you didn't know. Oh you kid!' she concluded, archly. And with this she withdrew the bodkin, dipped its prickle into a pot labelled *Poyson*

Moste Foule, and approached the supine young industrial alchemist with the tip of her tongue held between her teeth.

'Slip me the Formula for the Transmutation of Borax Without the Use of Cockatrice-egg,' she said (speaking with some difficulty, her tongue, as we have already noted, you clod, being between her teeth), 'and we'll be back in the Grand Ballroom of the Mayflower in lots of time to see Ed Finnegan made a KTV; afterwards we can tiptoe up to any of the thirty-odd double rooms which my Company keeps rented at all times, and you may have your wicked will o' me without fearing the House-Dick, because I'll put a Cheese-it spell on the door, see, which it's proof against Force, Force-Fields, Stealth, Mort-Main, Nigromancy, Mopery, and Gawk: so give, Cully, give.'

A cold sneer crossed George's hot lips. 'I say, what an absolutely rotten proposal!' he exclaimed. 'You know perfectly well that I have sworn by the most frightful oaths to remain true in mind and deed to Alchymy Ltd., of Canada, and to keep myself physically clean, mentally straight, and morally pure! I suppose you're one of these simply awful party girls which one hears that General Semantics, Inc., of Delaware, keeps on their payrolls to entrap, ensorcel, enviegle, enchaunt, enduce, endive, and endamage clean-living young chaps into betraying secrets. Well, I shan't, do you hear? Better I should die. So there!'

But the bim, far from being one whit abashed by this manly defiance, laughed as coarsely as the position of her tongue would permit. 'Well, if that don't take the cake,' she snickered. 'Gee, what a simp!' and made feint as though she would withdraw George's Plug, two Pips or no two Pips.

'No, really, don't touch me, do you hear?' George said, stoutly, trying to roll over on his stomach, 'I'm really most frightflie ticklish, and besides, without the Plug I should swell up with lint in simply no time; funny thing about me, I'm very susceptible to navel lint, always was, from a child.'

But the silken cords held him fast.

'The Formula for the Transmutation of Borax Without the Use of Cockatrice-egg,' she said, inexorably, making little jabs at him with the bodkin dip't in Venom.

George mimicked her: '"—Uthe of Cockatwithe-egg"!'

Unguardedly she laughed, releasing the tip of her tongue from between her teeth, and thus . . . Those who are Cupboard Certified Auditors of The Deep School will understand *thus*, and those who are *not* needn't imagine for one minute that we are going to reveal for free, secrets for which others have paid good money, no siree. Suffice it, then, to say that in a trice George had leapt out of his bonds, flung the bodkin from the bim's hands with such force that it pierced the door and hung quivering. This produced a startled cry from behind the door, which George flung open, revealing a man, a tape-recorder, and a flash-camera. The man first cringed, then assumed an expression combining both defiance and a falsely hearty air of goodwill.

'Weh-hell, Laine,' he burbled.

'What,' demanded George, sternly, 'is the Assistant Director of Research for the Middle Atlantic States Division of Alchymy Ltd., of Canada, doing cowering behind the door of a hovel on an alley off of Eye Street, with a tape-recorder and a flash-camera; what?'—a question which, put like that, might make any man pause before answering.

Mr Marcantonio Paracelcus (for such was his name), paused before answering. He swallowed. 'It was a Test, you see, George.'

'I fail to see.'

'Well, it was a *test.* The Company is considering you for an important new job. In order to find out how you would shape up under pressure, we have tested you. I am, um, happy to say that you have passed the Test.'

George said, 'Oh, good. Then I get the job. *What* job?'

Mr Marcantonio Paracelcus seemed to find some difficulty in answering this question. Whilst he stood there, came a buzz and a clatter, and that which George had hithertofore considered to be merely a tallboy-sized TV set opened up, revealing itself to be an Observation Armoire containing a microphone, *two* tape recorders, an automatic closed-circuit television camera, and Dr Roger Bacon Buxbaum, Chief Director of Research for the Middle Atlantic States Division of Alchymy, Ltd. of Canada. Marcantonio Paracelcus, on perceiving his superior, turned ashen, livid, and pale, in that order.

'The job in question, George,' said Dr Buxbaum, 'is that which until a moment ago was held by the gentleman you now see cowering behind the door; but which is no longer so held. On realizing that you were being considered for his position, he determined on this unworthy method of discrediting you: hence, the tape-recorder, on which he hoped to capture the sound of your voice as you revealed the Formula for the Transmutation of Borax without the Use of Cockatrice-egg; hence the flash-camera with which he hoped to capture the sight of you in a,' and here the benign, balding Buxbaum blushed a bit, 'compromising position with this young female person here. Little did he know,' the urbane researcher winked, and placed his right forefinger by the right side of his nose, 'that we were onto his jazz from the word Go . . .

'And to think that he would sully the semi-sacred season of the Coronation by his meretricious machinations; fie, sir, do you call yourself a Thiscountrean? But I forebear harshness; modern science has taught us that such a one as you is really sick, and needs help. Come along now—George! Expect to see you for lunch, day after tomorrow, at the Alembic, one sharp!'

George went pink with pleasure, for what was the Alembic but the most expensive eatery favoured by the upper echelons of the MASD of Alchymy Ltd. (Canada); and this invite betokened his full acceptance into the post previously held by his unfortunate predecessor, who even now, snivelling miserably, was being firmly guided out by the elbow. George's feelings of sorrow, which did him credit, were tempered by the reflection that, after suitable treatment at the Company's Rehabilitation Farm in North Baffin Land, the man might still prove capable of many years of devoted service; though, of course, in a minor capacity.

For a moment all was silent in the hovel on the alley off of Eye Street. George eyed the bim. The bim eyed the floor. After a while she spoke. 'I suppose you hate me,' she said.

'No, I—'

'I suppose you think I'm miserable and treacherous.'

'No, I—'

'I suppose you think I would really have stuck you with a poysoned bodkin, don't you? Well, the jar only contained a Sophronia Finkelstein preparation for the treatment of tired skin and subcutaneous tissues; so there.'

George said, 'No, I fully realize that as a bim, and as a sorceress under contract to General Semantics Inc., of Delaware, you were only carrying out your duty. And now, if you don't mind, I wonder if I might use your phone to call a taxi?'

Fancy his astonishment when she burst into tears.

'We have no phone,' she wept. 'I'm not a bim. I never worked for General Semantics. My parents couldn't afford to send me to Sorcery School. How I put you under that spell and brought you here, my old Auntie Eglantine was a white witch and I picked up some little piddly old spells from her, is all. I am really just a Drum Majorette, 1/c. Oh, I wish I were dead! A hoo, hoo, hoo!'

George, at first with awkwardness, then with growing appreciation for the task, patted her hands, her shoulders, and the general area of the small of her back. 'To tell you the truth, Miss Yancey-Courtney,' he said, 'I would just purely hate it if you were to be a bim. I mean, like, those hairy *feet*? And their toe-nails *glow* in the dark? Why, a man couldn't hardly relish his victuals, let alone keep his mind on his Transmutations . . . Of course, I'm just speaking speculatively, I mean; having always kept myself physically pure, mentally clean, and morally square, according to the terms of my Triune Oath to the Company, which I have never regretted,' he said, regretfully.

'Of course,' she murmured, wiping her eyes on his shirt tail.

'Listen,' she said, 'do you know when it was that I first felt a revulsion I was barely able to conquer at the infamous Marcantonio Paracelcus's proposal? It was when the Veep rode in. When he gave out the Challenge I could see you clench your fists until your knuckles went what I mean *white*; as if you were just *daring* any old Recounter to challenge the Selection!'

'Hm,' said George, grimly.

'I'll bet you must be awfully strong.'

George, modestly, said, well, shoving all that lead and gold around, *you* know. She said that she could well imagine. There was a pause. Then he asked what time it was. She said it was 7:45, why? He said that if they hurried, they could still get to see Ed Finnegan dubbed a KTV. She said, yes, they could, couldn't they? She asked if he was very fond of Ed Finnegan. There was a pause. He said that as a matter of fact he couldn't stand Ed Finnegan.

'Neither can I!'

'All those trained wombats!'

'And that incessant, hearty laugh!'

There was another pause. Then, 'My, those are handsome galoshes!' she said.

'Gift of the President.'

'Pipe the silvern buckles, will yuh?'

'Mmm.'

'But don't you think you'd be more comfortable if you took them off?'

'The buckles?'

'Oh you silly! The ga*loshes*!'

'I might at that.'

And he did. And he was.

Outside, the Northern Lights hissed and crackled (or, again, it might have been the dragons, which were so bad that year); outside, the noise of revelry continually rose and fell in the streets; but inside, all was quiet in the hovel on the alley off of Eye Street.

THOMAS BURNETT SWANN

THE SUDDEN WINGS

DRAGONFLY

'Gaius and Phoebe!' Mark called, pointing towards the coast of Greece. Above their heads a scarlet sail groaned in the wind and hurried them south toward Rhodes, Cyprus, and Gaza on their way to Petra, the rock-built capital of the Nabataeans. The voyage was long, the coastline dull, and Mark had spent the morning asleep in the sun or playing a lyre to his sister Phoebe, while their uncle, Gaius chatted with the sailors about the respective merits of Roman and Asiatic women. Towards noon Mark had laid aside the lyre and strolled to the side of the ship in time to see a remarkable sight.

'Look,' he called, his red, wind-blown hair a riot of curls. A short blue tunic outlined the young manliness of his body.

Phoebe came at once, but Gaius took his time, and Mark heard him call to his companions, 'But Etruscan women—they are the deep ones. Their slanted eyes hold secrets older than Babylon.' Casually he walked to Mark's side.

'You look as if you had seen a god,' he said, his lean tanned face seeming to say, 'What you have to show me had better be good.' Slender and straight as a battering ram, with eyes like onyx, he seemed more brother than uncle. Ten years ago, at the age of twenty, he had taken part with Titus (now emperor) in the sack of Jerusalem, and he always carried himself with martial erectness.

'In a sense I have,' Mark said. On a rocky headland, where the waves beat a mist of spray, a curious creature had paused to watch the vessel.

'A centaur,' cried Mark, an antiquarian at twenty. There was no race, he knew, more ancient than the centaurs, who had roamed the Greek mainland when the Cretans first waded ashore from their ships with purple sails. The centaur drummed a hoof on the rocks and his flanks glittered in the sun. His bearded face looked as old as the rocks, as mottled and brown, but somehow gentle.

'Hellooooo,' Mark called, his voice almost lost in the forward plunge of the ship. Tentatively the centaur raised an arm and returned the greeting. Even as they watched, a woman, nude except for her riotous blue hair, emerged from the cliffs and joined him on the headland.

'His nymph,' Mark whispered. 'The centaurs have no women of their own. She may be an Oread.' She placed her arm around his shoulder and stared after the ship. The whiteness of her body formed an eerie contrast to the blue of her hair. She looked as if she had spent her life in a stream or a tree, avoiding the sun. There was no friendliness in her eyes for the voyagers. Doubtless she remembered the centaurs slain by Greek and Roman warriors. At last she took her lord by the hand and led him away from the sea. They did not look back at the ship.

Mark felt a sharp loneliness as they disappeared; he hated the old cruelties which had made the nymph distrustful and wanted to shout to her: 'Wait, I won't hurt you!' Once as a child he had caught sight of a satyr on the slopes of Vesuvius and felt the same loss when it eluded him. Before Romulus, the entire Mediterranean world had abounded in satyrs and centaurs; Tritons had played in the sea-pools, and even the air, it was said, had throbbed with sudden wings as boys and girls had tumbled through the clouds. But now, thanks to the depredations of man, such creatures were almost extinct. Mark's eyes misted.

'Tears for a centaur,' Gaius reproached him, 'but never for a maiden. Twenty years old and heartfree still. You have set a bad example for your sister. She thinks she must remain single to look after you. And so she follows you into the further corners of the world.' Impulsively he hugged them against his chest; his crisp linen tunic was fragrant with bitter almonds, an essence much prized by Roman gentlemen. 'Children, children. You will find no mates among the rocks and scorpions.'

Mark laughed. 'There are always the Arabs.'

'With camel urine in their hair,' sniffed Phoebe.

Then they noticed that one of the sailors had joined them and was staring where the centaur and nymph had disappeared. The sailor looked perplexed and frightened. His calloused hands tightened on the gunwales.

'The Old Ones,' he muttered, 'It is a bad sign.'

'A bad sign?' asked Mark. 'How could they harm us?'

'Not those. Others.'

'You think we will meet others?'

'Aye, *you* will. At Petra. The Boy with Wings, the Dragonfly. His face is as young as your own, his heart, as old as evil. I have never seen him myself—never been beyond Gaza, nor want to go.'

Mark's interest mounted. A boy with wings, a creature more fabulous than centaurs. 'You make him sound dangerous.'

'The natives think so. They have offered a hundred camels to anyone who captures him.'

'Is he dangerous to Romans?'

'That I can't say. To your sister, I should think. She will draw him down like honey on idols. And you—your hair is the very colour of his wings. He may grow jealous.' He reached out and gripped Mark by the shoulder. 'Beware of him, boy. Beware of him! They say—'

'Decimus, came an angry cry from the stern, 'man the rudder oar!' The sailor broke off and hurried to his post.

Phoebe's eyes were round as honey cakes.

Mark took her hand. 'Frightened?' At nineteen she made him think of kittens and lambs and baby dolphins, of eiderdown and pink rose buds, of all things soft, vulnerable, and delicious. But she was more than soft and girlish; she was adolescence becoming womanhood. Even a voluminous stola could not conceal the delectable undulations of her body. A ripe fig, the young men of Rome had said, hoping in vain to pluck her.

She shivered and squeezed his hand. 'Not with you.'

Gaius laughed. 'You are babies, both of you. As pink and smooth as children, and as inexperienced. Look at Mark. Not the shadow of a beard! What does he know of Dragonflies or of cities in the desert, where the tribesmen sacrifice boys to the god Dusares? You ought to have married Romans, both of you, and stayed at home.'

'What is there left in Rome,' Mark wanted to know, 'with you away?' What was left in Italy, for that matter? Two years ago there had been a villa in Pompeii, devoted parents and friends beloved since childhood. Then came the rain of fire, the shaken earth, the frantic flight to the sea with Phoebe and the cat he had rescued from the burning temple of Venus, and finally escape in a fishing bark. They had sailed to Rome to live with their uncle Gaius. When the new emperor Titus, his general in the attack on Jerusalem, had sent him as unofficial emissary to prepare for a possible Roman occupation of Petra, they had asked to go with him.

Gaius returned to his conversation with the sailors and Mark and Phoebe strolled together around the deck of the ship. Mark looked at her with adoration. Her jonquil-yellow hair, curled on top, was drawn into a fillet behind her head. The fullness of her lips in another woman would have seemed a pout; in Phoebe it was girlhood grown voluptuous, and her blue enormous eyes were those of a maiden who, a little shocked by the licentiousness of Rome, had chosen to observe with wonder rather than participate.

Venus Cat, large, yellow, and, since Pompeii, almost tailless, trotted between them. Discriminating, he ignored most humans, but brother and sister he followed with a doglike constancy.

They paused beneath a bronze image of Portumnus, the harbour god for whom the ship was named.

'I should have married you to one of my friends in Rome,' Mark said mischievously, knowing that the subject annoyed her. 'That would have satisfied Gaius.'

'To a dandy in a silken toga, with myrrh in his curls? Who divides his time between the games and the baths? You would have done better to marry my friend, Cornelia.'

'A sweet thing, Cornelia. Her eyes reminded me of a ewe's. I always expected her to bleat.'

'You see,' she said, 'we both think up reasons not to marry.' It was not uncommon for maidens to wed at twelve and boys at fourteen; Mark and Phoebe, then, had long been marriageable.

Venus Cat mewed peremptorily, a signal for attention. He placed his shoulders on the deck and lifted his stub of a tail to indicate that he wished to be stroked where the tail joined the body.

'But are we enough, the three of us and Venus Cat? We have shut out so much since Pompeii.'

'The more we admit, the more we can lose,' Phoebe said with finality. He knew that she was remembering their villa ignited by lava and their parents trapped in the flames. They had both suffered nightmares for two years. 'We are admitting Petra. Surely that is enough.'

'It is Petrans we need, not Petra.' But he was no more eager than Phoebe to make room in his heart for the hurtful ties of friendship and love.

'There is always the Dragonfly,' she said lightly.

'He sounds unsociable, but I want to see him.' He yawned. 'But we still have a long journey and I am feeling drowsy.'

The wind sighed in the tall square sail and the painted emblem, a she-wolf suckling the twins, trembled as if to spring from the canvas. The little ship surged forward at five or more knots. Phoebe remained on deck while Mark retreated to the wicker cabin under the curving neck of the stern. Ensconcing Venus Cat beside him on the couch, he tried to sleep.

Usually sleep came soon and, except for nightmares, he slept long and soundly whatever the hour. But a winged boy danced in his brain. He rose, opened a chest of citrus wood, and took a small bronze image in his hand—Vaticanus, god of the Vatican Hill where Gaius owned his villa and where Mark and Phoebe had lived with him for two years; Mark's household god. In the immemorial attitude of prayer, he raised his arms and prayed:

'God of the hill, be with us on our journey, if you can. And the dragonfly boy—may we find him friendly.'

At the entrance to the Sik or ravine leading to Petra stood the Grand Portal, a fifty foot decorative arch with statues and niches. Beyond the portal precipitous walls overtowered the Romans and their guides with brown and yellow sandstone and made a twilight of morning. A paved road ran beside a dry river bed. Glens broke away to the side, brimming with olive and fig trees. His mouth was dry and dusty, and Mark would have paused to pick some figs, but the Arab guides, reeking with sweat and the camel urine with which they oiled their hair, hurried him forward and furtively peered at the sky. Even Venus Cat was shooed from the alluring byways which plunged among the vegetation. The Arabs seemed little acquainted with cats and avoided him as if he were a scorpion, but they did not hesitate, when he turned aside, to drive him into the path.

'They expect the Dragonfly,' Gaius whispered. 'They are speaking Aramaic, which I learned at the siege of Jerusalem. One of them said they must sacrifice to Dusares, because the Dragonfly has been seen in the area lately.'

'What will they sacrifice?' Phoebe whispered.

'A boy, most likely. With unblemished skin and the rosy innocence of youth.'

Phoebe looked horrified and tugged at Mark's tunic. 'Gaius, you are describing Mark.'

Gaius laughed. 'A Roman citizen should be safe enough. The Nabataean king, Rabel II, is said to be eager to conciliate the Romans—even to consolidate with them. That, of course, is why I am here. He is still very young—younger that Mark—and thought to be weary of his mother's domination.'

A shadow broke the ribbon of sunlight at their feet. They looked up. A bird-like creature wheeled above their heads.

The Arabs stopped and peered at the sky. They spoke rapidly. Diaphanous wings glinted in the sunlight. Mark stood very still; his heart-beat quickened. The Dragonfly sank into the gorge. His features materialized in the fitful light: firm but slender limbs, hair as red as Mark's but wild with a hundred winds, and—half again as tall as the body—pointed wings which opened and closed with the slow, deliberate strokes of anemones under the sea. He descended in lilting curves, like milk weed settling in a breeze. The Arabs had fled to the empty river bed, where they crouched among the boulders and hurled stones at the invader. He began to laugh, a deep laugh, musical yet chilling, that echoed down the gorge and set bulbuls springing from olive trees. Now he was almost within reach of the stones. He poised just above them, beating his wings to hold his position.

Mark waved his arms in a frantic warning. 'Go back,' he shouted. 'They will hurt you!' He imagined the delicate wings broken by rocks, the creature falling into the hands of the natives who would sacrifice him in their ghastly rituals. 'Go back!'

The Dragonfly laughed. He caught a rising stone and dropped it on his attackers. The Arabs who were bombarding him scrambled for cover. He looked at Mark and Phoebe and waved his hand.

'Red Hair and Yellow Hair,' he called. 'You with your golden cat. Welcome—and be my friends.' He fanned them with the rush of wings, and Mark looked directly into his face—high slender cheekbones, eyes as green and unfathomable as the sea beyond the Pillars of Hercules. A young face, a boy's face, but with eyes as old as—what had the sailor said?—evil. They stared after him as he beat his way up the gorge and vanished over the edge.

The Arabs emerged from the watercourse and resumed their journey. With every few steps they peered anxiously at the sky, muttered to each other, and glared at Mark as if he had summoned the creature with Roman magic.

Mark, Phoebe, and Gaius walked in silence, peering hopefully at the top of the gorge, now three hundred feet above their path. Even Gaius, the practical warrior, seemed shaken by the encounter. It was Phoebe who finally spoke.

'He was beautiful,' she said. 'Like a king. Like a—'

'God whose image we saw in Rome,' interrupted Mark.

'I am trying to think which one. But he looks so lonely. The Arabs hate him. Who can he find for a friend?'

'The birds,' said Phoebe. 'Would you be lonely with wings?'

'I would want my sister and uncle to have them too. Vultures make poor company. I should think the extent of their conversation would be directions for finding the latest carcass.'

'I suppose you are right. *We* shall have to become his friends.'

'You will do nothing of the kind,' snapped Gaius. 'For all we know he is deadly—not a boy at all, but a demon with a boy's face. Like the sirens who sang to Ulysses.'

They looked at him in surprise.

'Listen to me, my innocents. We will learn more about him in Petra. The Arabs may have good reason for fearing him.'

The gorge flared suddenly into a glen where, directly in their path, a temple loomed from the living rock, its red façade rising in columns, pediments, and, on the second level a rounded tower. The side walls were the unhewn rock itself. It was like a great poppy, flowering from the desert by some divine horticulture, a poppy on the scale of the giant hero Moses who, in local legend, was said to have channelled the Sik with a blow of his magic rod.

'It must be the Khazneh,' Gaius said. 'The Arabs call it the Treasury of Pharaoh, but it is really a temple to Isis.'

An image of the goddess stood in the open tower and smiled benignantly down on the petty humans who sought entrance to her valley. The Arabs greeted her with salaams and cries of praise. But Mark suspected that it was not Isis who controlled the valley.

'It might be called the temple of the Dragonfly,' he whispered to Phoebe.

She squeezed his hand. 'We will find him again?'

'Tomorrow,' said Mark. But a vague uneasiness gripped him. Her enthusiasm seemed excessive. What was she feeling? He had often encouraged her to fall in love. But a boy with wings—

He chided himself. Phoebe in love? Ridiculous. She was only curious like himself. Tomorrow they would satisfy their curiosity and the matter would end. Suddenly he remembered the god who had looked like the Dragonfly. His image stood in the Forum—Mors, the lord of death.

MOON-GIRL

Mark and Gaius had gone to a banquet in the palace, but Phoebe, a woman and therefore uninvited, remained in the house reserved for them by the king—an extraordinary house cut into the side of a mountain. Room after room plunged into the dark rock, the walls hung with tapestries which smouldered in the light of tall candelabra. But the rock-hewn walls were cold and damp in spite of the hangings, and Phoebe, bathed and scented by an Arab slave girl, had stretched on a couch and drawn a coverlet around her shoulders. A lamp was left burning, a crouching lion which seemed to spit fire from its jaws. She had hated the dark since Pompeii; besides, there were scorpions in Petra.

She fell asleep at once, with Venus Cat beside her. The old dream returned. It was autumn, and a great fire had broken from the mouth of Vesuvius. Giant shapes seemed to dance in the fire and fling it in streamers down the slopes. A heat wave gripped the city and the parched inhabitants gasped in the shade while stones and ashes possessed the sky. The sun vanished as if eclipsed, the ground shuddered and split—

Phoebe cried out and opened her eyes. The dream was gone but another had taken its place. The Dragonfly leaned above her, his wings as tall as the roof, and said, 'Yellow Hair'.

She sat up.

'I am glad you are finally awake,' he continued. 'I thought I was going to have to shake you. I have been here for a long time.'

It was not a dream; he sat beside her on the couch. She uttered a little cry and he looked at her with wide, serious eyes.

'Have you been on my couch all the time?' she asked.

He pointed to a chair with a curving back. 'The cathedra was hard. There was no room for my wings. Besides, I have been watching you sleep.'

The coverlet had fallen to the floor. She was wearing a lavender tunic, less than knee length, with a silken sash. A highly indecorous garment for any pursuit except sleeping. She drew a pillow over her knees.

'I have been watching the rise and fall of your bosom,' he continued without a smile. 'It is a pretty thing to watch—up and down, up and down, like a little swelling sea. It has made me want to kiss you.'

Phoebe stiffened and took Venus Cat in her arms. 'I must call my brother. He wanted to see you again.'

'He has not returned. I have already looked for him.' He reached up and touched her hair which, free of its fillet, had spilled in a saffron chaos over the couch.

'Yellow Hair,' he said. 'A pretty name. But you must have another.'

'Phoebe.' I am alone with the Dragonfly, she thought. I am dressed for sleeping, not company, and he has said that he wishes to kiss me. If I cry out, a slave will come at once. But wonder wrestled with fear, and she did not call. It was like those assignations her friends had told her about in Rome, when a young man came to their bedchambers, evading the watchful eye of parents and slaves, and made indelicate advances, which might or might not be encouraged. I will talk to him and learn his intentions, she told herself. Then perhaps I shall scream for a slave.

'Phoebe, the Moon,' he said. 'Not silver nor white, but a harvest moon, round and warm and golden. My name is Eros.'

'Eros. Love?'

'For you. And Red Hair. Tell me about him.'

She spoke quickly. It was always a pleasure to speak of Mark. 'He pretends not to care about things. He sleeps a great deal and laughs easily. He jokes and teases. He will tell you that he possesses no real ability. But he is a fine poet, and a brave and honourable man. When Pompeii was burning he carried me safely to a boat and went back into the city to rescue some children who were trapped in a temple.'

'His hair is red like mine. I want him to be my friend. Will he love me?'

'He wanted to see you again.'

'Will you love me too?'

She hesitated, uncertain how he meant the word. 'I am very fond of you.'

'Will you come with me?'

'Where?'

'To my home in the mountain.'

'Tomorrow, with Mark.'

His eyes darkened. 'You don't love me then.'

'It is too soon.'

'Soon? Is love the sum of minutes or days? No, it is a pine torch kindled in a second. I saw you in my valley and I loved you instantly.' His face was even younger than Mark's, and his artless, literal speech was that of a child. But his eyes were old. To look into them was to tumble down stairways of malachite with fireflies whirling around her.

He leaned forward and kissed her. She felt the brush of his wings, like butterflies, and wondered how anything so soft and tenuous could lift him from the ground. His breath was as sweet as nard, and she thought of the herbs and flowers he must gather from the mountaintops to season his delicate foods.

'Do you know now if you love me?'

'I am not sure.'

'Have you had many lovers?'

'No,' she said sharply.

'Not even one? A slave perhaps who caught your eye in the market-place, or a young friend of your brother?'

'Not even one.'

'I have heard that Roman women take husbands at twelve and lovers not later than fifteen. Sometimes I eavesdrop on caravans, and the great Roman ladies, the wives of consuls and governors—bound to join their husbands—do not always sleep alone.'

'I have neither a husband nor a lover, and I am a mature woman of nineteen.'

'Ah,' he said, 'That is why you are not sure if you love me. You have had no practice. I am glad. I will teach you how to love. But first we will talk.'

'Why do you never smile?' she asked him. She too could talk with the directness of a child. She felt as if he were a new playmate and she must get to know him as quickly as possible, without the polite evasions of adult society.

'Because there is nothing to smile about. I guess I have forgotten how.'

'Yet I heard you laugh this morning.'

'With scorn, not joy. Scorn for those dirty Nabataeans who were throwing rocks at me. It is joy I have forgotten.'

'You are lonely,' she said, 'and no wonder. How long have you been in Petra?'

'Before the Edomites, before the Kenites and Horites even, the valley belonged to my people. There were hundreds of us, and we slept in the caves and beat the air with our wings. But the Kenites were thousands, and they drove us to the cliffs and one by one the others died or were killed.'

She touched his wing in a gesture of sympathy and felt it shudder. 'All those centuries you have had no friends?'

'There have been a few,' he said. 'Once there was a Nabataean boy, I taught him to chisel animals out of the rocks—gazelle and jackal and coney with the face of a pig and the hands of a child. But they sacrificed him to Dusares. I found him at the Place of Sacrifice with a knife in his heart. That was a hundred years ago. He was the last.'

'A hundred years without a friend!' She would befriend him, she would love him. Too long she had held herself aloof in her brother's love. But Mark would be the last to begrudge her the Dragonfly. 'I will come to your mountain now, if you like. You must show it to Mark tomorrow.'

'Of course. You have said he will be my friend.' He handed her a cloak. 'Wrap yourself in your palla. The sky is cold.'

Her sandals made no sound on the carpeted floors, and Eros, barefoot, walked as quietly as a lizard. Venus Cat padded after them, puzzled by this tall winged boy who had come in the hush of night. In the vestibule at the entrance, the guard was sleeping, his head on his knees.

'He was easy to put to sleep,' said Eros. 'You ought to be better guarded. Someone might carry you away.'

She did not ask him what power he had used on the guard. It was a night of spells.

White broom flowers, so thick around their feet that neither stems nor leaves were visible, carpeted the roadside and a honied fragrance permeated the air. Phoebe stared at the rock-cut front of the house, a dusky purple in the light of torches which burned between the four entrances and cast the columns into a fitful dancing. Briefly she regretted the loss of this solid mansion, with its slaves and friendly torches and Venus Cat, and fought down an urge to wake the guard. Eros saw her shiver and drew the palla closely around her shoulders.

He lifted her in his arms and sprang into the air. The valley wheeled below them, the tall hills jutting toward the sky, their faces pitted with caves whose entrances were broken with columns and whose accesses were flights of stairs chiselled in the rock. He breathed heavily with his burden. Phoebe dared not question him; she lay like a cast-off cloak in his arms and scarcely dared to open her eyes.

But fear yielded to amazement. His slender, powerful arms assured her that he would not falter. His breath came regularly now. He pointed out hills and buildings. In the heart of the valley, a palace, a temple, a market place with stalls and a covered walkway crouched by a river like animals beside a water course. The temple and palace were elephants, the stalls, zebras, and the walkway, the curving length of a serpent. They crossed the Wady Moussa, the river which divided the town, and circled the rock Al Habi, where a thousand pigeons slept in the honeycombed façade of the Columbarium.

The torchlight of the city dwindled, the fragrance of white broom became the freshness of nearby rains, of closeness to the clouds, and Phoebe's heart expanded with the wonder of her flight. The stars burned as brightly as the torches they had left in the city. Were they going to the moon, that fretted amber palace? A fitting home for the Dragonfly. What were his wings but starlight materialized, his hair, a tangle of moonbeams?

Behind them the fires of the city blinked into darkness. Night is a raven, she thought, and his black hushed wings have seized us. But the Dragonfly will defend me—what darkness can shake his flight?

They approached the face of a great cliff, whose craggy expanse seemed to offer no entrance. But soon a small cavern loomed in their path. He settled on a ledge and led her into the blackness which, as they advanced, lessened and opened into a large chamber with lamps like swans. Carpets woven of rushes were strewn on the floor, while the walls held frescos where the birds of the land frolicked in poppies and oleanders—with white breasts and red patches on their tails. A couch stood at the rear, beside it a marble table like a truncated Doric column.

'My house,' he said proudly.

'You painted the walls yourself?'

'Yes, and gathered the furnishings from caravans. Some of them are very old—older than the Romans even.' He showed her a gaming board whose squares of shell were inlaid with lapis-lazuli and red limestone. 'That is from Ur. Once a Chaldean princess passed through this land. I came to her caravan at night and played with her. She would not come back with me, as you have, but she gave me the board. In return I gave her an Egyptian emerald as large as a pigeon's egg.'

She felt the desolation of a life in which his only friends were the travellers of the desert, shadows who paused and passed, leaving him all the lonelier for their momentary presence. She wanted to take him in her arms and be to him more than the Chaldean princess who had left behind her the hard, inadequate comfort of a gaming board.

She shivered. 'The flight has made me ill.'

'It is not the flight,' he said. 'You are afraid of me. Why? Surely I am not a monster. I have looked at myself in the mountain streams and seen a boy like your brother.'

She stared at his skin in the lamplight, reddish like the cliffs of Petra, and wondered if the wind had beaten the red of the rocks into his pores until he became the desert, as beautiful and unknowable. 'Your beauty troubles me,' she said at last.

He helped her to the couch. From a glass flagon he filled an amber goblet and placed it in her hand. 'Drink,' he said. 'It will make you at ease with me.'

She drank the liquid, part wine, part honey, with a curious tartness she did not recognize.

'You understand that I am going to kiss you,' he said. 'But I will wait until you have rested.'

She said nothing. Nausea pressed at her throat. His hair burned with rose and amber, and the light in its tangles made her dizzy.

'I—I am rested,' she said.

He kissed her and cupped her face between his hands. Still he did not smile. 'Phoebe, the moon,' he said. 'But the moon is pale tonight. What shall I do to kindle its yellow fires?'

'You are the moon, not I.'

'No, I am Eros, who calls to the moon. Do you love me, moon-girl, Phoebe?'

'Yes,' She did not hedge her words; she did not debate the multiple meanings of love. 'Yes.'

'Come.' He rose and held out his hand to her.

'Eros, must we fly again so soon?'

'The moon must learn how to fly,' he said. 'Was there ever a moon who looked up instead of down? The heavens should be your home.'

Again they stood on the ledge, with the cliff falling sheer below them, black and terrible. Her sandals dislodged a stone; its scraping descent faded into silence.

He let go of her.

'Take my hand,' she cried. 'I am afraid by myself.'

'I am here to give you strength,' he said, and he was smiling for the first time. 'You must do what I say. Do you trust me, moon-girl?'

'Yes.'

He touched her cheek and his strength sustained her. Yes, she could face the night again in his arms. But he did not take her in his arms.

He stepped back and said: 'Jump.'

EROS AND MORS

The banquet was small, intimate, and very Roman. Nine guests—the king, two Nabataean advisers, and six Romans including Gaius and Mark—reclined on couches in a semicircle around a table with feet like a camel's. A suckling pig revolved on a portable spit, and young Arab slaves, as noiseless as jackals, served the dates, cucumbers, melons, cheeses, kids boiled in milk, and locusts on skewers from the kitchen. Conversation soon turned to the benefits of union with Rome. The king and his ministers listened with enthusiasm as Gaius, august in his white toga, enumerated the advantages of the Pax Romana. Petra, said Gaius, harassed by an endless succession of wars, could secure her borders through alliance with Rome, and further, share in the benefits of Latin civilization—the roads, the aqueducts, the arts and games.

'A personable young man, the king,' Gaius said to Mark later in the evening, as they left the palace and headed for their villa. 'If he has his way, Petra will be the next Roman colony.'

'But his mother, Shaquilath, opposes annexation?'

'Bitterly. Till a year ago she ruled as regent, and her power is still formidable. She wishes to preserve the independence of her people at any price. I have not met the woman, but she is said to believe that the only good Roman is one whose bones have been picked by the vultures.'

The square, flat-roofed palace, with its façade of obelisks in half-relief, faded behind them and they entered a garden of scattered paths and summer houses roofed with the boughs of palm trees. A river barred their way, and oleanders rioted along its banks, their deep green leaves black in the moonlight. A frightened coney vanished on soundless hooves. They paused beside a bridge, talking of the banquet, the king, the implacable Shaquilath. Mark grew thirsty.

'Is the river water good?' he asked.

'All the water in Petra is good,' Gaius said. 'That is why so many armies have fought for the place.'

Mark stooped and filled his hands with water. He tasted it gingerly. Yes, it was clear and cold. He began to drink. He heard a footstep but, anticipating Gaius, did not turn his head. An object struck his temple and pain exploded into blackness.

He awoke shivering. He was nude; his feet and hands were bound and the rough thongs cut into his flesh. He seemed to be lying on a stone platform a little longer than himself and approached by shallow steps. The stone was jagged beneath his back. Cymbals clashed and tambourines jangled. Torches lurched in the darkness, as if their bearers were drunk. Figures leaped through the air, and he saw that they were nude like himself, their brown limbs glittering in the light of a moon so bright that it made him blink. They whistled and clapped their hands, but paid him little attention. He remembered the young men sacrificed to Dusares and hoped that he would continue to be ignored.

'You are awake,' said a woman's voice in uncertain Latin.

'It is well. I would talk with you.' She was the tallest woman he had ever seen, a giantess with arms that bulged like a Spartan's beneath her thin silk robe. Everything about her was massive and masculine except her voice, which was strangely soft.

Mark was frightened but he was also angry. 'I am a citizen of Rome,' he cried. 'This very night I have dined with the king. Release me at once.'

A tinkle of laughter greeted his command. 'I know you have dined with my son. My men had you under surveillance the entire evening.'

This, then, was Shaquilath. Surely the queen would not be a party to human sacrifice, however she hated the Romans.

'What do you intend to do with me?'

'Send you to Dusares, what else?'

'The Romans will send *you* an army.'

'Better an army than smooth-tongued ambassadors. The entrance to this valley is almost impregnable. But Roman ambassadors speak with silver tongues and my son listens. Why do you think I chose you for the sacrifice? To end this traitorous talk of annexation by Rome. The Romans will hardly negotiate with a country which has murdered—as they see it—one of their citizens. They will either ignore us or attack us. For either we shall be ready.' She looked at him appraisingly; nude and bound, he felt like a side of beef in the marketplace. 'You are beautiful enough for any god. When beauty is joined to political expediency—' She leaned very close to him. Her breath smelled of wine and olive oil.

'You are little older than my son. A boy like him, stubborn and foolish, but I could have loved you.' She touched his cheek with the fingertips; her bony fingers were surprisingly gentle. 'You are almost too beautiful to give to the god. I wish I could afford the luxury of saving you.'

The dance had become a bacchanalia; indeed, Mark recalled, the Arabs identified Dusares with the Greek Bacchus. Clouds had covered the moon, and the whirling figures, vivid by moonlight, looked now as shadowy and terrible as shades of the Underworld, dancing beside the Styx. He remembered a line which Gaius had read to him from a holy book of the Jews: how King David had 'danced before the Lord with all his might'. But the lord of the Arabs was not the lord of the Jews; this dance held a sinister difference. He breathed a prayer to his household god, Vaticanus:

'God of the Roman hill, be with me on this alien hill!'

The queen bent and kissed him on the cheek. 'Understand it is not that I wish you ill.' She turned and addressed the crowd in Aramaic. The dancing ceased, the dancers watched her with rapt attention, or rather they watched the knife in her upraised hand. Mark shuddered, no longer with cold, and thought of Phoebe. Beloved sister, I must die and leave you in a strange land, with Gaius and Venus Cat your only friends. I will come to you, if I can, as a shade, and watch over your sleep and your journeys, but when were the dead a comfort to the living?

He closed his eyes and whispered her name.

The arms which lifted him were quick and firm. He opened his eyes.

'Lie still,' said the Dragonfly, as he sprang from the altar. He dipped towards the ground and struggled above the heads of the astonished watchers before they could gather their wits. With the moon behind clouds, they had missed his approach, but now they cried out in rage and hurled insults in Aramaic whose meaning even Mark could guess.

With his heavy burden, the Dragonfly laboured like a galley in the Straits of Messina. He must surely sink, Mark thought. To increase their danger, the crowd had begun to throw stones, and the Dragonfly must dodge the sharp-edged missiles as well as keep to the air. At last they glided over the rim of the cliff and sank into darkness and safety.

They found shelter on a ledge midway down the side of the cliff and obscured from the Arabs by an overhanging rock; the sounds of their outraged pursuers came to them muffled as if by fathoms of water. The Dragonfly gasped; his strength was gone. But he lost no time in unbinding Mark and massaging his wrists and ankles.

It was hard for Mark to speak. What could he say? The Dragonfly's risk had been enormous. 'They would have killed you if they could.'

'Yes, the god would have gorged himself,' said the Dragonfly. 'Two victims instead of one. Now they will seize some poor Arab boy who came to watch the sacrifice.'

'There was every chance you would be caught. Yet you ran the risk for a stranger.'

'Stranger? We met in the Sik, did we not? I welcomed you. Could I let you die? My name, by the way, is Eros.'

'And mine is Mark.'

'I know.'

'But how did you learn of my danger?'

'I came to look for you in your house, and you had not returned. I had come earlier, and your sister had told me of the banquet. I flew to the palace, which was now dark. I retraced your route and found Gaius unconscious in the garden. When he revived and told me what had happened, I guessed where you were and came to the Place of Sacrifice. He was going to bring soldiers from the king, but I knew it would be too late by the time they had climbed the cliff. I crouched out of sight on this very ledge until the moon went behind a cloud and the people had hypnotized themselves with dancing and words.'

'You say you talked to Phoebe in the villa?'

'And took her to my mountain. She is waiting for you.'

'Can you take me there now?'

'I am not able to lift you so high. Phoebe was lighter.'

'To the villa then?'

'Perhaps. But rest now. It is too soon. We have things to talk about.'

'You have done me a great service,' said Mark, with a depth of feeling rare for him since Pompeii. His heart went out to this bold, miraculous being who had saved his life. He grasped his hand and Eros returned the pressure.

'Yes, I have helped you,' said Eros. 'Therefore I love you the more. But what have you done for me?'

Startled, Mark asked, 'What could I have done?'

'Nothing—yet.'

'Will there be something?'

'Soon. First we must talk. Phoebe loves me. She had hidden her heart in a little silver casket, but I found the key. There are hearts like agate in sunken galleys, and brave is the man who dives to recover them. There are hearts like an eagle's nest, on the tallest windy cliff. Where is your heart, Mark?'

'It is yours in friendship.'

'You do not know what I am going to ask of you.'

'Ask it then.'

'First you must serve me in little ways. See, I am hurt.' He held out his arm and disclosed a wound below the elbow. A flying stone had struck him. He tore a strip from his tunic.

'Bind my wound and thus draw close to me. A man loves where he lessens pain.'

Mark bound the wound, tenderly as if he were binding a child, and loved him. I have found a friend, he thought. In this wild land of rocks and tombs, the heart I lost in Pompeii has stirred, a little, and let me love. Phoebe will love him as I do. We will make him one with us; he will bring us his hurts and his loneliness.

'Have you other wounds?' he asked.

'Yes. They are not of the body.'

'Of the heart then. Loneliness perhaps?'

'Till tonight I was utterly alone. Like an ibex on a dark hill, with wolves in the valley. Half of me hungers still. For the love of a brother for brother, the clasp of hands; the field where comrades battle a common foe. Man's love for a woman has moods like the moon-drawn tide. It ebbs and flows, colours with the hour and the shape of the clouds. Man's love for his brother is constant. The clear hard burning of a diamond. Will you call me brother?'

'Yes, my brother.'

'And trust me?'

'You have saved my life.'

From a pouch hung at his side he drew a small vial. 'Drink,' he said. 'It will make you strong.'

Mark drank. The liquid was sweet with honey and bitter with a herb he did not recognize. It burned his throat and pulsed in his veins like lava.

'Now,' Eros said, 'you must jump from the ledge.'

'Jump?' he cried. 'I would die on the rocks! I have no wings like you.'

'I will give you wings. First you must trust me. Phoebe jumped.'

He looked at the Dragonfly with growing horror. 'From this cliff?'

'A taller one.'

'And—?'

'She fell on the rocks. I came to her and held out my hand and drew her beside me. Her body was whole, because she had trusted me. From her shoulders sprouted wings like my own. She said: "You must find Mark and give him wings."'

'I must see her at once!'

'After you have jumped. You should trust me without proof, as she did.'

Mark shook his head vehemently. 'I want to see her now.'

'I have spoken to you as a friend and brother. Your answer is no?'

'Until I have seen my sister.'

'Mark, Mark, what is love without trust? You have disappointed me. Come, I will take you back to your villa. Then I will send Phoebe to you. But you have lost her. She has chosen to remain with me.'

The Dragonfly rose to his feet and drew Mark beside him on the narrow ledge. 'I have loved you,' he said.

And Mark loved him. Dark, sweet currents flooded the tide pools of his body, those sun-parched pits of broken shell; a swift renewing tide freighted with sea grapes and the purple murex. Did he tell the truth, after all, this creature who seemed too beautiful to lie? This boy with amber in his hair and moonlight in his wings?

But the sirens, beautiful and sweet-tongued, had lured Ulysses. No, Mark dared not trust him.

A shadow sank from the moon. 'Mark, I have found you!'
'Phoebe,' he cried, 'it is true then. He has given you wings.'
'And you,' she said. 'You shall have them too. Trust him.'
'It is too late,' said the Dragonfly sadly.
'It is *not* too late,' Mark cried, and jumped.

He opened his eyes and found them standing above him, their faces anguished. They expected to find me dead, he thought, and truly I am dying. The pain is like a crucifixion, with a hundred nails piercing my body.

Eros knelt. 'Can you take my hand?' he asked doubtfully. Mark strained towards him, wrenching his body into a fury of pain. But the fingers of the Dragonfly eased him and somehow he struggled to his feet.

'Mark,' cried Phoebe. 'Your shoulders—they are sprouting wings!'

Eros examined them, marvelling. 'It will be a long time,' he said, 'before they are grown. You were not yet ready. But you were less unready than I feared.'

'You told me that love was a pine torch,' said Phoebe to Eros, 'kindled in a second. There is another kind, I think. A field set afire by the slow accumulating rays of a summer sun. Sudden wings or slow—either is love.'

MERVYN PEAKE

SAME TIME, SAME PLACE

That night, I hated father. He smelt of cabbage. There was cigarette ash all over his trousers. His untidy moustache was yellower and viler than ever with nicotine, and he took no notice of me. He simply sat there in his ugly armchair, his eyes half closed, brooding on the Lord knows what. I hated him. I hated his moustache. I even hated the smoke that drifted from his mouth and hung in the stale air above his head.

And when my mother came through the door and asked me whether I had seen her spectacles, I hated her too. I hated the clothes she wore; tasteless and fussy. I hated them deeply. I hated something I had never noticed before; it was the way the heels of her shoes were worn away on their outside edges—not badly, but appreciably. It looked mean to me, slatternly, and horribly human. I hated her for being human—like father.

She began to nag me about her glasses and the threadbare condition of the elbows of my jacket, and suddenly I threw my book down. The room was unbearable. I felt suffocated. I suddenly realized that I must get away. I had lived with these two people for nearly twenty-three years. I had been born in the room immediately overhead. Was this the life for a young man? To spend his evenings watching the smoke drift out of his father's mouth and stain that decrepit old moustache, year after year—to watch the worn away edges of my mother's heels—the dark-brown furniture and the familiar stains on the chocolate-coloured carpet? I would go away; I would shake off the dark, smug mortality of the place. I would forgo my birthright. What of my father's business into which I would step at his death? What of it? To hell with it.

I began to make my way to the door but at the third step I caught my foot in a ruck of the chocolate-coloured carpet and in reaching out my hand for support, I sent a pink vase flying.

Suddenly I felt very small and very angry. I saw my mother's mouth opening and it reminded me of the front door and the front door reminded me of my urge to escape—to where? To where?

I did not wait to find an answer to my own question, but, hardly knowing what I was doing, ran from the house.

The accumulated boredom of the last twenty-three years was at my back and it seemed that I was propelled through the garden gate from its pressure against my shoulder-blades.

The road was wet with rain, black and shiny like oilskin. The reflection of the streetlamps wallowed like yellow jellyfish. A bus was approaching—a bus to Piccadilly, a bus to the never-never land—a bus to death or glory.

I found neither. I found something which haunts me still.

The great bus swayed as it sped. The black street gleamed. Through the window a hundred faces fluttered by as though the leaves of a dark book were being flicked over. And I sat there, with a sixpenny ticket in my hand. What was I doing! Where was I going?

To the centre of the world, I told myself. To Piccadilly Circus, where anything might happen. What did I *want* to happen?

I wanted life to happen! I wanted adventure; but already I was afraid. I wanted to find a beautiful woman. Bending my elbow I felt for the swelling of my biceps. There wasn't much to feel. 'O hell,' I said to myself, 'O damnable hell: this is *awful*.'

I stared out of the window, and there before me was the Circus. The lights were like a challenge. When the bus had curved its way from Regent Street and into Shaftesbury Avenue, I alighted. Here was the jungle all about me and I was lonely. The wild beasts prowled around me. The wolf packs surged and shuffled. Where was I to go? How wonderful it would have been to have known of some apartment, dimly lighted; of a door that opened to the secret knock, three short ones and one long one—where a strawberry blonde was waiting—or perhaps, better still, some wise old lady with a cup of tea, an old lady, august and hallowed, and whose heels were not worn down on their outside edges.

But I knew nowhere to go either for glamour or sympathy. Nowhere except The Corner House.

I made my way there. It was less congested than usual. I had only to queue for a few minutes before being allowed into the great eating-palace on the first floor. Oh, the marble and the gold of it all! The waiters coming and going, the band in the distance—how different all this was from an hour ago, when I stared at my father's moustache.

For some while I could find no table and it was only when moving down the third of the long corridors between tables that I saw an old man leaving a table for two. The lady who had been sitting opposite him remained where she was. Had she left, I would have had no tale to tell. Unsuspectingly I took the place of the old man and in reaching for the menu lifted my head and found myself gazing into the midnight pools of her eyes.

My hand hung poised over the menu. I could not move for the head in front of me was magnificent. It was big and pale and indescribably proud—and what I would now call a greedy look, seemed to me then to be an expression of rich assurance; of majestic beauty.

I knew at once that it was not the strawberry blonde of my callow fancy that I desired for glamour's sake, nor the comfort of the tea-tray lady—but this glorious creature before me who combined the mystery and exoticism of the former with the latter's mellow wisdom.

Was this not love at first sight? Why else should my heart have hammered like a foundry? Why should my hand have trembled above the menu? Why should my mouth have gone dry?

Words were quite impossible. It was clear to me that she knew everything that was going on in my breast and in my brain. The look of love which flooded from her eyes all but unhinged me. Taking my hand in hers she returned it to my side of the table where it lay like a dead thing on a plate. Then she passed me the menu. It meant nothing to me. The hors-d'œuvres and the sweets were all mixed together in a dance of letters.

What I told the waiter when he came, I cannot remember, nor what he brought me. I know that I could not eat it. For an hour we sat there. We spoke with our eyes, with the pulse and stress of our excited breathing—and towards the end of this, our first meeting, with the tips of our fingers that in touching each other in the shadow of the teapot, seemed to speak a language richer, subtler and more vibrant than words.

At last we asked to go—and as I rose I spoke for the first time. 'Tomorrow?' I whispered. 'Tomorrow?' She nodded her magnificent head slowly. 'Same place? Same time?' She nodded again.

I waited for her to rise, but with a gentle yet authoritative gesture she signalled me away.

It seemed strange, but I knew I must go. I turned at the door and saw her sitting there, very still, very upright. Then I descended to the street and made my way to Shaftesbury Avenue, my head in a whirl of stars, my legs weak and trembling, my heart on fire.

I had not decided to return home, but found nevertheless that I was on my way back—back to the chocolate-coloured carpet, to my father in the ugly armchair—to my mother with her worn shoe heels.

When at last I turned the key it was near midnight. My mother had been crying. My father was angry. There were words, threats, and entreaties on all sides. At last I got to bed.

The next day seemed endless but at long last my excited fretting found some relief in action. Soon after tea I boarded the west-bound bus. It was already dark but I was far too early when I arrived at the Circus.

I wandered restlessly here and there, adjusting my tie at shop windows and filing my nails for the hundredth time.

At last, when waking from a day dream as I sat for the fifth time in Leicester Square, I glanced at my watch and found I was three minutes late for our tryst.

I ran all the way panting with anxiety but when I arrived at the table on the first floor I found my fear was baseless. She was there, more regal than ever, a monument of womanhood. Her large, pale face relaxed into an expression of such deep pleasure at the sight of me that I almost shouted for joy.

I will not speak of the tenderness of that evening. It was magic. It is enough to say that we determined that our destinies were inextricably joined.

When the time came for us to go I was surprised to find that the procedure of the previous night was once more expected of me. I could in no way make out the reason for it. Again I left her sitting alone at the table by the marble pillar. Again I vanished into the night alone, with those intoxicating words still on my lips. 'Tomorrow . . . tomorrow . . . same time . . . same place . . .'

The certainty of my love for her and hers for me was quite intoxicating. I slept little that night and my restlessness on the following day was an agony both for me and my parents.

Before I left that night for our third meeting, I crept into my mother's bedroom and opening her jewel box I chose a ring from among her few trinkets. God knows it was not worthy to sit upon my loved-one's finger, but it would symbolize our love.

Again she was waiting for me although on this occasion I arrived a full quarter of an hour before our appointed time. It was as though, when we were together, we were hidden in a veil of love—as though we were alone. We heard nothing else but the sound of our voices, we saw nothing else but one another's eyes.

She put the ring upon her finger as soon as I had given it to her. Her hand that was holding mine tightened its grip. I was surprised at its power. My whole body trembled. I moved my foot beneath the table to touch hers. I could find it nowhere.

When once more the dreaded moment arrived, I left her sitting upright, the strong and tender smile of her farewell remaining in my mind like some fantastic sunrise.

For eight days we met thus, and parted thus, and with every meeting we knew more firmly than ever, that whatever the difficulties that would result, whatever the forces against us, yet it was now that we must marry, now, while the magic was upon us.

On the eighth evening it was all decided. She knew that for my part it must be a secret wedding. My parents would never countenance so rapid an arrangement. She understood perfectly. For her part she wished a few of her friends to be present at the ceremony.

'I have a few colleagues,' she had said. I did not know what she meant, but her instructions as to where we should meet on the following afternoon put the remark out of my mind.

There was a registry office in Cambridge Circus, she told me, on the first floor of a certain building. I was to be there at four o'clock. She would arrange everything.

'Ah, my love,' she had murmured, shaking her large head slowly from side to side, 'how can I wait until then?' And with a smile unutterably bewitching, she gestured for me to go, for the great memorial hall was all but empty.

For the eighth time I left her there. I knew that women must have their secrets and must be in no way thwarted in regard to them, and so, once again I swallowed the question that I so longed to put to her. Why, O why had I always to leave her there—and why, when I arrived to meet her—was she always there to meet me?

On the following day, after a careful search. I found a gold ring in a box in my father's dressing table. Soon after three, having brushed my hair until it shone like sealskin I set forth with a flower in my buttonhole and a suitcase of belongings. It was a beautiful day with no wind and a clear sky.

The bus fled on like a fabulous beast, bearing me with it to a magic land.

But alas, as we approached Mayfair we were held up more than once for long stretches of time. I began to get restless. By the time the bus had reached Shaftesbury Avenue I had but three minutes in which to reach the Office.

It seemed strange that when the sunlight shone in sympathy with my marriage, the traffic should choose to frustrate me. I was on the top of the bus and having been given a very clear description of the building, was able, as we rounded at last in Cambridge Circus, to recognize it at once. When we came alongside my destination the traffic was held up again and I was offered the perfect opportunity of disembarking immediately beneath the building.

My suitcase was at my feet and as I stooped to pick it up I glanced at the windows on the first floor—for it was in one of those rooms that I was so soon to become a husband.

I was exactly on a level with the windows in question and commanded an unbroken view of the interior of a first floor room. It could not have been more than a dozen feet away from where I sat.

I remember that our bus was hooting away, but there was no movement in the traffic ahead. The hooting came to me as through a dream for I had become lost in another world.

My hand was clenched upon the handle of the suitcase. Through my eyes and into my brain an image was pouring. The image of the first floor room.

I knew at once that it was in that particular room that I was expected. I cannot tell you why, for during those first few moments I had not seen her.

To the right of the stage (for I had the sensation of being in a theatre) was a table loaded with flowers. Behind the flowers sat a small pin-striped registrar. There were four others in the room, three of whom kept walking to and fro. The fourth, an enormous bearded lady, sat on a chair by the window. As I stared, one of the men bent over to speak to her. He had the longest neck on earth. His starched collar was the length of a walking stick, and his small bony head protruded from its extremity like the skull of a bird. The other two gentlemen who kept crossing and re-crossing were very different. One was bald. His face and cranium were blue with the most intricate tattooing. His teeth were gold and they shone like fire in his mouth. The other was a well-dressed young man, and seemed normal enough until, as he came for a moment closer to the window I saw that instead of a hand, the cloven hoof of a goat protruded from the left sleeve.

And then suddenly it all happened. A door of their room must have opened for all at once all the heads in the room were turned in one direction and a moment later a something in white trotted like a dog across the room.

But it was no dog. It was vertical as it ran. I thought at first that it was a mechanical doll, so close was it to the floor. I could not observe its face, but I was amazed to see the long train of satin that was being dragged along the carpet behind it.

It stopped when it reached the flower-laden table and there was a good deal of smiling and bowing and then the man with the longest neck in the world placed a high stool in front of the table and, with the help of the young man with the goat-foot, lifted the white thing so that it stood upon the high stool. The long satin dress was carefully draped over the stool so that it reached to the floor on every side. It seemed as though a tall dignified woman was standing at the civic altar.

And still I had not seen its face, although I knew what it would be like. A sense of nausea overwhelmed me and I sank back on the seat, hiding my face in my hands.

I cannot remember when the bus began to move. I know that I went on and on and on and that finally I was told that I had reached the terminus. There was nothing for it but to board another bus of the same number and make the return journey. A strange sense of relief had by now begun to blunt the edge of my disappointment. That this bus would take me to the door of the house where I was born gave me a twinge of homesick pleasure. But stronger was my sense of fear. I prayed that there would be no reason for the bus to be held up again in Cambridge Circus.

I had taken one of the downstairs seats for I had no wish to be on an eye-level with someone I had deserted. I had no sense of having wronged her but she had been deserted nevertheless.

When at last the bus approached the Circus, I peered into the half darkness. A street lamp stood immediately below the registry office. I saw at once that there was no light in the office and as the bus moved past I turned my eyes to a group beneath the street lamp. My heart went cold in my breast.

Standing there, ossified as it were into a malignant mass—standing there as though they never intended to move until justice was done—were the five. It was only for a second that I saw them but every lamp-lit head is for ever with me—the long necked man with his bird skull head, his eyes glinting like chips of glass; to his right the small bald man, his tattooed scalp thrust forward, the lamplight gloating on the blue markings. To the left of the long-necked man stood the youth, his elegant body relaxed, but a snarl on his face that I still sweat to remember. His hands were in his pockets but I could see the shape of the hoof through the cloth. A little ahead of these three stood the bearded woman, a bulk of evil—and in the shadow that she cast before her I saw in that last fraction of a second, as the bus rolled me past, a big whitish head, very close to the ground.

In the dusk it appeared to be suspended above the kerb like a pale balloon with a red mouth painted upon it—a mouth that, taking a single diabolical curve, was more like the mouth of a wild beast than of a woman.

Long after I had left the group behind me—set as it were for ever under the lamp, like something made of wax, like something monstrous, long after I had left it I yet saw it all. It filled the bus. They filled my brain. They fill it still.

When at last I arrived home I fell weeping upon my bed. My father and mother had no idea what it was all about but they did not ask me. They never asked me.

That evening, after supper, I sat there, I remember, six years ago in my own chair on the chocolate-coloured carpet. I remember how I stared with love at the ash on my father's waistcoat, at his stained moustache, at my mother's worn away shoe heels. I stared at it all and I loved it all. I needed it all.

Since then I have never left the house. I know what is best for me.

KEITH ROBERTS

TIMOTHY

Anita was bored; and when she was bored odd things were liable to happen. Granny Thompson, who studied her granddaughter far more closely than she would have cared to admit, had been noticing a brooding look in her eyes for some days. She cast about for chores that would keep her mind off more exotic mischief for a time. 'There's the 'en run,' intoned the old lady. 'That wants a good gooin'-uvver fer a start. 'Arf the posts orl of a tip, 'oles everywheer . . . An' the path up ter the you-knows-wot. Nearly *went* on that, yisdey. Place gooin' orl of 'eap, an' yer sits there *moanin'* . . .'

Anita sneered, 'Chicken runs. Paths up to you-know-whats. I want to do something *interesting*, Gran. Like working a brand new spell. Can't we—'

'No we *kent!*' snapped the old lady irritably. 'Spells, spells, kent think o' nothink but *spells*. You want ter look a bit lively, my gel. Goo on out an' earn yer keep, sit there chopsin' . . . Goo on, git summat *done*. Git some o' that fat orf yer . . .'

Anita hissed furiously. She was very proud of her figure.

'Mackle up that there chair-back in the wosh'ouse,' snarled Granny, warming to her theme. 'Tek the truck down to ole Goody's place an' git them line props wot's bin cut an' waitin' 'arf a month. Git rid of orl that muck an' jollop yer chucked down by the copper 'ole a week larst *Toosdey*. Git the three o'clock inter Ket'rin', save my legs fer a change. 'Ole 'eap o' stuff we're run out on . . .'

'Oh *please*, Gran, not today . . .'

Granny Thompson softened a little. She didn't like going to Kettering either. 'Well goo on uvver to Aggie Everett's then an' git a couple of 'andfuls o' flour . . . an' watch she dunt put no chiblins o' nothink in *with* it. Aggie's sense o' wot's funny ent the same as anybody normal . . . An' when yer gits back yer kin goo up an' git orl that bird's-nest muck out o' the *thack*. I ent 'avin' that game agin, wadn't the same fer a month larst time I went up that there ladder . . .'

Anita fled, partly to escape her Granny's inventiveness, partly because there was some truth in the crack about her weight. In the winter she seemed to store fat like a dormouse, there was no answer to it; she'd tried a summer dress on only the day before and there had been too much Anita nearly everywhere. She decided to make a start on the chicken run. Levitation and spellraising were all very well in their way but there was something peculiarly satisfying once in a while in taking ordinary wood and nails and a perfectly normal hammer and lashing about as vigorously as possible. She rapidly tired of the job though. The rolls of wire netting were recalcitrant, possessed of a seemingly infinite number of hooks and snags that all but defied unravelment; once undone, they buried themselves gleefully in her palms. And the ground was soaked and nasty so that worms spurted out whenever she tried to drive a post. Anita leaned on the somewhat dishevelled end frame of the run and yawned. She probed the mind of the nearest of its occupants and got back the usual moronic burbling about the next feeding-time. Hens are easily the most boring of companions.

Anita snorted, pushed back her hair, wiped her hot face and decided to go to Aggie's for the flour. She knew her Granny still had a good stock of practically everything in the larder and that the errand was only an excuse to get her out from underfoot for a while, but that didn't matter. She could take the long path around the far side of Foxhanger; perhaps the wood creatures were waking up by now.

She walked between the trees, well muffled in jeans, boots, and donkey jacket. As she moved she scuffed irritably at twigs and leaves. She hated this time of the year with a peculiar loathing. February is a pointless sort of month: neither hot nor cold, neither winter nor spring. No animals, no birds, the sky a dull, uniform grey . . . Anita hung her head and frowned. If only things would get a move on . . . There were creatures in old tree stumps and deep in the ground but the few she was able to contact were dozy and grumpy and made it quite clear they wanted to be left alone for another six weeks, longer if possible. Anita decided she would like to hibernate, curled paws over nose in some brown crackling lair of leaves. Another year she really must try it; at least she might wake up feeling like doing something.

If she had expected any comfort from Aggie Everett she was disappointed. The old lady was morose; she had recently developed a head cold, had treated herself with a variety of ancient remedies and felt as she put it 'wuss in consiquence'. She was wearing a muffler knotted several times around her thin neck; her face was pale and even more scrinched-looking than usual while her nose, always a delicate member, glowed like a stop-light. She confided to Anita that things 'orl wanted a good shove, like'; her nephews would be coming down for the spring equinox and there were great plans for festivities but until then the Witches' Calendar was empty. The boys were away making cardboard boxes in far-off Northampton and there was nothing to do at all . . .

On the way back, weighed down by boredom and a bag of flour, Anita took a short cut across part of the Johnsons' land and saw Timothy on the horizon. Lacking anything better to do, she detoured so as to pass close by where he stood. She couldn't help noticing that Timothy looked as depressed as she felt. He had been made the previous spring to keep the birds off the new crops, so he was nearly a year old; and for nine months now he had had nothing to do but stand and be rained on and blown about by the wind and stare at the crown of Foxhanger wood away across the fields. Anita nodded mechanically as she trudged past. 'Afternoon, Timothy.' But it seemed he was too tired even to flap a ragged sleeve at her. She walked on.

Twenty yards away she stopped, struck by a thought. She stood still for a moment, weighing possibilities and feeling excited for the first time in weeks. Then she went back, stepping awkwardly on the chunky soil. She set the flour down, put her hands on her hips and looked at Timothy with her head on one side and her eyes narrowed appraisingly.

His face was badly weathered, of course, but that was unimportant; if anything, it tended to give him character. She walked up to him, brushed the lapels of his coat and tilted his old floppy hat to a more rakish angle. She made motions as if parting his wild straw hair. Timothy watched her enigmatically from his almond-shaped slits of eyes. He was a very well built scarecrow; the Johnson boys had put him together one weekend when they were home from college and Anita, who loved dolls and effigies, had watched the process with delight. She prodded and patted him, making sure his baling-wire tendons had not rotted from exposure. Timothy was still in good order; and although he was actually held up by a thick stake driven into the ground he had legs of his own, which was a great advantage. Anita walked around him, examining him with the air of a connoisseur. There were great possibilities in Timothy.

She moved back a few paces. Her boredom was forgotten now; she saw the chance of a brand new and very interesting spell. She squatted on her heels, folded her arms and rocked slightly to aid concentration.

Around her, winter-brown fields and empty sky waited silently; there was no breath of wind. Anita opened her eyes, ran through the incantation quickly to make sure she had it firmly set in her mind. Then she waved a hand and began to mutter rapidly.

A strange thing happened. Although the day remained still something like a breeze moved across the ground to Timothy. Had there been grass it might have waved; but there was no grass, and the soil twinkled and shifted and was still again. The wind touched the scarecrow and it seemed his shoulders stiffened, his head came up a trifle. One of his outstretched arms waved; a wisp of straw dropped from his cuff and floated to the ground. The stake creaked faintly to itself.

Anita was vastly pleased. She stood and did a little jig; then she looked around carefully. For a moment she was tempted to finish the job on the spot and activate Timothy; but the Johnson farmhouse was in sight and scarecrows that talk and walk, and sing maybe and dance, are best not seen by ordinary folk. Anita scurried off with her head full of plans. Twenty yards away she remembered the flour and went back for it. Timothy stirred impatiently on his post and a wind that was not a wind riffled the ragged tails of his coat. 'Sorry,' called Anita. 'I'll come back tonight, we can talk then. Besides, I'd better look up the rest of the trick, just to be sure.' She skipped away, not turning back again, and Timothy might or might not have waved . . .

The sky was deep grey when she returned, and the swell of land on which the scarecrow stood looked dark and rough as a dog's back. Timothy was silhouetted against the last of the light, a black drunken shape looking bigger than he really was. Anita breathed words over him, made passes; then she undid the wire and cord that held him to his stake and Timothy slid down and stood a little uncertainly on his curious feet. Anita held his arm in case he tumbled and broke himself apart. 'How do you feel?' she asked.

'Stiff,' said Timothy. His voice had a musty, earthy sort of quality and when he opened his mouth there was an old smell of dry soil and libraries. Anita walked slowly with him across the furrows; for a time he tottered and reeled like an old man or a sick one, then he began to get more assurance and strode out rapidly. At first his noseless round face looked odd in the twilight but Anita soon got used to it. After all Timothy was a personality, and personalities do not need to be conventionally handsome. She crossed the field with the scarecrow jolting beside her, headed for the cover of the nearest trees.

She found Timothy's mind was as empty as a thing could be; but that was part of his charm, because Anita could stock it with whatever she wanted him to know. At first the learning process was difficult because one question had a knack of leading to a dozen others and often the simplest things are hardest to explain. Thus,

'What's night?'

'Night is now. When it's dark.'

'What's dark?'

'When there isn't any light.'

'What's light?'

'Er . . . Light is when you can see Foxhanger across the fields. Dark is when you can't.'

'What's "see" . . . ?'

Anita was on firmer ground when it came to the question of scare-crows.

'What's a scarecrow?'

'A thing they put in a field when there are crops. The birds don't come because they think it's a man.'

'I was in a field. Am I a scarecrow?'

'No, you're not. Well maybe once on a time, but not any more. I changed you.'

'Am I a man?'

'You will be . . .' And Anita leaned on the arm of the giant, and felt the firmness of his wooden bones, and was very proud.

Timothy was back in his place by first light and Anita spent some time scuffing out tracks. When the scarecrow walked he had a way of plonking his feet down very hard so they sank deeply into the ground. If old Johnson saw the marks he might take it into his head to wait up and see what queer animal was on the prowl, and Anita hated the thought of Timothy being parted by a charge from a twelve bore. She was only just beginning to find out how interesting he could be.

During the following weeks Granny Thompson had little cause for complaint. She rarely saw her granddaughter; in the daytime Anita was usually mugging up fresh spellwork, or trying with the aid of a hugely battered *Britannica* to solve some of the more brilliant of Timothy's probings; and at night she was invariably and mysteriously absent. Her Granny finally raised the question of these absences.

'*Gallivantin'*,' snorted the elder Thompson. 'Yore got summat *on*, I knows that. The question is, *wot?*'

'But Gran, I don't know what you mean . . .'

'Kep me up 'arf the night larst night,' pronounced Granny. 'I could 'ear yer, gooin' on. Chelp chelp chelp, ev'ry night alike, but I kent 'ear nothink *answer* . . .' And then with a suddenly gimlet-like expression, 'Yore got a *bloke* agin, my gel, that's wot . . .'

'Really, Gran,' said Anita primly. 'The very *idea* . . .'

'Anita, what's a witch?'

'I've told you a dozen times, Timothy. A witch is somebody like me or Gran, or Aggie Everett I suppose. We can . . . talk to all sorts of people. Like yourself. Normal folk can't.'

'Why can't other people talk to me?'

'Well, they . . . it's hard to explain. It doesn't matter anyway, you've got me. I talk to you. I made you.'

'Yes, Anita . . .'

'I've got a new dress,' said Anita, pirouetting. Timothy stood stiffly by the gate and watched her. 'An' new shoes . . . but I'm not wearing them tonight because I don't want the damp to spoil them. I've got all new things because it's spring.' She held her hand out to Timothy and felt the brittle strength in him as he helped her over the gate. He had a sort of clumsy courtesy that was all his own. 'Anita, what's spring?'

Anita was exasperated. 'It's when . . . oh, the birds come back from Africa, don't ask me where's Africa because I shan't tell you . . . and there's nice scents in the air at night and the leaves come on the trees and you get new clothes and you can go out and everything feels different. I like spring.'

'What's "like"?'

Anita stopped, puzzled. 'Well, it's . . . I don't know. It's a feeling you have about people. I like you, for instance. Because you're gentle and you think about the things I think about.' Overhead a bat circled and dipped and the evening light showed redly through his wings and for a moment he almost spoke to Anita; then he saw the gauntness walking with her along the path and spun back up into the sky. 'I shall have to teach you about liking,' said Anita. 'There's still so many things you don't know.' She pelted ultrasonics after the noctule but if he was still in range he didn't answer. 'Come on, Timothy,' she said. 'I think we'll go to Deadman's Copse and see if the badgers are out yet.'

'Spells,' said Anita. 'Marjoram and wormsblood and quicksilver and cinnabar. Mandrakes and tar and honey. Divination by sieve and shears. Can you remember all that?'

'Yes, Anita.'

'You've got a very good brain, Timothy, you remember practically everything now. You've got most of the standard manual word for word, and I only read it through to you once. You really could be very useful . . . I think you're developing what they call a Balanced Personality. Though there's so much to put in; I still keep remembering bits I haven't done . . . Would you like to learn poetry?'

'What's poetry?'

Anita fumed momentarily, then started to laugh. 'I'm tired of defining things, it gets harder all the time. We shall just have to do some, that's all; I'll bring a book tomorrow.' And the day after she brought the book; it was one of her treasures, heavy and old and bound with leather. She opened Timothy's mind till he could read Shakespeare better than a man, then they went to Drawback to get a dramatic setting and Anita found Timothy's withered lips were just right for the ringing utterances of the old mad Lear. Next night they did a piece of *Tempest*, choosing for it the ghostly locale of Deadman's Copse. Anita read Ariel, though as she pointed out she was a little too well-developed for the part. Timothy made a fine Prospero; the cursing boomed out in great style though the bit about pegging people in oaks was if anything rather too realistic. When Timothy spoke the words Anita could see quite clearly how bad it would be to get mixed up with the knotty entrails of a tree as big as that.

The next day it rained, making the ground soggy and heavy. Mud covered Anita's ankles before she was halfway across the field. Timothy looked a little sullen and there was a pungent, rotting smell about his clothing that she found alarming. 'It's no good,' she said, 'we shall just have to get you under cover. I hate the idea of you standing out all the time; I don't expect you mind though.'

'Anita, what's "mind"?'

By mid-April Anita would normally have been busying herself about a hundred and one things connected with the field creatures and their affairs, but she was still mainly preoccupied with Timothy. Somehow she had stopped thinking of him as a scarecrow; the thing she had woken up was beginning to work by itself now and often when she came to release him he would bubble with notions of his own that had come to him in the grey time before the sun drained away his power. He asked her how she knew the names the bats called each other and why she was always sure when the weasel was too close for comfort; so she gave him a sixth sense, and portions of the seventh, eighth, and ninth for good measure. Then she could leave him standing on watch in his field and scurry off on her own business and Timothy would tattle and wheeze out the night's news when next he saw her. He found out where the fieldmice were building, and how the Hodges were faring on their rounds; then one of the hares under Drawback was taken by a lurcher and Timothy heard the scream and told Anita stiffly, making the death seem like a lab report; and Anita angrily gave him emotions and after that the tears would squeeze from somewhere and roll down his football face whenever he thought about killing.

A week later Anita came home with the dawn to find her Granny waiting for her. 'This,' said the old lady without preamble, ''as gotta *stop*.'

Anita flung herself down in one of the big armchairs and yawned. 'Wha', Gran . . .'

'Gallivantin',' said Granny Thompson sternly. 'Muckin' about wi' that gret thing uvver at the Johnsonses. *Ugghhh* . . . Giz me the creeps it does straight . . . Gret mucky thing orl straw an' stuff, sets yer teeth on edge ter *think* on it . . .' She crossed to one of the little windows and opened it. A breeze moved cold and sweet, ruffling Anita's hair. The room was shadowy but the sky outside was bright; somewhere a bird started to sing, all on his own. '*Gallivantin*',' said Granny again, as if to clinch matters.

Anita was nearly asleep; she'd used a lot of power that night and she was very tired. She said dreamily, 'He's not a thing, Gran. He's Timothy. He's very sweet. I invented him, he knows about *everything* . . .' Then a little more sharply, '*Gran!* How did you know—'

Granny Thompson sniffed. 'I knows wot I *knows* . . . There's ways an' means, my gel . . . Some as even you dunt know, artful though yer might be . . .'

Anita had a vision of something skulking in hedgerows, pouring itself across open ground like spilled jam. A very particular vision this, it lashed its tail and spat. She said reproachfully, 'You didn't play fair. You used a familiar . . .'

Granny looked virtuous. 'I ent sayin' I *did*, an' there agin I ent sayin' I *didn't* . . .'

'It was Vortigern,' said Anita, pouting. 'It must have been. None of the others would peach on me. But *him* . . .'

'Never mind 'ow I *knows*,' said Granny Thompson sternly. 'Or 'oo tole me. The thing is, yore gone fur *enough*. Any more an' I wunt be risponsible, straight I wunt . . .'

'But Gran, he's nice, And . . . well, I'm sorry for him. I don't like to think of him being left on his own now. It would be . . . well, like somebody dying almost. He's too clever now, can't just . . . *eeeooohhh* . . . jus' leave him li' that . . .'

'Clever,' muttered Granny, looking at the wall and not seeing it. 'That ent no call fer pity . . . You save yer pity fer the next world, my gel, there ent no place fer it 'ere . . . Brains, pah. Straw an' dirt an' muck orf the fields, that's brains. Same with 'im, same with 'em orl. You'll learn . . .'

But the homily was lost on Anita; she had incontinently fallen asleep.

She dreamed of Timothy that morning, woke and slept again to see if he would come back. He did; he was standing far away in his field and waving his arms to her and calling but his voice was so thick and distant she couldn't hear the words. But he wanted something, that was plain; and Anita woke and blinked, thought she knew what it was, and forgot again. She rubbed her eyes, saw the sunlight, felt the warmth of the air. It was lunchtime, and the day was as hot as June.

The fields were dark and rough and a full moon was rising. Anita crossed the open ground behind Foxhanger. A hunting bird called, close and low; she stopped and saw distant woods humped on their hills, looking like palls of smoke in the moonhaze. Timothy was waiting for her, a tiny speck a long way off in the night. When she reached him he looked gaunter than ever; his fingers stuck out in bundles from his sleeves, and his hat was askew. The night wind stirred his coat, moonlight oozed through the tatters and rags. Anita felt a queer stirring inside her; but she released him as usual and Timothy wriggled from the stake and dropped awkwardly to the ground. He said, 'It's a lovely night, Anita.' He took an experimental step or two. 'After you'd gone this morning my leg broke; but I mended it with wire and it's all right again now.' Anita nodded, her mind on other things. 'Good,' she said. 'Good, Timothy, that's fine . . .'

In February the ground had been bare and red-brown; now the harshness was lost under a new green hair. That was the corn Timothy had been made to protect. She took his arm. 'Timothy,' she said. 'Let's walk. I'm afraid I've got an awful lot to say.'

They paced the field, on the path that was beaten hard where the tractor came each day; and Anita told Timothy about the world. Everything she knew, about people dying, and living, and hoping; and how all things, even good things, get old and dirty and wornout, and the winds blow through them, and the rain washes them away. As it has always been, as it will be forever until the sun is cold. 'Timothy,' she said gently, 'one day . . . even my great Prince will be dust. It will be as though He had never been. He, and all the people of His house. Nobody knows why; nobody ever will. It's just the way things are.'

Timothy jolted gravely alongside; Anita held his thin arm and although he had no real face she could tell by his expression that he understood what she was saying. 'Timothy,' she said. 'I've got to go away . . .'

'Yes, Anita . . .'

She swallowed. 'It's right what Gran says. You're old now and nearly finished and there are so many other things to do. I haven't been fair, Timothy. You've just been a . . . well, a sort of toy. You know . . . I wasn't ever really interested in you. You were just something I made when I was bored. You sort of grew on me.'

'Yes, Anita . . .'

They turned at the furthest end of their walk. The air was wine-warm on her face and arms and Timothy smelled faintly of old brass spoons and what he was thinking about it was impossible to say. 'It's spring now,' said Anita. 'It's the time you put on a new dress and do your hair and find someone nice you can drive with or talk with or just walk along with and watch the night coming and the owls and the stars. They're the things that have to be done because they start right deep down inside you, in the blood. It's the same with animals nearly, they wake up and everything's fresh and green, and it's as if winter was the night and summer is one great long day . . .'

They had reached Timothy's stake. In the west the sky was still turquoise; an owl dropped down against the light like a black flake of something burned. Anita propped Timothy against his post. He seemed stiffer already and more lifeless somehow. She put his hat right; it was always flopping down. As she reached up she saw something shine silver on his wizened-turnip face. She was startled, until she remembered she had given him feelings. Timothy was crying.

She hugged him then, not knowing what to do. She felt the hardness of him and the crackling dryness, the knobs and angles of his bits-and-pieces body. 'Oh Timothy,' she said. 'Timothy, I'm sorry, but I just can't go with you any more. There won't be any spells for you after this, I've taken the power off . . .' She stepped back, not looking at him. 'I'll go now,' she said. 'This way's best, honestly. I won't tie you back on to your stick or anything, you can just stand here awhile and watch the bats and the owl. And in the morning you'll just sort of fade away; it won't hurt or anything . . .' She started to walk off down the slope, feeling the blades of new corn touch her calves. 'Goodbye, Timothy,' she called.

Something iron-hard snagged at her. She fell, rolled over horrified and tried to get up. Her ankles were caught; she wriggled and the night vanished, shut away by rough cloth that smelled of earth. 'Love,' croaked Timothy. 'Please, Anita, love . . .' And she felt his twiggy fingers move up and close over her breasts.

She looped like a caterpillar caught by the tail and her fists hit Timothy squarely, bang-bang. Dust flew, and the seeds of grass; then Anita was up and running down the hill, stumbling over the rough ground, and Timothy was close behind her, a flapping patch of darkness with his musty old head bobbing and his arms reaching out. His voice floated to her through the night. 'Anita . . . *love* . . .'

She reached the bottom of the field tousled and too shocked to defend herself at all, cut across the Johnsons' stackyard with Timothy still hard on her heels. A dog volleyed barks, subsided whimpering as he caught the strange scent on the air. Back up the hill, a doubling across Home Paddock; a horse bolted in terror as old cloth flapped at his eyes. Near the hedge Timothy gained once more, but he lost time climbing the gate. Anita spun around fifty yards away. 'Timothy, *go back! Timothy, no!*'

He came on again; she took three deep breaths, lifted her arm and flung something at him that crackled and fizzed and knocked a great lump of wadding from his shoulder. One arm flopped down uselessly but the rest of him still thumped towards her. Anita was angry now; her face was white in the moonlight and there was a little burning spot on each cheek and her mouth was compressed till her lips were hardly visible at all. 'Scarecrow!' she shouted. 'Old dirty thing made of straw! *Spider's home!*' She'd had time to aim; her next shot took Timothy full in the chest and bowled him backwards. He got up and came on again although he was much slower.

Anita waited for him on the little bridge over the Fyne-brook. She stood panting and pushing the hair out of her eyes with each hand in turn and the rage was white-hot now and choking her. Around her, brightnesses fizzed and sparked; as Timothy came within range she hit him again and again, arms and legs and head. Pieces flew from him and bounced across the grass. He reached the bridge but he was only a matchstick man now, his thin limbs glinting under tatters of cloth. Anita took a breath and held it, shut her eyes then opened them very wide, made a circle with her hands, thrust fire at Timothy. His wooden spine broke with a great sound; what was left of him folded in the middle, tumbled against the handrail of the bridge. He fell feet over head into the stream. The current seized him, whirling him off; he fetched up twenty yards away and lay quiet, humped in a reedbed like a heap of broken umbrellas.

Anita moved forward one foot at a time, ready to bolt again or throw more magic; but there was no need. Timothy was finished; he stayed still, the water rippling through his clothes. A little bright beetle shot from somewhere into his coatsleeve, came out at the elbow and sculled away down the stream. Timothy's face was pressed into mud so he could see nothing, but his voice still whispered in Anita's mind. '*Please . . . please . . .*'

She ran again, faster than ever. Along beside the brook, across the meadow, through Foxhanger, up the garden path. She burst into the kitchen of the cottage, spinning Granny Thompson completely around. Took the stairs three at a time and banged her bedroom door shut behind her. She flung herself on the bed and sobbed and wrapped blankets around her ears; but all night long, until the last of the power ran down, she could hear Timothy thinking old mouldy thoughts about rooks and winds, and worms in the thick red ground.

STERLING E. LANIER

THE KINGS OF THE SEA

I don't remember how magic came into the conversation at the club, but it had, somehow.

'Magic means rather different things to different people. To me . . .' Brigadier Donald Ffellowes, late of Her Majesty's forces, had suddenly begun talking. He generally sat, ruddy, very British and rather tired looking, on the edge of any circle. Occasionally he would add a date, a name, or simply nod, if he felt like backing up someone else's story. His own stories came at odd intervals and to many of us, frankly verged on the incredible, if not downright impossible. A retired artilleryman, Ffellowes now lived in New York, but his service had been all over the world, and in almost every branch of military life, including what seemed to be police or espionage work. That's really all there is to be said about either his stories or him, except that once he started one, no one ever interrupted him.

'I was attached to the embassy in Berlin in '38, and I went to Sweden for a vacation. Very quiet and sunny, because it was summer, and I stayed in Smaaland, on the coast, at a little inn. For a bachelor who wanted a rest, it was ideal, swimming every day, good food, and no newspapers, parades, crises, or Nazis.

'I had a letter from a Swedish pal I knew in Berlin to a Swedish nobleman, a local landowner, a sort of squire in those parts. I was so absolutely happy and relaxed I quite forgot about going to see the man until the second week of my vacation, and when I did, I found he wasn't at home in any case.

'He owned a largish, old house about three miles from the inn, also on the coast road, and I decided to cycle over one day after lunch. The inn had a bike. It was a bright, still afternoon, and I wore my bathing trunks under my clothes, thinking I might get a swim either at the house or on the way back.

'I found the place easily enough, a huge, dark-timbered house with peaked roofs, which would look very odd over here, and even at home. But it looked fine there, surrounded by enormous old pine trees, on a low bluff over the sea. There was a lovely lawn, close cut, spread under the trees. A big lorry—you'd say a moving van—was at the door, and two men were carrying stuff out as I arrived. A middle-aged woman, rather smartly dressed, was directing the movers, with her back to me so that I had a minute or two to see what they were moving. One of them had just manhandled a largish black chair, rather archaic in appearance, into the lorry and then had started to lift a long, carved wooden chest, with a padlock on it, in after the chair. The second man, who must have been the boss mover, was arguing with the lady. I didn't speak too much Swedish, although I'm fair at German, but the two items I saw lifted into the van were apparently the cause of the argument, and I got the gist of it, you know.

' "But Madame," the mover kept on saying, "Are you sure these pieces should be *destroyed*? They look very old."

' "You have been paid," she kept saying, in a stilted way. "Now get rid of it any way you like. Only take it away, now, at once."

'Then she turned and saw me, and believe it or not, blushed bright red. The blush went away quickly, though, and she asked me pretty sharply what I wanted.

'I answered in English, that I had a letter to Baron Nyderstrom. She switched to English, which she spoke pretty well, and appeared a bit less nervous. I showed her the letter, which was a simple note of introduction, and she read it and actually smiled at me. She wasn't a bad-looking woman—about forty-five, forty-eight, somewhere in there, anyway—but she was dressed to the nines, and her hair was dyed an odd shade of metallic brown. Also, she had a really hard mouth and eyes.

' "I'm so sorry," she said, "but the baron, who is my nephew, is away for a week and a half. I know he would have been glad to entertain an English officer friend of Mr—" here she looked at the letter "—of Mr Sorendson, but I'm afraid he is not around, while as you see, I am occupied. Perhaps another time?" She smiled brightly, and also rather nastily, I thought. "Be off with you," but polite.

'Well, really there was nothing to do except bow, and I got back on my bike and went wheeling off down the driveway.

'Halfway down the drive, I heard the lorry start, and I had just reached the road when it passed me, turning left, away from the direction of the inn, while I turned to the right.

'At that point something quite appalling happened. Just as the van left the drive, and also—as I later discovered—the estate's property line, something, a great weight, seemed to start settling over my shoulders, while I was conscious of a terrible cold, a cold which almost numbed me and took my wind away.

'I fell off the bike and half stood, half knelt over it, staring back after the dust of the lorry and completely unable to move. I remember the letters on the licence and on the back of the van, which was painted a dark red. They said *Solvaag and Mechius, Stockholm.*

'I wasn't scared, mind you, because it was all too quick. I stood staring down the straight dusty road in the hot sun, conscious only of a terrible weight and the freezing cold, the weight pressing me down and the icy cold numbing me. It was as if time had stopped. And I felt utterly depressed, too, sick and, well, *hopeless.*

'Suddenly, the cold and the pressure stopped. They were just gone, as if they had never been, and I was warm, in fact, covered with sweat, and feeling like a fool there in the sunlight. Also, the birds started singing among the birches and pines by the road, although actually, I suppose they had been all along. I don't think the whole business took over a minute, but it seemed like hours.

'Well, I picked up the bike, which had scraped my shins, and started to walk along, pushing it. I could think quite coherently, and I decided I had had either a mild coronary or a stroke. I seemed to remember that you felt cold if you had a stroke. Also, I was really dripping with sweat by now and felt all swimmy; you'd say dizzy. After about five minutes, I got on the bike and began to pedal, slowly and carefully, back to my inn, deciding to have a doctor check me out at once.

'I had only gone about a third of a mile, numbed still by shock—after all I was only twenty-five, pretty young to have a heart attack or a stroke, either—when I noticed a little cove, an arm of the Baltic, on my right, which came almost up to the road, with tiny blue waves lapping at a small beach. I hadn't noticed it on the way to the baron's house, looking the other way, I guess, but now it looked like heaven. I was soaked with sweat, exhausted by my experience, and now had a headache. That cool sea-water looked really marvellous, and as I said earlier, I had my trunks on under my clothes. There was even a towel in the bag strapped to the bike.

'I undressed behind a large pine tree ten feet from the road, and then stepped into the water. I could see white sand for about a dozen feet out, and then it appeared to get deeper quickly. I sat down in the shallow water, with just my neck sticking out, and began to feel human again.

Even the headache receded into the background. There was no sound but the breeze soughing in the trees and the chirping of a few birds, plus the splash of little waves on the shore behind me. I felt at peace with everything and shut my eyes, half sitting, half floating in the water. The sun on my head was warm.

'I don't know what made me open my eyes, but I must have felt something watching, some presence. I looked straight out to sea, the entrance of the little cove, as I opened them, and stared into a face which was looking at me from the surface of the water about eight feet away, right where it began to get deeper.'

No one in the room had moved or spoken once the story had started, and since Ffellowes had not stopped speaking since he began, the silence as he paused now was oppressive, even the muted sound of traffic outside seeming far off and unreal.

He looked around at us, then lit a cigarette and continued steadily.

'It was about two feet long, as near as I could tell, with two huge, oval eyes of a shade of amber yellow, set at the corners of its head. The skin looked both white and vaguely shimmery; there were no ears or nose that I could see, and there was a big, wide, flat mouth, opened a little, with blunt, shiny, rounded teeth. But what struck me most was the rage in the eyes. The whole impression of the face was vaguely—only vaguely, mind you—serpentine, snakelike, except for those eyes. They were mad, furious, raging, and not like an animal's at all, but like a man's. I could see no neck. The face "sat" on the water, so to speak.

'I had only a split second to take all this in, mind you, but I was conscious at once that whatever this was, it was livid at *me* personally, not just at people. I suppose it sounds crazy, but I *knew* this right off.

'I hadn't even moved, hadn't had a chance, when something flickered under the head, and a grip like a steel cable clamped onto my hip. I dug my heels in the sand and grabbed down, pushing as hard as I could, but I couldn't shake that grip. As I looked down, I saw what had hold of me and damn near fainted, because it was a hand. It was double the size of mine, dead white, and had only two fingers and a thumb, with no nails, but it was a hand. Behind it was a boneless-looking white arm like a giant snake or an eel, stretching away back towards the head, which still lay on the surface of the water. At the same time I felt the air was cold, almost freezing, as if a private iceberg was following me again, although not to the point of making me numb. Oddly enough, the cold didn't seem to be *in* the water, though I can't explain this very well.

'I pulled back hard, but I might as well have pulled at a tree trunk for all the good it did. Very steadily the pressure on my hip was increasing, and I knew that in a minute I was going to be pulled out to that head. I was kicking and fighting, splashing the water and clawing at that hand, but in the most utter silence. The hand and arm felt just like rubber, but I could feel great muscles move under the hard skin.

'Suddenly I began to scream. I knew my foothold on the bottom sand was slipping and I was being pulled loose so that I'd be floating in a second. I don't remember what I screamed, probably just yelling with no words. I knew for a certainty that I would be dead in thirty seconds, you see.' He paused, then resumed.

'My vision began to blur, and I seemed to be slipping, mentally, not physically, into a blind, cold world of darkness. But still I fought, and just as I began to be pulled loose from my footing, I heard two sounds. One was something like a machine-gun, but ringing through it I heard a human voice shouting and, I thought, shouting one long word. The shout was very strong, ringing and resonant, so resonant that it pierced through the strange mental fog I was in, but the word was in no language I knew. Then I blacked out, and that was that.

'When I opened my eyes, I was in a spasm of choking. I was lying face down on the little beach, my face turned sideways on my crossed arms, and was being given artificial respiration. I vomited up more water and then managed to choke out a word or two, probably obscene. There was a deep chuckle, and the person who had been helping me turned me over, so that I could see him. He pulled me up to a sitting position and put a tweed-clad arm around my shoulders, giving me some support while I recovered my senses.

'Even kneeling as he was, when I turned to look at him, I could see he was a very tall man, in fact, a giant. He was wearing a brown tweed suit with knickerbockers, heavy wool knee socks, and massive buckled shoes. His face was extraordinary. He was what's called an ash-blond, almost white-haired, and his face was very long, with high cheekbones, and also very white, with no hint of color in the cheeks. His eyes were green and very narrow, almost Chinese looking, and terribly piercing. Not a man you would ever forget if you once got a look at him. He looked about thirty-five, and was actually thirty, I later found out.

'I was so struck by his appearance, even though he was smiling gently, that I almost forgot what had happened to me. Suddenly I remembered though, and gave a convulsive start and tried to get up. As I did so, I turned to look at the water, and there was the cove, calm and serene, with no trace of that thing, or anything else.

'My new acquaintance tightened his grip on my shoulders and pulled me down to a sitting position, speaking as he did so.

'"Be calm, my friend. You have been through a bad time, but it is gone now. You are safe."

'The minute I heard his voice, I knew it was he who had shouted as I was being pulled under. The same timbre was in his speech now, so that every word rang like a bell, with a concealed purring under the words.

'I noticed more about him now. His clothes were soaked to the waist, and on one powerful hand he wore an immense ring set with a green seal stone, a crest. Obviously he had pulled me out of the water, and equally obviously, he was no ordinary person.

'"What was it," I gasped finally, "and how did you get me loose from it?"

'His answer was surprising. "Did you get a good look at it?" He spoke in pure, unaccented "British" English, I might add.

'"I did," I said with feeling. "It was the most frightful, bloody thing I ever saw, and people ought to be warned about this coast! When I get to a phone, every paper in Sweden *and* abroad will hear about it. They ought to fish this area with dynamite!"

'His answer was a deep sigh. Then he spoke. "Face-to-face, you have seen one of Jormungandr's Children," he said, "and that is more than I or any of my family have done for generations." He turned to face me directly and continued, "And I must add, my friend, that if you tell a living soul of what you have seen, I will unhesitatingly pronounce you a liar or a lunatic. Further, I will say I found you alone, having a seeming fit in this little bay, and saved you from what appeared to me to be a vigorous attempt at suicide."

'Having given me this belly-punch, he lapsed into a brooding silence, staring out over the blue water, while I was struck dumb by what I had heard. I began to feel I had been saved from a deadly sea monster only to be captured by an apparent madman.

'Then he turned back to me, smiling again. "I am called Baron Nyderstrom," he said, "and my house is just a bit down the road. Suppose we go and have a drink, change our clothes and have a bit of a chat."

'I could only stammer, "But your aunt said you were away, away for more than a week. I came to see you because I have a letter to you." I fumbled in my bathing suit, and then lurched over to my clothes under the trees. I finally found the letter, but when I gave it to him, he stuck it in his pocket. "In fact I was just coming from your house when I decided to have a swim here. I'd had a sick spell as I was leaving your gate, and I thought the cool water would help."

'"As you were leaving my gate?" he said sharply, helping me to get into my clothes. "What do you mean 'a sick spell', and what was that about my aunt?"

'As he assisted me, I saw for the first time a small, blue sports car, of a type unfamiliar to me, parked on the road at the head of the beach. It was in this, then, my rescuer had appeared. Half carrying, half leading me up the gentle slope, he continued his questioning, while I tried to answer him as best I could. I had just mentioned the lorry and the furniture as he got me into the left-hand bucket seat, having detailed in snatches my fainting and belief that I had had a mild stroke or heart spasm, when he got really stirred up.

'He levered his great body, and he must have been six-foot five, behind the wheel like lightning, and we shot off in a screech of gears and spitting of gravel. The staccato exhaust told me why I thought I had heard a machine-gun while fighting that incredible thing in the water.

'Well, we tore back up the road, into and up his driveway, and without a word, he slammed on the brakes and rushed into the house as if all the demons of hell were at his heels. I was left sitting stupefied in the car. I was not only physically exhausted and sick, but baffled and beginning again to be terrified. As I looked around the pleasant green lawn, the tall trees and the rest of the sunny landscape, do you know I wondered if through some error in dimensions, I had fallen out of my own proper space and landed in a world of monsters and lunatics!

'It could only have been a moment when the immense figure of my host appeared in the doorway. On his fascinating face was an expression which I can only describe as being mingled half sorrow, half anger. Without a word, he strode down his front steps and over to the car where, reaching in, he picked me up in his arms as easily as if I had been a doll instead of 175 pounds of British subaltern.

'He carried me up the steps and as he walked, I could hear him murmuring to himself in Swedish. It sounded to me like gibberish, with several phrases I could just make out being repeated over and over. "What could they do, what else could they do! She would not be warned. What else could they do?'

'We passed through a vast dark hall, with great beams high overhead, until we came to the back of the house, and into a large sunlit room, overlooking the sea, which could only be the library or study. There were endless shelves of books, a huge desk, several chairs, and a long, low padded window seat on which the baron laid me down gently.

'Going over to a closet in the corner, he got out a bottle of aquavit and two glasses, and handed me a full one, taking a more modest portion for himself. When I had downed it—and I never needed a drink more—he pulled up a straight-backed chair and set it down next to my head. Seating himself, he asked my name in the most serious way possible, and when I gave it, he looked out of the window a moment.

'"My friend," he said finally, "I am the last of the Nyderstroms. I mean that quite literally. Several rooms away, the woman you met earlier today is dead, as dead as you yourself would be, had I not appeared on the road, and from the same, or at least a similar cause. The only difference is that she brought this fate on herself, while you, a stranger, were almost killed by accident, and simply because you were present at the wrong time." He paused and then continued with the oddest sentence, although, God knows, I was baffled already. "You see," he said, "I am a kind of game warden and some of my charges are loose."

'With that, he told me to lie quiet and started to leave the room. Remembering something, however, he came back and asked if I could remember the name of the firm which owned the mover's lorry I had seen. Fortunately I could, for as I told you earlier, it was seared on my brain by the strange attack I had suffered while watching it go up the road. When I gave it to him, he told me again not to move and left the room for another, from which I could hear him faintly using a telephone. He was gone a long time, perhaps half an hour, and by the time he came back, I was standing looking at his books. Despite the series of shocks I had gone through, I now felt fairly strong, but it was more than that. This strange man, despite his odd threat, had saved my life, and I was sure that I was safe from *him* at least. Also, he was obviously enmeshed in both sorrow and some danger, and I felt strongly moved to try and give him a hand.

'As he came back into the room, he looked hard at me, and I think he read what I was thinking, because he smiled, displaying a fine set of teeth.

'"So—once again you are yourself. If your nerves are strong, I wish you to look on my late aunt. The police have been summoned and I need your help."

'Just like that! A dead woman in the house and he needed my help!

'Well, if he was going to get rid of me, why call the police? Anyway, I felt safe as I told you, and you'd have to see the man, as I did, to know why.

'At any rate, we went down the great hall to another room, much smaller, and then through that again until we found ourselves in a little sewing room, full of women's stuff and small bits of fancy furniture. There in the middle of the room lay the lady whom I had seen earlier telling the movers to go away. She certainly appeared limp, but I knelt and felt her wrist because she was lying face down. Sure enough, no pulse at all and quite cold. But when I started to turn her over, a huge hand clamped on my shoulder and the baron spoke. "I don't advise it," he said warningly. "Her face isn't fit to look at. She was frightened to death, you see."

'I simply told him I had to, and he just shrugged his shoulders and stepped back. I got my hands under one shoulder and started to turn the lady, but my God, as the profile came into view, I dropped her and stood up like a shot. From the little I saw, her mouth was drawn back like an animal's, showing every tooth, and her eye was wide open and glaring in a ghastly manner. That was enough for me.

'Baron Nyderstrom led me from the room and back into the library, where we each had another aquavit in silence.

'I started to speak, but he held up his hand in a kind of command, and started talking.

'"I shall tell the police that I passed you bathing on the beach, stopped to chat, and then brought you back for a drink. We found my aunt dead of heart failure and called the police. Now, sir, I like you, but if you will not attest to this same story, I shall have to repeat what I told you I would say at the beach, and I am well known in these parts. Also, the servants are away on holiday, and I think you can see that it would look ugly for you."

'I don't like threats, and it must have showed, because although it would have looked bad as all hell, still I wasn't going to be a party to any murders, no matter how well-planned. I told him so, bluntly, and he looked sad and reflective, but not particularly worried.

'"Very well", he said at length, "I can't really blame you, because you are in a very odd position." His striking head turned towards the window in brief thought, and then he turned back to face me directly and spoke.

'"I will make a bargain with you. Attest my statement to the police, and then let me have the rest of the day to talk to you. If, at the end of the day, I have not satisfied you about my aunt's death, you have my word, solemnly given, that I will go to the police station and attest *your* story, the fact that I have been lying and anything else you choose to say."

'His words were delivered with great gravity, and it never for one instant occurred to me to doubt them. I can't give you any stronger statement to show you how the man impressed me. I agreed straightaway.

'In about ten minutes the police arrived, and an ambulance came with them. They were efficient enough, and very quick, but there was one thing that showed through the whole of the proceedings, and it was that the Baron Nyderstrom was *somebody*! All he did was state that his aunt had died of a heart attack and that was that! I don't mean the police were serfs, or crooks either for that matter. But there was an attitude of deference very far removed from servility or politeness. I doubt if royalty gets any more nowadays, even in England. When he had told me earlier that his name was "known in these parts", it was obviously the understatement of the decade.

'Well, the police took the body away in the ambulance, and the baron made arrangements for a funeral parlour and a church with local people over the telephone. All this took awhile, and it must have been 4.30 when we were alone again.

'We went back into the library. I should mention that he had got some cold meat, bread, and beer from a back pantry, just after the police left, and so now we sat down and made ourselves some sandwiches. I was ravenous, but he ate quite lightly for a man of his size, in fact only about a third of what I did.

'When I felt full, I poured another glass of an excellent beer, lit a cigarette, sat back and waited. With this man, there was no need of unnecessary speech.

'He was sitting behind his big desk facing me, and once again that singularly attractive smile broke through.

'"You are waiting for your story, my friend, if I may call you so. You shall have it, but I ask your word as a man of honour that it not be for repetition." He paused briefly. "I know it is yet a further condition, but if you do not give it, there is no recourse except the police station and jail for me. If you do, you will hear a story and perhaps—perhaps, I say, because I make no promises—see and hear something which no man has seen or heard for many, many centuries, save only for my family and not many of them. What do you say?"

'I never hesitated for a second. I said "yes", and I should add that I've never regretted it. No, never.'

Ffellowes's thoughts seemed far away, as he paused and stared out into the murky New York night, dimly lit by shrouded street lamps, and the fog lights on passing cars. No one spoke, and no sound broke the silence of the room but a muffled cough. He continued.

'Nyderstrom next asked me if I knew anything about Norse mythology. Now this question threw me for an absolute loss. What did a dangerous animal and an awful death have to do with Norse mythology, to say nothing of a possible murder?

'However, I answered I'd read of Odin, Thor, and a few other gods as a child in school, the Valkyries, of course, and that was about it.

'"Odin, Thor, the Valkyries, and a few others?" My host smiled, "You must understand that they are rather late Norse and even late German adaptions of something much older. Much, much older, something with its roots in the dawn of the world.

'" Listen," he went on, speaking quietly but firmly, "and when I have finished we will wait for that movers' truck to return. I was able to intercept it, and what it took, because of that very foolish woman, must be returned."

'He paused as if at a loss how to begin, and then went on. His bell-like voice remained muted, but perfectly audible, while he detailed one of the damnedest stories I've ever heard. If I hadn't been through what I had that day, and if he hadn't been what he was, I could have thought I was listening to the Grand Master of all the lunatics I'd ever met.

'"Long ago," he said, "my family came from inner Asia. They were some of the people the latercomers called *Aesir*, the gods of Valhalla, but they were not gods, only a race of wandering conquerors. They settled here, on this spot, despite warnings from the few local inhabitants, a small, dark, shore-dwelling folk. This house is built on the foundations of a fortress, a very old one, dating at the very least back to the second century BC. It was destroyed later in the wars of the sixteenth century, but that is modern history.

'"At any rate, my remote ancestors began soon to lose people. Women bathing, boys fishing, even full-grown warriors out hunting, they would vanish and never return. Children had to be guarded and so did the livestock, which had a way of disappearing also, although that of course was preferable to the children.

'"Finally, for no trace of the mysterious marauders could be found, the chief of my family decided to move away. He had prayed to his gods and searched zealously, but the reign of silent, stealthy terror never ceased, and no human or other foe could be found.

'"But before he gave up, the chief had an idea. He sent presents and a summons to the shaman, the local priest, not of our own people, but of the few, furtive, little shore folk, the strand people, who had been there when we came. We despised and avoided them, but we had never harmed them. And the bent little shaman came and answered the chief's questions.

'"What he said amounted to this. We, that is my people, had settled on the land made sacred in the remote past to Jormungandr. Now Jormungandr in the standard Norse sagas and myths is the great, world-circling sea serpent, the son of the Aesir renegade Loki and a giantess. He is a monster who on the day of Ragnarok will arise to assault Asgard. But actually, these myths are based on something quite, quite different. The ancient Jormungandr was a god of the sea all right, but he was here before any Norsemen, and he had children, who were semi-mortal and very, very dangerous. All the Asgard business was invented later, by people who did not remember the reality, which was both unpleasant and a literal, living menace to ancient men.

'"My ancestor, the first of our race to rule here, asked what he could do to abate the menace. Nothing, said the shaman, except go away. Unless, if the chief were brave enough, he, the shaman, could summon the Children of the God, and the chief could ask *them* how *they* felt!

'"Well, my people were anything but Christians in those days, and they had some rather nasty gods of their own. Also, the old chief, my ancestor, was on his mettle, and he liked the land he and his tribe had settled. So—he agreed, and although his counsellors tried to prevent him, he went alone at night to the shore with the old shaman of the shore people. And what is more, he returned.

'"From that day to this we have always lived here on this stretch of shore. There is a vault below the deepest cellar where certain things are kept and a ceremony through which the eldest son of the house of Nyderstrom must pass. I will not tell you more about it save to say that it involves an oath, one we have never broken, and that the other parties to the oath would not be good for men to see. You should know, for you have seen one!"

'I had sat spellbound while this rigmarole went on, and some of the disbelief must have showed in my eyes, because he spoke rather sharply all at once.

'"What do you think the Watcher in the Sea was, the 'animal' that seized you? If it had been anyone else in that car but myself—!"

'I nodded, because after recalling my experience on my swim, I was less ready to dismiss his story, and I had been in danger of forgetting my adventure. I also apologized and he went on talking.

'"The woman you spoke to was my father's much younger sister, a vain and arrogant woman of no brainpower at all. She lived a life in what is now thought of as society, in Stockholm, on a generous allowance from me, and I have never liked her. Somewhere, perhaps as a child, she learned more than she should about the family secret, which is ordinarily never revealed to our women.

'"She wished me to marry and tried ceaselessly to entrap me with female idiots of good family whom she had selected.

'"It is true that I must someday marry, but my aunt irritated me beyond measure, and I finally ordered her out of the house and told her that her allowance would cease if she did not stop troubling me. She was always using the place for house parties for her vapid friends, until I put a stop to it.

'"I knew when I saw her body what she had done. She must have found out that the servants were away and that I would be gone for the day. She sent men from Stockholm. The local folk would not obey such an order from her, in my absence. She must have had duplicate keys, and she went in and down and had moved what she should never have seen, let alone touched. It was sacrilege, no less, and of a very real and dangerous kind. The fool thought the things she took held me to the house, I imagine.

'"You see," he went on, with more passion in his voice than I had previously heard. "*They* are not responsible. They do not see things as we do. They regarded the moving of those things as the breaking of a trust, and they struck back. You appeared, because of the time element, to have some connection, and they struck at you. You do see what I mean, don't you?"

'His green eyes fixed themselves on me in an open appeal. He actually wanted sympathy for what, if his words were true, must be the damnedest set of beings this side of madness. And even odder, you know, he had got it. I had begun to make a twisted sense of what he said, and on that quiet evening in the big shadowed room, I seemed to feel an ancient and undying wrong, moreover one which badly needed putting right.

'He seemed to sense this and went on, more quietly.

'"You know, I still need your help. Your silence later, but more immediate help now. Soon that lorry will be here and the things it took must be restored.

'"I am not now sure if I can heal the breach. It will depend on the Others. If they believe me, all will go as before. If not—well, it was my family who kept the trust, but also who broke it. I will be in great danger, not only to my body but also to my soul. Their power is not all of the body.

'"We have never known," he went on softly, "why they love this strip of coast. It is not used, so far as we know, for any of their purposes, and they are not subject to our emotions or desires in any case. But they do, and so the trust is honoured."

'He looked at his watch and murmured "six o'clock". He got up and went to the telephone, but as his hand met the receiver, we both heard something.

'It was a distant noise, a curious sound, as if, far away somewhere, a wet piece of cloth were being dragged over stone. In the great silent house, the sound could not be localized, but it seemed to me to come from deep below us, perhaps in a cellar. It made my hair stiffen.

'"Hah," he muttered. "They are stirring. I wonder—"

'As he spoke, we both became conscious of another noise, one which had been growing upon us for some moments unaware, that of a powerful motor engine. Our minds must have worked together for as the engine noise grew, our eyes met and we both burst into simultaneous gasps of relief. It could only be the furniture van, returning at last.

'We both ran to the entrance. The hush of evening lay over the estate, and shadows were long and dark, but the twin lights turning into the drive cast a welcome luminance over the entrance.

'The big lorry parked again in front of the main entrance, and the two workmen I had seen earlier got out. I could not really understand the rapid gunfire Swedish, but I gathered the baron was explaining that his aunt had made a mistake. At one point both men looked appalled, and I gathered that Nyderstrom had told them of his aunt's death. (He told me later that he had conveyed the impression that she was unsound mentally: it would help quiet gossip when they saw a report of the death.)

'All four of us went around to the rear of the van, and the two men opened the doors. Under the baron's direction they carried out and deposited on the gravel the two pieces of furniture I had seen earlier. One was the curious chair. It did not look terribly heavy, but it had a box bottom, solid sides instead of legs and no arm rests. Carved on the oval-topped head was a hand grasping a sort of trident, and when I looked closely, I got a real jolt. The hand had only two fingers and a thumb, all without nails, and I suddenly felt in my bones the reality of my host's story.

'The other piece was the small, plain, rectangular chest, a bit like a large toy chest, with short legs ending in feet like a duck's. I mean three-toed and *webbed*, not the conventional "duck foot" of the antique dealers.

'Both the chair and the chest were made of a dark wood, so dark it looked oily, and they had certainly not been made yesterday.

'Nyderstrom had the two men put the two pieces in the front hall and then paid them. They climbed back into their cab, so far as I could make out, apologizing continuously for any trouble they might have caused. We waved from the porch and then watched the lights sweep down the drive and fade into the night. It was fully dark now, and I suddenly felt a sense of plain old-fashioned fright as we stood in silence on the dark porch.

'"Come," said the baron, suddenly breaking the silence, "we must hurry. I assume you will help?"

'"Certainly," I said. I felt I had to, you see, and had no lingering doubts at all. I'm afraid that if he'd suggested murdering someone, by this time I'd have agreed cheerfully. There was a compelling, hypnotic power about him. Rasputin was supposed to have had it and Hitler also, although I saw *him* plenty, and never felt it. At any rate, I just couldn't feel that anything this man wanted was wrong.

'We manhandled the chair and the chest into the back of the house, stopping at last in a back hall in front of a huge oaken door, which appeared to be set in a stone wall. Since the house was made of wood, this stone must have been part of the original building, the ancient fort, I guess, that he'd mentioned earlier.

'There were three locks on the door, a giant old padlock, a smaller newer one and a very modern-looking combination. Nyderstrom fished out two keys, one of them huge, and turned them. Then, with his back to me, he worked the combination. The old house was utterly silent, and there was almost an atmospheric hush, the kind you get when a bad thunderstorm is going to break. Everything seemed to be waiting, waiting for something to happen.

'There was a click and Nyderstrom flung the great door open. The first thing I noticed was that it was lined with steel on the other, inner side, and the second, that it opened on a broad flight of shallow steps leading down on a curve out of sight into darkness. The third impression was not visual at all. A wave of odour, strong but not unpleasant, of tide pools, seaweed, and salt air poured out of the opening. And there were several large patches of water on the highest steps, large enough to reflect the light.

'Nyderstrom closed the door again gently, not securing it, and turned to me. He pointed, and I now saw on one wall of the corridor to the left of the door, about head height, a steel box, also with a combination lock. A heavy cable led from it down to the floor. Still in silence, he adjusted the combination and opened the box. Inside was a knife switch, with a red handle. He left the box open and spoke, solemnly and slowly.

'"I am going down to a confrontation. You must stay right here, with the door open a little, watching the steps. I may be half an hour, but at most three quarters. If I come up *alone*, let me out. If I come up *not* alone, slam the door, turn the lock and throw that switch. Also if anything *else* comes up, do so. This whole house, under my direction, and at my coming of age, was extensively mined and you will have exactly two and a half minutes to get as far as possible from it. Remember, at *most*, three-quarters of an hour. At the end of that time, even if nothing has happened, you will throw that switch and run . . . !"

'I could only nod. There seemed to be nothing to say, really.

'He seemed to relax a little, patted me on the shoulders, and turned to unlock the strange chest. Over his shoulder he talked to me as he took things out. "You are going to see one thing at any rate, a true Sea King in full regalia. Something, my friend, no one has seen who is not a member of my family since the late Bronze Age."

'He stood up and began to undress quickly, until he stood absolutely naked. I have never seen a more wonderful figure of a man, pallid as an ivory statue, but huge and splendidly formed. On his head, from out of the stuff in the chest, he had set a narrow coronet, only a band in the back, but rising to a flanged peak in front. Mounted in the front peak was a plaque on which the three-fingered hand and trident were outlined in purple gems. The thing was solid gold. Nyderstrom then stooped and pulled on a curious, short kilt, made of some scaly hide, like a lizard's and coloured an odd green-gold. Finally, he took in his right hand a short, curved, gold rod, ending in a blunt, stylized trident.

'We looked at each other a moment and then he smiled. "My ancestors were very successful Vikings," he said, still smiling. "You see, they always could call on *help*."

'With that, he swung the door open and went marching down the steps. I half shut it behind him and settled down to watch and listen.

'The sound of his footsteps receded into the distance, but I could still hear them in the utter silence for a long time. His family vault, which I was sure connected somehow with the sea, was a long way down. I crouched, tense, wondering if I would ever see him again. The whole business was utterly mad, and I believed every word of it. I still do.

'The steps finally faded into silence. I checked my watch and found ten minutes had gone by.

'Suddenly, as if out of an indefinite distance, I heard his voice. I recognized it instantly, for it was a long quavering call, sonorous and bell-like, very similar to what I had heard when he rescued me in the afternoon. The sound came from far down in the earth, echoing faintly up the dank stairs and died into silence. Then it came again, and when it died, yet again.

'My heart seemed to stop. I knew that this brave man was summoning something no man had a right to see and calling a council in which no one with human blood in his veins should sit.

'Silence, utter and complete, followed. I could hear nothing, save for an occasional faint drop of water falling somewhere out of my range of vision.

'I glanced at my watch. Twenty-one minutes had gone by. The minutes seemed to crawl endlessly, meaninglessly. I felt alone and in a strange dream, unable to move, frozen, an atom caught in a mesh beyond my comprehension.

'Then far away, I heard it, a faint sound. It was faint but regular, and increasing in volume, measured and remorseless. It was a tread, and it was coming up the stair in my direction.

'I glanced at my watch, thirty-four minutes. It could be my friend, still within his self-appointed limits of time. The step came nearer, nearer still. It was, so far as my straining ear could judge, a single step. It progressed further, and suddenly into the circle of light stepped Nyderstrom.

'He was alone and as he came up he waved in greeting. He was dripping wet and the light gleamed on his shining body. I threw the door wide and he stepped through.

'As his head emerged into the light, I stepped back, almost involuntarily. There was a look of exaltation and wonder on it, such as I have never seen on a human face. The strange green eyes flashed, and there was a faint flush on the high cheekbones. He looked like a man who has seen a vision of Paradise.

'He walked rather wearily, but firmly, over to the switch box, which he closed and locked. Then he turned to me, still with that blaze of radiance on his face.

' "All is well, my friend. They are again at peace with men. They have accepted me and the story of what has happened. All will be well now, with my house, and with me."

'I stared at him hard, but he said no more and began to divest himself of his incredible regalia. He had one more thing to say, and I can hear it still as if it were yesterday, spoken almost as an afterthought.

' "They say the blood of the guardians is getting too thin again. But that also is settled. I have seen my bride." '

LARRY NIVEN

NOT LONG BEFORE THE END

A swordsman battled a sorcerer, once upon a time.

In that age such battles were frequent. A natural antipathy exists between swordsmen and sorcerers, as between cats and small birds, or between rats and men. Usually the swordsman lost, and humanity's average intelligence rose some trifling fraction. Sometimes the swordsman won, and again the species was improved; for a sorcerer who cannot kill one miserable swordsman is a poor excuse for a sorcerer.

But this battle differed from the others. On one side, the sword itself was enchanted. On the other, the sorcerer knew a great and terrible truth.

We will call him the Warlock, as his name is both forgotten and impossible to pronounce. His parents had known what they were about. He who knows your name has power over you, but he must speak your name to use it.

The Warlock had found his terrible truth in middle age.

By that time he had travelled widely. It was not from choice. It was simply that he was a powerful magician, and he used his power, and he needed friends.

He knew spells to make people love a magician. The Warlock had tried these, but he did not like the side effects. So he commonly used his great power to help those around him, that they might love him without coercion.

He found that when he had been ten to fifteen years in a place, using his magic as whim dictated, his powers would weaken. If he moved away, they returned. Twice he had had to move, and twice he had settled in a new land, learned new customs, made new friends. It happened a third time, and he prepared to move again. But something set him to wondering.

Why should a man's powers be so unfairly drained out of him?

It happened to nations too. Throughout history, those lands which had been richest in magic had been overrun by barbarians carrying swords and clubs. It was a sad truth, and one that did not bear thinking about, but the Warlock's curiosity was strong.

So he wondered, and he stayed to perform certain experiments.

His last experiment involved a simple kinetic sorcery set to spin a metal disc in midair. And when that magic was done, he knew a truth he could never forget.

So he departed. In succeeding decades he moved again and again. Time changed his personality, if not his body, and his magic became more dependable, if less showy. He had discovered a great and terrible truth, and if he kept it secret, it was through compassion. His truth spelled the end of civilization, yet it was of no earthly use to anyone.

So he thought. But some five decades later (the date was on the order of 12000 BC) it occurred to him that all truths find a use somewhere, sometime. And so he built another disc and recited spells over it, so that (like a telephone number already dialled but for one digit) the disc would be ready if ever he needed it.

The name of the sword was Glirendree. It was several hundred years old, and quite famous.

As for the swordsman, his name is no secret. It was Belhap Sattlestone Wirldess ag Miracloat roo Cononson. His friends, who tended to be temporary, called him Hap. He was a barbarian, of course. A civilized man would have had more sense than to touch Glirendree, and better morals than to stab a sleeping woman. Which was how Hap had acquired his sword. Or vice versa.

The Warlock recognized it long before he saw it. He was at work in the cavern he had carved beneath a hill, when an alarm went off. The hair rose up, tingling, along the back of his neck. 'Visitors,' he said.

'I don't hear anything,' said Sharla, but there was an uneasiness to her tone. Sharla was a girl of the village who had come to live with the Warlock. That day she had persuaded the Warlock to teach her some of his simpler spells.

'Don't you feel the hair rising on the back of your neck? I set the alarm to do that. Let me just check . . .' He used a sensor like a silver hula hoop set on edge. 'There's trouble coming. Sharla, we've got to get you out of here.'

'But . . .' Sharla waved protestingly at the table where they had been working.

'Oh, that. We can quit in the middle. That spell isn't dangerous.' It was a charm against love-spells, rather messy to work, but safe and tame and effective. The Warlock pointed at the spear of light glaring through the hoop-sensor. 'That's dangerous. An enormously powerful focus of mana power is moving up the west side of the hill. You go down the east side.'

'Can I help? You've taught me *some* magic.'

The magician laughed a little nervously. 'Against that? That's Glirendree. Look at the size of the image, the colour, the shape. No. You get out of here, and right now. The hill's clear on the eastern slope.'

'Come with me.'

'I can't. Not with Glirendree loose. Not when it's already got hold of some idiot. There are obligations.'

They came out of the cavern together, into the mansion they shared. Sharla, still protesting, donned a robe and started down the hill. The Warlock hastily selected an armload of paraphernalia and went outside.

The intruder was halfway up the hill: a large but apparently human being carrying something long and glittering. He was still a quarter of an hour downslope. The Warlock set up the silver hula hoop and looked through it.

The sword was a flame of mana discharge, an eye-hurting needle of white light. Glirendree, right enough. He knew of other, equally powerful mana foci, but none were portable, and none would show as a sword to the unaided eye.

He should have told Sharla to inform the Brotherhood. She had that much magic. Too late now.

There was no coloured borderline to the spear of light.

No green fringe effect meant no protective spells. The swordsman had not tried to guard himself against what he carried. Certainly the intruder was no magician, and he had not the intelligence to get the help of a magician. Did he know *nothing* about Glirendree?

Not that that would help the Warlock. He who carried Glirendree was invulnerable to any power save Glirendree itself. Or so it was said.

'Let's test that,' said the Warlock to himself. He dipped into his armload of equipment and came up with something wooden, shaped like an ocarina. He blew the dust off it, raised it in his fist and pointed it down the mountain. But he hesitated.

The loyalty spell was simple and safe, but it did have side effects. It lowered its victim's intelligence.

'Self-defence,' the Warlock reminded himself, and blew into the ocarina.

The swordsman did not break stride. Glirendree didn't even glow; it had absorbed the spell that easily.

In minutes the swordsman would be here. The Warlock hurriedly set up a simple prognostics spell. At least he could learn who would win the coming battle.

No picture formed before him. The scenery did not even waver.

'Well, now,' said the Warlock. '*Well*, now!' And he reached into his clutter of sorcerous tools and found a metal disc. Another instant's rummaging produced a double-edged knife, profusely inscribed in no known language, and very sharp.

At the top of the Warlock's hill was a spring, and the stream from that spring ran past the Warlock's house. The swordsman stood leaning on his sword, facing the Warlock across that stream. He breathed deeply, for it had been a hard climb.

He was powerfully muscled and profusely scarred. To the Warlock it seemed strange that so young a man should have found time to acquire so many scars. But none of his wounds had impaired motor functions. The Warlock had watched him coming up the hill. The swordsman was in top physical shape.

His eyes were deep blue and brilliant, and half an inch too close together for the Warlock's taste.

'I am Hap,' he called across the stream. 'Where is she?'

'You mean Sharla, of course. But why is that your concern?'

'I have come to free her from her shameful bondage, old man. Too long have you—'

'Hey, hey, hey. Sharla's my *wife*.'

'Too long have you used her for your vile and lecherous purposes. Too—'

'She stays of her own free will, you nit!'

'You expect me to believe that? As lovely a woman as Sharla, could she love an old and feeble warlock?'

'Do I look feeble?'

The Warlock did not look like an old man. He seemed Hap's age, some twenty years old, and his frame and his musculature were the equal of Hap's. He had not bothered to dress as he left the cavern. In place of Hap's scars, his back bore a tattoo in red and green and gold, an elaborately curlicued pentagramic design, almost hypnotic in its extradimensional involutions.

'Everyone in the village knows your age,' said Hap. 'You're two hundred years old, if not more.'

'Hap,' said the Warlock. 'Belhap something-or-other roo Cononson. Now I remember. Sharla told me you tried to bother her last time she went to the village. I should have done something about it then.'

'Old man, you lie. Sharla is under a spell. Everybody knows the power of a warlock's loyalty spell.'

'I don't use them. I don't like the side effects. Who wants to be surrounded by friendly morons?' The Warlock pointed to Glirendree. 'Do you know what you carry?'

Hap nodded ominously.

'Then you ought to know better. Maybe it's not too late. See if you can transfer it to your left hand.'

'I tried that. I can't let go of it.' Hap cut at the air, restlessly, with his sixty pounds of sword. 'I have to sleep with the damned thing clutched in my hand.'

'Well, it's too late then.'

'It's worth it,' Hap said grimly. 'For now I can kill you. Too long has an innocent woman been subjected to your lecherous—'

'I know, I know.' The Warlock changed languages suddenly, speaking high and fast. He spoke thus for almost a minute, then switched back to Rynaldese. 'Do you feel any pain?'

'Not a twinge,' said Hap. He had not moved. He stood with his remarkable sword at the ready, glowering at the magician across the stream.

'No sudden urge to travel? Attacks of remorse? Change of body temperature?' But Hap was grinning now, not at all nicely. 'I thought not. Well, it had to be tried.'

There was an instant of blinding light.

When it reached the vicinity of the hill, the meteorite had dwindled to the size of a baseball. It should have finished its journey at the back of Hap's head. Instead, it exploded a millisecond too soon. When the light had died, Hap stood within a ring of craterlets.

The swordsman's unsymmetrical jaw dropped, and then he closed his mouth and started forward. The sword hummed faintly.

The Warlock turned his back.

Hap curled his lip at the Warlock's cowardice. Then he jumped three feet backward from a standing start. A shadow had pulled itself from the Warlock's back.

In a lunar cave with the sun glaring into its mouth, a man's shadow on the wall might have looked that sharp and black. The shadow dropped to the ground and stood up, a humanoid outline that was less a shape than a window view of the ultimate blackness beyond the death of the universe. Then it leapt.

Glirendree seemed to move of its own accord. It hacked the demon once lengthwise and once across, while the demon seemed to batter against an invisible shield, trying to reach Hap even as it died.

'Clever,' Hap panted. 'A pentagram on your back, a demon trapped inside.'

'That's clever,' said the Warlock, 'but it didn't work. Carrying Glirendree works, but it's not clever. I ask you again, do you know what you carry?'

'The most powerful sword ever forged.' Hap raised the weapon high. His right arm was more heavily muscled than his left, and inches longer, as if Glirendree had been at work on it. 'A sword to make me the equal of any warlock or sorceress, and without the help of demons, either. I had to kill a woman who loved me to get it, but I paid that price gladly. When I have sent you to your just reward, Sharla will come to me—'

'She'll spit in your eye. Now will you listen to me? Glirendree *is* a demon. If you had an ounce of sense, you'd cut your arm off at the elbow.'

Hap looked startled. 'You mean there's a demon imprisoned in the metal?'

'Get it through your head. *There is no metal.* It's a demon, a bound demon, and it's a parasite. It'll age you to death in a year unless you cut it loose. A warlock of the northlands imprisoned it in its present form, then gave it to one of his bastards, Jeery of Something-or-other. Jeery conquered half this continent before he died on the battlefield, of senile decay. It was given into the charge of the Rainbow Witch a year before I was born, because there never was a woman who had less use for people, especially men.'

'That happens to have been untrue.'

'Probably Glirendree's doing. Started her glands up again, did it? She should have guarded against that.'

'A year,' said Hap. 'One year.'

But the sword stirred restlessly in his hand. 'It will be a glorious year,' said Hap, and he came forward.

The Warlock picked up a copper disc. 'Four,' he said, and the disc spun in midair.

By the time Hap had sloshed through the stream, the disc was a blur of motion. The Warlock moved to keep it between himself and Hap, and Hap dared not touch it, for it would have sheared through anything at all. He crossed around it, but again the Warlock had darted to the other side. In the pause he snatched up something else: a silvery knife, profusely inscribed.

'Whatever that is,' said Hap, 'it can't hurt me. No magic can affect me while I carry Glirendree.'

'True enough,' said the Warlock. 'The disc will lose its force in a minute anyway. In the meantime, I know a secret that I would like to tell, one I could never tell to a friend.'

Hap raised Glirendree above his head and, two-handed, swung it down on the disc. The sword stopped jarringly at the disc's rim.

'It's protecting you,' said the Warlock. 'If Glirendree hit the rim now, the recoil would knock you clear down to the village. Can't you hear the hum?'

Hap heard the whine as the disc cut the air. The tone was going up and up the scale.

'You're stalling,' he said.

'That's true. So? Can it hurt you?'

'No. You were saying you knew a secret.' Hap braced himself, sword raised, on one side of the disc, which now glowed red at the edge.

'I've wanted to tell someone for such a long time. A hundred and fifty years. Even Sharla doesn't know.' The Warlock still stood ready to run if the swordsman should come after him. 'I'd learned a little magic in those days, not much compared to what I know now, but big, showy stuff. Castles floating in the air. Dragons with golden scales. Armies turned to stone, or wiped out by lightning, instead of simple death spells. Stuff like that takes a lot of power, you know?'

'I've heard of such things.'

'I did it all the time, for myself, for friends, for whoever happened to be king, or whomever I happened to be in love with. And I found that after I'd been settled for a while, the power would leave me. I'd have to move elsewhere to get it back.'

The copper disc glowed bright orange with the heat of its spin. It should have fragmented, or melted, long ago.

'Then there are the dead places, the places where a warlock dares not go. Places where magic doesn't work. They tend to be rural areas, farmlands and sheep ranges, but you can find the old cities, the castles built to float which now lie tilted on their sides, the unnaturally aged bones of dragons, like huge lizards from another age.

'So I started wondering.'

Hap stepped back a bit from the heat of the disc. It glowed pure white now, and it was like a sun brought to earth. Through the glare Hap had lost sight of the Warlock.

'So I built a disc like this one and set it spinning. Just a simple kinetic sorcery, but with a constant acceleration and no limit point. You know what mana is?'

'What's happening to your voice?'

'Mana is the name we give to the power behind magic.' The Warlock's voice had gone weak and high.

A horrible suspicion came to Hap. The Warlock had slipped down the hill, leaving his voice behind! Hap trotted around the disc, shading his eyes from its heat.

An old man sat on the other side of the disc. His arthritic fingers, half-crippled with swollen joints, played with a rune-inscribed knife. 'What I found out—oh, there you are. Well, it's too late now.'

Hap raised his sword, and his sword changed.

It was a massive red demon, horned and hooved, and its teeth were in Hap's right hand. It paused, deliberately, for the few seconds it took Hap to realize what had happened and to try to jerk away. Then it bit down, and the swordsman's hand was off at the wrist.

The demon reached out, slowly enough, but Hap in his surprise was unable to move. He felt the taloned fingers close his windpipe.

He felt the strength leak out of the taloned hand, and he saw surprise and dismay spread across the demon's face.

The disc exploded. All at once and nothing first, it disintegrated into a flat cloud of metallic particles and was gone, flashing away as so much meteorite dust. The light was as lightning striking at one's feet. The sound was its thunder. The smell was vapourized copper.

The demon faded, as a chameleon fades against its background. Fading, the demon slumped to the ground in slow motion, and faded further, and was gone. When Hap reached out with his foot, he touched only dirt.

Behind Hap was a trench of burnt earth.

The spring had stopped. The rocky bottom of the stream was drying in the sun.

The Warlock's cavern had collapsed. The furnishings of the Warlock's mansion had gone crashing down into that vast pit, but the mansion itself was gone without trace.

Hap clutched his messily severed wrist, and he said, 'But what happened?'

'Mana,' the Warlock mumbled. He spat out a complete set of blackened teeth. 'Mana. What I discovered was that the power behind magic is a natural resource, like the fertility of the soil. When you use it up, it's gone.'

'But—'

'Can you see why I kept it a secret? One day all the wide world's mana will be used up. No more mana, no more magic. Do you know that Atlantis is tectonically unstable? Succeeding sorcerer-kings renew the spells each generation to keep the whole continent from sliding into the sea. What happens when the spells don't work any more? They couldn't possibly evacuate the whole continent in time. Kinder not to let them know.'

'But . . . that disc.'

The Warlock grinned with his empty mouth and ran his hands through snowy hair. All the hair came off in his fingers, leaving his scalp bare and mottled. 'Senility is like being drunk. The disc? I told you. A kinetic sorcery with no upper limit. The disc keeps acceleration until all the mana in the locality has been used up.'

Hap moved a step forward. Shock had drained half his strength. His foot came down jarringly, as if all the spring were out of his muscles.

'You tried to kill me.'

The Warlock nodded. 'I figured if the disc didn't explode and kill you while you were trying to go around it, Glirendree would strangle you when the constraint wore off. What are you complaining about? It cost you a hand, but you're free of Glirendree.'

Hap took another step, and another. His hand was beginning to hurt, and the pain gave him strength. 'Old man,' he said thickly. 'Two hundred years old. I can break your neck with the hand you left me. And I will.'

The Warlock raised the inscribed knife.

'That won't work. No more magic.' Hap slapped the Warlock's hand away and took the Warlock by his bony throat.

The Warlock's hand brushed easily aside, and came back, and up. Hap wrapped his arms around his belly and backed away with his eyes and mouth wide open. He sat down hard.

'A knife always works,' said the Warlock.

'Oh,' said Hap.

'I worked the metal myself, with ordinary blacksmith's tools, so the knife wouldn't crumble when the magic was gone. The runes aren't magic. They only say—'

'Oh,' said Hap. 'Oh.' He toppled sideways.

The Warlock lowered himself on to his back. He held the knife up and read the markings, in a language only the Brotherhood remembered.

AND THIS, TOO, SHALL PASS AWAY. It was a very old platitude, even then.

He dropped his arm back and lay looking at the sky.

Presently the blue was blotted by a shadow.

'I told you to get out of here,' he whispered.

'You should have known better. What's *happened* to you?'

'No more youth spells. I knew I'd have to do it when the prognostics spell showed blank.' He drew a ragged breath. 'It was worth it. I killed Glirendree.'

'Playing hero, at your age! What can I do? How can I help?'

'Get me down the hill before my heart stops. I never told you my true age—'

'I knew. The whole village knows.' She pulled him to sitting position, pulled one of his arms around her neck. It felt dead. She shuddered, but she wrapped her own arm around his waist and gathered herself for the effort. 'You're so thin! Come on, love. We're going to stand up.' She took most of his weight on to her, and they stood up.

'Go slow. I can hear my heart trying to take off.'

'How far do we have to go?'

'Just to the foot of the hill, I think. Then the spells will work again, and we can rest.' He stumbled. 'I'm going blind,' he said.

'It's a smooth path, and all downhill.'

'That's why I picked this place. I knew I'd have to use the disc someday. You can't throw away knowledge. Always the time comes when you use it, because you have to, because it's there.'

'You've changed so. So—so ugly. And you smell.'

The pulse fluttered in his neck, like a hummingbird's wings. 'Maybe you won't want me, after seeing me like this.'

'You can change back, can't you?'

'Sure. I can change to anything you like. What colour eyes do you want?'

'I'll be like this myself someday,' she said. Her voice held cool horror. And it was fading; he was going deaf.

'I'll teach you the proper spells, when you're ready. They're dangerous. Blackly dangerous.'

She was silent for a time. Then: 'What colour were *his* eyes? You know, Belhap Sattlestone whatever.'

'Forget it,' said the Warlock, with a touch of pique.

And suddenly his sight was back.

But not forever, thought the Warlock as they stumbled through the sudden daylight. When the mana runs out, I'll go like a blown candle flame, and civilization will follow. No more magic, no more magic-based industries. Then the whole world will be barbarian until men learn a new way to coerce nature, and the swordsmen, the damned stupid swordsmen will win after all.

JOHN BRUNNER

THE WAGER LOST BY WINNING

Down the slopes of a pleasant vale an army marched in good order: colours at the head fluttering on the warm summer breeze, drummers beating a lively stroke for the men behind perspiring in their brass-plated cuirasses and high thonged boots. Each man except the officers wore a baldric with an axe and a knife in leather frogs, and carried a spear and a wide square shield. The officers rode horses draped with fine light mail, wore shirts and breeches of velvet sewn with little steel plates, and carried straight swords in decorated sheaths, the pommels bright with enamel and gilt.

Leaning on his staff of curdled light, the traveller in black stood in the shade of a chestnut tree and contemplated them as they went by. Directly he clapped eyes on them, the banners had told him whence they hailed; no city but Teq employed those three special hues in its flag—gold, and silver, and the red of new-spilled blood. They symbolized the moral of a legend which the traveller knew well, and held barbarous, to the effect that all treasure must be bought by expending life.

In accordance with that precept, the Lords of Teq before they inherited their fathers' estates must kill all challengers and did so by any means to hand, whether cleanly by the sword or subtly by drugs and venoms. Consequently some persons had come to rule in Teq who were less than fit—great only in their commitment to evil.

'That,' said the traveller to the leaves on the chestnut tree, 'is a highly disturbing spectacle!'

However, he stood as and where he was, neither hidden nor conspicuous, and as ever allowed events to follow their natural course. Few of the rank-and-file soldiery noticed him as they strode along, being preoccupied with the warmth of the day and the weight of their equipment, but two or three of the officers gave him curious glances. However, they paid no special attention to the sight of an old man in a black cloak, and likely, a mile or two beyond, would have dismissed him altogether from their minds.

That was customary and to be expected. Few folk recognized the traveller in black, unless they were enchanters of great skill and could detect the uniqueness of one who had many names yet but a single nature, or perhaps if they were learned in curious arts and aware of the significance of the season: that epoch after the conjunction of four specific planets when, it was said, he to whom the task had been given of bringing forth order from chaos went out by ordinary roads to oversee that portion of the All which lay in his charge.

And this saying was true. Uncomplaining—for it was not in his nature to resent the inescapable—at this particular season the traveller did go forth to meet and talk with many, many people, and (for this too was in his single nature) to grant to those who craved such the fulfilment of their heart's desire.

The journeys he had made were far beyond counting. Most of them, anyhow, were indistinguishable—not because the same events transpired during each or all, but because in chaos they were so unalike as to be similar. None the less, a little by a little earnests were coming of his eventual triumph. Not only could he recall that here beneath a hill lay prisoned Laprivan of the Yellow Eyes, or there in a volcano Fegrim was pent: now there had been a change in his habits which he had not wished. Formerly he had been used to setting forth from, and returning to, the city Ryovora on his travels.

But Ryovora had passed into Time.

Soon, as the black-garbed traveller counted soonness, all things would have but one nature. He would be unique no more, and time would have a stop. Whereupon . . .

Release.

Watching the purposeful progress of the army, the traveller considered the idea with a faint sense of surprise. It had never previously crossed his mind. But, clearly, it was a wise and kindly provision that his nature should include the capacity to become weary, so that, his mission over, he might surrender to oblivion with good grace.

The rearguard of the army passed, slow commissary-wagons drawn by mules bumping on the rough track. The drumbeats died in the distance, their last faint reverberation given back by the hillsides like the failing pulse of a giant's heart, and he stirred himself to continue on his way.

It was not until he came, much later, to Erminvale that he realized, weary or no, he must still contend with vastly subtle forces arrayed against him.

For a little while, indeed, he could almost convince himself that this was to be the last of his journeys, and that his next return would find the places he had known tight in the grip of Time. The borderland between rationality and chaos seemed to be shrinking apace, as the harsh constraint of reason settled on this corner of the All. Reason implies logic, logic requires memory, and memory Time, not the randomness existent in eternity.

Thus, beyond Leppersley, the folk remembered Farchgrind, and that being's chiefest attribute had once been that no one should recall his deceits, but fall prey to them again and again. Where once there had been a monstrous pile of follies, each a memento to some new-minted prank—'Build thus and worship me and I will give you more wealth than you can carry!' or 'Build thus and worship me and I will give you health and vigour like a man of twenty!'—there were sober families in small neat timber houses, framed with beams pilfered from the ancient temples, who said, 'Yes, we hear Farchgrind when he speaks to us, but we think what became of Grandfather when he believed what he was told, and carry on about our daily business.'

The traveller spoke to Farchgrind almost in sorrow, saying it was not his doing that had made men so sceptical, but had to suffer without denial the expected retort.

'You too,' said the elemental, 'are part of the way things are, and I—I am only part of the way things were!'

Moreover, at Acromel, the place where honey was bitter, that tall black tower like a pillar of onyx crowned with agate where once the dukes had made sacrifice to the Quadruple God was broken off short, snapped like a dry stick. In among the ruins fools made ineffectual attempts to revive a dying cult, but their folly was footling compared to the grand insanities of the enchanter Manuus who once had taken a hand in the affairs of this city, or even of the petty tyrant Vengis, whose laziness and greed almost brought a fearful calamity on the innocent citizens of Ys.

'Ah, if only I could find the key to this mystery!' said one of them, who had bidden the traveller to share the warmth of a fire fed with leather-bound manuscripts from the ducal library.

'Then should I have men come to me and bow the knee, offer fine robes to bar the cold instead of these shabby rags, savoury dishes to grace my table instead of this spitted rat I'm toasting on a twig, and nubile virgins from the grandest families to pleasure me, instead of that old hag I was stupid enough to take to wife!'

'As you wish, so be it,' said the traveller, and knocked his staff on the altar-stone the man was using as a hearth.

In the cold dawn that followed, the wife went running to her neighbours to report a miracle: her husband was struck to a statue, unmoving yet undead. And, because no other such wonder had occurred since the departure of the Quadruple God, things transpired as he had wished. Men set him up on the stump of the great black tower and wrapped their best robes about him, burned expensive delicacies on a brazier that the scent might waft to his nostrils, and sought beautiful girls that their throats might be cut and their corpses hung before him on gallows strung with chains—all this in strict accordance with the ancient customs.

But after a while, when their adulation failed to bring them the favours that they begged, they forgot him and left him helpless to watch the robes fade and the fire die to ashes and the girls' bodies feed the maggots until nothing was left but their bare white bones.

Likewise, a packman met at Gander's Well spoke in the shade of brooding Yorbeth, whose tap-root fed his branches with marvellous sap from an underground spring, and said, 'Each year when the snows melt I come back here and with the proper precautions contrive to pluck the fruit and leaves from these long boughs. Such growths no sun ever shone on before! See here: a furry ball that cries in a faint voice when my hand closes on it! And here too: a leaf transparent as glass that yet shows, when you peer through, a scene beyond that no man can swear to identifying! And these I take, and sell to wealthy strangers in cities far abroad.

'But what irks me'—and he leaned forward, grimacing—'is this simple injustice. I make the trip to Gander's Well, risk death or worse to garner my wares, and tramp the hot dusty summer roads with a heavy pack on my shoulders, and then I must sell for a pittance to some amateur enchanter who doubtless botches the conjuration he plans to work with what I supplied . . . Would that I knew beyond a peradventure what can be wrought by using the means I'm making marketable!'

'As you wish,' sighed the traveller, 'so be it.' He knocked three times with his staff on the coping of the well and went aside to speak of release to Yorbeth—that release which he himself was coming so unexpectedly to envy. For there was one way only to comprehend the possibilities of what grew on this tall tree, and that was to take Yorbeth's place within its trunk.

Where, trapped and furious, the packman shortly found himself, possessed of all the secret lore he had suspected down to the use that might be made of a sheet of the bark in conjuring Ogram-Vanvit from his lair, and powerless to exploit it for his gain.

Yorbeth himself ceased to be. Heavy-hearted, the traveller went on.

In Eyneran, a mountainous land where folk kept sheep and goats, he had once incarcerated a chilly elemental named Karth, thanks to whose small remaining power one strange valley stayed frozen beneath a mask of ice when all around the summer flowers grew bright and jangly music came from the bell-wethers of the grazing flocks. Here the traveller came upon a fellow who with flint and steel sought to ignite the ice, grim-visaged and half blue with cold.

'Why,' enquired the traveller, 'do you spend so long in this unprofitable pastime?'

'Are you a simpleton like all the rest?' countered the man, frenziedly striking spark after spark. 'Is it not in the nature of ice, by ordinary, to melt when the hot sun falls on it? Since what is in this valley does not melt, it cannot be ice. Certainly it is not stone—rock-crystal or quartz is wholly different from this. Therefore it must be of an amberous nature, and amber is congealed resin, and resin burns well, as any drudge knows who has lighted fires with pine-knots. Accordingly this so-called "ice" must burn. Sooner or later,' he concluded in a more dispirited tone, and wiped his brow. The gesture made a little crackling noise, for so bitter was the wind in this peculiar valley that the sweat of his exertion turned at once to a layer of verglas on his skin.

The traveller thought sadly of the scrivener Jacques of Ys, who also had been persuaded that he alone of all the world was perfectly right, and held back his opinion of the would-be ice-burner's logic. Sensing disagreement, though, the fellow gave him a sharp and hostile glare.

'I'm sick of being mocked by everyone!' he exclaimed. 'Would that the true nature of this imagined ice could this moment become clear for you and all to see!'

'As you wish, so be it,' said the traveller, and knew that the time of release had come also to Karth. With the cessation of his dwindled ancient power, sunlight thawed the glaciers and warm zephyrs fathered water from their edge.

The man looked and touched, and paddled his hands in it, and cried out in dismay.

'If this is water, that must have been ice—but that was not ice, therefore this is not water!'

Spray lashed him; rivulets formed around his ankles.

'It is not water,' he declared, and stood his ground. But when the pent-up floods broke loose they swept him with his flint and steel far down the hillside and dashed him to death on a rock.

Aloof, the black-clad traveller stood on a rock and watched the whirling waters, thinking that he, so aged that there was not means to measure his duration, knew now what it meant to say, 'I am old.'

So too in Gryte, a fair city and a rich one, there was a lady who could have had her choice of fifty husbands, but kept her heart whole, as she claimed, for one man who would not look at her though he had wooed and conquered maidens for a hundred miles around.

'Why does he scorn me?' she cried. 'He must be hunting for a wife who will give him surcease from this endless quest—can he not come to me who hunger for him?'

'As you wish, so be it,' said the traveller, and next day the man she dreamed of came a-courting her, so that she imagined all her hopes fulfilled and made him free of her body and household. And the day after he treated her as he had treated all the rest: rose from the couch where he had taken his pleasure, without a kind look or a kiss, and left her to wring her hands and moan that she was undone.

Likewise there stood a gravestone in the cemetery of Barbizond, under the arch of rainbow signalling the presence of the bright being Sardhin which shed a gentle never-ceasing rain. The traveller visited it because he owed a particular private debt to the man beneath, a former merchant enchanter of Ryovora who had performed a service for him unknowingly, and full of years and honour had gone to his repose.

Turning away, the traveller was addressed by a person in a cape of leaves who might have passed at a glance for seven years of age, either boy or girl.

'Good morrow, sir! Think you to brace yourself for death by contemplating it, or have you cause to wish it might overtake some other before yourself?'

'In that case, what?' the traveller said.

'Why, then I could be of service,' the person said slyly. 'I have been for thirty-one years as you see me now—dwarfed, sexless and agile. What better end could I turn such a gift to than to become the finest assassin ever known in Barbizond? You stand among testimonials to my skill; here a miserly old ruffian whose children paid me half his coffer-load, there an eldest son who blocked his brother's way to an inheritance—'

'You speak openly of it!'

'Why, sir, no one is to hear me save yourself, and would folk not think you deranged were you to say a child had boasted of such matters to you?'

'In truth, a childish form is a deep disguise,' the traveller conceded. 'But did you speak to me to solicit my custom, or more because that disguise is perfectly efficient?'

The person scowled. 'Why, frankly, from time to time the very secrecy which benefits my trade does gall me. I gain my living, yes, but no one knows except those whom I have served that I'm the master above all masters. Would that I might enjoy the fame my brilliance deserves!'

'As you wish, so be it,' said the traveller, and struck his staff against the nearest tomb. That very evening rumours took their rise in Barbizond, and everyone who had lost a relative in suspicious circumstances, to a poison subtler than the enchanters could define, or a silent noose, or a knife hissing from the shadows, began to realize how well the appearance of a child of tender years might mask a killer.

The traveller passed the body next morning, sprawled on a dung-heap by the road to Teq.

Will it be now? The question haunted the traveller as he went. With half his being he was apprehensive, for all he had ever known through innumerable eons was the task allotted him; with the balance, he yearned for it. Karth gone, Yorbeth gone—would there shortly also be an end for Tuprid, and Caschalanva, and Laprivan of the Yellow Eyes?

On impulse, when he came to the grove of ash-trees at Segrimond which was one of the places where such things were possible, he constrained Wolpec in the proper fashion to enter a candle, but when he tried to smoke a piece of glass over its flame and read the three truths that would appear on it, it cracked across. With resignation he concluded that this was not for him to learn, and went his way.

In Kanish-Kulya the wall that once had divided Kanishmen from Kulyamen, decked along its top with the skulls of those dead in a long-ago war, had crumbled until it was hardly more than a bank enshrouded with ivy and convolvulus, and roads pierced it along which went the gay carts of pedlars and the tall horses of adventure-seeking knights. Yet in the minds of men the barrier stood firm as ever.

'Not only,' groused a certain Kanish merchant to the traveller, 'does my eldest daughter decline to accept her fate and be sacrificed in traditional manner to Fegrim, pent in that volcano yonder—she adds insult to injury and proposes to wed a Kulyan brave!'

The traveller, who knew much about the elemental called Fegrim, held his peace.

'This I pledge on my life!' the merchant fumed. 'If my daughter carries on the way she's going, I shall never want to speak to her again, nor let her in my house!'

'As you wish, so be it,' said the traveller. From that moment forward the merchant uttered never another word; dumb, he stood by to watch the fine procession in which the girl went to claim her bridegroom, and before she returned home with him her father was dead, so that the house was no longer his.

But nothing in this was remarkable. Greed, hate, jealousy—they were commonplace, and it was not to be questioned that they should defeat themselves.

Onward again, therefore, and now at last to Erminvale.

In that land of pleasant rolling downs and copses of birch and maple, there stood the village Wantwich, of small white farms parted by tidy hedgerows, radiating out from a central green where of a summer evening the young people would gather with a fiddler and a harpist to dance and court in bright costumes of pheasants' feathers and fantastical jingling bangles. At one side of it was a pond of sweet water which the traveller in black had given into the charge of the being Horimos, for whom he had conceived a strange affection on discovering that this one alone among all known elementals was too lazy to be harmful, desiring chiefly to be left in peace. While others older than themselves danced, the village children would splash in the pond with delighted cries, or paint their bare bodies with streaks of red and blue clay from the bank, proudly writing each other's names if they knew how. In winter, moreover, it served for them to skate, and well wrapped in the whole hides of goats they slid across on double wooden runners strapped to their soles.

Good things were plentiful in Erminvale: creamy milk, fat cheeses, turnips so firm and sweet you might carve a slice and eat it with a dressing of salt, berries and nuts of every description, and bearded barley for nutritious bread. Also they brewed good beer, and on a festival day they would wheel three barrels on to the green from which anyone, resident or traveller, might swig at will, the first mug always being poured, of course, to Horimos. Content with that small token of esteem, he slumbered at the bottom of his mud.

All this was what the girl named Viola had known since a child, and from reports she had heard through visitors she felt well satisfied. Where else was there that offered a better life? Great cities were crowded and full of smoke and stinks; moreover, they had more demanding deities than

Horimos, like Hnua-Threl of Barbizond who must be invoked with combats to the death before his altar, or that blind Lady Luck who smiled randomly on the folk of Teq and might tomorrow turn her back for good on one she favoured today.

She had heard of Teq from the finely clad rider who had come, a while ago, on a tall roan stallion, twirling long fair mustachios and spilling gold from his scrip like sand for all his complaints about the size of his room at Wantwich's only inn, and the inferiority of beer compared to the wines of his home city.

He had arrived on the first fine evening of spring, when Viola and her betrothed man Leluak joined all the other young people in a giddy whirling dance across the green, and because he was a stranger and it was courteous—and also, she admitted to herself, because all the other girls would be impressed—she had accepted his request to join him in executing some new-fashionable steps from Teq. Instruction took a moment only; she was a skilful dancer, light on her slim legs that not even winter's dark had worn to paleness from their summer brown. After that they talked.

She learned his name was Achoreus, and that he served one of the great lords of Teq. She learned further that he thought her beautiful, which she granted, for everyone had always told her so; she had long sleek tresses, large liquid eyes like opals, and skin of satin smoothness. He declared further that such loveliness was wasted in a backwater hamlet and should be displayed to the nobility and gentry of a place like Teq. She thanked him but said she was already spoken for. Thereupon he demonstrated that for all his arrogant airs he lacked common civility, for he tried to fondle her inside her bodice, and she walked away.

Had he acted civilly, asking her to walk in the woods with him and find a bed of moss, she would have agreed, of course; it was the custom of Wantwich to receive all strangers as one would one's friends. But as things were—so she told Leluak when bidding him goodnight—he seemed to expect that the mere sight of him would make her forget the boy she had grown up with all her life, a foolishness which she could not abide.

Accordingly, all plans for her marriage went ahead in the ancient manner, until at sunset on the day before the festival her father, her mother, her two sisters and her aunt prepared her in the proper fashion for her night alone, in which she must pass to the five high peaks enclosing Erminvale and there plant the prescribed seeds against the time of bearing children: an apple, a sloe, a cob, an acorn and a grain of barley.

With a leather wallet and a flask of water, carrying a torch of sweet-scented juniper, and followed by the cries of well-wishing from the assembled villagers who tomorrow would attend her wedding on the green, she set forth into the gathering dusk.

The tramp was a long one, and difficult in the dark, but she had wandered through Erminvale since she was old enough to be allowed out of her mother's sight, and though she must clamber up rocky slopes and thread her way through thickets where night-birds hooted and chattered, she gained each peak in turn with no worse injury than thorn-scratches on her calves. As dawn began to pale the sky, she set in place the final seed, the grain of barley, and watered it from her body to give it a healthy start in life. Then, singing, she turned back, weary but excited, on the road towards her home. By about noon she would be safe in Leluak's embrace, and the feasting and merry-making could begin.

Still a mile off, however, she began to sense that something was amiss. Smoke drifted to her on the breeze, but it lacked the rich scent of baking which she had expected. A little closer, and she started to wonder why there was no shrill music audible, for no one had ever been able to prevent Fiddler Jarge from striking up directly his instrument was tuned whether or no the bride had come back from the hills.

Worst of all, at the Meeting Rock that marked the last bend in the road, the huge granite slab at which the groom traditionally seized his bride by the hand to lead her into Wantwich, there was no sign of Leluak.

She broke into a run, terrified, and rounded the rock. Instantly she saw the furthest outlying house, that of the Remban family which she remembered seeing built when she was a toddler, and almost fainted with the shock. Its fine clean walls were smeared with a grime of smoke, its gate was broken and their finest plough-ox lay bellowing silent suffering through a bubble of blood.

Worse still beyond: the Harring house afire—source of the smoke she'd smelled! Her own home with its shutters ripped off the hinges, the front door battered down with an axe from the kindling pile! Leluak's, unmarked, but the door ajar and on the threshold a smear of hour-old blood!

Wildly she ran onward to the village green, and there was the worst of all—Jarge's fiddle broken on the ground, the beer-barrels set out for the wedding drained, a patch of scorched grass she could not account for near to them, and all the water of the pond fouled with the blood of the ducks which daily quacked there, joyously.

Crouched in her chair, from which for longer than Viola could recall she had watched and grinned at the countless weddings she had witnessed, sat the only remaining villager of Wantwich: Granny Anderland, who was in fact a great-great-grandmother.

'What happened?' Viola shrieked at her. 'What *happened*?'

But all that Granny could do—all that she had ever been able to do since Viola was a baby—was to expose her toothless grin and rock back and forth in her chair.

Helpless, Viola screamed Leluak's name a few times, but after that she collapsed on the ground from weariness and horror, and that was how the traveller found her when he chanced that way.

He barely checked his pace as he came into Wantwich, along another road than that which Viola had followed on her return from the five peaks. But his expression grew sterner with every stride he took until when finally he could survey the full measure of the calamity from the centre of the green his brow was dark as a thundercloud.

His footsteps were too soft upon the sward for the weeping girl to hear them through her sobs, and it was plain that the old woman had either been so shocked by what she'd seen that she had lost her reason, or else was long ago too senile to understand the world. Accordingly he addressed the girl.

At the sound of his voice she cringed away, her face wet with tears displaying mindless terror. But there was little in the sight of an old man leaning on a staff to suggest that he could be connected with the rape of Wantwich. And, for all that he looked angry beyond description, there was no reason why that anger should be aimed at her.

'Who are you, child?' the traveller enquired.

'I—I'm Viola, sir,' the girl forced out against her sobs.

'And what has happened here today?'

'I don't know, I don't know!' Wringing her hands, she rose. 'Why should anyone want to do this to us? Monsters must have done it—evil beings!'

'There are few such creatures left hereabouts,' the traveller said. 'More likely it was men, if one can dignify them with that name. You were away from the village?'

'I was to be married today!' Viola choked.

'I see. Therefore you were walking the five peaks and planting seeds on them.'

'You—you know our customs, sir?' Viola was regaining control of herself, able to mop away her blinding tears and look at the newcomer. 'Yet I don't remember that I saw you here before.'

'This is not the first time that I've been to Wantwich,' the traveller said, refraining from any reference to the number of earlier visits. 'But to pursue the important matter: did not this old lady witness what occurred?'

'Granny Anderland has been like this for many years,' Viola said dully. 'She likes to be talked to, nods and sometimes giggles, but beyond that . . .' She gave a hopeless shrug.

'I see. In that case we must resort to other means to determine what went on. Girl, are you capable of being brave?'

'If you can do something, sir, to help me get back my man—and to right the wrong that's been done to all these good people—I'll be as brave as you require of me.' She stared at him doubtfully. 'But *can* you do anything?'

'You wish me to?'

'Sir, I *beg* you to!'

'As you wish, so be it,' said the traveller, and took her hand, leading her across the green, past the patch of grass scorched black—at which he cast a puzzled glance—and to the very rim of the sweetwater pond.

'Stand firm,' he commanded. 'Do not be afraid of what you see.'

'I—I don't understand . . .'

'Better for you that you should not,' the traveller muttered, and thrust his staff into the water. He dissolved one of the forces binding the light of which it was composed, and a shaft of brilliance lanced downward to the bottom.

'Horimos!' he said. 'Come forth!'

The girl's eyes grew round with wonder, and then her mouth also, with dismay, for the water heaved and bubbled sluggishly as pitch, and from the plopping explosions a thick voice seemed to take form, uttering words.

'*Le-e-ave me-e a-a-lo-o-one . . .*'

'Move!' rapped the traveller. 'Stir yourself—you've slumbered for eons in that soft bed of mud! Shall I remove you to Kanish-Kulya, make you share the pit of a volcano with Fegrim?'

A noise between a grumble and a scream.

'Yes, he'd be a restless companion for you, wouldn't he?' the traveller rasped. 'Up! Up! I need to speak with you!'

Beside him, Viola had gone down on her knees, all colour vanished from her cheeks. Too petrified even to blink, she saw the water from which she had so often drunk, in which she had so often bathed, rise into tumult—yet absurdly slowly, as though time had been extended to double length. More bubbles burst, and she could watch their surface part; waves and ripples crossed the surface so slowly, one would have thought to push them aside into new directions without wetting one's palm.

And ultimately—

'You may prefer to close your eyes now,' the traveller said didactically, and added, 'Horimos! Speak! And be quick—the sooner you tell me what I need to know, the sooner you can sink back into your ooze. What's become of all the people from this village of yours?'

'Been taken away,' Horimos mumbled. It was not exactly a mouth he used to shape the words, but then, like all elementals, his physical form was mostly accidental.

'How and by whom? Come on, tell me the whole story!' The traveller rapped the bank impatiently with his staff.

'Army marched in this morning,' Horimos sighed. 'Went around the village, drove everyone to the green—most of them were there already anyway. Set up a forge there where the grass is blackened, welded fetters for everyone on to a chain. Killed some ducks and hens for supper, drank the beer in the barrels, herded everyone away. Good riddance, say I—never had a moment's peace since you put me here, what with fiddling and laughing and swimming and skating and all the rest of it!'

'Whose was the army? What colours did they fly?'

'Should I know who uses a flag of silver, red and gold?'

The traveller gave a thoughtful nod.

'But you made no attempt to interfere?'

'Told you—glad to see the back of them.' Horimos made the whole surface of the pond yawn in a colossal expression of weariness. 'And but for you I'd have enjoyed a decent sleep for a while, now I'm alone!'

'For that,' said the traveller softly, 'until the folk of Wantwich are again in their homes, you shall itch so much you can enjoy no rest. Begone with you. Hope that the matter is speedily set to rights.'

'But—!'

'You argue with me?'

Horimos declined. When once again he had subsided to the bottom of his pond, the water was no longer pellucid and still as before, but roiled continually without a breeze to stir it.

'Who are you?' Viola whispered from beside him. 'I'd always thought Horimos was—was . . .'

'Was imaginary?' The traveller chuckled. 'Not exactly. But his chief fault is the very human one of being a fool. As for my own identity: you may call me Mazda, or what you will. I have many names, but only one nature.'

He waited to see if the information, which might be given only to someone who directly requested it, meant anything to her. With relief he noted that it did, for on the instant a blend of hope and awe transfigured her pretty face.

But then he took a second look, and his heart sank. For, in along with the rest, that expression betrayed a trace of selfishness.

'Is it true, then,' she demanded fiercely, 'that I can require of you my heart's desire?'

'Why not? Is not the heart's desire the enforcement of will upon chaos? Am I not occupied with reducing chaos to order? But think well!' He raised his staff sternly. 'I cannot know what is in your secret mind, only what you convey to me in words. Reflect! Ponder!'

'I don't have to,' she said with terrible directness. 'I want to be reunited with my man!'

The traveller sighed, but as always was resigned to the natural course of events. 'As you wish, so be it,' he said.

'What shall I do?' she cried, suddenly overcome with a sense of the finality of what she had said.

'Wait.'

'No more than wait? Here?' She turned frantically around, surveying the ravished homes, the slaughtered livestock, the smoke that still drifted over the burning house. 'But——'

And when she looked again for the traveller in black, he was gone.

A little after, when the sun was still high in the sky, there were clopping noises on the road by which the army had arrived, and she stirred from her torpor and made to flee. But the horseman easily ran her down, bowing from his saddle to sweep her off her feet and park her on the withers of his steed, laughing at her vain attempts to break away.

'I missed you when they rounded up the rest of them,' said Achoreus of Teq. 'I couldn't forget a pretty face like yours. Even less can I forget an insult like the one you offered me when I came before! So I dawdled, thinking you'd be back eventually, and here you are. Not for long, though! You're going to rejoin your family, and friends, and this country bumpkin you preferred to me!'

He set spurs to his horse, and away they galloped in the wake of the miserable gang of captives strung with chains.

Laughter rang loud and shrill under the gorgeous canopy that shaded Lord Fellian of Teq from the naked rays of the sun. The canopy was of pleated dragon-hide, bought at the cost of a man's life in a distant land where chaos and reason had once been less evenly matched and strange improbable beasts went about with lion's claws and eagle's beaks and wings of plated bronze. Report held there were no more such creatures to be found; even their bones had been rejected by reality.

'But I have my canopy!' Lord Fellian would say.

Its shade fell on a floor of patterned stone: marble was the commonest of the types of tile composing it, outnumbered by chalcedony, jasper, sardonyx, chrysoberyl, and others yet so rare that they had no name save 'one of the tiles in Lord Fellian's gallery'. This was on the very apex of the grand high tower from which Lord Fellian might survey his domain: lands from here to the skyline and beyond which bled wealth into his coffers.

But on the houseward side was a high wall, that when he sat in his throne of state—made of the bones of a beast of which the enchanters declared no more than one could ever have existed, translucent as water but harder than steel—not even an absent glance over his shoulder might reveal to him the sole building in all of Teq which outreached his tower. Atop that mighty edifice reposed the figure of Lady Luck, the goddess blind in one eye, masked over the other, whose smile or frown dictated the fortunes of those who ruled in Teq.

It was not the custom to look on her. It was said that those who secretly tried to, in order to discover which way her gaze was bent, would die a fearful death. And indeed, the agents of Lords Fellian, Yuckin, and Nusk did occasionally deposit in the chief market-square the bodies of men who had clearly undergone some repulsive torment, that they might be a warning and a caution to the common folk.

More often than not, these corpses belonged to persons who had boasted of their looking on the Lady. It was taken for granted that the others belonged to those who had not even enjoyed the pleasure of boasting as reward for their supreme gamble.

It was the one gamble no Lord of Teq would take. Why should they? Was not their very affluence proof that the Lady bent that enigmatic smile continually on them?

Lord Fellian on his chair of inexplicable bones cramped with pure gold, robed in cloth dyed with the purple of the genuine murex, shod with sandals of the softest kidskin on which had been stamped, again in gold, a series of runes to guard the path he took; his foppish locks entwined with ribbons, his nails painted with ground pearls, his weak eyes aided with lenses not of rock-crystal such as his rivals must make do with but of very diamond, his lobes hung with amber, his fingers glittering with sapphires; he, Lord Fellian, the greatest winner among all the past and present Lords of Teq, laughed and laughed and laughed.

The noise drowned out the soft rattling from the table on which a trained monkey, tethered with a velvet leash, kept spilling and gathering up a set of ivory dice, their values after each throw being recorded by a slave on sheets of parchment; likewise the humming of a gaming-wheel turned by an idiot—both these, with bias eliminated, to determine whether after fifty thousand throws or spins there would be some subtle preference to help him in his ceaseless conflict against Lords Yuckin and Nusk. Furthermore, it drowned out the chirruping of the gorgeous songbirds in a gilded cage against the high screening wall, which he had won last week from Nusk in a bout at shen fu, and the drone of the musicians playing on a suite of instruments he had won—along with their players—from Yuckin a year or more past. Those instruments were of eggshells, ebony, and silver, and their sound was agonizingly sweet.

Before the chair of bones, Achoreus, who had committed himself to the service of Lord Fellian when he was but seventeen grinned from ear to ear at the brilliant inspiration of his master.

'Before those fools learn that winning from me costs me nothing,' Fellian declared, 'I shall have taken the very roofs from over their heads! They will be shamed if they refuse to match my stakes, and I may climb as high as I wish, while they—poor fools!—struggle to clamber after me. Oh, how I look forward to seeing Yuckin's face when tonight I bet him a hundred skilful servants, including girls fit for a royal bed! You've done well, Achoreus. Torquaida, come you here!'

From among the gaggle of retainers who hourly by day and night attended Fellian, subservient to his slightest whim, there shuffled forward the elderly treasurer whose mind enclosed, as he boasted, even such detail of his master's coffers as how many of the copper coins used to pay off tradesmen had been clipped around the edge, instead of honestly worn.

In no small part, Fellian acknowledged, his victories in the endless betting-matches with his rivals were due to Torquaida instructing him what they could or could not stake to correspond with his own wagers. He had rewarded the treasurer suitably, while those who served his rivals were more likely punished for letting go irreplaceable wonders on lost bets, and grew daily bitterer by consequence.

'Young Achoreus here,' declared the lord, 'has performed a service for us. We have now, thanks to him, one hundred or more skilled servants surplus to the needs of the household, and additionally some children who can doubtless be trained up in a useful skill. How, say you, should this service be repaid?'

'This is a difficult estimate,' frowned Torquaida. His ancient voice quavered; Fellian scowled the musicians into silence that he might hear better. 'There are two aspects of the matter to consider. First, that he has brought us a hundred servants—this is easy enough. Give him a hundred golden dirhans to increase his stake in the wager he has made with Captain Ospilo of Lord Yuckin's train; that bet is won, and the odds are nine to five, so that he will gain one hundred eighty dirhans.'

Achoreus preened his moustachios, very conscious of the envious gaze of all the other household officers who were standing around the gallery.

'Beyond that, however,' Torquaida continued in his reedy tones, 'it remains to be established what the nature of these new arrivals is. As one should not wager on a horse without inspecting both it and the competition, so too one must begin by looking over the captives.'

'Well said,' approved Fellian, and clapped his hands. 'Let them be brought before me! Clear a space large enough for them!'

'Sir,' ventured Achoreus, 'there were not a few among them who appeared to resent the—ah—invitation I extended to enter your lordship's service. It will be best to make space also for the escort I detailed to accompany them.'

'What?' Fellian leaned forward in his chair, scowling terribly. 'Say you that a man on whom Lady Luck smiles so long and so often is to be injured by some—some stupid peasant? Or have you not disarmed them?'

Seeing all his new fortunes vanishing, Achoreus replied placatingly. 'My lord, there was hardly a weapon in the whole village save rustic implements whose names I scarcely know, not having had truck with such subjects—scythes, perhaps, or hatchets. These we deprived them of. But a person without spirit makes a poor servant, and a person with spirit still possesses—well—feet, and fingernails!'

'Hmmm . . .' Fellian rubbed his chin. 'Yes, I remember well a gladiator whom Lord Yuckin set against my champion in a year gone by, who lost both net and trident and still won the bout by some such underhand trick as clawing out his opponent's vitals with his nails! On that I lost—but no matter.' He gave an embarrassed cough. 'Good, then! Bring them up, but make sure they're under guard.'

Relieved, Achoreus turned to issue the necessary orders. Accordingly, in a little while, to the music of their fetters clanking, a sorry train of captives wended their way out of the grand courtyard of the palace, up the lower stages of the ramp leading to the gallery, which were of common granite, and stage by stage on to the higher levels, where the parapets were of garnets in their natural matrix, and the floor of cat's-eye, peridot, and tourmaline. Refused food on the long trudge from Erminvale to discourage the energy needed for escape, accorded barely enough water to moisten their lips, they found the gradual incline of the ramp almost too much for them, and their escorts had to prod them forward with the butts of spears.

At last, however, they were ranged along the gallery, out of the shade of the dragon-skin awning, blinking against sunlight at their new and unlooked-for master. At one end of the line was Leluak, one eye swollen shut from a blow and testifying to his vain resistance; as far distant as possible from him, Viola, nearly naked from the struggle that had led to Achoreus ripping most of her clothes. And between them, every villager from Wantwich bar Granny Anderland, from babes in arms to the aged.

Accompanied by the proud Achoreus, Torquaida went along the line peering into face after face, occasionally poking to test the hardness of a muscle, his finger sharp as his own styli. He halted before one bluff middle-aged fellow in a red jerkin who looked unutterably weary.

'Who are you?' he croaked.

'Uh . . .' The man licked his lips. 'My name is Harring.'

'Say "so please you"!' Achoreus rasped, and made a threatening move towards his sword.

Harring mumbled the false civility.

'And what can you do?' Torquaida pursued.

'I'm a brewer.' And, reluctantly after a brief mental debate: 'Sir!'

'You learn swiftly,' Achoreus said with mock approval, and accompanied Torquaida down the line.

'I'm a baker, sir.'

'And I, a sempstress!'

'And I a bodger, turner, and mender of ploughs!'

The answers came pat upon the question, as though by naming their trades the captives could reassure themselves they still retained some dignity by virtue of their skill. At Torquaida's direction a clerk made lists of all the names and crafts, leaving aside the children under twelve, and finally presented them with a flourish to Lord Fellian.

Scrutinizing them through his diamond lenses, he addressed Achoreus.

'And what standard are these folk in their professions? Competent, or makeshift?'

'As far as I could judge, sir,' Achoreus answered, 'they are competent. Their houses seemed sturdy, their fences were well mended, and they had sound byres and folds for their livestock.'

'Hmmm!' Fellian rubbed his chin delicately with the flat of a gemstone ringed to his middle finger. 'Then they might be better kept than staked. We have no brewer in the household, have we? Some scullery drab or turnspit would be less useful than that man—what's his peasant's name? Harring! Therefore do thus, Torquaida: take away their children and put them to nurse or to be apprenticed, then sort the rest and for each one you put into the household take one servant we already have and set him at my disposal to be staked tonight. Why, was this not an inspiration that I had?' He rubbed his hands and gave a gleeful chuckle.

'Oh, how I long to see the faces of those dunderheads when I bet fifty servants against each of them tonight! How can I lose? If they win, they will merely clutter up their households with those we find most of a nuisance, while I gain tradesmen in their place, and should I win—which I shall, of course—I shall have spare overseers to cope with what they choose to get rid of! We must do this again, Achoreus!'

Achoreus bowed low and once more stroked his moustachios.

'Take them away now,' Fellian commanded, and leaned back in his throne, reaching with long pale fingers for the mouthpiece of a jade hookah on a lacquered table nearby. An alert slave darted forward and set a piece of glowing charcoal on the pile of scented herbs contained in the bowl.

Frightened, but too weak and exhausted to resist, the folk of Wantwich turned under the goading of the soldiers to wend their way back to the courtyard below. Fellian watched them. As the end of the line drew level with his throne, he snapped his fingers and all looked expectantly towards him.

'That girl at the end,' he murmured. 'She's not unattractive in a country way. Set her apart, bathe, perfume, and dress her, and let her attend me in my chamber.'

'But—!' Achoreus took a pace forward.

'You wish to comment?' Fellian purred dangerously.

Achoreus hesitated, and at last shook his head.

'Let it be done, then,' Fellian smiled, and sucked his hookah with every appearance of content.

Furious, Achoreus turned to superintend the final clearance of the gallery, and thought the task was done, but when he glanced around there was one stranger remaining: a man in a black cloak, leaning on a staff and contemplating Fellian with an expression of interest.

'Achoreus!' Fellian rasped. 'Why have you not taken that man with the rest?'

Staring, Achoreus confessed, 'I have not seen him before; he was not with the villagers we assembled and—Ah, but I have! Now I recall that I saw him watch our army pass, when we were on the outward leg from Teq. He stood beneath a tree with that same staff in hand.'

'And came to join the captives of his own accord?' Fellian said with a laugh. An answering ripple of amusement at what passed for his great wit echoed from his sycophants. 'Well then! We shall not deny him the privilege he craves.'

Faces brightened everywhere. Fellian was a capricious master, but when he spoke in this jovial fashion it was certain he felt merry, and in that mood he might well distribute favours and gifts at random, saying that this was to impress on his retinue the supreme importance of luck.

'So, old man!' he continued. 'What brings you hither, if it was not the long chain linking those who have been here a moment back?'

'A need to know,' said the traveller, and paced forward on the jewelled floor, his staff going tap-tap.

'To know what? When the gaming-wheel of life will spin to a halt for you against the still dark pointer of death? Why, go ask Lady Luck face to face, and she will tell you instantly!'

At that, certain of his companions blanched; it was not in good taste to say such things.

'To know,' said the traveller unperturbed, 'why you sent armed raiders to rape the village Wantwich.'

'Ah yes,' Fellian said ironically. 'Strangers do ask questions of that order, lacking the understanding we have of the realities of life. They think all they need do in their lives is act reasonably, meet obligations, pay their debts. And then some random power intrudes on their silly calm existence, perhaps with a leash, perhaps with a sword, and all their reasoning is futile in face of a superior! And from that they learn the truth. Not sense but luck is what rules the cosmos—do you hear me? Luck!'

He leaned forward, uttering the last word with such venom that a spray of spittle danced down to the floor.

'See, there's an idiot who turns a gaming-wheel for me. Bring the creature here!'

Retainers rushed to obey. Fellian peeled rings from his fingers, holding stones that might bring the price of a small farm and vineyard, and threw them on the skirt of the idiot's robe, soiled with testaments of animality.

'Turn her free! Luck has smiled her way today!'

'Not so,' said the traveller.

'What? You contradict?'

'Say disagree, rather,' the traveller murmured. 'Is it not greater luck for an idiot to be fed, housed, and clothed by a great lord than to be given some pretty baubles and left to fend alone? She'll starve, and that's not luck but misfortune.'

Fellian began to redden as the validity of the point sank in, and glared fiercely at someone to his right whom he suspected of being about to giggle.

'You chop logic, do you?' he rasped. 'You're a schoolman of the kind we take to gaze on Lady Luck's statue, and who thereupon die rather horribly!'

'To persuade the dead to accept one's point of view is a somewhat fruitless task, is it not?' said the traveller mildly. 'You say, as I understand you, that life is one long wager. This may be so, but if it is, why then does one need to make more wagers? As I gather you do, in that you propose to stake human beings against your rival lords tonight, and for this purpose kidnapped the inoffensive folk of Wantwich.'

'What else gives spice to life but winning wagers?' Fellian countered. 'I sit here, and by that token it is plain that I already won my greatest gamble. I staked my very existence on the right to be a lord of Teq, and that I am here proves that the lady on the tower smiles my way!'

The traveller cocked his head sardonically. 'This is why they term you a great winner?' he enquired.

'Of course it is!' Fellian raged.

'But I will name you a bet you will not accept.'

'What?' Fellian howled, and all around there were looks of shocked dismay. 'Think you you can insult a lord of Teq with impunity? Guards, seize and bind him! He has offered me a mortal affront and he must pay for it!'

'Have I affronted you? How? To say that I can name a bet you will not accept is not to insult you—unless you would decline the gamble!' The traveller straightened with a stern expression and fixed Fellian with hard eyes.

'Am I to bet with a nobody? I bet only against my coevals, who can match my stakes!' Fellian snorted. 'I do not even know who you are! Some bumpkin, were I to accept your conditions, could come to me and say, "I wager my rags and clogs, all I possess against all that *you* possess."'

'But there is one thing any man may bet against any other,' said the traveller. 'For no man can have more than one of it.'

There was silence for the space of several heartbeats. 'My lord,' Torquaida said at last, 'he means life.'

'Precisely,' said the traveller. 'Life . . .'

'It is true, my lord,' Torquaida said at length in a rusty voice. 'To stake one man's life against another is a fair bet.'

Fellian licked his lips. He blustered, 'But even so . . . ! A life that may have fifty years to run, like mine—for I'm young yet—against one which may snuff out tomorrow, or next week?'

'But we have not yet agreed on the bet,' Torquaida said. 'Is it not over-soon to name the stake?'

Fellian flashed him a grateful smile; this was the outlet he had been unable to spot himself. He said loudly, 'Yes, a very important point! What bet is this that you wish to make with me, old man?'

'I bet you,' said the traveller into a universal hush, 'that the statue of Lady Luck has her back turned to your throne.'

There was an instant of appalled shock. But with a great effort of recovery, Fellian forced out a booming laugh.

'Why, that bet's lost already!' he exclaimed. 'Is it not proof in itself of her favour towards me that I sit here amid unparalleled riches?'

'No!' said the traveller sharply. 'How can you determine the outcome until you have concluded the bet? Or *any* bet!'

'Enough of this nonsense,' Fellian grunted. 'Take the old fool away. It will be proof enough in a month from now, when I have won still more bets against my coevals, that she smiles on me. And if he is still alive by then, after the diet of my dungeons, I will claim from him the satisfaction of his owings.'

'A month is too long,' said the traveller. 'A day will suffice. I will see you again sooner than that; at dawn tomorrow, let us say. For now, farewell.'

'Seize him!' Fellian bellowed, and the soldiers who had remained behind, on Achoreus's signal, when the party of captives were led away, dashed forward to obey. But in some fashion which none present could decipher, they went crashing against one another, as though they had sought to seize an armful of the empty air.

And on seeing that, Fellian's face grew grey.

In the great cave-like kitchens of the tower, a cook sweated with ladle and skewer at a cauldron of half a hog's-head capacity. The fire roaring beneath scorched his skin, the smoke blinded his eyes with tears, but still he attended to his duty.

From the dark corner of the hearth, a voice enquired for whom the savoury-smelling broth was being prepared.

'Why, for Lord Fellian,' sighed the cook.

'But no man could engulf such a vast deal of soup, surely.'

'Oh no, sir.' The cook grimaced. 'There will be much left over.'

'And you will get your share?'

'I, sir?' The cook gave a rueful chuckle. 'No, tonight I shall sup as ever off a crust of dry bread and a piece of mouldy bacon-rind. But if I'm lucky I shall have the dregs of the wine-goblets from the high table, and the liquor will soothe my grumbling belly enough to let me sleep.'

Among the fierce ammonia stench of guano, the falconer worked by an unglazed window, tooling with gnarled but delicate hands a design of rhythmical gold leaf to the hood and jesses of a peregrine falcon.

'This leather is beautiful,' said a soft voice from over his shoulder. 'But doubtless you yourself put on far finer stuffs when you go forth of an evening to enjoy yourself at a tavern?'

'I, sir?' grunted the falconer, not turning around. 'Why, no, I have no time to amuse myself. And had I the time, I'd be constrained to wear what you see upon me now—old canvas breeches, bound around the waist with fraying rope. Besides, with what would I purchase a mug of ale? With a scoop of fewmets?'

In the stables, a groom passed a soft cloth caressingly over the fitments of the stalls; they were of jacynth and ivory, and the mangers were filled with new sweet hay, fine oats that might have baked bread, and warm-scented bran.

'Palatial!' said an admiring voice from beyond the partition. 'And this only for a horse?'

'Aye, sir,' muttered the groom. 'For Western Wind, Lord Fellian's favourite steed.'

'Upon what do you then sleep—high pillows filled with swansdown, beneath a coverlet of silk?'

'On straw, sir! Do not jest with me! And if I have time to gather clay and stop the chinks in my hovel against the night's cold, I count myself lucky.'

Beside a marble bath, which ran scented water from a gargoyle's mouth, a slender girl measured out grains of rare restorative spices on to a sponge, a loofah and the bristles of a brush made from the bristles of a wild boar.

'With such precautions,' said the traveller from within the cloud of rising steam, 'beauty must surely be preserved far beyond the normal span!'

'Think you, sir, that I would dare to waste one grain of this precious essence on my own skin?' retorted the girl, and brushed back a tress of hair within which—though she could at most be aged twenty—there glinted a betraying thread of silver. 'I'd be lucky, when they detected my fault, to be thrown over the sill of that window! Beneath it there is at least a kitchen-midden to give me a soft landing. No, my whole capital is my youth, and it takes the powers of an elemental and the imagination of a genius to spread it thin enough to satisfy Lord Fellian from spring to autumn.'

'Then,' said the traveller, 'why do you endure it?'

'Because he is a winner in the game of life.'

'And how do you know that?'

'Why,' sighed the girl, 'everyone says so.'

In the high-vaulted banqueting hall, as the sun went down, the rival lords Yuckin and Nusk came with their retinue to feast at the expense of the current greatest winner prior to the onset of the night's gambling. They had had to come to his tower too often of late; there was no friendly chat between them, but instead they sat apart at tables of chrysoprase and tourmaline, supping from emerald mugs and plates of gold.

Lord Fellian, who should have been delighted at the discomfiture of his rivals, was downcast, and the talk at his long table was all of the strange intruder who had laid him so threatening a bet.

'It's nonsense!' roundly declared Achoreus, who had at seventeen pledged himself to the lord he believed likeliest to win and win. 'As you rightly said, sir, no lord of Teq bets with a fundless nobody—and, moreover, the bet he named is by definition incapable of being resolved!'

But his brow was pearled with sweat, and when he spoke his voice was harsh with a hoarseness no amount of wine could relieve.

'And how say you, Torquaida?' demanded Fellian, hungry for reassurance.

'To me,' wheezed the old treasurer reedily, 'what I just heard from Captain Achoreus makes good sense. A bet cannot be resolved on an outcome which by definition is inestablishable. There is no need to worry. Lord Yuckin and Lord Nusk, like or dislike, would have to concede the correctness of cancelling such a gamble—else might they challenge you, or themselves be challenged, on an equally arbitrary base!'

Even that counsel, however, did not set Fellian's mind at ease as he toyed with rare delicacies that the stewards set before him.

'Ah, would that I might know the outcome of this wager, however foolish!' he grumbled, and at that the black-clad traveller, standing apart in the secrecy of an embrasure, gave a heavy sigh.

'As you wish,' he murmured, 'so be it. You have won your bet with me, Lord Fellian. And in the same instant when you won, you lost beyond all eternal hope.'

That question settled, he went away.

Shortly, they cleared the dishes from the hall, bringing in their place the hand-carved dominoes required for the game shen fu, the lacquered plaques destined for match-me-mine and mark-me-well, the tumbling gilded cages full of coloured balls known as The Lady's Knucklebones, the gaming-wheels, the blind songbirds trained to peck out one and only one of three disparately dyed grains of wheat, the jumping beans, the silver-harnessed fleas, the baby toads steeped in strong liquor, and all the other appurtenances on which the Lords of Teq were accustomed to place their bets. Additionally, from among the respective trains, the lords ordered to come forth their current champions at wrestling, boxing with cestae, and gladiatorial combat, not to mention tumblers, jumpers, imbeciles armed with brushes full of paint, dice-throwing monkeys, and whatever else they had lately stumbled across upon the outcome of whose acts a bet might be made.

It was the custom of the challengers to name the games, and of the challenged to declare the stakes. Thus, in strict accordance with routine, Lord Yuckin as the last loser to Lord Fellian, cleared his throat and began with a single hand of shen fu, to which Lord Fellian consented, and won a cage of desert hoppers—the typical small stake of the early hours.

Then Lord Nusk bet on a jumping toad, and won a purse of coins from Barbizond, to which Lord Fellian replied with a spin of the wheel and won a bag of sapphires. He nudged his companions and whispered that the old fool on the gallery had been wrong.

Thus too he won five bouts, on toads, on fleas, more on shen fu, and lastly on the pecking birds. After that he lost a second spin of the wheel and with it a chased enamelled sword that Torquaida dismissed as pretty but not practical—no special loss.

'Now, I think,' murmured the pleased Lord Fellian, and on Lord Nusk naming shen fu as the next bout, declared his stake: fifty hale servants on this single toss.

The impact was all he could have wished. Though they might scornfully disdain knowledge of such mundane matters, none knew better than the Lords of Teq how many folk were kept employed to ensure their affluence, through what different and varied skills. To bet one servant was occasionally a last-resort gesture after a bad night; to bet fifty at one go was unprecedented.

Captain Achoreus chortled at the dismay that overcame the rival lords, and nudged old Torquaida in his fleshless ribs. 'The greatest winner!' he murmured, and signalled for another mug of wine.

Yet, when the dominoes were dealt, the Star of Night fell to Lord Nusk and only the Inmost Planet to Lord Fellian.

Lord Nusk, who was a fat man with a round bald pate fringed with black, grinned from ear to ear and rubbed his enormous paunch. Scowling, Lord Fellian trembled and made challenge to Lord Yuckin at the same game.

Lord Yuckin, thin and gaunt, eyes blank behind lenses of white crystal, named as much gold as a man might carry, and won, and challenged back, and Lord Fellian staked the other fifty servants.

Whereupon he displayed the chief prize of shen fu, the Crown of Stars, and mocked Lord Yuckin's petty deal of Planets Conjoined.

Next, on a hopping toad, he won back from Lord Nusk the former fifty servants, and again from Lord Yuckin a fresh batch, including three armourers that lord could ill afford to lose, and beyond that a farm in the Dale of Vezby, and a whole year's vintage of sparkling wine, and three trade-galleys with all their crews, and then from Lord Nusk the High Manor of Coper's Tor, with the right to make the celebrated ewes'-milk cheese according to a secret recipe; then lost for five short minutes the Marches of Gowth with all four fortresses and the Shrine of Fire, but won them back on a spin of the wheel and along with them the Estate of Brywood, the Peak of Brenn, and all the territory from Haggler's Mound to Cape Dismay.

Then began the calculated process of attrition: the cook who knows how to make sorbets without ice, the kitchen enchanter you display so proudly when we come to dine—the charmer who brings out game from barren ground by playing a whistle—the eight-foot-tall swordsman who beat my own champion in the last public games . . .

Torquaida grew harried trying to keep track of the winnings and match what was now in hand against what remained to the rival lords. Often and often now he instructed Lord Fellian that such a stake was unworthy, for the girl had suffered the smallpox and her skin was scarred, or else that guardsman had a palsy and his sword-blow was unreliable, or else that coffer of coins had an enchantment on it and touched by the winner would turn to pebbles . . .

Lord Fellian awarded him free of feoff the Estate of Brywood as reward for the memory he brought to bear. And he laughed joyously at the disarray of his opponents.

Far down below that ringing laughter, cast back by the high vaults of the banquet-hall, reached the ears of those miserable deportees from Wantwich who were still awake. Some were asleep—on straw if they were lucky, on hard tiles if they were not, but at least asleep.

One who was wakeful, though not on a hard floor but on a mattress of eider's feathers, and in a diaphanous robe of finest lawn embroidered with seed pearls, was the girl Viola, among all the other female objects of pleasure kept for Lord Fellian's delight. At a footstep on the floor beside her couch, she started and peered into dark, seeing only a black form outlined on greater blackness.

'Is someone there?' she whimpered.

'I,' said the traveller, and added, 'I go where I will, though there be locks and bars.'

'Why did you do this to me?' she complained, and began to weep.

'You did it,' said the traveller, contradicting her. 'Will you not be reunited with your man Leluak? You are both here in the same city, captive in the same building; when Lord Fellian's patience expires you will be cast forth together to share the same dank alleyway and the same fevers, chills, and pestilences. Can there be a closer reunion?'

'I chose wrongly,' said Viola after a while. 'Tell me what I should have chosen.'

The traveller brightened a little; at least this cruel experience inflicted by Achoreus and Fellian had imbued some sense in the girl's skull. He said, 'You knew, I believe, Captain Achoreus before the rape of Wantwich.'

'I did so. I companioned him when he joined us for a spring dance.'

'Out of courtesy?'

'Of course.' In the dark, Viola bridled.

'Or because he was a stranger, and handsome, and every other girl in the village would have changed places with you?'

'A little of that too,' she admitted meekly when she had considered the question.

'Is it not true, daughter, that you were more concerned to regain the handsomest, most eligible bachelor in Wantwich, against whom you had competed with all the other girls less lovely than yourself, than to right the wrong done to your family and friends coincidentally upon the day of your projected wedding?'

'I must have been!' Viola moaned. 'Would that that mistake of mine could be undone!'

'As you wish, so be it,' said the traveller with a chuckle. He knocked his staff against the carved head of the couch on which she lay to take her uneasy rest. 'Sleep, my daughter, and wake at dawn.'

Which, the traveller added for himself alone, *brings the matter to a highly satisfactory conclusion.*

Dazed with elation, when the returning sun began to gild the turrets of the city of Teq with the promise of a new day, Lord Fellian struggled to the high gallery of his tower in order to witness the departure of his defeated rivals. No one in all the history of this city had won so fantastic a victory in a single night's gambling! Stripped of their wealth to such a degree that it would take a month or more simply to account the balancing of the debts incurred, Lords Yuckin and Nusk were creeping like whipped dogs into the morning twilight.

Lord Fellian leaned drunkenly over the parapet of the gallery and whooped like a falconer sighting quarry; when some cowed member of Lord Yuckin's train glanced up to see what the noise was, he spilled the contents of his latest beaker of wine on the upturned face and roared with laughter again.

'So much for the old fool who bet that Lady Luck's face was turned away from me!' he bellowed, and laughed again as the racket of his boasting was reflected from the nearby rooftops.

'Are you sure?'

On the heels of his politely voiced question the traveller in black emerged from shadow and crossed the jewelled floor of the gallery, his cloak making a faint swishing as he went.

'Why, you . . . !' Lord Fellian gasped and made to draw back, but the parapet was hard against him and there was nowhere to retreat save into insubstantial air. 'Where are my guards? Hey—*guards*!'

'None of them came up here with you,' said the traveller gravely. 'They are persuaded that upon a winner like yourself Lady Luck smiles so long and so favourably that no harm will come to you.'

'Ah-hah!' Fellian began to regain his composure. 'And for that reason you have lost your bet with me, have you not?'

'Why, no,' said the traveller with genuine regret, for it had always seemed a shame to him that a person of real intelligence should be seduced into the paths of action chosen by Lord Fellian. 'I have won.'

He smote with his staff against the wall that screened the gallery from sight of the tallest tower in Teq, and a slice fell away like a wedge cut from a cheese. Beyond, there where Fellian's reflex-driven gaze fled before he could check himself, Lady Luck's pinnacle loomed on the easterly blueness of the dawn.

A scream died still-born in Lord Fellian's throat. He stared and stared, and after a while he said, 'But . . . But there's only a stump!'

And it was true; against the sky, a broken jagged edge of stone marked the former location of the celebrated statue.

He began to giggle. 'Why, you've lost after all!' he chuckled. 'You did not make me the wager that Lady Luck had ceased to smile on me, which would be a fair victory—only that her back was turned to me.'

'True,' said the traveller sadly.

'Then—'

'Then I have won.' He gestured with his staff. 'Go forward; examine those chunks of stone I have broken from the wall of your gallery.'

Uncertain, but cowed, Fellian obeyed. His fingertips fumbled across rough plaster while he coughed in dust, and found smooth chased stones not conformable to the flat surface of the broken wall. A knot of hair-ribbon interpreted in sculpture; the slope of a gown, petrified, slanting over shoulder-blades of granite . . .

'There was a storm,' said the traveller didactically. 'The figure tumbled and landed in the street. It has, has it not, always been the custom that any who looked upon the statue of Lady Luck should die? Save the breath you'd waste on an answer; I've seen the bones of some whom your agents dumped in the market-place on precisely that excuse, regardless of whether the charge was true.

'Accordingly, none recognized the fragments. When you commanded stonemasons to assemble the necessary material and build a wall atop your handsome tower here, they gathered up what they could, and into the wall they set the broken pieces of the statue, with the back to your throne. My bet with you is won, and you are done for.

'During the night, you have bankrupted your rival lords; shame and custom will combine to compel them to honour their debt, and they will cede to you all the wealth they have filched from the people of this land.

But you will have no joy of it; you placed the greatest stake you knew on the bet I made, and now I claim my winnings.'

He stretched out his staff. One hand clutching the mocking back of the statue, the other clawing at air as though to take handfuls of it and stuff it into his choking lungs, Lord Fellian rolled his eyes upwards in his sockets and departed into nowhere.

A while later, when they came upon the corpse, those who had pledged themselves to his service began to quarrel about partitioning what he had left behind—in sum, the total wealth of the city and its environs.

'I will have the treasury!' cried Torquaida. 'It is due to me!' But a younger and more vigorous man, a clerk, struck him down with a candlestick. His old pate cracked across like the shell of an egg.

'If I can have nothing more, I'll take the booty Lord Fellian cheated me of!' vowed Captain Achoreus, and set off in search of the girl Viola. But he tripped on the slippery marble steps of the entrance to the women's quarters, and by the time he recovered from the blow on the head which resulted she was awake and away.

On hearing that his lord was dead, a loser in the game of life after all, the groom with whom the traveller had spoken saddled up Western Wind, sighing.

'At least this small recompense is due me,' he muttered, and opened the door of the stall. Later, in Barbizond, he offered the stallion at a livery stable to cover some mares on heat, and from the foals which resulted built up a fine string of horses of his own.

Likewise the falconer, on being told the news, gathered his prize merlin and went out into the countryside to get what living he could; he lost the merlin by flying it at an eagle that had stolen a child, a match the eagle was bound to win. But the child was the only son of a wealthy landholder, and in gratitude he made the falconer bailiff of his estate, second only to himself, until his son came of age.

Also the cook, being informed, gathered a brand from under his cauldron and went forth by a secret passage he knew of, leading from the back of the ox-roasting hearth. There he turned his ankle on a square object lying in the dust of the passageway, and the light of the brand showed him that it was the lost Book of Knightly Vigour, from which—legend claimed—the Count of Hyfel, founder of Teq, had gained the amorous skill to woo and wed his twenty-seven brides. With recipes from it he opened a cookshop, and defeated lovers from a score of lands trudged over hill and dale to savour his unique dishes.

Bewildered amid the confusion, the captives from Wantwich, however, were content to be able to find their way to freedom in the warm morning sun.

On their first return, the villagers were a trifle puzzled to discover that the pond beside the green, which for as long as they could recall had been placid, now roiled unaccountably. However, as the repairs proceeded—new roofs and shutters, new gates and fences, to replace those broken by the troops from Teq—that disturbance ceased. Before the new beer was brewed, the new barrels were coopered, and a new fiddle made for Fiddler Jarge, the water had regained its normal state.

And, on the day when—belatedly—Leluak led out his bride to start the dancing among the assembled people, a stranger in a black cloak stood with a benign smile under the shelter of a sycamore.

'Was it not clever, Horimos?' he said under his breath to the prisoned elemental beneath the water. 'Was it not ingenious to pervert the thinking of rational man into the random path of a gambler, who lacks even the dangerous knowledge of an enchanter when he tampers with the forces of chaos?'

Unnoticed except by the traveller, the pond gave off a bubble full of foul marshy gas, which might have been intended for an answer.

'*Shut—brr-up!*'

'By all means, Horimos,' murmured the traveller, and drained the mug of Brewer Harring's good beer which he, like all passers-by on a festival day, had been offered. He set the vessel on a handy stump, and the music rose to a frantic, gay crescendo.

When, a little diffidently, the new bride came to greet him and make him welcome among the other company, there was no trace of his presence except the empty mug.

PETER S. BEAGLE

LILA THE WEREWOLF

Lila Braun had been living with Farrell for three weeks before she found out she was a werewolf. They had met at a party when the moon was a few nights past the full, and by the time it had withered to the shape of a lemon Lila had moved her suitcase, her guitar, and her Ewan MacColl records two blocks north and four blocks west to Farrell's apartment on Ninety-eighth Street. Girls sometimes happened to Farrell like that.

One evening Lila wasn't in when Farrell came home from work at the bookstore. She had left a note on the table, under a can of tuna fish. The note said that she had gone up to the Bronx to have dinner with her mother, and would probably be spending the night there. The coleslaw in the refrigerator should be finished up before it went bad.

Farrell ate the tuna fish and gave the coleslaw to Grunewald. Grunewald was a half-grown Russian wolfhound, the colour of sour milk. He looked like a goat, and had no outside interests except shoes. Farrell was taking care of him for a girl who was away in Europe for the summer. She sent Grunewald a tape recording of her voice every week.

Farrell went to a movie with a friend, and to the West End afterward for beer. Then he walked home alone under the full moon, which was red and yellow. He reheated the morning coffee, played a record, read through a week-old 'News Of The Week In Review' section of the Sunday *Times*, and finally took Grunewald up to the roof for the night, as he always did. The dog had been accustomed to sleep in the same bed with his mistress, and the point was not negotiable. Grunewald mooed and scrabbled and butted all the way, but Farrell pushed him out among the looming chimneys and ventilators and slammed the door. Then he came back downstairs and went to bed.

He slept very badly. Grunewald's baying woke him twice; and there was something else that brought him half out of bed, thirsty and lonely, with his sinuses full and the night swaying like a curtain as the figures of his dream scurried offstage. Grunewald seemed to have gone off the air—

perhaps it was the silence that had awakened him. Whatever the reason, he never really got back to sleep.

He was lying on his back, watching a chair with his clothes on it becoming a chair again, when the wolf came in through the open window. It landed lightly in the middle of the room and stood there for a moment, breathing quickly, with its ears back. There was blood on the wolf's teeth and tongue, and blood on its chest.

Farrell, whose true gift was for acceptance, especially in the morning, accepted the idea that there was a wolf in his bedroom and lay quite still, closing his eyes as the grim, black-lipped head swung towards him. Having once worked at a zoo, he was able to recognize the beast as a Central European subspecies: smaller and lighter-boned than the northern timber wolf variety, lacking the thick, ruffy mane at the shoulders and having a more pointed nose and ears. His own pedantry always delighted him, even at the worst moments.

Blunt claws clicking on the linoleum, then silent on the throw rug by the bed. Something warm and slow splashed down on his shoulder, but he never moved. The wild smell of the wolf was over him, and that did frighten him at last—to be in the same room with that smell and the Miró prints on the walls. Then he felt the sunlight on his eyelids, and at the same moment he heard the wolf moan softly and deeply. The sound was not repeated, but the breath on his face was suddenly sweet and smoky, dizzyingly familiar after the other. He opened his eyes and saw Lila. She was sitting naked on the edge of the bed, smiling, with her hair down.

'Hello, baby,' she said. 'Move over, baby. I came home.'

Farrell's gift was for acceptance. He was perfectly willing to believe that he had dreamed the wolf; to believe Lila's story of boiled chicken and bitter arguments and sleeplessness on Tremont Avenue; and to forget that her first caress had been to bite him on the shoulder, hard enough so that the blood crusting there as he got up and made breakfast might very well be his own. But then he left the coffee perking and went up to the roof to get Grunewald. He found the dog sprawled in a grove of TV antennas, looking more like a goat than ever, with his throat torn out. Farrell had never actually seen an animal with its throat torn out.

The coffeepot was still chuckling when he came back into the apartment, which struck him as very odd. You could have either werewolves or Pyrex nine-cup percolators in the world, but not both, surely. He told Lila, watching her face. She was a small girl, not really pretty, but with good eyes and a lovely mouth, and with a curious sullen gracefulness that had been the first thing to speak to Farrell at the party. When he told her how Grunewald had looked, she shivered all over, once.

'Ugh!' she said, wrinkling her lips back from her neat white teeth. 'Oh baby, how awful. Poor Grunewald. Oh, poor Barbara.' Barbara was Grunewald's owner.

'Yeah,' Farrell said. 'Poor Barbara, making her little tapes in Saint-Tropez.' He could not look away from Lila's face.

She said, 'Wild dogs. Not really wild, I mean, but with owners. You hear about it sometimes, how a pack of them get together and attack children and things, running through the streets. Then they go home and eat their Dog Yummies. The scary thing is that they probably live right around here. Everybody on the block seems to have a dog. God, that's scary. Poor Grunewald.'

'They didn't tear him up much,' Farrell said. 'It must have been just for the fun of it. And the blood. I didn't know dogs killed for the blood. He didn't have any blood left.'

The tip of Lila's tongue appeared between her lips, in the unknowing reflex of a fondled cat. As evidence, it wouldn't have stood up even in old Salem; but Farrell knew the truth then, beyond laziness or rationalization, and went on buttering toast for Lila. Farrell had nothing against werewolves, and he had never liked Grunewald.

He told his friend Ben Kassoy about Lila when they met in the Automat for lunch. He had to shout it over the clicking and rattling all around them, but the people sitting six inches away on either hand never looked up. New Yorkers never eavesdrop. They hear only what they simply cannot help hearing.

Ben said, 'I told you about Bronx girls. You better come stay at my place for a few days.'

Farrell shook his head. 'No, that's silly. I mean, it's only Lila. If she were going to hurt me, she could have done it last night. Besides, it won't happen again for a month. There has to be a full moon.'

His friend stared at him. 'So what? What's that got to do with anything? You going to go on home as though nothing had happened?'

'Not as though nothing had happened,' Farrell said lamely. 'The thing is, it's still only Lila, not Lon Chaney or somebody. Look, she goes to her psychiatrist three afternoons a week, and she's got her guitar lesson one night a week, and her pottery class one night, and she cooks eggplant maybe twice a week. She calls her mother every Friday night, and one night a month she turns into a wolf. You see what I'm getting at? It's still Lila, whatever she does, and I just can't get terribly shook about it. A little bit, sure, because what the hell. But I don't know. Anyway, there's no mad rush about it. I'll talk to her when the thing comes up in conversation, just naturally. It's okay.'

Ben said, 'God damn. You see why nobody has any respect for liberals anymore? Farrell, I know you. You're just scared of hurting her feelings.'

'Well, it's that too,' Farrell agreed, a little embarrassed. 'I hate confrontations. If I break up with her now, she'll think I'm doing it because she's a werewolf. It's awkward, it feels nasty and middle-class. I should have broken up with her the first time I met her mother, or the second time she served the eggplant. Her mother, boy, there's the real werewolf, there's somebody I'd wear wolfsbane against, that woman. Damn, I wish I hadn't found out. I don't think I've ever found out anything about people that I was the better for knowing.'

Ben walked all the way back to the bookstore with him, arguing. It touched Farrell, because Ben hated to walk. Before they parted, Ben suggested, 'At least you could try some of that stuff you were talking about, the wolfsbane. There's garlic, too—you put some in a little bag and wear it around your neck. Don't laugh, man. If there's such a thing as werewolves, the other stuff must be real, too. Cold iron, silver, oak, running water—'

'I'm not laughing at you,' Farrell said, but he was still grinning. 'Lila's shrink says she has a rejection thing, very deep-seated, take us years to break through all that scar tissue. Now if I start walking around wearing amulets and mumbling in Latin every time she looks at me, who knows how far it'll set her back? Listen, I've done some things I'm not proud of, but I don't want to mess up anyone's analysis. That's the sin against God.' He sighed and slapped Ben lightly on the arm. 'Don't worry about it. We'll work it out, I'll talk to her.'

But between that night and the next full moon, he found no good, casual way of bringing the subject up. Admittedly, he did not try as hard as he might have: it was true that he feared confrontations more than he feared werewolves, and he would have found it almost as difficult to talk to Lila about her guitar playing, or her pots, or the political arguments she got into at parties. 'The thing is,' he said to Ben, 'it's sort of one more little weakness not to take advantage of. In a way.'

They made love often that month. The smell of Lila flowered in the bedroom, where the smell of the wolf still lingered almost visibly, and both of them were wild, heavy zoo smells, warm and raw and fearful, the sweeter for being savage. Farrell held Lila in his arms and knew what she was, and he was always frightened; but he would not have let her go if she had turned into a wolf again as he held her. It was a relief to peer at her while she slept and see how stubby and childish her fingernails were, or that the skin around her mouth was rashy because she had been snacking on chocolate. She loved secret sweets, but they always betrayed her.

It's only Lila after all, he would think as he drowsed off. Her mother used to hide the candy, but Lila always found it. Now she's a big girl, neither married nor in a graduate school, but living in sin with an Irish musician, and she can have all the candy she wants. What kind of a werewolf is that. Poor Lila, practising *Who killed Davey Moore? Why did he die?* . . .

The note said that she would be working late at the magazine, on layout, and might have to be there all night. Farrell put on about four feet of Telemann laced with Django Reinhardt, took down *The Golden Bough*, and settled into a chair by the window. The moon shone in at him, bright and thin and sharp as the lid of a tin can, and it did not seem to move at all as he dozed and woke.

Lila's mother called several times during the night, which was interesting. Lila still picked up her mail and most messages at her old apartment, and her two roommates covered for her when necessary, but Farrell was absolutely certain that her mother knew she was living with him. Farrell was an expert on mothers. Mrs Braun called him Joe each time she called and that made him wonder, for he knew she hated him. Does she suspect that we share a secret? Ah, poor Lila.

The last time the telephone woke him, it was still dark in the room, but the traffic lights no longer glittered through rings of mist, and the cars made a different sound on the warming pavement. A man was saying clearly in the street, 'Well, *I*'d shoot'm. *I*'d shoot'm.' Farrell let the telephone ring ten times before he picked it up.

'Let me talk to Lila,' Mrs Braun said.

'She isn't here.' What if the sun catches her, what if she turns back to herself in front of a cop, or a bus driver, or a couple of nuns going to early Mass? 'Lila isn't here, Mrs Braun.'

'I have reason to believe that's not true.' The fretful, muscular voice had dropped all pretence of warmth. 'I want to talk to Lila.'

Farrell was suddenly dry-mouthed and shivering with fury. It was her choice of words that did it. 'Well, I have reason to believe you're a suffocating old bitch and a bourgeois Stalinist. How do you like them apples, Mrs B?' As though his anger had summoned her, the wolf was standing two feet away from him. Her coat was dark and lank with sweat, and yellow saliva was mixed with the blood that strung from her jaws. She looked at Farrell and growled far away in her throat.

'Just a minute,' he said. He covered the receiver with his palm. 'It's for you,' he said to the wolf. 'It's your mother.'

The wolf made a pitiful sound, almost inaudible, and scuffed at the floor. She was plainly exhausted. Mrs Braun pinged in Farrell's ear like a bug against a lighted window. 'What, what? Hello, what is this? Listen, you put Lila on the phone right now. Hello? I want to talk to Lila. I know she's there.'

Farrell hung up just as the sun touched a corner of the window. The wolf became Lila. As before, she only made one sound. The phone rang again, and she picked it up without a glance at Farrell. 'Bernice?' Lila always called her mother by her first name. 'Yes—no, no—yeah, I'm fine. I'm all right, I just forgot to call. No, I'm all right, will you listen? Bernice, there's no law that says you have to get hysterical. Yes, you are.' She dropped down on the bed, groping under her pillow for cigarettes. Farrell got up and began to make coffee.

'Well, there was a little trouble,' Lila was saying. 'See, I went to the zoo, because I couldn't find—Bernice, I know, I *know*, but that was, what, three months ago. The thing is, I didn't think they'd have their horns so soon. Bernice, I had to, that's all. There'd only been a couple of cats and a—well, sure they chased me, but I—well, Momma, Bernice, what did you want me to do? Just what did you want me to do? You're always so dramatic—why do I shout? I shout because I can't get you to listen to me any other way. You remember what Dr Schechtman said—what? No, I told you, I just forgot to call. No, that is the reason, that's the real and only reason. Well, whose fault is that? What? Oh, Bernice. Jesus Christ, Bernice. All right, *how* is it Dad's fault?'

She didn't want the coffee, or any breakfast, but she sat at the table in his bathrobe and drank milk greedily. It was the first time he had ever seen her drink milk. Her face was sandy-pale, and her eyes were red. Talking to her mother left her looking as though she had actually gone ten rounds with the woman. Farrell asked, 'How long has it been happening?'

'Nine years,' Lila said. 'Since I hit puberty. First day, cramps; the second day, this. My introduction to womanhood.' She snickered and spilled her milk. 'I want some more,' she said. 'Got to get rid of that taste.'

'Who knows about it?' he asked. 'Pat and Janet?' They were the two girls she had been rooming with.

'God, no. I'd never tell them. I've never told a girl. Bernice knows, of course, and Dr Schechtman—he's my head doctor. And you now. That's all.' Farrell waited. She was a bad liar, and only did it to heighten the effect of the truth. 'Well, there was Mickey,' she said. 'The guy I told you about the first night, you remember? It doesn't matter. He's an acidhead in Vancouver, of all the places. He'll never tell anybody.'

He thought: I wonder if any girl has ever talked about me in that sort of voice. I doubt it, offhand. Lila said, 'It wasn't too hard to keep it secret. I missed a lot of things. Like I never could go to the riding camp, and I still want to. And the senior play, when I was in high school. They picked me to play the girl in *Liliom*, but then they changed the evening, and I had to say I was sick. And the winter's bad, because the sun sets so early. But actually, it's been a lot less trouble than my goddamn allergies.' She made a laugh, but Farrell did not respond.

'Dr Schechtman says it's a sex thing,' she offered. 'He says it'll take years and years to cure it. Bernice thinks I should go to someone else, but I don't want to be one of those women who runs around changing shrinks like hair colours. Pat went through five of them in a month one time. Joe, I wish you'd say something. Or just go away.'

'Is it only dogs?' he asked. Lila's face did not change, but her chair rattled, and the milk went over again. Farrell said, 'Answer me. Do you only kill dogs, and cats, and zoo animals?'

The tears began to come, heavy and slow, bright as knives in the morning sunlight. She could not look at him; and when she tried to speak she could only make creaking, cartilaginous sounds in her throat. '*You* don't know,' she whispered at last. 'You don't have any idea what it's like.'

'That's true,' he answered. He was always very fair about that particular point.

He took her hand, and then she really began to cry. Her sobs were horrible to hear, much more frightening to Farrell than any wolf noises. When he held her, she rolled in his arms like a stranded ship with the waves slamming into her. I always get the criers, he thought sadly. My girls always cry, sooner or later. But never for me.

'Don't leave me!' she wept. 'I don't know why I came to live with you—I knew it wouldn't work—but don't leave me! There's just Bernice and Dr Schechtman, and it's so lonely. I want somebody else, I get so lonely. Don't leave me, Joe. I love you, Joe. I love you.'

She was patting his face as though she were blind. Farrell stroked her hair and kneaded the back of her neck, wishing that her mother would call again. He felt skilled and weary, and without desire. I'm doing it again, he thought.

'I love you,' Lila said. And he answered her, thinking, I'm doing it again. That's the great advantage of making the same mistake a lot of times. You come to know it, and you can study it and get inside it, really make it yours. It's the same good old mistake, except this time the girl's hang-up is different. But it's the same thing. I'm doing it again.

The building superintendent was thirty or fifty: dark, thin, quick, and shivering. A Lithuanian or a Latvian, he spoke very little English. He smelled of black friction tape and stale water, and he was strong in the twisting way that a small, lean animal is strong. His eyes were almost purple, and they bulged a little, straining out—the terrible eyes of a herald angel stricken dumb. He roamed in the basement all day, banging on pipes and taking the elevator apart.

The superintendent met Lila only a few hours after Farrell did; on that first night, when she came home with him. At the sight of her the little man jumped back, dropping the two-legged chair he was carrying. He promptly fell over it, and did not try to get up, but cowered there, clucking and gulping, trying to cross himself and make the sign of the horns at the same time. Farrell started to help him up, but he screamed. They could hardly hear the sound.

It would have been merely funny and embarrassing, except for the fact that Lila was equally as frightened of the superintendent, from that moment. She would not go down to the basement for any reason, nor would she enter or leave the house until she was satisfied that he was nowhere near. Farrell had thought then that she took the superintendent for a lunatic.

'I don't know how he knows,' he said to Ben. 'I guess if you believe in werewolves and vampires, you probably recognize them right away. I don't believe in them at all, and I live with one.'

He lived with Lila all through the autumn and the winter. They went out together and came home, and her cooking improved slightly, and she gave up the guitar and got a kitten named Theodora. Sometimes she wept, but not often. She turned out not to be a real crier.

She told Dr Schechtman about Farrell, and he said that it would probably be a very beneficial relationship for her. It wasn't, but it wasn't a particularly bad one either. Their lovemaking was usually good, though it bothered Farrell to suspect that it was the sense and smell of the Other that excited him. For the rest, they came near being friends. Farrell had known that he did not love Lila before he found out that she was a werewolf, and this made him feel a great deal easier about being bored with her.

'It'll break up by itself in the spring,' he said, 'like ice.'

Ben asked, 'What if it doesn't?' They were having lunch in the Automat again. 'What'll you do if it just goes on?'

'It's not that easy.' Farrell looked away from his friend and began to explore the mysterious, swampy innards of his beef pie. He said, 'The trouble is that I know her. That was the real mistake. You shouldn't get to know people if you know you're not going to stay with them, one way or another. It's all right if you come and go in ignorance, but you shouldn't know them.'

A week or so before the full moon, she would start to become nervous and strident, and this would continue until the day preceding her transformation. On that day, she was invariably loving, in the tender, desperate manner of someone who is going away; but the next day would see her silent, speaking only when she had to. She always had a cold on the last day, and looked grey and patchy and sick, but she usually went to work anyway.

Farrell was sure, though she never talked about it, that the change into wolf shape was actually peaceful for her, though the returning hurt. Just before moonrise she would take off her clothes and take the pins out of her hair, and stand waiting. Farrell never managed not to close his eyes when she dropped heavily down on all fours; but there was a moment before that when her face would grow a look that he never saw at any other time, except when they were making love. Each time he saw it, it struck him as a look of wondrous joy at not being Lila anymore.

'See, I know her,' he tried to explain to Ben. 'She only likes to go to colour movies, because wolves can't see colour. She can't stand the Modern Jazz Quartet, but that's all she plays the first couple of days afterward. Stupid things like that. Never gets high at parties, because she's afraid she'll start talking. It's hard to walk away, that's all. Taking what I know with me.'

Ben asked, 'Is she still scared of the super?'

'Oh, God,' Farrell said. 'She got his dog last time. It was a Dalmatian—good-looking animal. She didn't know it was his. He doesn't hide when he sees her now, he just gives her a look like a stake through the heart. That man is a really classy hater, a natural. I'm scared of him myself.' He stood up and began to pull on his overcoat. 'I wish he'd get turned on to her mother. Get some practical use out of him. Did I tell you she wants me to call her Bernice?'

Ben said, 'Farrell, if I were you, I'd leave the country. I would.'

They went out into the February drizzle that sniffled back and forth between snow and rain. Farrell did not speak until they reached the corner where he turned towards the bookstore. Then he said very softly, 'Damn, you have to be so careful. Who wants to know what people turn into?'

May came, and a night when Lila once again stood naked at the window, waiting for the moon. Farrell fussed with dishes and garbage bags, and fed the cat. These moments were always awkward. He had just asked her, 'You want to save what's left of the rice?' when the telephone rang.

It was Lila's mother. She called two and three times a week now. 'This is Bernice. How's my Irisher this evening?'

'I'm fine, Bernice,' Farrell said. Lila suddenly threw back her head and drew a heavy, whining breath. The cat hissed silently and ran into the bathroom.

'I called to inveigle you two uptown this Friday,' Mrs Braun said. 'A couple of old friends are coming over, and I know if I don't get some young people in we'll just sit around and talk about what went wrong with the Progressive Party. The Old Left. So if you could sort of sweet-talk our girl into spending an evening in Squaresville—'

'I'll have to check with Lila.' She's *doing* it, he thought, that terrible woman. Every time I talk to her, I sound married. I see what she's doing, but she goes right ahead anyway. He said, 'I'll talk to her in the morning.' Lila struggled in the moonlight, between dancing and drowning.

'Oh,' Mrs Braun said. 'Yes, of course. Have her call me back.' She sighed. 'It's such a comfort to me to know you're there. Ask her if I should fix a fondue?'

Lila made a handsome wolf: tall and broad-chested for a female, moving as easily as water sliding over stone. Her coat was dark brown, showing red in the proper light, and there were white places on her breast. She had pale green eyes, the colour of the sky when a hurricane is coming.

Usually she was gone as soon as the changing was over, for she never cared for him to see her in her wolf form. But tonight she came slowly towards him, walking in a strange way, with her hindquarters almost dragging. She was making a high, soft sound, and her eyes were not focusing on him.

'What is it?' he asked foolishly. The wolf whined and skulked under the table, rubbing against the leg. Then she lay on her belly and rolled and as she did so the sound grew in her throat until it became an odd, sad, thin cry; not a hunting howl, but a shiver of longing turned into breath.

'Jesus, don't do that!' Farrell gasped. But she sat up and howled again, and a dog answered her from somewhere near the river. She wagged her tail and whimpered.

Farrell said, 'The super'll be up here in two minutes flat. What's the matter with you?' He heard footsteps and low frightened voices in the apartment above them. Another dog howled, this one nearby, and the wolf wriggled a little way towards the window on her haunches, like a baby, scooting. She looked at him over her shoulder, shuddering violently. On an impulse, he picked up the phone and called her mother.

Watching the wolf as she rocked and slithered and moaned, he described her actions to Mrs Braun. 'I've never seen her like this,' he said. 'I don't know what's the matter with her.'

'Oh, my God,' Mrs Braun whispered. She told him.

When he was silent, she began to speak very rapidly. 'It hasn't happened for such a long time. Schechtman gives her pills, but she must have run out and forgotten—she's always been like that, since she was little. All the thermos bottles she used to leave on the school bus, and every week her piano music—'

'I wish you'd told me before,' he said. He was edging very cautiously towards the open window. The pupils of the wolf's eyes were pulsing with her quick breaths.

'It isn't a thing you tell people!' Lila's mother wailed in his ear. 'How do you think it was for me when she brought her first little boyfriend—' Farrell dropped the phone and sprang for the window. He had the inside track, and he might have made it, but she turned her head and snarled so wildly that he fell back. When he reached the window, she was already two fire-escape landings below, and there was eager yelping waiting for her in the street.

Dangling and turning just above the floor, Mrs Braun heard Farrell's distant yell, followed immediately by a heavy thumping on the door. A strange, tattered voice was shouting unintelligibly beyond the knocking. Footsteps crashed by the receiver and the door opened.

'My dog, my dog!' the strange voice mourned. 'My dog, my dog, my dog!'

'I'm sorry about your dog,' Farrell said. 'Look, please go away. I've got work to do.'

'I got work,' the voice said. 'I know my work.' It climbed and spilled into another language, out of which English words jutted like broken bones. 'Where is she? Where is she? She kill my dog.'

'She's not here.' Farrell's own voice changed on the last word. It seemed a long time before he said, 'You'd better put that away.

Mrs Braun heard the howl as clearly as though the wolf were running beneath her own window: lonely and insatiable, with a kind of gasping laughter in it. The other voice began to scream. Mrs Braun caught the phrase *silver bullet* several times. The door slammed; then opened and slammed again.

Farrell was the only man of his own acquaintance who was able to play back his dreams while he was having them: to stop them in mid-flight, no matter how fearful they might be—or how lovely—and run them over and over studying them in his sleep, until the most terrifying reel became at once utterly harmless and unbearably familiar. This night that he spent running after Lila was like that.

He would find them congregated under the marquee of an apartment house, or romping around the moonscape of a construction site: ten or fifteen males of all races, creeds, colours, and previous conditions of servitude; whining and yapping, pissing against tyres, inhaling indiscriminately each other and the lean, grinning bitch they surrounded. She frightened them, for she growled more wickedly than coyness demanded, and where she snapped, even in play, bone showed. Still they tumbled on her and over her, biting her neck and ears in their turn; and she snarled but she did not run away.

Never, at least, until Farrell came charging upon them, shrieking like any cuckold, kicking at the snuffling lovers. Then she would turn and race off into the spring dark, with her thin, dreamy howl floating behind her like the train of a smoky gown. The dogs followed, and so did Farrell, calling and cursing. They always lost him quickly, that jubilant marriage procession, leaving him stumbling down rusty iron ladders into places where he fell over garbage cans. Yet he would come upon them as inevitably in time, loping along Broadway or trotting across Columbus Avenue towards the Park; he would hear them in the tennis courts near the river, breaking down the nets over Lila and her moment's Ares. There were dozens of them now, coming from all directions. They stank of their joy, and he threw stones at them and shouted, and they ran.

And the wolf ran at their head, on sidewalks and on wet grass; her tail waving contentedly, but her eyes still hungry, and her howl growing ever more warning than wistful. Farrell knew that she must have blood before sunrise, and that it was both useless and dangerous to follow her. But the night wound and unwound itself, and he knew the same things over and over, and ran down the same streets, and saw the same couples walk wide of him, thinking he was drunk.

Mrs Braun kept leaping out of a taxi that pulled up next to him; usually at corners where the dogs had just piled by, knocking over the crates stacked in market doorways and spilling the newspapers at the subway kiosks. Standing in broccoli, in black taffeta, with a front like a ferry-boat—yet as lean in the hips as her wolf-daughter—with her plum-coloured hair all loose, one arm lifted, and her orange mouth pursed in a bellow, she was no longer Bernice but a wronged fertility goddess getting set to blast the harvest. 'We've got to split up!' she would roar at Farrell, and each time it sounded like a sound idea. Yet he looked for her whenever he lost Lila's trail, because she never did.

The superintendent kept turning up too, darting after Farrell out of alleys or cellar entrances, or popping from the freight elevators that load through the sidewalk. Farrell would hear his numberless passkeys clicking on the flat piece of wood tucked into his belt.

'You see her? You see her, the wolf, kill my dog?' Under the fat, ugly moon, the Army .45 glittered and trembled like his own mad eyes.

'Mark with a cross.' He would pat the barrel of the gun and shake it under Farrell's nose like a maracas. 'Mark with a cross, bless by a priest. Three silver bullets. She kill my dog.'

Lila's voice would come sailing to them then, from up in Harlem or away near Lincoln Center, and the little man would whirl and dash down into the earth, disappearing into the crack between two slabs of sidewalk. Farrell understood quite clearly that the superintendent was hunting Lila underground, using the keys that only superintendents have to take elevators down to the black sub-sub-basements, far below the bicycle rooms and the wet, shaking laundry rooms, and below the furnace rooms, below the passages walled with electricity meters and roofed with burly steam pipes; down to the realms where the great dim water mains roll like whales, and the gas lines hump and preen, down where the roots of the apartment houses fade together, and so along under the city, scrabbling through secret ways with silver bullets, and his keys rapping against the piece of wood. He never saw Lila, but he was never very far behind her.

Cutting across parking lots, pole-vaulting between locked bumpers, edging and dancing his way through fluorescent gaggles of haughty children; leaping uptown like a salmon against the current of the theatre crowds; walking quickly past the random killing faces that floated down the night tide like unexploded mines, and especially avoiding the crazy faces that wanted to tell him what it was like to be crazy—so Farrell pursued Lila Braun, of Tremont Avenue and CCNY, in the city all night long. Nobody offered to help him, or tried to head off the dangerous-looking bitch bounding along with the delirious gaggle of admirers streaming after her; but then, the dogs had to fight through the same clenched legs and vengeful bodies that Farrell did. The crowds slowed Lila down, but he felt relieved whenever she turned towards the emptier streets. *She must have blood soon, somewhere.*

Farrell's dreams eventually lost their clear edge after he played them back a certain number of times, and so it was with the night. The full moon skidded down the sky, thinning like a tatter of butter in a skillet, and remembered scenes began to fold sloppily into each other. The sound of Lila and the dogs grew fainter whichever way he followed. Mrs Braun blinked on and off at longer intervals; and in dark doorways and under subway gratings, the superintendent burned like a corposant, making the barrel of his pistol run rainbow. At last he lost Lila for good, and with that it seemed that he woke.

It was still night, but not dark, and he was walking slowly home on Riverside Drive through a cool, grainy fog. The moon had set, but the river was strangely bright: glittering grey as far up as the Bridge, where headlights left shiny, wet paths like snails. There was no one else on the street.

'Dumb broad,' he said aloud. 'The hell with it. She wants to mess around, let her mess around.' He wondered whether werewolves could have cubs, and what sort of cubs they might be. Lila must have turned on the dogs by now, for the blood. Poor dogs, he thought. They were all so dirty and innocent and happy with her.

'A moral lesson for all of us,' he announced sententiously. 'Don't fool with strange, eager ladies, they'll kill you.' He was a little hysterical. Then, two blocks ahead of him, he saw the gaunt shape in the grey light of the river; alone now, and hurrying. Farrell did not call to her, but as soon as he began to run, the wolf wheeled and faced him. Even at that distance, her eyes were stained and streaked and wild. She showed all the teeth on one side of her mouth, and she growled like fire.

Farrell trotted steadily towards her, crying, 'Go home, go home! Lila, you dummy, get on home, it's morning!' She growled terribly, but when Farrell was less than a block away she turned again and dashed across the street, heading for West End Avenue. Farrell said, 'Good girl, that's it,' and limped after her.

In the hours before sunrise on West End Avenue, many people came out to walk their dogs. Farrell had done it often enough with poor Grunewald to know many of the dawn walkers by sight, and some to talk to. A fair number of them were whores and homosexuals, both of whom always seem to have dogs in New York. Quietly, almost always alone, they drifted up and down the Nineties, piloted by their small, fussy beasts, but moving in a kind of fugitive truce with the city and the night that was ending. Farrell sometimes fancied that they were all asleep, and that this hour was the only true rest they ever got.

He recognized Robie by his two dogs, Scone and Crumpet. Robie lived in the apartment directly below Farrell's, usually unhappily. The dogs were horrifying little homebrews of Chihuahua and Yorkshire terrier, but Robie loved them.

Crumpet, the male, saw Lila first. He gave a delighted yap of welcome and proposition (according to Robie, Scone bored him, and he liked big girls anyway) and sprang to meet her, yanking his leash through Robie's slack hand. The wolf was almost upon him before he realized his fatal misunderstanding and scuttled desperately in retreat, meowing with utter terror.

Robie wailed, and Farrell ran as fast as he could, but Lila knocked Crumpet off his feet and slashed his throat while he was still in the air. Then she crouched on the body, nuzzling it in a dreadful way.

Robie actually came within a step of leaping upon Lila and trying to drag her away from his dead dog. Instead, he turned on Farrell as he came panting up, and began hitting him with a good deal of strength and accuracy. 'Damn you, damn you!' he sobbed. Little Scone ran away around the corner, screaming like a mandrake.

Farrell put up his arms and went with the punches, all the while yelling at Lila until his voice ripped. But the blood frenzy had her, and Farrell had never imagined what she must be like at those times. Somehow she had spared the dogs who had loved her all night, but she was nothing but thirst now. She pushed and kneaded Crumpet's body as though she were nursing.

All along the avenue, the morning dogs were barking like trumpets. Farrell ducked away from Robie's soft fists and saw them coming; tripping over their trailing leashes, running too fast for their stubby legs. They were small, spoiled beasts, most of them, overweight and shortwinded, and many were not young. Their owners cried unmanly pet names after them, but they waddled gallantly towards their deaths, barking promises far bigger than themselves, and none of them looked back.

She looked up with her muzzle red to the eyes. The dogs did falter then, for they knew murder when they smelled it, and even their silly, nearsighted eyes understood vaguely what creature faced them. But they knew the smell of love too, and they were all gentlemen.

She killed the first two to reach her—a spitz and a cocker spaniel—with two snaps of her jaws. But before she could settle down to her meal, three Pekes were scrambling up to her, though they would have had to stand on each other's shoulders. Lila whirled without a sound, and they fell away, rolling and yelling but unhurt. As soon as she turned, the Pekes were at her again, joined now by a couple of valiant poodles. Lila got one of the poodles when she turned again.

Robie had stopped beating on Farrell, and was leaning against a traffic light, being sick. But other people were running up now: a middle-aged black man, crying; a plump youth in a plastic car coat and bedroom slippers, who kept whimpering, 'Oh God, she's eating them, look at her, she's really eating them!'; two lean, ageless girls in slacks, both with foamy beige hair. They all called wildly to their unheeding dogs, and they all grabbed at Farrell and shouted in his face. Cars began to stop.

The sky was thin and cool, rising pale gold, but Lila paid no attention to it. She was ramping under the swarm of little dogs; rearing and spinning in circles, snarling blood. The dogs were terrified and bewildered, but they never swerved from their labour. The smell of love told them that they were welcome, however ungraciously she seemed to receive them. Lila shook herself, and a pair of squealing dachshunds, hobbled in a double harness, tumbled across the sidewalk to end at Farrell's feet. They scrambled up and immediately towed themselves back into the maelstrom. Lila bit one of them almost in half, but the other dachshund went on trying to climb her hindquarters, dragging his ripped comrade with him. Farrell began to laugh.

The black man said, 'You think it's funny?' and hit him. Farrell sat down, still laughing. The man stood over him, embarrassed, offering Farrell his handkerchief. 'I'm sorry, I shouldn't have done that,' he said. 'But your dog killed my dog.'

'She isn't my dog,' Farrell said. He moved to let a man pass between them, and then saw that it was the superintendent, holding his pistol with both hands. Nobody noticed him until he fired; but Farrell pushed one of the foamy-haired girls, and she stumbled against the superintendent as the gun went off. The silver bullet broke a window in a parked car.

The superintendent fired again while the echoes of the first shot were still clapping back and forth between the houses. A Pomeranian screamed that time, and a woman cried out, 'Oh, my God, he shot Borgy!' But the crowd was crumbling away, breaking into its individual components like pills on television. The watching cars had sped off at the sight of the gun, and the faces that had been peering down from windows disappeared. Except for Farrell, the few people who remained were scattered halfway down the block. The sky was brightening swiftly now.

'For God's sake, don't let him!' the same woman called from the shelter of a doorway. But two men made shushing gestures at her, saying, 'It's all right, he knows how to use that thing. Go ahead, buddy.'

The shots had at last frightened the little dogs away from Lila. She crouched among the twitching splotches of fur, with her muzzle wrinkled back and her eyes more black than green. Farrell saw a plaid rag that had been a dog jacket protruding from under her body. The superintendent stooped and squinted over the gun barrel, aiming with grotesque care, while the men cried to him to shoot. He was too far from the werewolf for her to reach him before he fired the last silver bullet, though he would surely die before she died. His lips were moving as he took aim.

Two long steps would have brought Farrell up behind the superintendent. Later he told himself that he had been afraid of the pistol, because that was easier than remembering how he had felt when he looked at Lila. Her tongue never stopped lapping around her dark jaws; and even as she set herself to spring, she lifted a bloody paw to her mouth. Farrell thought of her padding in the bedroom, breathing on his face. The superintendent grunted and Farrell closed his eyes. Yet even then he expected to find himself doing something.

Then he heard Mrs Braun's unmistakable voice. '*Don't you dare!*' She was standing between Lila and the superintendent: one shoe gone, and the heel off the other one; her knit dress torn at the shoulder, and her face tired and smudgy. But she pointed a finger at the startled superintendent, and he stepped quickly back, as though she had a pistol, too.

'Lady, that's a wolf,' he protested nervously. 'Lady, you please get, get out of the way. That's a wolf, I go shoot her now.'

'I want to see your licence for that gun.' Mrs Braun held out her hand. The superintendent blinked at her, muttering in despair. She said, 'Do you know that you can be sent to prison for twenty years for carrying a concealed weapon in this state? Do you know what the fine is for having a gun without a licence? The fine is Five. Thousand. Dollars.' The men down the street were shouting at her, but she swung around to face the creature snarling among the little dead dogs.

'Come on, Lila,' she said. 'Come on home with Bernice. I'll make tea and we'll talk. It's been a long time since we've really talked, you know? We used to have nice long talks when you were little, but we don't anymore.' The wolf had stopped growling, but she was crouching even lower, and her ears were still flat against her head. Mrs Braun said, 'Come on, baby. Listen, I know what—you'll call in sick at the office and stay for a few days. You'll get a good rest, and maybe we'll even look around a little for a new doctor, what do you say? Schechtman hasn't done a thing for you, I never liked him. Come on home, honey. Momma's here, Bernice knows.' She took a step towards the silent wolf, holding out her hand.

The superintendent gave a desperate, wordless cry and pumped forward, clumsily shoving Mrs Braun to one side. He levelled the pistol point-blank, wailing, 'My dog, my dog!' Lila was in the air when the gun went off, and her shadow sprang after her, for the sun had risen. She crumpled down across a couple of dead Pekes. Their blood dabbled her breasts and her pale throat.

Mrs Braun screamed like a lunch whistle. She knocked the superintendent into the street and sprawled over Lila, hiding her completely from Farrell's sight. 'Lila, Lila,' she keened her daughter, 'poor baby, you never had a chance. He killed you because you were different, the way they kill everything different.' Farrell approached her and stooped down, but she pushed him against a wall without looking up. 'Lila, Lila, poor baby, poor darling, maybe it's better, maybe you're happy now. You never had a chance, poor Lila.'

The dog owners were edging slowly back, and the surviving dogs were running to them. The superintendent squatted on the curb with his head in his arms. A weary, muffled voice said, 'For God's sake, Bernice, would you get up off me? You don't have to stop yelling, just get off.'

When she stood up, the cars began to stop in the street again. It made it very difficult for the police to get through.

Nobody pressed charges, because there was no one to lodge them against. The killer dog—or wolf, as some insisted—was gone; and if she had an owner, he could not be found. As for the people who had actually seen the wolf turn into a young girl when the sunlight touched her; most of them managed not to have seen it, though they never really forgot. There were a few who knew quite well what they had seen, and never forgot it either, but they never said anything. They did, however, chip in to pay the superintendent's fine for possessing an unlicensed handgun. Farrell gave what he could.

Lila vanished out of Farrell's life before sunset. She did not go uptown with her mother, but packed her things and went to stay with friends in the Village. Later he heard that she was living on Christopher Street; and later still, that she had moved to Berkeley and gone back to school. He never saw her again.

'It had to be like that,' he told Ben once. 'We got to know too much about each other. See, there's another side to knowing. She couldn't look at me.'

'You mean because you saw her with all those dogs? Or because she knew you'd have let that little nut shoot her?' Farrell shook his head.

'It was that, I guess, but it was more something else, something I know. When she sprang, just as he shot at her that last time, she wasn't leaping at him. She was going straight for her mother. She'd have got her too, if it hadn't been sunrise.'

Ben whistled softly. 'I wonder if her old lady knows.'

'Bernice knows everything about Lila,' Farrell said.

Mrs Braun called him nearly two years later to tell him that Lila was getting married. It must have cost her a good deal of money and ingenuity to find him (where Farrell was living then, the telephone line was open for four hours a day), but he knew by the spitefulness in the static that she considered it money well spent.

'He's at Stanford,' she crackled. 'A research psychologist. They're going to Japan for their honeymoon.'

'That's fine,' Farrell said. 'I'm really happy for her, Bernice.' He hesitated before he asked, 'Does he know about Lila? I mean, about what happens—?'

'Does he know?' she cried. 'He's proud of it—he thinks it's wonderful! It's his field!'

'That's great. That's fine. Goodbye, Bernice. I really am glad.'

And he was glad, and a little wistful, thinking about it. The girl he was living with here had a really strange hang-up.

JANE YOLEN

JOHANNA

The forest was dark and the snow-covered path was merely an impression left on Johanna's moccasined feet.

If she had not come this way countless daylit times, Johanna would never have known where to go. But Hartwood was familiar to her, even in the unfamiliar night. She had often picnicked in the cool, shady copses and grubbed around the tall oak trees. In a hard winter like this one, a family could subsist for days on acorn stew.

Still, this was the first night she had ever been out in the forest, though she had lived by it all her life. It was tradition—no, more than that—that members of the Chevril family did not venture into the midnight forest. 'Never, never go to the woods at night,' her mother said, and it was not a warning so much as a command. 'Your father went though he was told not to. He never returned.'

And Johanna had obeyed. Her father's disappearance was still in her memory, though she remembered nothing else of him. He was not the first of the Chevrils to go that way. There had been a great-uncle and two girl cousins who had likewise 'never returned'. At least, that was what Johanna had been told. Whether they had disappeared into the maw of the city that lurked over several mountains to the west, or into the hungry jaws of a wolf or bear, was never made clear. But Johanna, being an obedient girl, always came into the house with the setting sun.

For sixteen years she had listened to that warning. But tonight, with her mother pale and sightless, breathing brokenly in the bed they shared, Johanna had no choice. The doctor, who lived on the other side of the wood, must be fetched. He lived in the cluster of houses that rimmed the far side of Hartwood, a cluster that was known as 'the village', though it was really much too small for such a name. The five houses of the Chevril family that clung together, now empty except for Johanna and her mother, were not called a village, though they squatted on as much land.

Usually the doctor himself came through the forest to visit the

Chevrils. Once a year he made the trip. Even when the grandparents and uncles and cousins had been alive, the village doctor came only once a year. He was gruff with them and called them 'strong as beasts' and went away, never even offering a tonic. They needed none. They were healthy.

But the long, cruel winter had sapped Johanna's mother's strength. She lay for days silent, eyes cloudy and unfocused, barely taking in the acorn gruel that Johanna spooned for her. And at last Johanna had said: 'I will fetch the doctor.'

Her mother had grunted 'no' each day, until this evening. When Johanna mentioned the doctor again, there had been no answering voice. Without her mother's no, Johanna made up her own mind. She *would* go.

If she did not get through the woods and back with the doctor before dawn, she felt it would be too late. Deep inside she knew she should have left before, even when her mother did not want her to go. And so she ran as quickly as she dared, following the small, twisting path through Hartwood by feel.

At first Johanna's guilt and the unfamiliar night were a burden, making her feel heavier than usual. But as she continued running, the crisp night air seemed to clear her head. She felt unnaturally alert, as if she had suddenly begun to discover new senses.

The wind moulded her short dark hair to her head. For the first time she felt graceful and light, almost beautiful. Her feet beat a steady tattoo on the snow as she ran, and she felt neither cold nor winded. Her steps lengthened as she went.

Suddenly a broken branch across the path tangled in her legs. She went down heavily on all fours, her breath caught in her throat. As she got to her feet, she searched the darkness ahead. Were there other branches waiting?

Even as she stared, the forest seemed to grow brighter. The light from the full moon must be finding its way into the heart of the woods. It was a comforting thought.

She ran faster now, confident of her steps. The trees seemed to rush by. There would be plenty of time.

She came at last to the place where the woods stopped, and cautiously she ranged along the last trees, careful not to be silhouetted against the sky. Then she halted.

She could hear nothing moving, could see nothing that threatened. When she was sure, she edged out onto the short meadow that ran in a downward curve to the back of the village.

Once more she stopped. This time she turned her head to the left and right. She could smell the musk of the farm animals on the wind, blowing faintly up to her. The moon beat down upon her head and, for a moment, seemed to ride on her broad, dark shoulder.

Slowly she paced down the hill towards the line of houses that stood like teeth in a jagged row. Light streamed out of the rear windows, making threatening little earthbound moons on the graying snow.

She hesitated.

A dog barked. Then a second began, only to end his call in a whine.

A voice cried out from the house furthest on the right, a woman's voice, soft and soothing. 'Be quiet, Boy.'

The dog was silenced.

She dared a few more slow steps towards the village, but her fear seemed to proceed her. As if catching its scent, the first dog barked lustily again.

'Boy! Down!' It was a man this time, shattering the night with authority.

She recognized it at once. It was the doctor's voice. She edged towards its sound. Shivering with relief and dread, she came to the backyard of the house on the right and waited. In her nervousness, she moved one foot restlessly, pawing the snow down to the dead grass. She wondered if her father, her great-uncle, her cousins had felt this fear under the burning eye of the moon.

The doctor, short and too stout for his age, came out of the back door, buttoning his breeches with one hand. In the other he carried a gun. He peered out into the darkness.

'Who's there?'

She stepped forward into the yard, into the puddle of light. She tried to speak her name, but she suddenly could not recall it. She tried to tell why she had come, but nothing passed her closed throat. She shook her head to clear the fear away.

The dog barked again, excited, furious.

'My God,' the doctor said, 'it's a deer.'

She spun around and looked behind her, following his line of sight. There was nothing there.

'That's enough meat to last the rest of this cruel winter,' he said. He raised the gun, and fired.

ANGELA CARTER

THE ERL-KING

The lucidity, the clarity of the light that afternoon was sufficient to itself; perfect transparency must be impenetrable, these vertical bars of a brass-coloured distillation of light coming down from sulphur-yellow interstices in a sky hunkered with grey clouds that bulge with more rain. It struck the wood with nicotine-stained fingers, the leaves glittered. A cold day of late October, when the withered blackberries dangled like their own dour spooks on the discoloured brambles. There were crisp husks of beechmast and cast acorn cups underfoot in the russet slime of dead bracken where the rains of the equinox had so soaked the earth that the cold oozed up through the soles of the shoes, lancinating cold of the approach of winter that grips hold of your belly and squeezes it tight. Now the stark elders have an anorexic look; there is not much in the autumn wood to make you smile but it is not yet, not quite yet, the saddest time of the year. Only, there is a haunting sense of the imminent cessation of being; the year, in turning, turns in on itself. Introspective weather, a sickroom hush.

The woods enclose. You step between the first trees and then you are no longer in the open air; the wood swallows you up. There is no way through the wood any more, this wood has reverted to its original privacy. Once you are inside it, you must stay there until it lets you out again for there is no clue to guide you through in perfect safety; grass grew over the track years ago and now the rabbits and the foxes make their own runs in the subtle labyrinth and nobody comes. The trees stir with a noise like taffeta skirts of women who have lost themselves in the woods and hunt round hopelessly for the way out. Tumbling crows play tig in the branches of the elms they clotted with their nests, now and then raucously cawing. A little stream with soft margins of marsh runs through the wood but it has grown sullen with the time of the year; the silent, blackish water thickens, now, to ice. All will fall still, all lapse.

A young girl would go into the wood as trustingly as Red Riding

Hood to her granny's house but this light admits of no ambiguities and, here, she will be trapped in her own illusion because everything in the wood is exactly as it seems.

The woods enclose and then enclose again, like a system of Chinese boxes opening one into another; the intimate perspectives of the wood changed endlessly around the interloper, the imaginary traveller walking towards an invented distance that perpetually receded before me. It is easy to lose yourself in these woods.

The two notes of the song of a bird rose on the still air, as if my girlish and delicious loneliness had been made into a sound. There was a little tangled mist in the thickets, mimicking the tufts of old man's beard that flossed the lower branches of the trees and bushes; heavy bunches of red berries as ripe and delicious as goblin or enchanted fruit hung on the hawthorns but the old grass withers, retreats. One by one, the ferns have curled up their hundred eyes and curled back into the earth. The trees threaded a cat's cradle of half-stripped branches over me so that I felt I was in a house of nets and though the cold wind that always heralds your presence, had I but known it then, blew gentle around me, I thought that nobody was in the wood but me.

Erl-King will do you grievous harm.

Piercingly, now, there came again the call of the bird, as desolate as if it came from the throat of the last bird left alive. That call, with all the melancholy of the failing year in it, went directly to my heart.

I walked through the wood until all its perspectives converged upon a darkening clearing; as soon as I saw them, I knew at once that all its occupants had been waiting for me from the moment I first stepped into the wood, with the endless patience of wild things, who have all the time in the world.

It was a garden where all the flowers were birds and beasts; ash-soft doves, diminutive wrens, freckled thrushes, robins in their tawny bibs, huge, helmeted crows that shone like patent leather, a blackbird with a yellow bill, voles, shrews, fieldfares, little brown bunnies with their ears laid together along their backs like spoons, crouching at his feet. A lean, tall, reddish hare, up on its great hind legs, nose a-twitch. The rusty fox, its muzzle sharpened to a point, laid its head upon his knee. On the trunk of a scarlet rowan a squirrel clung, to watch him; a cock pheasant delicately stretched his shimmering neck from a brake of thorn to peer at him. There was a goat of uncanny whiteness, gleaming like a goat of snow, who turned her mild eyes towards me and bleated softly, so that he knew I had arrived.

He smiles. He lays down his pipe, his elder bird-call. He lays upon me his irrevocable hand.

His eyes are quite green, as if from too much looking at the wood.

There are some eyes can eat you.

The Erl-King lives by himself all alone in the heart of the wood in a house which has only the one room. His house is made of sticks and stones and has grown a pelt of yellow lichen. Grass and weeds grow in the mossy roof. He chops fallen branches for his fire and draws his water from the stream in a tin pail.

What does he eat? Why, the bounty of the woodland! Stewed nettles; savoury messes of chickweed sprinkled with nutmeg; he cooks the foliage of shepherd's purse as if it were cabbage. He knows which of the frilled, blotched, rotted fungi are fit to eat; he understands their eldritch ways, how they spring up overnight in lightless places and thrive on dead things. Even the homely wood blewits, that you cook like tripe, with milk and onions, and the egg-yolk yellow chanterelle with its fan-vaulting and faint scent of apricots, all spring up overnight like bubbles of earth, unsustained by nature, existing in a void. And I could believe that it has been the same with him; he came alive from the desire of the woods.

He goes out in the morning to gather his unnatural treasures, he handles them as delicately as he does pigeons' eggs, he lays them in one of the baskets he weaves from osiers. He makes salads of the dandelion that he calls rude names, 'bum-pipes' or 'piss-the-beds', and flavours them with a few leaves of wild strawberry but he will not touch the brambles, he says the Devil spits on them at Michaelmas.

His nanny goat, the colour of whey, gives him her abundant milk and he can make soft cheese that has a unique, rank, amniotic taste. Sometimes he traps a rabbit in a snare of string and makes a soup or stew, seasoned with wild garlic. He knows all about the wood and the creatures in it. He told me about the grass snakes, how the old ones open their mouths wide when they smell danger and the thin little ones disappear down the old ones' throats until the fright is over and out they come again, to run around as usual. He told me how the wise toad who squats among the kingcups by the stream in summer has a very precious jewel in his head. He said the owl was a baker's daughter; then he smiled at me. He showed me how to thread mats from reeds and weave osier twigs into baskets and into the little cages in which he keeps his singing birds.

His kitchen shakes and shivers with birdsong from cage upon cage of singing birds, larks and linnets, which he piles up one on another against the wall, a wall of trapped birds. How cruel it is, to keep wild birds in cages! But he laughs at me when I say that; laughs, and shows his white, pointed teeth with the spittle gleaming on them.

He is an excellent housewife. His rustic home is spick and span. He puts his well-scoured saucepan and skillet neatly on the hearth side by side, like a pair of polished shoes. Over the hearth hang bunches of drying mushrooms, the thin, curling kind they call jew's-ears, which have grown on the elder trees since Judas hanged himself on one; this is the kind of lore he tells me, tempting my half-belief. He hangs up herbs in bunches to dry, too—thyme, marjoram, sage, vervain, southernwood, yarrow. The room is musical and aromatic and there is always a wood fire crackling in the grate, a sweet, acrid smoke, a bright, glancing flame. But you cannot get a tune out of the old fiddle hanging on the wall beside the birds because all its strings are broken.

Now, when I go for walks, sometimes in the mornings when the frost has put its shiny thumbprint on the undergrowth or sometimes, though less frequently, yet more enticingly, in the evenings when the cold darkness settles down, I always go to the Erl-King and he lays me down on his bed of rustling straw where I lie at the mercy of his huge hands.

He is the tender butcher who showed me how the price of flesh is love; skin the rabbit, he says! Off come all my clothes.

When he combs his hair that is the colour of dead leaves, dead leaves fall out of it; they rustle and drift to the ground as though he were a tree and he can stand as still as a tree, when he wants the doves to flutter softly, crooning as they come, down upon his shoulders, those silly, fat, trusting woodies with the pretty wedding rings round their necks. He makes his whistles out of an elder twig and that is what he uses to call the birds out of the air—all the birds come; and the sweetest singers he will keep in cages.

The wind stirs the dark wood; it blows through the bushes. A little of the cold air that blows over graveyards always goes with him, it crisps the hairs on the back of my neck but I am not afraid of him; only, afraid of vertigo, of the vertigo with which he seizes me. Afraid of falling down.

Falling as a bird would fall through the air if the Erl-King tied up the winds in his handkerchief and knotted the ends together so they could not get out. Then the moving currents of the air would no longer sustain them and all the birds would fall at the imperative of gravity, as I fall down for him, and I know it is only because he is kind to me that I do not fall still further. The earth with its fragile fleece of last summer's dying leaves and grasses supports me only out of complicity with him, because his flesh is of the same substance as those leaves that are slowly turning into earth.

He could thrust me into the seed-bed of next year's generation and I would have to wait until he whistled me up from my darkness before I could come back again.

Yet, when he shakes out those two clear notes from his bird call, I come, like any other trusting thing that perches on the crook of his wrist.

I found the Erl-King sitting on an ivy-covered stump winding all the birds in the wood to him on a diatonic spool of sound, one rising note, one falling note; such a sweet piercing call that down there came a soft, chirruping jostle of birds. The clearing was cluttered with dead leaves, some the colour of honey, some the colour of cinders, some the colour of earth. He seemed so much the spirit of the place I saw without surprise how the fox laid its muzzle fearlessly upon his knee. The brown light of the end of the day drained into the moist, heavy earth; all silent, all still and the cool smell of night coming. The first drops of rain fell. In the wood, no shelter but his cottage.

That was the way I walked into the bird-haunted solitude of the Erl-King, who keeps his feathered things in little cages he has woven out of osier twigs and there they sit and sing for him.

Goat's milk to drink, from a chipped tin mug; we shall eat the oat-cakes he has baked on the hearthstone. Rattle of the rain on the roof. The latch clanks on the door; we are shut up inside with one another, in the brown room crisp with the scent of burning logs that shiver with tiny flame, and I lie down on the Erl-King's creaking palliasse of straw. His skin is the tint and texture of sour cream, he has stiff, russet nipples ripe as berries. Like a tree that bears bloom and fruit on the same bough together, how pleasing, how lovely.

And now—ach! I feel your sharp teeth in the subaqueous depths of your kisses. The equinoctial gales seize the bare elms and make them whizz and whirl like dervishes; you sink your teeth into my throat and make me scream.

The white moon above the clearing coldly illuminates the still tableaux of our embracements. How sweet I roamed, or, rather, used to roam; once I was the perfect child of the meadows of summer, but then the year turned, the light clarified and I saw the gaunt Erl-King, tall as a tree with birds in its branches, and he drew me towards him on his magic lasso of inhuman music.

If I strung that old fiddle with your hair, we could waltz together to the music as the exhausted daylight founders among the trees; we should have better music than the shrill prothalamions of the larks stacked in their pretty cages as the roof creaks with the freight of birds you've lured to it while we engage in your profane mysteries under the leaves.

He strips me to my last nakedness, that underskin of mauve, pearlized satin, like a skinned rabbit; then dresses me again in an embrace so lucid and encompassing it might be made of water. And shakes over me dead leaves as if into the stream I have become.

Sometimes the birds, at random, all singing, strike a chord.

His skin covers me entirely; we are like two halves of a seed, enclosed in the same integument. I should like to grow enormously small, so that you could swallow me, like those queens in fairy tales who conceive when they swallow a grain of corn or a sesame seed. Then I could lodge inside your body and you would bear me.

The candle flutters and goes out. His touch both consoles and devastates me; I feel my heart pulse, then wither, naked as a stone on the roaring mattress while the lovely, moony night slides through the window to dapple the flanks of this innocent who makes cages to keep the sweet birds in. Eat me, drink me; thirsty, cankered, goblin-ridden, I go back and back to him to have his fingers strip the tattered skin away and clothe me in his dress of water, this garment that drenches me, its slithering odour, its capacity for drowning.

Now the crows drop winter from their wings, invoke the harshest season with their cry.

It is growing colder. Scarcely a leaf left on the trees and the birds come to him in even greater numbers because, in this hard weather, it is lean pickings. The blackbirds and thrushes must hunt the snails from hedge bottoms and crack the shells on stones. But the Erl-King gives them corn and when he whistles to them, a moment later you cannot see him for the birds that have covered him like a soft fall of feathered snow. He spreads out a goblin feast of fruit for me, such appalling succulence; I lie above him and see the light from the fire sucked into the black vortex of his eye, the omission of light at the centre, there, that exerts on me such a tremendous pressure, it draws me inwards.

Eyes green as apples. Green as dead sea fruit.

A wind rises; it makes a singular, wild, low, rushing sound.

What big eyes you have. Eyes of an incomparable luminosity, the numinous phosphorescence of the eyes of lycanthropes. The gelid green of your eyes fixes my reflective face. It is a preservative, like a green liquid amber; it catches me. I am afraid I will be trapped in it for ever like the poor little ants and flies that stuck their feet in resin before the sea covered the Baltic. He winds me into the circle of his eye on a reel of birdsong. There is a black hole in the middle of both your eyes; it is their still centre, looking there makes me giddy, as if I might fall into it.

Your green eye is a reducing chamber. If I look into it long enough, I will become as small as my own reflection, I will diminish to a point and vanish. I will be drawn down into that black whirlpool and be consumed by you. I shall become so small you can keep me in one of your osier cages and mock my loss of liberty. I have seen the cage you are weaving for me; it is a very pretty one and I shall sit, hereafter, in my cage among the other singing birds but I—I shall be dumb, from spite.

When I realized what the Erl-King meant to do to me, I was shaken with a terrible fear and I did not know what to do for I loved him with all my heart and yet I had no wish to join the whistling congregation he kept in his cages although he looked after them very affectionately, gave them fresh water every day and fed them well. His embraces were his enticements and yet, oh, yet! they were the branches of which the trap itself was woven. But in his innocence he never knew he might be the death of me, although I knew from the first moment I saw him how Erl-King would do me grievous harm.

Although the bow hangs beside the old fiddle on the wall, all the strings are broken so you cannot play it. I don't know what kind of tunes you might play on it, if it were strung again; lullabies for foolish virgins, perhaps, and now I know the birds don't sing, they only cry because they can't find their way out of the wood, have lost their flesh when they were dipped in the corrosive pools of his regard and now must live in cages.

Sometimes he lays his head on my lap and lets me comb his lovely hair for him; his combings are leaves of every tree in the wood and dryly susurrate around my feet. His hair falls down over my knees. Silence like a dream in front of the spitting fire while he lies at my feet and I comb the dead leaves out of his languorous hair. The robin has built his nest in the thatch again, this year; he perches on an unburnt log, cleans his beak, ruffles his plumage. There is a plaintive sweetness in his song and a certain melancholy, because the year is over—the robin, the friend of man, in spite of the wound in his breast from which Erl-King tore out his heart.

Lay your head on my knee so that I can't see the greenish inward-turning suns of your eyes any more.

My hands shake.

I shall take two huge handfuls of his rustling hair as he lies half dreaming, half waking, and wind them into ropes, very softly, so he will not wake up, and, softly, with hands as gentle as rain, I shall strangle him with them.

Then she will open all the cages and let the birds free; they will change back into young girls, every one, each with the crimson imprint of his love-bite on their throats.

She will carve off his great mane with the knife he uses to skin the rabbits; she will string the old fiddle with five single strings of ash-brown hair.

Then it will play discordant music without a hand touching it. The bow will dance over the new strings of its own accord and they will cry out: 'Mother, mother, you have murdered me!'

JAMES TIPTREE JR.

BEYOND THE DEAD REEF

A love that is not sated
Calls from a poisoned bed;
Where monsters half-created
Writhe, unliving and undead.
None knows for what they're fated;
None knows on what they've fed.

My informant was, of course, spectacularly unreliable.

The only character reference I have for him comes from the intangible nuances of a small restaurant-owner's remarks, and the only confirmation of his tale lies in the fact that an illiterate fishing-guide appears to believe it. If I were to recount all the reasons why no sane mind should take it seriously, we could never begin. So I will only report the fact that today I found myself shuddering with terror when a perfectly innocent sheet of seaworn plastic came slithering over my snorkelling-reef, as dozens have done for years—and get on with the story.

I met him one evening this December at the Cozumel *Buzo*, on my first annual supply trip. As usual, the *Buzo*'s outer rooms were jammed with tourist divers and their retinues and gear. That's standard. *El Buzo* means, roughly, The Diving, and the *Buzo* is their place. Marcial's big sign in the window reads 'DIVVERS UELCOME! BRING YR FISH WE COK WITH CAR. FIRST DRINK FREE!'

Until he went in for the 'Divvers', Marcial's had been a small quiet place where certain delicacies like stone-crab could be at least semi-legally obtained. Now he did a roaring trade in snappers and groupers cooked to order at outrageous fees, with a flourishing sideline in fresh fish sales to the neighbourhood each morning.

The 'roaring' was quite literal. I threaded my way through a crush of burly giants and giantesses of all degrees of nakedness, hairiness, age, proficiency, and inebriation—all eager to share their experiences and plans in voices powered by scuba-deafened ears and Marcial's free drink, beneath which the sound-system could scarcely be heard at full blast.

(Marcial's only real expense lay in first-drink liquor so strong that few could recall whether what they ultimately ate bore any resemblance to what they had given him to cook.) Only a handful were sitting down yet and the amount of gear underfoot and on the walls would have stocked three sports shops. This was not mere exhibitionism; on an island chronically short of washers, valves, and other spare parts the diver who lets his gear out of his sight is apt to find it missing in some vital.

I paused to allow a young lady to complete her massage of the neck of a youth across the aisle who was deep in talk with three others, and had time to notice the extraordinary number of heavy spear-guns racked about. Oklahomans, I judged, or perhaps South Florida. But then I caught clipped New England from the centre group. Too bad; the killing mania seems to be spreading yearly, and the armament growing ever more menacing and efficient. When I inspected their platters, however, I saw the usual array of lavishly garnished lobsters and common fish. At least they had not yet discovered what to eat.

The mermaiden blocking me completed her task—unthanked—and I continued on my way in the little inner sanctum Marcial keeps for his old clientele. As the heavy doors cut off the uproar, I saw that this room was full too—three tables of dark-suited Mexican businessmen and a decorous family of eight, all quietly intent on their plates. A lone customer sat at the small table by the kitchen door, leaving an empty seat and a child's chair. He was a tall, slightly balding Anglo some years younger than I, in a very decent sports jacket. I recalled having seen him about now and then on my banking and shopping trips to the island.

Marcial telegraphed me a go-ahead nod as he passed through laden with more drinks, so I approached.

'Mind if I join you?'

He looked up from his stone-crab and gave me a slow, owlish smile.

'Welcome. A *diverse* welcome,' he enunciated carefully. The accent was vaguely British, yet agreeable. I also perceived that he was extremely drunk, but in no common way.

'Thanks.'

As I sat down I saw that he was a diver too, but his gear was stowed so unobtrusively I hadn't noticed it. I tried to stack my own modest snorkel outfit as neatly, pleased to note that like me, he seemed to carry no spear-guns. He watched me attentively, blinking once or twice, and then returned to an exquisitely exact dissection of his crab.

When Marcial brought my own platter of crab—unasked—we engaged in our ritual converse. Marcial's English is several orders of magnitude better than my Spanish, but he always does me the delicate courtesy of allowing me to use his tongue. How did I find my rented casita on the coco ranch this year? Fine. How goes the tourist business this year? Fine. I learn from Marcial: the slight pause before his answer with a certain tone, meant that in fact the tourist business was lousy so far, but would hopefully pick up; I used the same to convey that in fact my casa was in horrible shape but reparable. I tried to cheer him by saying that I thought the *Buzo* would do better than the general *turismo*, because the diving enthusiasm was spreading in the States. 'True,' he conceded. 'So long as they don't discover other places—like Belizé.' Here he flicked a glance at my companion, who gave his solemn blink. I remarked that my country's politics were in disastrous disarray, and he conceded the same for his; the Presidente and his pals had just made off with much of the nation's treasury. And I expressed the hope that Mexico's new oil would soon prove a great boon. 'Ah, but it will be a long time before it gets to the little people like us,' said Marcial, with so much more than his normal acerbity that I refrained from my usual joke about his having a Swiss bank account. The uproar from the outer rooms had risen several decibels but just before Marcial had to leave he paused and said in a totally different voice, 'My grandson Antonito Vincente has four teeth!'

His emotion was so profound that I seized his free hand and shook it lightly, congratulating him in English. And then he was gone, taking on his 'Mexican waiter' persona quite visibly as he passed the inner doors.

As we resumed our attention to the succulence before us, my companion said in his low, careful voice, 'Nice chap, Marcial. He likes you.'

'It's mutual,' I told him between delicate mouthfuls. Stone-crab is not to be gulped. 'Perhaps because I'm old enough to respect the limits where friendship ends and the necessities of life take over.'

'I say, that's rather good,' my companion chuckled, 'Respect for the limits where friendship ends and the necessities of life take over, eh? Very few Yanks do, you know. At least the ones we see down here.'

His speech was almost unslurred, and there were no drinks before him on the table. We chatted idly a bit more. It was becoming apparent that we would finish simultaneously and be faced with the prospect of leaving together, which could be awkward, if he, like me, had no definite plans for the evening.

The dilemma was solved when my companion excused himself momentarily just as Marcial happened by.

I nodded to his empty chair. 'Is he one of your old customers, Senor Marcial?'

As always Marcial understood the situation at once. 'One of the oldest,' he told me, and added low-voiced, 'muy bueno gentes—a really good guy. Un poco de difficultades—' he made an almost imperceptible gesture of drinking—'But controllado. And he has also négocios—I do not know all, but some are important for his country.—So you really like the crab?' he concluded in his normal voice. 'We are honoured.'

My companion was emerging from the rather dubious regions that held the excusado.

Marcial's recommendation was good enough for me. Only one puzzle remained: what was his country? As we both refused dulce and coffee, I suggested that he might care to stroll down to the Marina with me and watch the sunset.

'Good thought.'

We paid up Marcial's outrageous bills, and made our way through the exterior Bedlam, carrying our gear. One of the customers was brandishing his spear-gun as he protested his bill. Marcial seemed to have lost all his English except the words 'Police', and cooler heads were attempting to calm the irate one. 'All in a night's work,' my companion commented as we emerged into a blaze of golden light.

The marina to our left was a simple L-shaped *muelle*, or pier, still used by everything from dinghies to commercial fisherman and baby yachts. It will be a pity when and if the town decides to separate the sports tourist trade from the more interesting working craft. As we walked out towards the pier in the last spectacular colour of the tropic sunset over the mainland, the rigging lights of a cruise ship standing out in the channel came on, a fairyland illusion over the all-too-dreary reality.

'They'll be dumping and cleaning out their used bunkers tonight,' my companion said, slurring a trifle now. He had a congenial walking gait, long-strided but leisurely. I had the impression that his drunkenness had returned slightly; perhaps the fresh air. 'Damn crime.'

'I couldn't agree more,' I told him. 'I remember when we used to start snorkelling and scuba-diving right off the shore here—you could almost wade out to untouched reefs. And now—'

There was no need to look; one could smell it. The effluvia of half a dozen hotels and the town behind ran out of pipes that were barely covered at low tide; only a few parrot-fish, who can stand anything, remained by the hotel-side restaurants to feed on the crusts the tourists threw them from their tables. And only the very ignorant would try out—once—the dilapidated Sunfish and water-ski renters who plied the small stretches of beach between hotels.

We sat down on one of the near benches to watch a commercial trawler haul net. I had been for some time aware that my companion, while of largely British culture, was not completely Caucasian. There was a minute softness to the voice, a something not quite dusky about hair and fingernails—not so much as to be what in my youth was called 'A touch of the tar-brush', but nothing that originated in Yorkshire, either. Nor was it the obvious Hispano-Indian. I recollected Marcial's earlier speech and enlightenment came.

'Would I be correct in taking Marcial's allusion to mean that you are a British Honduran—forgive me, I mean a Belizéian, or Belizan?'

'Nothing to forgive, old chap. We haven't existed long enough to get our adjectives straight.'

'May god send you do.' I was referring to the hungry maws of Guatemala and Honduras, the little country's big neighbours, who had the worst of intentions towards her. 'I happen to be quite a fan of your country. I had some small dealings there after independence which involved getting all my worldly goods out of your customs on a national holiday, and people couldn't have been finer to me.'

'Ah yes. Belizé the blessed, where sixteen nationalities live in perfect racial harmony. The odd thing is, they do.'

'I could see that. But I couldn't quite count all sixteen.'

'My own grandmother was a Burmese—so called. I think it was the closest grandfather could come to Black. Although the mix *is* extraordinary.'

'My factor there was a very dark Hindu with red hair and a Scottish accent, named Robinson. I had to hire him in seven minutes. He was a miracle of efficiency. I hope he's still going.'

'Robinson . . . Used to work for customs?'

'Why, yes, now you recall it.'

'He's fine . . . Of course, we felt it when the British left. Among other things, half the WCs in the hotels broke down the first month. But there are more important things in life than plumbing.'

'That I believe . . . But you know, I've never been sure how much help the British would have been to you. Two years before your independence I called the British Embassy with a question about your immigration laws, and believe it or not I couldn't find one soul who even knew there *was* a British Honduras, let alone that they owned it. One child finally denied it flatly and hung up. And this was their main embassy in Washington, DC. I realized then that Britain was not only sick, but crazy.'

'Actually denied our existence, eh?' My companion's voice held a depth and timbre of sadness such as I have heard only from victims of better-known world wrongs. Absently his hand went under his jacket, and he pulled out something gleaming.

'Forgive me.' It was a silver flask, exquisitely plain. He uncapped and drank, a mere swallow, but, I suspected, something of no ordinary power. He licked his lips as he recapped it, and sat up straighter while he put it away.

'Shall we move along out to the point?'

'With pleasure.'

We strolled on, passing a few late sports-boats disgorging hungry divers.

'I'm going to do some modest exploring tomorrow,' I told him. 'A guide named Jorge'—in Spanish it's pronounced Hor-hay—'Jorge Chuc is taking me out to the end of the North reef. He says there's a pretty little untouched spot there. I hope so. Today I went South, it was so badly shot over I almost wept. Cripples—and of course shark everywhere. Would you believe I found a big she-turtle, trying to live with a steel bolt through her neck? I managed to catch her, but all I could do for her was pull it out. I hope she makes it.'

'Bad . . . Turtles are tough, though. If it wasn't vital you may have saved her. But did you say that Jorge Chuc is taking you to the end of the North reef?'

'Yes, why. Isn't it any good?'

'Oh, there is one pretty spot. But there's some very bad stuff there too. If you don't mind my advice, don't go far from the boat. I mean, a couple of metres. And don't follow anything. And above all be very sure it *is* Jorge's boat.'

His voice had become quite different, with almost military authority.

'A couple of metres!' I expostulated. 'But—'

'I know, I know. What I don't know is why Chuc is taking you there at all.' He thought for a moment. 'You haven't by any chance offended him, have you? In any way?'

'Why no—we were out for a long go yesterday, and had a nice chat on the way back. Yes . . . although he is a trifle changeable, isn't he? I put it down to fatigue, and gave him some extra dinero for being only one party.'

My companion made a untranslatable sound, compounded of dubiety, speculation, possible enlightenment, and strong suspicion.

'Did he tell you the name of that part of the reef? Or that it's out of sight of land?'

'Yes, he said it was far out. And that part of it was so poor it's called dead.'

'And you chatted—forgive me, but was your talk entirely in Spanish?'

I chuckled deprecatingly. 'Well, yes—I know my Spanish is pretty horrible, but he seemed to get the drift.'

'Did you mention his family?'

'Oh yes—I could draw you the whole Chuc family tree.'

'H'mmm. . . .' My companion's eyes had been searching the pier-side where the incoming boats were being secured for the night.

'Ah. There's Chuc now. This is none of my business, you understand—but do I have your permission for a short word with Jorge?'

'Why yes. If you think it necessary.'

'I do, my friend. I most certainly do.'

'Carry on.'

His long-legged stride had already carried him to Chuc's big skiff, the *Estrellita.* Chuc was covering his motors. I had raised my hand in greeting, but he was apparently too busy to respond. Now he greeted my companion briefly, but did not turn when he clambered into the boat uninvited. I could not hear the interchange. But presently the two men were standing, faces somewhat averted from each other as they conversed. My companion made rather a long speech, ending with questions. There was little response from Chuc, until a sudden outburst from him took me by surprise. The odd dialogue went on for some time after that; Chuc seemed to calm down. Then the tall Belizian waved me over.

'Will you say exactly what I tell you to say!'

'Why, yes, if you think it's important.'

'It is. Can you say in Spanish, "I ask your pardon, Mr Chuc. I mistook myself in your language. I did not say anything of what you thought I said. Please forgive my error. And please let us be friends again."'

'I'll try.'

I stumbled through the speech, which I will not try to reproduce here, as I repeated several phrases with what I thought was better accent, and I'm sure I threw several verbs into the conditional future. Before I was through, Chuc was beginning to grin. When I came to the 'friends' part he had relaxed, and after a short pause, said in very tolerable English, 'I see, so I accept your apology. We will indeed be friends. It was a regrettable error . . . And I advise you, do not again speak in Spanish.'

We shook on it.

'Good,' said my companion. 'And he'll take you out tomorrow, but not to the dead reef. And keep your hands off your wallet tonight, but I suggest liberality tomorrow eve.'

We left Chuc to finish up, and paced down to a bench at the very end of the *muelle*. The last colours of evening, peaches and rose shot with unearthly green, were set off by a few low-lying clouds already in grey shadow, like sharks of the sky passing beneath a sentimental vision of bliss.

'Now what was all *that* about!' I demanded of my new friend. He was just tucking the flask away again, and shuddered lightly.

'I don't wish to seem overbearing but *that* probably saved your harmless life, my friend. I repeat Jorge's advice—stay away from that Spanish of yours unless you are absolutely sure of being understood.'

'I know it's ghastly.'

'That's not actually the problem. The problem is that it isn't ghastly enough. Your pronunciation is quite fair, and you've mastered some good idioms, so people who don't know you think you speak much more fluently than you do. In this case the trouble came from your damned rolled rrrs. Would you mind saying the words for "but" and "dog"?'

'*Pero . . . perro*. Why?'

'The difference between a rolled and a single r, particularly in Maya Spanish, is very slight. The upshot of it was that you not only insulted his boat in various ways, but you ended by referring to his mother as a dog . . . He was going to take you out beyond the Dead Reef and leave you there.'

'*What*?'

'Yes. And if it hadn't been I who asked—he knows I know the story—you'd never have understood a thing. Until you turned up as a statistic.'

'Oh Jesus Christ . . .'

'Yes,' he said dryly.

'I guess some thanks are in order,' I said finally. 'But words seem a shade inadequate. Have you any suggestions?'

My companion suddenly turned and gave me a highly concentrated look.

'You were in World War Two, weren't you? And afterwards you worked around a bit.' He wasn't asking me, so I kept quiet. 'Right now, I don't see anything,' he went on. 'But just possibly I might be calling on you,' he grinned, 'with something you may not like.'

'If it's anything I can do from a wheelchair, I won't forget.'

'Fair enough. We'll say no more about it now.'

'Oh yes we will,' I countered. 'You may not know it, but you owe *me* something. I can smell a story when one smacks me in the face. What I want from you is the story behind this Dead Reef business, and how it is that Jorge knows you know something special about it. If I'm not asking too much? I'd really like to end our evening with your tale of the Dead Reef.'

'Oho. My error—I'd forgotten Marcial telling me you wrote . . . Well, I can't say I enjoy reliving it, but maybe it'll have a salutary effect on your future dealings in Spanish. The fact is, I was the one it happened to, and Jorge was driving a certain boat. You realize, though, there's not a shred of proof except my own word? And my own word—' he tapped the pocket holding his flask '—is only as good as you happen to think it is.'

'It's good enough for me.'

'Very well then. Very well,' he said slowly, leaning back. 'It happened about three, no four years back—by god, you know this is hard to tell, though there's not much to it.' He fished in another pocket, and took out, not a flask, but the first cigarette I'd seen him smoke, a *Petit Caporal*. 'I was still up to a long day's scuba then, and, like you, I wanted to explore North. I'd run into this nice, strong, young couple who wanted the same thing. Their gear was good, they seemed experienced and sensible. So we got a third tank apiece, and hired a trustable boatman—not Jorge, Victor Camul—to take us north over the worst of the reef. It wasn't so bad then, you know.

'We would be swimming North with the current until a certain point, where if you turn East, you run into a long reverse eddy that makes it a lot easier to swim back to Cozumel. And just to be extra safe, Victor was to start out up the eddy in two hours sharp to meet us and bring us home. I hadn't one qualm about the arrangements. Even the weather co-operated—not a cloud, and the forecast perfect. Of course, if you miss up around here, the next stop is four hundred miles to Cuba, but you know that; one gets used to it . . . By the way, have you heard they're still looking for that girl who's been gone two days on a Sunfish with no water?'

I said nothing.

'Sorry.' He cleared his throat. 'Well, Victor put us out well in sight of shore. We checked watches and compasses and lights. The plan was for the lad Harry to lead, Ann to follow, and me to bring up the rear. Harry had dayglo-red shorts you could see a mile, and Ann was white-skinned with long black hair and a brilliant neon-blue and orange bathing suit on her little rump—you could have seen her in a mine at midnight. Even I got some yellow water safety tape and tied it around my arse and tanks.

'The one thing we didn't have then was a radio. At the time they didn't seem worth the crazy cost, and were unreliable besides. I had no way of guessing I'd soon give my life for one—and very nearly did.

'Well, when Victor let us out and we got organized and started North single file over the dead part of the reef, we almost surfaced and yelled for him to take us back right then. It was purely awful. But we knew there was better stuff ahead, so we stuck it out and flippered doggedly along—actually doing pretty damn fair time, with the current—and trying not to look too closely at what lay below.

'Not only was the coral dead, you understand—that's where the name got started. We think now it's from oil and chemical wash, such as that pretty ship out there is about to contribute—but there was tons and tons of litter, *basura* of all description, crusted there. It's everywhere, of course—you've seen what washes on to the mainland beach—but here the current and the reef produce a particularly visible concentration. Even quite large heavy things—bedsprings, auto chassis in addition to things you'd expect, like wrecked skiffs. Cozumel, *Basurera del Caribe*!'

He gave a short laugh, mocking the Gem-of-the-Caribbean ads, as he lit up another Caporal. The most polite translation of *basurera* is garbage can.

'A great deal of the older stuff was covered with that evil killer algae—you know, the big coarse red-brown hairy kind, which means that nothing else can ever grow there again. But some of the heaps were too new.

'I ended by getting fascinated and swimming lower to look. Always keeping one eye on that blue-and-orange rump above me with her white legs and black flippers. And the stuff—I don't mean just Chlorox and *detergente* bottles, beer cans, and netting—but weird things like about ten square metres of butchered pink plastic baby-dolls—arms and legs wiggling, and rose-bud mouths—it looked like a babies' slaughter-house. Syringes, hypos galore. Fluorescent tubes on end, waving like drowned orchestra conductors. A great big red sofa with a skeletonized banana-stem or *something* sitting in it—when I saw that, I went back up and followed right behind Ann.

'And then the sun dimmed unexpectedly, so I surfaced for a look. The shoreline was fine, we had plenty of time, and the cloud was just one of a dozen little thermals that form on a hot afternoon like this. When I went back down Ann was looking at me, so I gave her the All's fair sign. And with that we swam over a pair of broken dories and found ourselves in a different world—the beauty patch we'd been looking for.

'The reef was live here—whatever had killed the coral hadn't reached yet, and the damned basura had quit or been deflected, aside from a beer bottle or two. There was life everywhere; anemones, sponges, conches, fans, stars—and fish, oh my! No one ever came here, you see. In fact, there didn't seem to have been any spearing, the fish were as tame as they used to be years back.

'Well, we began zigzagging back and forth, just revelling in it. And every time we'd meet head on we'd make the gesture of putting our fingers to our lips, meaning Don't tell anyone about this, ever!

'The formation of the reef was charming, too. It broadened into a sort of big stadium, with allées and cliffs and secret pockets, and there were at least eight different kinds of coral. And most of it was shallow enough so the sunlight brought out the glorious colours—those little black and yellow fish—butterflies, or I forget their proper name—were dazzling. I kept having to brush them off my mask, they wanted to look in.

'The two ahead seemed to be in ecstasies; I expect they hadn't seen much like this before. They swam on and on, investigating it all—and I soon realized there was real danger of losing them in some coral pass. So I stuck tight to Ann. But time was passing. Presently I surfaced again to investigate—and, my god, the shoreline was damn near invisible and the line-up we had selected for our turn marker was all but passed! Moreover, a faint hazy overcast was rising from the West.

'So I cut down again, intending to grab Ann and start, which Harry would have to see. So I set off after the girl. I used to be a fair sprint-swimmer, but I was amazed how long it took me to catch her. I recall vaguely noticing that the reef was going a bit bad again, dead coral here and there. Finally, I came right over her, signed to her to halt, and kicked up in front of her nose for another look.

'To my horror, the shoreline was gone and the overcast had overtaken the sun. We would have to swim East by compass, and swim hard. I took a moment to hitch my compass around where I could see it well—it was the old-fashioned kind—and then I went back down for Ann. And the damn fool girl wasn't there. It took me a minute to locate that blue bottom and white legs; I assumed she'd gone after Harry, having clearly no idea of the urgency of our predicament.

I confess the thought crossed my mind that I could cut out of there, and come back for them later with Victor, but this was playing a rather iffy game with someone else's lives. And if they were truly unaware, it would be fairly rotten to take off without even warning them. So I went after Ann again—my god, I can still see that blue tail and the white limbs and black feet and hair with the light getting worse every minute and the bottom now gone really rotten again. And as bad luck would have it she was going in just the worst line—north-north-west.

'Well I swam and I swam and I *swam*. You know how a chase takes you, and somehow being unable to overtake a mere girl made it worse. But I was gaining, age and all, until just as I got close enough to sense something was wrong, she turned sidewise above two automobile tyres—and I saw it wasn't a girl at all.

'I had been following a god-damned great fish—a fish with a bright blue and orange band around its belly, and a thin white body ending in a black, flipperlike tail. Even its head and nape were black, like her hair and mask. It had a repulsive catfish-like mouth, with barbels.

'The thing goggled at me, and then swam awkwardly away, just as the light went worse yet. But there was enough for me to see that it was no normal fish, either, but a queer archaic thing that looked more tacked together than grown. This I can't swear to, because I was looking elsewhere by then, but it was my strong impression that as it went out of my line of sight its whole tail broke off.

'But, as I say, I was looking elsewhere. I had turned my light on, although I was not deep but only dim, because I had to read my watch and compass. It had just dawned on me that I was very probably a dead man. My only chance, if you can call it that, was to swim East as long as I could, hoping for that eddy and Victor. And when my light came on, the first thing I saw was the girl, stark naked and obviously stone cold dead, lying in a tangle of nets and horrid stuff on the bottom ahead.

'Of Harry or anything human there was no sign at all. But there was a kind of shining, like a pool of moonlight, around her, which was so much stronger than my lamp that I clicked it off and swam slowly toward her, through the nastiest mess of *basura* I had yet seen. The very water seemed vile. It took longer to reach her than I had expected, and soon I saw why.

'They speak of one's blood running cold with horror, y'know. Or people becoming numb with horror piled on horrors. I believe I experienced both those effects. It isn't pleasant, even now.' He lit a third Caporal, and I could see that the smoke column trembled. Twilight had fallen while he'd been speaking. A lone mercury lamp came on at the shore end of the pier; the one near us was apparently out, but we sat in what would ordinarily have been a pleasant tropic evening, sparkling with many moving lights—whites, reds, and greens, of late-moving incomers and the rainbow lighting from the jewel-lit cruise ship ahead, all cheerfully reflected in the unusually calm waters.

'Again I was mistaken, you see. It wasn't Ann at all; but the rather more distant figure of a young woman, of truly enormous size. All in this great ridge of graveyard luminosity, of garbage in phosphorescent decay. The current was carrying me slowly, inexorably, right towards her—as it had carried all that was there now. And perhaps I was also a bit hypnotized. She grew in my sight metre by metre as I neared her. I think six metres—eighteen feet—was about it, at the end . . . I make that guess later, you understand, as an exercise in containing the unbearable—by recalling the size of known items in the junkpile she lay on. One knee, for example, lay alongside an oil drum. At the time she simply filled my world. I had no doubt she was dead, and very beautiful. One of her legs seemed to writhe gently.

'The next stage of horror came when I realized that she was not a gigantic woman at all—or rather, like the fish, she was a woman-shaped construction. The realization came to me first, I think, when I could no longer fail to recognize that her 'breasts' were two of those great net buoys with the blue knobs for nipples.

'After that it all came with a rush—that she was a made-up body—all sorts of pieces of plastic, rope, styrofoam, netting, crates, and bolts—much of it clothed with that torn translucent white polyethylene for skin. Her hair was a dreadful tangle of something, and her crotch was explicit and unspeakable. One hand was a torn, inflated rubber glove, and her face—well, I won't go into it except that one eye was a traffic reflector and her mouth was partly a rusted can.

'Now you might think this discovery would have brought some relief, but quite the opposite. Because simultaneously I had realized the very worst thing of all—

'She was alive.'

He took a long drag on his cigarette.

'You know how things are moved passively in water? Plants waving, a board see-sawing and so on? Sometimes enough almost to give an illusion of mobile life. What I saw was nothing of this sort.

'It wasn't merely that as I floated over her horrible eyes "opened" and looked at me, and her rusted-can mouth *smiled*. Oh, no.

'What I mean is that as she smiled, first one whole arm, shedding junk, stretched up and reached for me *against the current*, and then the other did the same.

'And when I proved to be out of reach, this terrifying figure, or creature, or unliving life, actually sat up, again *against the current*, and reached up towards me with both arms at full extension.

'And as she did so, one of her "breasts"—the right one—came loose and dangled by some sort of tenuous thready stuff.

'All this seemed to pass in slow motion—I even had time to see that there were other unalive yet living things moving near her on the pile. Not fish, but more what I should have taken, on land, for rats or vermin—and I distinctly recall the paper-flat skeleton of something like a chicken, running and pecking. And other moving things like nothing in this world. I have remembered all this very carefully, y'see, from what must have been quick glimpses, because in actual fact I was apparently kicking like mad in a frenzied effort to get away from those dreadful, reaching arms.

'It was not till I shot to the surface with a mighty splash that I came somewhere near my senses. Below and behind me I could still see faint cold light. Above was twilight and the darkness of an oncoming small storm.

'At that moment the air in my last tank gave out—or rather that splendid Yank warning buzz, which means you have just time to get out of your harness, sounded off.

'I had, thank god, practised the drill. Despite being a terror-paralysed madman, habit got me out of the harness before the tanks turned into lethal deadweight. In my panic of course, the headlight went down too. I was left unencumbered in the night, free to swim towards Cuba, or Cozumel, and to drown as slow or fast as fate willed.

'The little storm had left the horizon stars free. I recall that pure habit made me take a sight on what seemed to be Canopus, which should be over Cozumel. I began to swim in that direction. I was appallingly tired, and as the adrenalin of terror which had brought me this far began to fade out of my system, I realized I could soon be merely drifting, and would surely die in the next day's sun if I survived till then. Nevertheless, it seemed best to swim whilst I could.

'I rather resented it when some time after a boat motor passed nearby. It forced me to attempt to yell and wave, nearly sinking myself. I was perfectly content when the boat passed on. But someone had seen—a spotlight wheeled blindingly, motors reversed, I was forcibly pulled from my grave and voices from what I take to be your Texas demanded, roaring with laughter,'—here he gave quite a creditable imitation—' "Whacha doin out hyar, boy, this time of night? Ain't no pussy out hyar, less'n y'all got a date with a mermaid." They had been trolling for god knows what, mostly beer.'

'The driver of that boat claimed me as a friend and later took me home for the night, where I told to him—and to him alone—the whole story. He was Jorge Chuc.

'Next day I found that the young couple, Harry and Ann, had taken only a brief look at the charming unspoiled area, and then started East, exactly according to plan, with me—or something very much like me—following behind them all the way. They had been a trifle surprised at my passivity and uncommunicativeness, and more so when, on meeting Victor, I was no longer to be found. But they had taken immediate action, even set a full-scale search in progress—approximately seventy kilometres from where I then was. As soon as I came to myself I had to concoct a wild series of lies about cramps and heart trouble to get them in the clear and set their minds at ease. Needless to say, my version included no mention of diver-imitating fish-life.'

He tossed the spark of his cigarette over the rail before us.

'So now, my friend, you know the whole story of all I know of what is to be found beyond the Dead Reef. It may be that others know of other happenings and developments there. Or of similar traps elsewhere. The sea is large . . . Or it may be that the whole yarn comes from neuroses long abused by stuff like this.'

I had not seen him extract his flask, but he now took two deep, shuddering swallows.

I sighed involuntarily, and then sighed again. I seemed to have been breathing rather inadequately during the end of his account.

'Ordinary thanks don't seem quite appropriate here,' I finally said. 'Though I do thank you. Instead I am going to make two guesses. The second is that you might prefer to sit quietly here alone, enjoying the evening, and defer the mild entertainment I was about to offer you to some other time. I'd be glad to be proved wrong . . . ?'

'No. You're very perceptive, I welcome the diverse—the deferred offer.' His tongue stumbled a bit now more from fatigue than anything he'd drunk. 'But what was your first guess?'

I rose and slowly paced a few metres to and fro, remembering to pick up my absurd snorkel bag. Then I turned and gazed out to the sea.

'I can't put it into words. It has something to do with the idea that the sea is still, well, strong. Perhaps it can take revenge? No, that's too simple. I don't know. I have only a feeling that our ordinary ideas of what may be coming on us may be—oh—not deep, or broad enough. I put this poorly. But perhaps the sea, or nature, will not die passively at our hands, . . . perhaps death itself may turn or return in horrible life upon us, besides the more mechanical dooms. . . .'

'Our thoughts are not so far apart,' the tall Belizan said. 'I welcome them to my night's agenda.'

'To which I now leave you, unless you've changed your mind?'

He shook his head. I hoisted his bag to the seat beside him. 'Don't forget this. I almost left mine.'

'Thanks. And don't you forget about dogs and mothers,' he grinned faintly.

'Goodnight.'

My footsteps echoed on the now deserted *muelle* left him sitting there. I was quite sure he was no longer smiling.

Nor was I.

PHYLLIS EISENSTEIN

SUBWORLD

You've seen the blind man. You've passed him a hundred times in the tunnel between subway lines. He sits on a campstool and sells peanuts and gum out of a burlap sack. He keeps a cigar box on his lap for money. I've always thought that he sells more peanuts than gum, but maybe that's because I never liked gum. Donny didn't either.

Donny was five when I started taking him on the subway. Sheila had custody during the week, but on Saturdays he was mine, and we would go to museums or the zoo or downtown movies. And the blind man was always in his spot, just as he was on the days I went to work. Donny was curious from the start, turning to stare as we passed. I suppose I should have given him a swat on the behind and a quick lecture on rudeness, but that would have been embarrassing there in the crowd. That was his mother's job anyway—discipline. I was the parent who didn't have to say No.

For a few weeks Donny only stared, never saying anything. At last, though, we happened to pass the blind man as someone was buying a bag of peanuts. Donny watched her drop a quarter in the box, take a small white bag, and tear it open. The following week, he tugged at my hand as we approached. 'Peanuts, Daddy?' he said. So the ritual began, and from then on the blind man received a quarter from me every Saturday.

Almost a year went by before Donny spoke to him. I would never have thought of doing it, but—like most adults—I had had a lot of practice in stifling my curiosity. Donny stood there, clutching his bag of peanuts, looking at the blind man with those wide blue eyes, those guileless eyes, and he said, 'Why do you wear sunglasses indoors?'

I felt my face flushing. I didn't know what to say; I was too embarrassed even to mumble an apology. I wanted to drag Donny away and pretend he hadn't said anything, but he had let go of my hand and stepped towards the blind man, and I would have to move to grab him. And just for the moment, I was too flustered to move.

The man said, 'I'm blind, little boy. Don't you see my white cane?'

'I see it,' said Donny. 'Are you crippled, too?'

He shook his head. 'The cane is like a long arm. I wave it in front of me to keep from walking into things, because I can't see them like you can.'

'If you can't see things, how come you know I'm a little boy?'

'That's easy,' said the blind man. 'I'm not deaf.'

I found my voice. 'Come on, Donny—we'll be late for the movie.' I stretched my hand towards him, and he took it. And then we were walking, a little faster than usual, and Donny was calling over his shoulder, 'See you later.'

We did see him, of course, on the homeward leg of the trip. He and Donny exchanged greetings as if they had known each other forever, and so their friendship was sealed. Peanuts and conversation, every Saturday. And if the blind man wasn't there, Donny wanted to know why, as if I had to answer for him. Where was he; was he sick, was he angry with us, had he moved to another country? 'I'm sure he'll be here next week,' I always said, and I was usually right.

Donny loved the subway. The older he got, the more he loved it. When he was seven, he started wanting to ride the trains from one end of the line to the other. The zoo, I would say; the new movie, the museum with all the push buttons, but no—he wanted to ride the subway. He would sit by the front window, his nose pressed to the glass, watching the lights and tracks rush towards us.

He also loved the platforms. He would wander around them, reading the advertisements, the system maps, the graffiti on the walls. He especially liked to look down on the tracks, at the places between the rails where mesh garbage traps were set—low, squarish things, open at one end so that the rushing wind of an approaching train would sweep trash into them. At first he wanted to know why they weren't squashed by the trains . . . and then he had to lean over in front of a waiting train to prove to himself that the cars really did sit high enough on the tracks to pass over them. That was the first time I ever yelled at him, when I yanked him away from the edge. He looked up at me in total innocence and said, 'But the train wasn't *moving*.'

'Do you see that sign?' I said, pointing to the wall across the tracks. Of course he had seen it: PLEASE STAND BEHIND THE YELLOW LINE. The yellow line marked the edge of the platform.

'I was careful,' he said.

'I don't care how careful you were. What if the motorman had started the train while you were leaning out in front of it?'

He looked down at his feet. 'I'm sorry.'

We watched the train roar off down the tunnel, leaving a string of red lights in its wake. The lights turned yellow, then green.

'Look, Daddy,' Donny said, tugging at my sleeve. 'The paper. It's *moving*.'

In the garbage trap, crumpled napkins and paper cups and newspapers were rustling. When a tiny grey body shot out of the trap, Donny squealed and jumped, clutching me with both hands. 'What is it?'

'A mouse. See, there's another.'

'A mouse!' He took a step towards the edge of the platform, but I held him back by one elbow.

'You'll watch them from back here,' I said.

'But it's a mouse!'

'From back here.'

He was fascinated by their scrambling about the garbage trap, by their dodging under the rails and across the ties. 'What are they doing, Daddy?'

'Looking for food, I suppose. Maybe some candy left in those wrappers.'

He took the white bag from his pocket. There were a few peanuts left in it. He threw one down on the tracks. It lay there, pale against the grime of the tunnel floor. When no mouse approached it, he turned a disappointed face to me. 'Don't they want it, Daddy?'

'They're not like squirrels in the park, Donny. They're afraid of people. I'm sure they'll come and get it sometime, though. They probably don't get peanuts very often.'

'What *do* they eat then?'

'Things people don't want. Things people drop accidentally. Look at all that garbage in the trap. People throw a lot of things on to the tracks, even though they're not supposed to.'

'I wouldn't want to eat garbage,' he said, and he took out another peanut and tossed it on to the tracks. A few minutes later, a mouse did come out of some hiding place too close to the platform for us to see and cautiously approached one of the peanuts. But then the tunnel began to vibrate from an oncoming train, and the mouse skittered away.

'They'll be hurt!' Donny cried. 'They'll be run over! Daddy! Oh, Daddy! The poor mice!'

I don't think they'll be hurt,' I said, my hand tightening on his shoulder just to be sure that he stayed behind the yellow line. 'They're so small that the train will go right over them without ever touching them.'

He was very relieved to see them reappear when the train was gone. And this time, one picked a peanut up in its mouth and scooted away with it.

'Maybe it's a mother mouse,' Donny said, 'taking it home to her babies.'

'Maybe,' I said. 'And how about us going home, too? Aren't *you* hungry?'

He nodded reluctantly. 'But can't we stay just a *little* while longer? Look, there's another one!'

'A little while,' I said. To be perfectly honest, I rather enjoyed watching the creatures, too. There had been mice in the subway for a long time. I had seen them when I was a kid. They were cute; not like cartoon mice, but still, in their own way, cute. Next, I guessed, Donny would ask for a mouse of his very own. I was already trying to formulate a good answer for that one, although I expected the cold truth would be best: his mother would never allow it. That would pass the buck, and the questions, to her. *The tough ones are yours, Sheila,* I thought; *you didn't even want me to have him on weekends.*

But he never asked me. Maybe he knew better, even at seven. He remained fascinated by the mice, though. He talked to them, as he talked to the animals in the zoo, to the squirrels in the park. He really needed a pet, I thought, and I tried to make it up to him by thinking up new and marvellous places to take him. Children's plays, puppet shows, circuses, amusement parks. But still, though he took pleasure in it all, he found a special delight in standing on the subway platform watching the mice.

'I wish I could go down there and play with them,' he said.

'They'd be afraid of you.'

'Not *me*. I'd be their friend.'

'They might bite you.'

'I'd be gentle, Daddy.'

I held his hand tightly. 'The tracks are electrified. If you touched them, you'd be burned badly. Remember when you burned yourself on the frying pan? Remember how it hurt?'

He looked up at me doubtfully. 'The mice run across the tracks.'

'They run across the ties. The ties are made of wood, and they're safe. The rails are made of metal and they're electrified. The mice are small enough to crawl underneath the rails without touching them. But you're not.'

'I'd be careful.'

'And while you were trying to play, a train could run you over.'

He pointed to one of the wall niches where workmen could retreat to let the trains pass by. 'I'd stand there every time one came through.'

'It's too dangerous,' I said firmly.

He looked up at me. 'I was just wishing, Daddy,' he said in a very small voice. 'Don't be angry.'

And then I had to hug him to show that I wasn't.

I never told his mother about those 'wishes'. I knew that she'd throw a fit about my not looking after her son properly. Her way would be to smack him a couple of times when he talked about playing with the mice, to knock the idea out of his head. She was always the one to say a flat No, without reason, without discussion. To him. To me. Her answers were always sharp, clean-cutting; and if you didn't jump *her* way . . . well, the best thing to do was leave. Or you'd be thrown out. I couldn't say any more which it had been with me. I only knew that I hadn't been the husband she'd wanted, or needed. Nor the proper sort of father to her son. She probably spent the week telling Donny that everything I said on Saturday was wrong. Poor kid, I thought. And poor me. Someday she was going to force him into making a choice between her and me, and I figured I'd be the one to lose. So I didn't say No to him, and I didn't yell at him, and I didn't punish him, because I wanted the time that we had to be cheerful, to be full of good memories for both of us. I spoiled him. She hated me for it.

We started buying two bags of peanuts after that day, one for Donny and one for the mice. Donny explained the situation very carefully to the blind man, who smiled and took two quarters from us. He didn't say—and I didn't say—that the subway authorities probably left poison for the mice and wouldn't appreciate anyone feeding them.

Sometimes I felt a little nervous about Donny throwing peanuts on to the tracks. I kept looking over my shoulder, knowing there was an embarrassed grin on my face, looking for some cop who would sneak up on us with a lecture on littering. There were signs everywhere, of course, on using the trash bins. People mostly ignored them, tossing lit cigarettes down on the ties, or crumpled bags with the remnants of lunch, or unwanted newspapers. But Donny had read the signs—why should I have doubted it?—and understood that when the police came by, the peanuts stayed in his pocket. Seven years old and he already knew that getting caught was bad, not the act itself.

By the time Donny was eight, he was telling the blind man all about his week, about school and his friends and his mother. The blind man never said much in return, just smiled and made encouraging noises as Donny babbled. I finally stopped being embarrassed by these chats, stopped rocking from foot to foot as people hurried by me to catch their trains. I came to feel that I knew him, too, as I knew my regular checker at the supermarket or my regular bank teller. And yet I knew nothing about him except that he was blind and he sold peanuts for a quarter in the subway.

Their favourite topic was the mice. Donny would talk enthusiastically about them, describing their latest activities, their scrambling through the trash trap or dashing off with peanuts, and the blind man would nod. The ninth or tenth time that Donny said, in that wistful little voice, 'I wish I could play with them,' the blind man cocked his head to one side, almost as if he could see Donny. 'Do you really wish it?' he said.

'Yes. Oh, yes.'

'Well, then, maybe I can do something about that.'

I shook my head automatically, then remembered whom I was talking to and said, 'Please don't make any promises to the boy; I don't think his mother would approve.'

He tilted his head up to me, and the overhead lights glinted off his dark glasses. 'We don't have to tell his mother,' he said. He turned to Donny. 'Can you keep a secret, son?'

'Oh, yes,' said Donny.

'And can your Daddy keep a secret?'

Donny looked up at me. 'Can you?'

'That depends,' I said. 'I won't keep a dangerous secret.' I didn't like the idea of Donny near the mice. Mice were dirty, diseased, maybe even rabid. These weren't laboratory mice, after all. 'Mice have very sharp teeth.'

'There's no danger,' said the blind man.

I put my hand on Donny's shoulder. 'I thought we were going to the ball game.'

'It doesn't matter, Daddy,' he said.

'I thought you were excited about this game.'

'But it's a *secret*, Daddy. We can always go to another ball game.'

I looked at the blind man. 'What are you talking about, mister?'

He shook his head. 'I can't tell you anything in a public place. But I'll show you.'

'What do you want to show us, and where do you want to show it?'

'You'll see when we get there. It isn't far away. Will you come?'

'Oh, Daddy, *please.*'

'This is silly,' I said. 'We're not going anywhere but the ball game.'

'I think you'll find it interesting, too,' said the blind man. 'Most folks do.'

'Daddy, *please.*'

'How far?'

'A few minutes' walk.'

I looked down at Donny's eager face. I couldn't imagine what the blind man wanted to show us. Or rather, I *could* imagine all sorts of things, things I would prefer Donny didn't see. 'What is it?' I said. 'Some sort of private menagerie?' I thought of cages full of mice, shredded newspapers heaped about, droppings, insects infesting the whole thing.

'I promise you it's interesting,' said the blind man. 'And pleasant.'

I frowned, trying to look stern for Donny's benefit. 'How much will it cost us?'

'It's free.'

'Oh, Daddy, *please.*'

I looked at my son once more. Well, I didn't need to see a ball game. 'All right,' I said. 'If it isn't far.'

The blind man dropped his camp stool and cigar box into his burlap sack and slung it over a shoulder. He gestured with his cane. 'This way.' He started down the tunnel at a brisk pace, swinging his cane in a shallow arc before him. We followed. The crowd was thin for the moment, a burst of travellers having just passed, and so the rest were able to give us a wide berth. About a dozen yards down the tunnel, he turned right through a narrow archway and descended a flight of steps.

The stairway was steep, and I held Donny's hand to keep him from falling. The light was bad, too, naked bulbs behind heavy mesh, and I was none too sure of my own footing in the shadows. The stairway led to another tunnel about twenty yards long, and then another, shorter stairway at the end of that.

The blind man paused at the top step. 'Stay close to me,' he said. 'It's easy to get lost.'

We descended into an area of muted light. It might have been a room, but I couldn't see any walls or ceiling. Instead there was a great, jagged, convoluted mass of material before me, some of it white, some grey, with here and there splotches of red or yellow or blue. It appeared to be a rigid substance, like sheets of styrofoam folded, broken, crumpled into a vast and unfathomable sculpture. Like cardboard mashed and twisted into a jungle. And it was all translucent, scattering and mellowing light that came from unseen sources.

An open space lay before us, floored and roofed and walled by the stuff. The blind man stepped into it, his feet sinking up to the ankles with a rustling sound. He motioned to us to follow. My feet sank, too, and it was like walking on mounded pillows, treacherous, playing havoc with my sense of balance. The stuff shifted, and I felt myself floundering; when I reached out to a wall for support, it also yielded under my hands. I fell over.

When I climbed to my feet, the blind man had vanished.

'I think we should go to the ball game,' I said to Donny. 'This doesn't look interesting at all.'

Donny pushed at one of the walls, and it moved under his fingers with a crinkling sound. 'I think it's kind of nice,' he said. 'Like a cave.' He had seen pictures of caves. He tugged at my arm. 'Come on, Daddy; let's see the rest.'

I glanced over my shoulder. We were only a few paces from the stairway. I wondered what was supporting all this stuff, whether it would fall in on us any minute. On the other hand, the blind man had gone ahead; he wasn't afraid. And he had obviously been here before.

I let Donny pull me forward.

We moved through a tunnel of irregular size, really a string of open spaces, rising, dipping, twisting, turning—a tortuous path through a crazy forest. I felt like we had stumbled into an abandoned funhouse. I had to watch every step, and still I slipped and staggered over the haphazard floor. But Donny seemed to adapt to it easily, maybe because his slight weight kept him from sinking as deep in the stuff as I did.

We came to a fork in the tunnel.

'Which way shall we go, Daddy?' asked Donny.

'I don't know. I don't see any signposts.' Then I shouted, 'Hello!'

'Hello yourself!' came the blind man's voice. 'Take the right fork. We're waiting for you.'

Donny pulled at my hand, but I wouldn't move.

We?

Uneasiness struck me. I didn't know who *we* might be, and I wasn't sure I wanted to find out. The whole place suddenly gave me a chill. If nothing else, it was a firetrap, and if it burned we'd have a hell of a time finding the stairway again. And who could be waiting for us? Muggers? After all, we didn't really *know* the blind man, not even his name. I looked affluent enough. Why shouldn't he and his friends want to knock me over the head, take my wallet, and leave me here, where nobody would find me, except maybe the mice. And Donny; what would they do with Donny?

'Come on!' called the blind man.

Donny let go of my hand and scrambled away.

'Hey!' I yelled and dived after him. But he was faster in that crazy tunnel. He turned a corner and was gone, all except the rustling sound of his feet against the yielding floor. 'Donny!' I shouted. And then I rounded the same corner and found him.

He stood in a large open space with the blind man and a woman. She was rather pretty; a little older than Sheila, I guessed, and a little plumper. She wore jeans and a sweatshirt. As I watched, she held out her hand to Donny, smiling, and he shook it. She glanced up at me then. 'Welcome,' she said. 'We've heard so much about the two of you.'

'Hello,' I said. 'We haven't heard anything about you.'

'No, of course not. Wilbur is very good about that.'

'Wilbur?'

She nodded at the blind man. 'He's been our friend for quite a long time.'

'Who are you?' I asked. 'And what *is* this place?'

'Would you care for some lunch?'

Donny looked back at me. 'I'm hungry, Daddy.' At the ball game he would have had a hot dog and ice cream by now.

'All right,' I said.

She clapped her hands three times, and in less than a minute, three people entered the open space from a tunnel opposite the one we had come by, each carrying a covered tray. They set the trays down on the floor, pressing them firmly into the surface so that they were level and stable. Then they withdrew.

'There's an art to eating here,' said the woman, settling herself cross-legged on the floor. 'Don't lean on the trays or put your weight too close to them, or they'll tip over.' To demonstrate, she leaned back, stretching one arm out to lift the cover from the nearest tray. On the platter was a mound of raw vegetables. She selected a carrot. The other trays held bread, lettuce, and sliced chicken. Carefully, we made sandwiches.

'Wilbur has been observing you for a long time,' the woman said. 'He had to know if you were the right sort of people.'

'What sort is the right sort?' I asked.

She looked at Donny. 'You are.'

He grinned.

'Thanks for the compliment,' I said. 'At least I hope it's a compliment. But the right sort for *what*?'

'We have a very pleasant life down here,' she said. 'Good food, good company. You'd like it.'

I had to smile. 'What kind of salary are you paying?'

She shook her head. 'I'm not kidding.'

'I don't know what you're talking about.'

'This isn't a job,' she said. 'It's a life. I'm suggesting that you join us. There's plenty of room for both of you.'

'Where?'

'Here. Right here.'

I looked around, at the wild sculptures we were in the middle of. 'You *live* here?'

'Yes.'

'Inside *this*?'

'It's very large. There's plenty of privacy, if you want it.'

'Inside this stuff? I can't believe it. It's ridiculous. We're in the subway, aren't we? In some storage area?'

'Something like that.'

'Are you sure this kind of thing is legal?'

'The police don't bother us.'

'Do they know you're here?'

'Yes. Of course.'

At that moment, four young faces peeped into the room from the other tunnel. Leaning against the wall, they made it rustle. The woman heard the noise and gestured them in. 'Come,' she said. 'No need to be shy.'

They surged forward, three boys and a girl, pushing each other in their eagerness, and they clustered about her like young puppies. They chattered all at once, so fast that I could hardly make out a word.

'Yes,' she said to them, laughing. 'If he wants to.' She looked at Donny. 'Are you interested in playing with my children?'

Donny crammed the remnants of his sandwich into his mouth and nodded vigorously. Then he glanced at me, the question in his eyes. I didn't know what to say.

'It'll give you and me a chance to talk,' said the woman.

I didn't want to let him go, but he was so eager, and the other children smiled so much . . . they were clean kids, neatly dressed, no more rambunctious than a new playmate usually made kids.

'Please, Daddy,' said Donny.

So I nodded, and the fivesome scrambled off together. I heard them laughing for quite a time after I couldn't see them any more.

I turned to the woman. 'What do you want to talk about?'

'You're welcome here,' she said.

'What is this place? A commune?'

She nodded.

'I don't know anything about you,' I said. 'I don't even know your name.'

'Clarissa.'

I pointed to the blind man. 'Does he live here?'

'No,' she said. 'Wilbur is happy on the outside.'

'The outside,' I repeated.

'He's our contact. We don't go out there, but he does.'

'You don't go out?'

'Never.'

'You don't ever go out of this place?'

'We don't go out of the subway.'

'I find this hard to believe. Where do you get your food and clothing? Where do you get your money?'

'We don't have any money. And the food is brought in to us.'

I looked at her sceptically. 'What is this—some religious cult?'

'No.'

'Then . . . why don't you go out? Are you afraid of the sun or something?'

She smiled. 'No, we're not afraid of the sun. But everything we need is here. We don't have to go out any more. And we don't want to. Do you?'

'Of course I do! My job is out there. My life is out there.'

'Is it a good job?'

'Yes, a very good job.'

'Do you like it?'

'Well enough.'

'And do you like your life well enough?'

'I don't want to give it up for . . . for this.'

She leaned back, sinking into the floor as into a soft couch. 'I thought that way at first. But after I'd visited down here a few times, I realized that I'd be happy giving up the job and the people and the harassment that's up there. The responsibility. The demands. I had a husband . . . But I was glad to leave him. It's quieter down here. I like that.' She looked sidelong at me. 'You might keep that in mind. You don't have to decide now.'

'I'm not going to,' I said. 'I'm not about to abandon everything and move in with you. However many of you there are.'

'Quite a few,' she said. 'There's room for more.'

'It's a nice life, is it?'

'Very.'

'But you didn't convince Wilbur of that.'

Wilbur smiled. 'I didn't have anything to run away from.'

'Well, I don't either,' I said.

The woman smiled. 'It's nice to know there's a place to go, just in case.'

I shook my head. 'I'm sorry. I don't consider living in the subway a viable alternative. It's crazy. It's *got* to be illegal.'

'Not for mice,' she said.

I stared at her. 'What do mice have to do with it?'

'We're mice,' she said.

Hesitantly, I chuckled. 'You think of yourself as mice?'

She nodded. 'This is a mouse warren. It's made of crumpled paper bags, and old newspapers, and all the rest of the garbage that people toss on to the tracks.'

'What are you talking about?'

'I'm a mouse,' she said. 'Well, I'm really a human being, but as far as the people who ride the subway are concerned, I'm a mouse. You've seen us among the tracks.'

I shook my head slowly. 'You're a real kidder, aren't you?'

'No.'

'Funny—you don't *look* like a mouse.'

'Yes I do,' she said. 'When I want to.'

That feeling of unease came back. Muggers would have been easier to cope with than this. I glanced at the blind man, wondering if he was as crazy as the woman. At that moment, I wanted very much to find Donny and get out of there. But I had let them take Donny away from me.

'Don't be frightened,' said the woman, as if reading my mind, though my mental state probably showed on my face. 'We won't hurt you. We're not crazy. In fact, we may be the sanest people in the world. We don't have any ulcers.' She stretched languorously, then drew her arms and legs in to her chest. 'All you need is the proper sort of illusion,' she said, 'and all the senses can be fooled—sight, smell, touch, everything. Stale candy bars can be transformed into a banquet. Mice into people. And people into mice.' As I watched, her form . . . wavered. Her arms and legs thinned, her torso fattened, her head flattened, her face protruded into a snout. Her clothes melted into gray fur, and a long, thin tail took shape at the base of her spine. She was a mouse—a human-sized, furry, bright-eyed mouse.

She rolled over to her stomach and stepped slowly towards me. I eased myself backward. Her mouth opened, showing sharp incisors, and the human voice issued from her throat: 'I won't hurt you; don't be afraid.' She closed her mouth, hiding those teeth, and came very close to me. I sat there tensely, ready to scramble away, but caught, fascinated by the transformation I had just witnessed. She came so near that her twitching whiskers brushed me, and they were like springy, coarse wires. She put her snout against my face, her cool moist snout, and licked my cheek. She put her paws up on my shoulders. She hugged me. Her fur was sleek and glossy. 'You see,' she whispered, 'there's nothing to be afraid of.' She hugged me again and then settled down close beside me, her head resting in my lap. Like a dog. Like a great collie dog. 'It's very nice,' she said, 'being a mouse.'

Somehow I found my voice. 'But you're not really a mouse. I mean, you're too big to be a mouse.'

'No,' she said. 'I'm mouse-sized. And so are you, right now. You're four inches tall.'

'That's ridiculous!' I gasped.

'Nothing is ridiculous down here.'

'You've hypnotized me!'

'Not really.'

"Yes, you have, and I don't like it. I'm going to leave now. Where's my son?'

'Safe,' she said. 'Playing with the other mice.'

'I want him here, now.'

'He won't want to leave. Children love it down here.'

'I don't care what he wants. He's going home with me if I have to tear this place apart to find him.' I pushed her head off my lap and stood up, staggering to find my balance.

'Children love it down here,' repeated the mouse who had been a woman. 'They adapt quickly.'

'I want my son!'

She looked up at me, and then she rose to her hind legs and the transformation reversed itself, the mouse melting back into a woman, the fur fading into jeans and sweatshirt. She crossed her arms over her chest. 'Very well,' she said. She clapped her hands four times, and a couple of moments later five laughing children burst into view, one of them mine.

Donny scrambled up to me, breathless, 'Come see the pirate treasure!' he said.

I took his hand. 'We're going home now, Donny.'

His face fell, laughter stifled. 'Do we *have* to?'

'Yes. It's getting late.'

His face screwed up, and for a moment he looked like he was going to cry. But he conquered it, saying, in a stiff little voice, 'Can we come back soon?'

'We'll see,' I said.

He looked back at the four children. He raised one hand to wave. 'See you later.'

I pulled him towards the tunnel that we had come in by. Sliding and stumbling, we made our way towards the outside world. Only when I reached the first stairway did I realize that the blind man was following.

'You need me,' he called.

I looked back sharply. 'For what?'

'You can't get out without me. I have the key.'

'I didn't notice any doors,' I said.

He grinned. 'Well, go on then—go on without me.'

And we did, until we came to the top of the second stairway and found ourselves facing a blank wall. I rapped on the concrete, and it sounded solid. I turned around. The blind man stood on the bottom step. 'All right,' I said. 'How do we get out?'

He seemed to be looking straight at me as he climbed. 'They don't want to keep you prisoner, you know. Everybody who joins them does it out of choice.'

'Then where's the exit?'

He put his hand against the wall. 'You really were the size of a mouse down there, you know. A grown man couldn't fit in that space.'

'Sure I was.'

'Some of them really are mice. After a while, you get to know the difference.'

Donny tugged at my hand. 'Three of the children were mice, Daddy. They showed me. They said I could be one, too, if I wanted.'

I looked down at him. 'People are not mice,' I told him.

'Sure they are, Daddy. Just like in Cinderella.'

'That was a fairy tale, Donny.'

'But I saw it.'

I looked at Wilbur. 'Get us out of here.'

He tapped the wall, and an arched section faded into mist. Beyond the mist I could see people walking in the tunnel. As the traffic thinned momentarily, the mist cleared.

'Go on,' said the blind man.

Donny and I stepped through. I turned back to thank our guide for the adventure, but he was gone, and in the wall of the tunnel was only a shadowed alcove, and no door at all. I fought the chill that ran up my back and started walking, pulling Donny along.

At the end of the tunnel, I glanced at the overhead clock, then at my watch. Both seemed to have stopped. Both said we had plenty of time to make the ball game. I stopped at a news stand to ask the time, and the vendor gave the same answer.

'I'm hungry, Daddy,' said Donny.

'But you just had a big lunch.'

'I know, but I'm still hungry.'

And so was I. Fairy food, I thought. It doesn't stay with you.

So we went to the ball game and had hot dogs and ice cream, and we didn't discuss the warren down in the subway.

I didn't know what to think. I didn't want to think about it. The best approach seemed to be to ignore the whole incident, treat it like a strange dream. But Donny wasn't about to let me do that. The following Saturday, when his mother dropped him off at my place, the first thing he wanted to do was visit the mice. 'The ones in the funny place all made of paper.'

'Did you tell your mother about them?' I asked.

He made a sour face, like he had just bitten into a lemon. '*She'd* never understand. Besides, it's a secret, remember? Did *you* tell anybody?'

'No, not a soul. Of course it's a secret.' I frowned. 'I don't think we should visit them today. We really won't have time. We're going to the zoo—they have two new lion cubs.'

'Just for a little while, Daddy? It was so much fun.'

'Not today, Donny. Maybe next week.'

He looked up into my face. 'Next week? Promise?'

'There are so many other things to do, Donny.'

'Promise me, Daddy. You never break your promises.'

'That's why I *won't* promise, Donny. I don't know when we can see the . . . the mice again. If ever.'

'Why not?'

I searched for some plausible excuse, but I couldn't find one and finally just blurted out the truth. 'Because people who turn into mice scare me.'

He took my hand and held it tight. 'You shouldn't be scared of them, Daddy. They won't hurt you.'

I squatted down beside him. 'You're sure of that, are you?'

'They're nice. They wouldn't hurt *anybody*. *Please* can't we visit them again? It's a wonderful place, and the mouse children say there's so much to explore and so many things to find. We found pirate treasure last time—jewels and gold and silver. It was so much fun!'

I shook my head. 'I don't know Donny. We'll talk about it another time, okay? I've got so many other plans for us. . . .'

He hung his head. 'Okay, Daddy. If you say so.'

We went to the zoo. He seemed to enjoy himself.

The next day I found out about the secret that Sheila had asked him to keep from me. She called. She was moving out to the West Coast, taking Donny with her. She had a fiancé waiting there, a new father for her son; he would have a complete family again. The court had approved. I could have him for a couple of weeks during the summer if I paid his air fare.

The following Saturday Donny was very subdued. He knew that I knew.

'Everything will be all right,' I said as we walked hand in hand towards the subway. We were going to a museum, just a typical Saturday outing. 'You'll like your new home. The weather is beautiful out there.'

He looked up at me. 'It isn't fair, is it? For her to take me so far away.'

'She's trying to do her best for you. She loves you.'

'But you love me, too.'

'You know I do.'

'Will you come visit me?'

'We'll make some kind of arrangement.'

'Promise?'

'Yes, I promise.'

Down in the subway, we bought our peanuts from the blind man, and Donny stood there for a long moment, looking at him silently. And the blind Wilbur knew, he *felt* the difference in my son.

'Is something wrong?' he asked.

Donny tucked the peanuts into his pocket. 'I'm going away,' he said. 'I won't come here any more. My Mom is taking me away.'

'Ah,' said the blind man.

Then Donny turned, tugging at my hand, and we walked off with a surge in the crowd. But he was in more of a rush than I; he walked too fast, lost my hand, and disappeared among the hurrying people. I called his name. I dodged among the walkers, craning my neck for a glimpse of him. I let myself be swept to the platform; I crisscrossed it, calling, calling. He wasn't there.

I ran back to the tunnel, to the blind man.

He was gone. His peanuts were gone, his cigar box, his campstool. All gone.

I went through the motions. What else could I do? I found a cop, and he called some more cops, and they made a methodical search of the station. They couldn't find him. They suggested he might have been kidnapped, and a bulletin went out on him. Too soon I had to call Sheila and listen to her scream at me over the phone. She had always known it would happen, she said. Always. Shortly after that, the police arrested me on a kidnapping charge. There were court dates and lawyers then, but no one could prove anything. No ransom note ever came. No body was ever found. Sheila finally moved to the Coast with her new husband. He was angry, too, but only for her sake. He had a couple of children by a previous marriage and couldn't get worked up over a stepchild.

Of course I knew where Donny was. Though the shadowed alcove contained only a blank wall when I passed it, I knew. But I couldn't very well tell anybody. *My son has turned into a mouse.* No.

Things blew over at last. Sheila was gone, and the police stopped following me around, and life settled back to its old routine, except for Saturdays. I still passed the blind man every day on my way to work. He had been there all the time, hadn't missed a day. I stopped for peanuts sometimes. I said Hello. He knew who I was. He looked towards me in a special way; as if he could see, as if he were waiting for me to say something, to do something. I didn't, not while I was being watched, and not for a long time afterward. But a day came, a Saturday when I was going to a ball game by myself—a day came when the crowds weren't quite as thick as usual, when I stopped for peanuts and I said, 'I'd like to see him.'

The blind man smiled. 'He always asks about you.'

'Today?' I said. 'Now?'

'I think we can manage that.'

The doorway was there, the stairs, the tunnel, the other stairs, and then we were in that mass of crumpled paper that was the mouse people's haven. I felt disoriented as I walked through that strange forest. The pathway seemed unfamiliar, twisting and turning confusingly, as if the whole thing had been re-engineered while I was gone. The open space at the end was different, too—long and narrow. There was a rise in the floor on the far side, and two couches set on the rise. They looked comfortable, upholstered in some soft material, with low backs and curving arms. They looked incongruous—furniture amid crumpled paper.

On one couch was a young woman, a lovely creature with pale skin and long fair hair; she wore a sleek blue gown and great masses of jewellery, a pirate's trove of gold and gems. On the other couch was a young man, a sturdy, muscular fellow in red velvet. On his head was a gold coronet. He looked familiar. He looked a little like the face I saw in the mirror every morning.

I remembered the lunch that had taken no time at all. No *outside* time.

He stood as I approached. He knew me right away. 'Daddy,' he said softly.

I tried to smile. My voice wasn't quite steady as I said, 'Hello, Donny.' He held out his hand and I shook it. A man's hand, not the small paw I had known. It was hard to believe that my little boy was gone, replaced by this man who answered to his name. And then he was hugging me, and I was hugging him, and it was strange, very strange, that he could reach all the way around me. We loosened our grip on each other at last, and I looked up at the crown on his head. 'You're king around here?'

He grinned and took the crown off and tossed it on the couch, the throne. 'It's a game,' he said. 'Some of the young people play it. It's just illusion anyway; the crown's the ring from a pop-top can.' He waved at the woman, still sitting on her own couch, watching us with great dark eyes. 'This is Mila, Daddy. My friend.'

'Hello, Mila,' I said.

She smiled.

'She's really a mouse,' said Donny, 'so she doesn't talk. But she makes a terrific-looking human.'

I stared at her, trying to imagine that slender form as a plump grey mouse, failing. But I didn't ask him to dispel the illusion. I wasn't ready for that.

'Have you come to stay with us?' Donny asked me. 'You'll like it here. The people are marvellous. The mice, too.'

I looked into his eyes. I seemed to see him as child and adult all at once, as son and stranger. 'How long has it been for you, Donny?'

'Almost twenty years.'

I shook my head. 'Only two for me.'

'They explained it to me pretty early, but I didn't understand till one day I noticed that Wilbur never got any older. Then I knew why you hadn't come to visit.'

'It was to protect you,' I said. 'The police were following me. They thought I had . . . done something with you.'

He looked away from me. 'I'm sorry, Daddy. I never meant to make trouble for you.'

Softly, I said, 'Why did you do it, Donny? You frightened your mother terribly.'

'You didn't tell her?'

'How could I?'

He shrugged. 'I don't know if I can explain how I felt. Pulled in two directions. Helpless. I loved you both, really. You *do* believe that?'

I nodded.

'But I was frightened. Everything was going wrong. I wanted it all to *stop*. There was only one way out, don't you see?'

Looking at him, I thought I could see. It was our fault, really, Sheila's and mine, because neither of us was willing to let go of him. 'Tell me about the last twenty years,' I said. 'We have so much to catch up on.'

We spent hours together, though my watch said it was only minutes. He told me about his life and his friends, both mice and people. They had gotten other recruits since Donny's arrival, and some of the human women had had children; but their human population was still low, and they would be happy to take in more. Only a few real mice were allowed among them, as special pets. The sole danger to all was the poison occasionally left by the city workers, but it was so obvious in appearance that the humans, at least, had no trouble avoiding it. There was plenty of safe food in the subway; they all took turns going out to the tunnel floor as mice and foraging. Time resumed a more ordinary pace for them out there, so they always hurried, not wanting to leave their friends waiting long.

'From the edge of the warren,' he said, 'you can see the trains pass in slow motion. *Very* slow. But you can't hear them: the sound is too low-pitched for our ears, they tell us.'

'They?'

'We've got some smart people down here. They've given me a pretty fair education.'

A bit self-consciously, I put my arm around his shoulders. 'You could get a better one on the outside. Why don't you come back with me?'

He looked surprised. 'Are you going back?'

'Of course.'

'You'd like it here, Daddy. Why don't you stay?'

I shook my head. 'It's not for me. It's too . . . different. It's easy for a kid to adjust, but I'm not a kid.'

'Most of our recruits come to us as adults.'

'Donny, I know what I've got up there, and I don't want to give it up.'

'Did you remarry?'

'No.' I laughed. 'It really hasn't been very long for me.'

'Sorry. I keep thinking of you as . . . old.'

'Do I look old?'

'No. You look just like I remember. Just like you should. And when you come back for your next visit . . . *I'll* be the one who's old.'

'I won't wait that long.'

'Even if you come back every one of *your* weeks, you'll see me age; and from my point of view, you'll always stay the same.' He shook his head, but smiled. 'It's like relativity, a little.'

'Have they taught you about *that* down here?'

'We have a college physics teacher with us. I think you'd like him.'

'I'm sure I would. But . . . why don't you come back with me?'

His smile faded. 'No. No, I can't do that.'

'Oh, come on. Just for a visit. We'll catch a movie—you don't get movies down here, do you? Maybe a ball game. Don't you miss them?'

He shrugged. 'It's an interesting life without all that stuff.'

'You could have both. Your mother isn't going to take you back. She wouldn't even know who you are. A guy I hang around with sometimes—that's not her little boy. She'd never guess.'

He looked at my eyes. 'No, Daddy,' he said. 'I can't.'

'Can't?'

'It's physically impossible.'

'What do you mean?'

'The amount of food I eat down here can't support the mass of a human body. I lost mass, lived off it, until I reached the point where my intake *could* support me. I only weigh a few ounces now, like an ordinary mouse. I can't go back to my old size. If I went out with you, I'd be a six-inch-high human. It's been tried, believe me. Once you've been here a while, you have to stay.'

Shocked, I whispered, 'Did they tell you that at the beginning?'

He nodded. 'They told me that if I stayed, it would be forever. And here I am.'

'But . . . but what if you change your mind?'

'The process takes a few months of subjective time. I could have gone back and just been . . . thin. I had my chance. But I wanted to stay. I love these people, Daddy. They're kind, good. They're happy.'

I looked hard at him. 'And *you're* happy?'

'Yes.' He smiled. 'Yes.'

He introduced me to the other adult humans—to the physics teacher and the advertising executive and the commodities broker; to the insurance agent and the lawyer and the Ph.D. candidate; to the woman I had met before, whose hair was now grey. It took her a minute to remember me. We had a party, and everyone laughed and joked. They seemed easy going and genuinely cheerful, and no one had to leave because of work or an early appointment. But at last I began to yawn.

They walked me to the stairway, all of them. I had lost track of the blind man sometime in that endless afternoon, but now he was waiting for me, standing on the second step.

'This is where we leave you,' said Donny. 'Take good care of him, Wilbur?'

The blind man nodded and began to tap his way up the stairs with his cane. I followed. I looked back once, to wave, but they were all gone, vanished into the crazy paper forest.

The routine of my life is pretty well set these days. On Saturdays I visit Donny, and he tries to convince me to stay. He's older than I am now, and I know that I'll lose him in a few more years. But I'd have lost him anyway if Sheila had taken him with her. He made his choice, and I've made mine, and my feeling for him isn't going to make me give up everything and become a mouse. After all, we *are* both adults. We have our separate lives. I like my work . . . well enough . . . and there's a strong possibility that I'll get my boss's job when he retires. That will mean more responsibility, but I look forward to it; I think I can handle it.

And if ever I can't . . . if ever it gets to be too much, and I find myself as lost and helpless and desperate as Donny was, I'll know, as he did, that there's a way out. Meanwhile, I buy peanuts from Wilbur every day.

Not for myself, of course. Not for myself.

BITE-ME-NOT OR FLEUR DE FUR

I

In the tradition of young girls and windows, the young girl looks out of this one. It is difficult to see anything. The panes of the window are heavily leaded, and secured by a lattice of iron. The stained glass of lizard-green and storm-purple is several inches thick. There is no red glass in the window. The colour red is forbidden in the castle. Even the sun, behind the glass, is a storm sun, a green-lizard sun.

The young girl wishes she had a gown of palest pastel rose—the nearest affinity to red, which is never allowed. Already she has long dark beautiful eyes, a long white neck. Her long dark hair is however hidden in a dusty scarf and she wears rags. She is a scullery maid. As she scours dishes and mops stone floors, she imagines she is a princess floating through the upper corridors, gliding to the dais in the Duke's hall. The Cursed Duke. She is sorry for him. If he had been her father, she would have sympathized and consoled him. His own daughter is dead, as his wife is dead, but these things, being to do with the cursing, are never spoken of. Except, sometimes, obliquely.

'*Rohise!*' dim voices cry now, full of dim scolding soon to be actualized.

The scullery maid turns from the window and runs to have her ears boxed and a broom thrust into her hands.

Meanwhile, the Cursed Duke is prowling his chamber, high in the East Turret carved with swans and gargoyles. The room is lined with books, swords, lutes, scrolls, and has two eerie portraits, the larger of which represents his wife, and the smaller his daughter. Both ladies look much the same with their pale, egg-shaped faces, polished eyes, clasped hands. They do not really look like his wife or daughter, nor really remind him of them.

There are no windows at all in the turret, they were long ago bricked up and covered with hangings. Candles burn steadily. It is always night in the turret. Save, of course, by night there are particular *sounds* all about it, to which the Duke is accustomed, but which he does not care for. By night, like most of his court, the Cursed Duke closes his ears with softened tallow. However, if he sleeps, he dreams, and hears in the dream the beating of wings. . . . Often, the court holds loud revel all night long.

The Duke does not know Rohise the scullery maid has been thinking of him. Perhaps he does not even know that a scullery maid is capable of thinking at all.

Soon the Duke descends from the turret and goes down, by various stairs and curving passages, into a large, walled garden on the east side of the castle.

It is a very pretty garden, mannered and manicured, which the gardeners keep in perfect order. Over the tops of the high, high walls, where delicate blooms bell the vines, it is just possible to glimpse the tips of sun-baked mountains. But by day the mountains are blue and spiritual to look at, and seem scarcely real. They might only be inked on the sky.

A portion of the Duke's court is wandering about in the garden, playing games or musical instruments, or admiring painted sculptures, or the flora, none of which is red. But the Cursed Duke's court seems vitiated this noon. Nights of revel take their toll.

As the Duke passes down the garden, his courtiers acknowledge him deferentially. He sees them, old and young alike, all doomed as he is, and the weight of his burden increases.

At the furthest, most eastern end of the garden, there is another garden, sunken and rather curious, beyond a wall with an iron door. Only the Duke possesses the key to this door. Now he unlocks it and goes through. His courtiers laugh and play and pretend not to see. He shuts the door behind him.

The sunken garden, which no gardener ever tends, is maintained by other, spontaneous, means. It is small and square, lacking the hedges and the paths of the other, the sundials and statues and little pools. All the sunken garden contains is a broad paved border, and at its centre a small plot of humid earth. Growing in the earth is a slender bush with slender velvet leaves.

The Duke stands and looks at the bush only a short while.

He visits it every day. He has visited it every day for years. He is waiting for the bush to flower. Everyone is waiting for this. Even Rohise, the scullery maid, is waiting, though she does not, being only sixteen, born in the castle and uneducated, properly understand why.

The light in the little garden is dull and strange, for the whole of it is roofed over by a dome of thick smoky glass. It makes the atmosphere somewhat depressing, although the bush itself gives off a pleasant smell, rather resembling vanilla.

Something is cut into the stone rim of the earth-plot where the bush grows. The Duke reads it for perhaps the thousandth time. *O, fleur de feu—*

When the Duke returns from the little garden into the large garden, locking the door behind him, no one seems truly to notice. But their obeisances now are circumspect.

One day, he will perhaps emerge from the sunken garden leaving the door wide, crying out in a great voice. But not yet. Not today.

The ladies bend to the bright fish in the pools, the knights pluck for them blossoms, challenge each other to combat at chess, or wrestling, discuss the menagerie lions; the minstrels sing of unrequited love. The pleasure garden is full of one long and weary sigh.

'Oh flurda fur

'Pourma souffrance—'

Sings Rohise as she scrubs the flags of the pantry floor.

'Ned ormey par,

'May say day mwar—'

'What are you singing, you slut?' someone shouts, and kicks over her bucket.

Rohise does not weep. She tidies her bucket and soaks up the spilled water with her cloths. She does not know what the song, because of which she seems, apparently, to have been chastised, means. She does not understand the words that somehow, somewhere—perhaps from her own dead mother—she learned by rote.

In the hour before sunset, the Duke's hall is lit by flambeaux. In the high windows, the casements of oil-blue and lavender glass and glass like storms and lizards, are fastened tight. The huge window by the dais was long ago obliterated, shut up, and a tapestry hung of gold and silver tissue with all the rubies pulled out and emeralds substituted. It describes the subjugation of a fearsome unicorn by a maiden, and huntsmen.

The court drifts in with its clothes of rainbow from which only the colour red is missing.

Music for dancing plays. The lean pale dogs pace about, alert for tidbits as dish on dish comes in. Roast birds in all their plumage glitter and die a second time under the eager knives. Pastry castles fall. Pink and amber fruits, and green fruits and black, glow beside the goblets of fine yellow wine.

The Cursed Duke eats with care and attention, not with enjoyment. Only the very young of the castle still eat in that way, and there are not so many of those.

The murky sun slides through the stained glass. The musicians strike up more wildly. The dances become boisterous. Once the day goes out, the hall will ring to *chanson*, to drum and viol and pipe. The dogs will bark, no language will be uttered except in a bellow. The lions will roar from the menagerie. On some nights the cannons are set off from the battlements, which are now all of them roofed in, fired out through narrow mouths just wide enough to accommodate them, the charge crashing away in thunder down the darkness.

By the time the moon comes up and the castle rocks to its own cacophony, exhausted Rohise has fallen fast asleep in her cupboard bed in the attic. For years, from sunset to rise, nothing has woken her. Once, as a child, when she had been especially badly beaten, the pain woke her and she heard a strange silken scratching, somewhere over her head. But she thought it a rat, or a bird. Yes, a bird, for later it seemed to her there were also wings. . . . But she forgot all this half a decade ago. Now she sleeps deeply and dreams of being a princess, forgetting, too, how the Duke's daughter died. Such a terrible death, it is better to forget.

'The sun shall not smite thee by day, neither the moon by night,' intones the priest, eyes rolling, his voice like a bell behind the Duke's shoulder.

'Ne moi mords pas,' whispers Rohise in her deep sleep. 'Ne mwar mor par, ne par mor mwar. . . .'

And under its impenetrable dome, the slender bush has closed its fur leaves also to sleep. O flower of fire, oh fleur de fur. Its blooms, though it has not bloomed yet, bear the ancient name *Nona Mordica*. In light parlance they call it Bite-Me-Not. There is a reason for that.

II

He is the Prince of a proud and savage people. The pride they acknowledge, perhaps they do not consider themselves to be savages, or at least believe that savagery is the proper order of things.

Feroluce, that is his name. It is one of the customary names his kind give their lords. It has connotations with diabolic royalty and, too, with a royal flower of long petals curved like scimitars. Also the name might be the partial anagram of another name. The bearer of that name was also winged.

For Feroluce and his people are winged beings. They are more like a nest of dark eagles than anything, mounted high among the rocky pilasters and pinnacles of the mountain. Cruel and magnificent, like eagles, the sombre sentries motionless as statuary on the ledge-edges, their sable wings folded about them.

They are very alike in appearance (less a race or tribe, more a flock, an unkindness of ravens). Feroluce also, black-winged, black-haired, aquiline of feature, standing on the brink of star-dashed space, his eyes burning through the night like all the eyes along the rocks, depthless red as claret.

They have their own traditions of art and science. They do not make or read books, fashion garments, discuss God or metaphysics or men. Their cries are mostly wordless and always mysterious, flung out like ribbons over the air as they wheel and swoop and hang in wicked cruciform, between the peaks. But they sing, long hours, for whole nights at a time, music that has a language only they know. All their wisdom and theosophy, and all their grasp of beauty, truth or love, is in the singing.

They look unloving enough, and so they are. Pitiless fallen angels. A travelling people, they roam after sustenance. Their sustenance is blood. Finding a castle, they accepted it, every bastion and wall, as their prey. They have preyed on it and tried to prey on it for years.

In the beginning, their calls, their songs, could lure victims to the feast. In this way, the tribe or unkindness of Feroluce took the Duke's wife, somnambulist, from a midnight balcony. But the Duke's daughter the first victim, they found seventeen years ago, benighted on the mountain side. Her escort and herself they left to the sunrise, marble figures, the life drunk away.

Now the castle is shut, bolted, and barred. They are even more attracted by its recalcitrance (a woman who says 'No'). They do not intend to go away until the castle falls to them.

By night, they fly like huge black moths round and round the carved turrets, the dull-lit leaded windows, their wings invoking a cloudy tindery wind, pushing thunder against thundery glass.

They sense they are attributed to some sin, reckoned a punishing curse, a penance, and this amuses them at the level whereon they understand it.

They also sense something of the flower, the *Nona Mordica*. Vampires have their own legends.

But tonight Feroluce launches himself into the air, speeds down the sky on the black sails of his wings, calling, a call like laughter or derision. This morning, in the tween-time before the light began and the sun-to-be drove him away to his shadowed eyrie in the mountain-guts, he saw a chink in the armour of the beloved refusing-woman-prey. A window, high in an old neglected tower, a window with a small eyelet which was cracked.

Feroluce soon reaches the eyelet and breathes on it, as if he would melt it. (His breath is sweet. Vampires do not eat raw flesh, only blood, which is a perfect food and digests perfectly, while their teeth are sound of necessity.) The way the glass mists at breath intrigues Feroluce. But presently he taps at the cranky pane, taps, then claws. A piece breaks away, and now he sees how it should be done.

Over the rims and upthrusts of the castle, which is only really another mountain with caves to Feroluce, the rumble of the Duke's revel drones on.

Feroluce pays no heed. He does not need to reason, he merely knows, *that* noise masks *this*—as he smashes in the window. Its panes were all faulted and the lattice rusty. It is, of course, more than that. The magic of Purpose has protected the castle, and, as in all balances, there must be, or come to be, some balancing contradiction, some flaw. . . .

The people of Feroluce do not notice what he is at. In a way, the dance with their prey has debased to a ritual. They have lived almost two decades on the blood of local mountain beasts, and bird-creatures like themselves brought down on the wing. Patience is not, with them, a virtue. It is a sort of foreplay, and can go on, in pleasure, a long, long while.

Feroluce intrudes himself through the slender window. Muscularly slender himself, and agile, it is no feat. But the wings catch, are a trouble. They follow him because they must, like two separate entities. They have been cut a little on the glass, and bleed.

He stands in a stony small room, shaking bloody feathers from him, snarling, but without sound.

Then he finds the stairway and goes down.

There are dusty landings and neglected chambers. They have no smell of life. But then there comes to be a smell. It is the scent of a nest, a colony of things, wild creatures, in constant proximity. He recognizes it. The light of his crimson eyes precedes him, deciphering blackness. And then other eyes, amber, green and gold, spring out like stars all across his path.

Somewhere an old torch is burning out. To the human eye, only mounds and glows would be visible, but to Feroluce, the Prince of the vampires, all is suddenly revealed. There is a great stone area, barred with bronze and iron, and things stride and growl behind the bars, or chatter and flee, or only stare. And there, without bars, though bound by ropes of brass to rings of brass, three brazen beasts.

Feroluce, on the steps of the menagerie, looks into the gaze of the Duke's lions. Feroluce smiles, and the lions roar. One is the king, its mane like war-plumes. Feroluce recognizes the king and the king's right to challenge, for this is the lions' domain, their territory.

Feroluce comes down the stair and meets the lion as it leaps the length of its chain. To Feroluce, the chain means nothing, and since he has come close enough, very little either to the lion.

To the vampire Prince the fight is wonderful, exhilarating and meaningful, intellectual even, for it is coloured by nuance, yet powerful as sex.

He holds fast with his talons, his strong limbs wrapping the beast which is almost stronger than he, just as its limbs wrap him in turn. He sinks his teeth in the lion's shoulder, and in fierce rage and bliss begins to draw out the nourishment. The lion kicks and claws at him in turn. Feroluce feels the gouges like fire along his shoulders, thighs, and hugs the lion more nearly as he throttles and drinks from it, loving it, jealous of it, killing it. Gradually the mighty feline body relaxes, still clinging to him, its cat teeth bedded in one beautiful swanlike wing, forgotten by both.

In a welter of feathers, stripped skin, spilled blood, stray semen, the lion and the angel lie in embrace on the menagerie floor. The lion lifts its head, kisses the assassin, shudders, lets go.

Feroluce glides out from under the magnificent deadweight of the cat. He stands. And pain assaults him. His lover has severely wounded him.

Across the menagerie floor, the two lionesses are crouched. Beyond them, a man stands gaping in simple terror, behind the guttering torch. He had come to feed the beasts, and seen another feeding, and now is paralyzed. He is deaf, the menagerie-keeper, previously an advantage saving him the horror of nocturnal vampire noises.

Feroluce starts towards the human animal swifter than a serpent, and checks. Agony envelops Feroluce and the stone room spins. Involuntarily, confused, he spreads his wings for flight, there in the confined chamber. But only one wing will open. The other, damaged and partly broken, hangs like a snapped fan. Feroluce cries out, a beautiful singing note of despair and anger. He drops fainting at the menagerie keeper's feet.

The man does not wait for more. He runs away through the castle, screaming invective and prayer, and reaches the Duke's hall and makes the whole hall listen.

All this while, Feroluce lies in the ocean of almost-death that is sleep or swoon, while the smaller beasts in the cages discuss him, or seem to.

And when he is raised, Feroluce does not wake. Only the great drooping bloody wings quiver and are still. Those who carry him are more than ever revolted and frightened, for they have seldom seen blood. Even the food for the menagerie is cooked almost black. Two years ago, a gardener slashed his palm on a thorn. He was banished from the court for a week.

But Feroluce, the centre of so much attention, does not rouse. Not until the dregs of the night are stealing out through the walls. Then some nervous instinct invests him. The sun is coming and this is an open place, he struggles through unconsciousness and hurt, through the deepest most bladed waters, to awareness.

And finds himself in a huge bronze cage, the cage of some animal appropriated for the occasion. Bars, bars all about him, and not to be got rid of, for he reaches to tear them away and cannot. Beyond the bars, the Duke's hall, which is only a pointless cold glitter to him in the maze of pain and dying lights. Not an open place, in fact, but too open for his kind. Through the window-spaces of thick glass, muddy sunglare must come in. To Feroluce it will be like swords, acids, and burning fire—

Far off he hears wings beat and voices soaring. His people search for him, call and wheel and find nothing.

Feroluce cries out, a gravel shriek now, and the persons in the hall rush back from him, calling on God. But Feroluce does not see. He has tried to answer his own. Now he sinks down again under the coverlet of his broken wings, and the wine-red stars of his eyes go out.

III

'And the Angel of Death,' the priest intones, 'shall surely pass over, but yet like the shadow, not substance—'

The smashed window in the old turret above the menagerie tower has been sealed with mortar and brick. It is a terrible thing that it was for so long overlooked. A miracle that only one of the creatures found and entered by it. God, the Protector, guarded the Cursed Duke and his court. And the magic that surrounds the castle, that too held fast. For from the possibility of a disaster was born a bloom of great value: Now one of the monsters is in their possession. A prize beyond price.

Caged and helpless, the fiend is at their mercy. It is also weak from its battle with the noble lion, which gave its life for the castle's safety (and will be buried with honour in an ornamented grave at the foot of the Ducal family tomb). Just before the dawn came, the Duke's advisers advised him, and the bronze cage was wheeled away into the darkest area of the hall, close by the dais where once the huge window was but is no more. A barricade of great screens was brought, and set around the cage, and the top of it covered. No sunlight now can drip into the prison to harm the specimen. Only the Duke's ladies and gentlemen steal in around the screens and see, by the light of a candlebranch, the demon still lying in its trance of pain and bloodloss. The Duke's alchemist sits on a stool nearby, dictating many notes to a nervous apprentice. The alchemist, and the apothecary for that matter, are convinced the vampire, having drunk the lion almost dry, will recover from its wounds. Even the wings will mend.

The Duke's court painter also came. He was ashamed presently, and went away. The beauty of the demon affected him, making him wish to paint it, not as something wonderfully disgusting, but as a kind of superlative man, vital and innocent, or as Lucifer himself, stricken in the sorrow of his colossal Fall. And all that has caused the painter to pity the fallen one, mere artisan that the painter is, so he slunk away. He knows, since the alchemist and the apothecary told him, what is to be done.

Of course much of the castle knows. Though scarcely anyone has slept or sought sleep, the whole place rings with excitement and vivacity. The Duke has decreed, too, that everyone who wishes shall be a witness. So he is having a progress through the castle, seeking every nook and cranny, while, let it be said, his architect takes the opportunity to check no other windowpane has cracked.

From room to room the Duke and his entourage pass, through corridors, along stairs, through dusty attics and musty storerooms he has never seen, or if seen has forgotten. Here and there some retainer is come on. Some elderly women are discovered spinning like spiders up under the eaves, half-blind and complacent. They curtsy to the Duke from a vague recollection of old habit. The Duke tells them the good news, or rather, his messenger, walking before, announces it. The ancient women sigh and whisper, are left, probably forget. Then again, in a narrow courtyard, a simple boy, who looks after a dovecote, is magnificently told. He has a fit from alarm, grasping nothing, and the doves who love and understand him (by not trying to) fly down and cover him with their soft wings as the Duke goes away. The boy comes to under the doves as if in a heap of warm snow, comforted.

It is on one of the dark staircases above the kitchen that the gleaming entourage sweeps round a bend and comes on Rohise the scullery maid, scrubbing. In these days, when there are so few children and young servants, labour is scarce, and the scullerers are not confined to the scullery.

Rohise stands up, pale with shock, and for a wild instant thinks that, for some heinous crime she has committed in ignorance, the Duke has come in person to behead her.

'Hear then, by the Duke's will,' cries the messenger. 'One of Satan's night-demons, which do torment us, has been captured and lies penned in the Duke's hall. At sunrise tomorrow, this thing will be taken to that sacred spot where grows the bush of the Flower of the Fire, and here its foul blood shall be shed. Who then can doubt the bush will blossom, and save us all, by the Grace of God.'

'And the Angel of Death,' intones the priest, on no account to be omitted, 'shall surely—'

'Wait,' says the Duke. He is as white as Rohise. 'Who is this?' he asks. 'Is it a ghost?'

The court stare at Rohise, who nearly sinks in dread, her scrubbing rag in her hand.

Gradually, despite the rag, the rags, the rough hands, the court too begins to see.

'Why, it is a marvel.'

The Duke moves forward. He looks down at Rohise and starts to cry. Rohise thinks he weeps in compassion at the awful sentence he is here to visit on her, and drops back on her knees.

'No, no,' says the Duke tenderly. 'Get up. Rise. You are so like my child, my daughter—'

Then Rohise, who knows few prayers, begins in panic to sing her little song as an orison:

Oh fleur de feu
Pour ma souffrance—

'Ah!' says the Duke. 'Where did you learn that song?'

'From my mother,' says Rohise. And, all instinct now, she sings again:

O flurda fur,
Pourma souffrance
Ned ormey par
May say day mwar—

It is the song of the fire-flower bush, the *Nona Mordica*, called Bite-Me-Not. It begins, and continues: *O flower of fire, For my misery's sake. Do not sleep but aid me; wake!* The Duke's daughter sang it very often. In those days the shrub was not needed, being just a rarity of the castle. Invoked as an amulet, on a mountain road, the rhyme itself had besides proved useless.

The Duke takes the dirty scarf from Rohise's hair. She is very, very like his lost daughter, the same pale smooth oval face, the long white neck and long dark polished eyes, and the long dark hair. (Or is it that she is very, very like the painting?)

The Duke gives instructions and Rohise is borne away.

In a beautiful chamber, the door of which has for seventeen years been locked, Rohise is bathed and her hair is washed. Oils and scents are rubbed into her skin. She is dressed in a gown of palest most pastel rose, with a girdle sewn with pearls. Her hair is combed, and on it is set a chaplet of stars and little golden leaves. 'Oh, your poor hands,' say the maids, as they trim her nails. Rohise has realized she is not to be executed. She has realized the Duke has seen her and wants to love her like his dead daughter. Slowly, an uneasy stir of something, not quite happiness, moves through Rohise. Now she will wear her pink gown, now she will sympathize with and console the Duke. Her daze lifts suddenly.

The dream has come true. She dreamed of it so often it seems quite normal. The scullery was the thing which never seemed real.

She glides down through the castle and the ladies are astonished by her grace. The carriage of her head under the starry coronet is exquisite. Her voice is quiet and clear and musical, and the foreign tone of her mother, long unremembered, is quite gone from it. Only the roughened hands give her away, but smoothed by unguents, soon they will be soft and white.

'Can it be she is truly the princess returned to flesh?'

'Her life was taken so early—yes, as they believe in the Spice-Lands, by some holy dispensation, she might return.'

'She would be about the age to have been conceived the very night the Duke's daughter d—That is, the very night the bane began—'

Theosophical discussion ensues. Songs are composed.

Rohise sits for a while with her adoptive father in the East Turret, and he tells her about the books and swords and lutes and scrolls, but not about the two portraits. Then they walk out together, in the lovely garden in the sunlight. They sit under a peach tree, and discuss many things, or the Duke discusses them. That Rohise is ignorant and uneducated does not matter at this point. She can always be trained. She has the basic requirements: docility, sweetness. There are many royal maidens in many places who know as little as she.

The Duke falls asleep under the peach tree. Rohise listens to the love-songs her own (her very own) courtiers bring her.

When the monster in the cage is mentioned, she nods as if she knows what they mean. She supposes it is something hideous, a scaring treat to be shown at dinner time, when the sun has gone down.

When the sun moves towards the western line of mountains just visible over the high walls, the court streams into the castle and all the doors are bolted and barred. There is an eagerness tonight in the concourse.

As the light dies out behind the coloured windows that have no red in them, covers and screens are dragged away from a bronze cage. It is wheeled out into the centre of the great hall.

Cannons begin almost at once to blast and bang from the roofholes. The cannoneers have had strict instructions to keep up the barrage all night without a second's pause.

Drums pound in the hall. The dogs start to bark. Rohise is not surprised by the noise, for she has often heard it from far up, in her attic, like a sea-wave breaking over and over through the lower house.

She looks at the cage cautiously, wondering what she will see. But she sees only a heap of blackness like ravens, and then a tawny dazzle, torchlight on something like human skin. 'You must not go down to look,' says the Duke protectively, as his court pours about the cage. Someone pokes between the bars with a gemmed cane, trying to rouse the nightmare which lies quiescent there. But Rohise must be spared this.

So the Duke calls his actors, and a slight, pretty play is put on throughout dinner, before the dais, shutting off from the sight of Rohise the rest of the hall, where the barbaric gloating and goading of the court, unchecked, increases.

IV

The Prince Feroluce becomes aware between one second and the next. It is the sound—heard beyond all others—of the wings of his people beating at the stones of the castle. It is the wings which speak to him, more than their wild orchestral voices. Besides these sensations, the anguish of healing and the sadism of humankind are not much.

Feroluce opens his eyes. His human audience, pleased, but afraid and squeamish, backs away, and asks each other for the two thousandth time if the cage is quite secure. In the torchlight the eyes of Feroluce are more black than red. He stares about. He is, though captive, imperious. If he were a lion or a bull, they would admire this 'nobility'. But the fact is, he is too much like a man, which serves to point up his supernatural differences unbearably.

Obviously, Feroluce understands the gist of his plight. Enemies have him penned. He is a show for now, but ultimately to be killed, for with the intuition of the raptor he divines everything. He had thought the sunlight would kill him, but that is a distant matter, now. And beyond all, the voices and the voices of the wings of his kindred beat the air outside this room-caved mountain of stone.

And so, Feroluce commences to sing, or at least, this is how it seems to the rabid court and all the people gathered in the hall. It seems he sings. It is the great communing call of his kind, the art and science and religion of the winged vampires, his means of telling them, or attempting to tell them, what they must be told before he dies. So the sire of Feroluce sang, and the grandsire, and each of his ancestors. Generally they died in flight, falling angels spun down the gulches and enormous stairs of distant peaks, singing. Feroluce, immured, believes that his cry is somehow audible.

To the crowd in the Duke's hall the song is merely that, a song, but how glorious. The dark silver voice, turning to bronze or gold, whitening in the higher registers. There seem to be words, but in some other tongue. This is how the planets sing, surely, or mysterious creatures of the sea.

Everyone is bemused. They listen, astonished.

No one now remonstrates with Rohise when she rises and steals down from the dais. There is an enchantment which prevents movement and coherent thought. Of all the roomful, only she is drawn forward. So she comes close, unhindered, and between the bars of the cage, she sees the vampire for the first time.

She has no notion what he can be. She imagined it was a monster or a monstrous beast. But it is neither. Rohise, starved for so long of beauty and always dreaming of it, recognizes Feroluce inevitably as part of the dream-come-true. She loves him instantly. Because she loves him, she is not afraid of him.

She attends while he goes on and on with his glorious song. He does not see her at all, or any of them. They are only things, like mist, or pain. They have no character or personality or worth; abstracts.

Finally, Feroluce stops singing. Beyond the stone and the thick glass of the siege, the wing-beats, too, eddy into silence.

Finding itself mesmerized, silent by night, the court comes to with a terrible joint start, shrilling and shouting, bursting, exploding into a compensation of sound. Music flares again. And the cannons in the roof, which have also fallen quiet, resume with a tremendous roar.

Feroluce shuts his eyes and seems to sleep. It is his preparation for death.

Hands grasp Rohise. 'Lady—step back, come away. So close! It may harm you—'

The Duke clasps her in a father's embrace. Rohise, unused to this sort of physical expression, is unmoved. She pats him absently.

'My lord, what will be done?'

'Hush, child. Best you do not know.'

Rohise persists.

The Duke persists in not saying.

But she remembers the words of the herald on the stair, and knows they mean to butcher the winged man. She attends thereafter more carefully to snatches of the bizarre talk about the hall, and learns all she needs. At earliest sunrise, as soon as the enemy retreat from the walls, their captive will be taken to the lovely garden with the peach trees. And so to the sunken garden of the magic bush, the fire-flower. And there they will hang him up in the sun through the dome of smoky glass, which will be slow murder to him, but they will cut him, too, so his blood, the stolen blood of the vampire, runs down to water the roots of the fleur de feu. And who can doubt that, from such nourishment, the bush will bloom? The blooms are salvation. Wherever they grow it is a safe place. Whoever wears them is safe from the draining bite of demons. Bite-Me-Not, they call it; vampire-repellent.

Rohise sits the rest of the night on her cushions, with folded hands, resembling the portrait of the princess, which is not like her.

Eventually the sky outside alters. Silence comes down beyond the wall, and so within the wall, and the court lifts its head, a corporate animal scenting day.

At the intimation of sunrise the black plague has lifted and gone away, and might never have been. The Duke, and almost all his castle full of men, women, children, emerge from the doors. The sky is measureless and bluely grey, with one cherry rift in the east that the court refers to as 'mauve', since dawns and sunsets are never any sort of red here.

They move through the dimly lightening garden as the last stars melt. The cage is dragged in their midst.

They are too tired, too concentrated now, the Duke's people, to continue baiting their captive. They have had all the long night to do that, and to drink and opine, and now their stamina is sharpened for the final act.

Reaching the sunken garden, the Duke unlocks the iron door. There is no room for everyone within, so mostly they must stand outside, crammed in the gate, or teetering on erections of benches that have been placed around, and peering in over the walls through the glass of the dome. The places in the doorway are the best, of course; no one else will get so good a view. The servants and lower persons must stand back under the trees and only imagine what goes on. But they are used to that.

Into the sunken garden itself there are allowed to go the alchemist and the apothecary, and the priest, and certain sturdy soldiers attendant on the Duke, and the Duke. And Feroluce in the cage.

The east is all 'mauve' now. The alchemist has prepared sorcerous safeguards which are being put into operation, and the priest, never to be left out, intones prayers. The bulge-thewed soldiers open the cage and seize the monster before it can stir. But drugged smoke has already been wafted into the prison, and besides, the monster has prepared itself for hopeless death and makes no demur.

Feroluce hangs in the arms of his loathing guards, dimly aware the sun is near. But death is nearer, and already one may hear the alchemist's apprentice sharpening the knife an ultimate time.

The leaves of the *Nona Mordica* are trembling, too, at the commencement of the light, and beginning to unfurl. Although this happens every dawn, the court points to it with optimistic cries. Rohise, who has claimed a position in the doorway, watches it too, but only for an instant. Though she has sung of the fleur de fur since childhood, she had never known what the song was all about. And in just this way, though she has dreamed of being the Duke's daughter most of her life, such an event was never really comprehended either, and so means very little.

As the guards haul the demon forward to the plot of humid earth where the bush is growing, Rohise darts into the sunken garden, and lightning leaps in her hands. Women scream and well they might. Rohise has stolen one of the swords from the East Turret, and now she flourishes it, and now she has swung it and a soldier falls, bleeding red, red, *red*, before them all.

Chaos enters, as in yesterday's play, shaking its tattered sleeves. The men who hold the demon rear back in horror at the dashing blade and the blasphemous gore, and the mad girl in her princess's gown. The Duke makes a pitiful bleating noise, but no one pays him any attention.

The east glows in and like the liquid on the ground.

Meanwhile, the ironically combined sense of impending day and spilled hot blood have penetrated the stunned brain of the vampire. His eyes open and he sees the girl wielding her sword in a spray of crimson as the last guard lets go. Then the girl has run to Feroluce. Though, or because, her face is insane, it communicates her purpose, as she thrusts the sword's hilt into his hands.

No one has dared approach either the demon or the girl. Now they look on in horror and in horror grasp what Feroluce has grasped.

In that moment the vampire springs, and the great swanlike wings are reborn at his back, healed and whole. As the doctors predicted, he has mended perfectly, and prodigiously fast. He takes to the air like an arrow, unhindered, as if gravity does not any more exist. As he does so, the girl grips him about the waist, and slender and light, she is drawn upward too. He does not glance at her. He veers towards the gateway, and tears through it, the sword, his talons, his wings, his very shadow beating men and bricks from his path.

And now he is in the sky above them, a black star which has not been put out. They see the wings flare and beat, and the swirling of a girl's dress and unbound hair, and then the image dives and is gone into the shade under the mountains, as the sun rises.

V

It is fortunate, the mountain shade in the sunrise. Lion's blood and enforced quiescence have worked wonders, but the sun could undo it all. Luckily the shadow, deep and cold as a pool, envelops the vampire, and in it there is a cave, deeper and colder. Here he alights and sinks down, sloughing the girl, whom he has almost forgotten. Certainly he fears no harm from her. She is like a pet animal, maybe, like the hunting dogs or wolves or lammergeyers that occasionally the unkindness of vampires have kept by them for a while. That she helped him is all he needs to know. She will help again. So when, stumbling in the blackness, she brings him in her cupped hands water from a cascade at the poolcave's back, he is not surprised. He drinks the water, which is the only other substance his kind imbibe. Then he smooths her hair, absently, as he would pat or stroke the pet she seems to have become. He is not grateful, as he is not suspicious. The complexities of his intellect are reserved for other things. Since he is exhausted he falls asleep, and since Rohise is exhausted she falls asleep beside him, pressed to his warmth in the freezing dark. Like those of Feroluce, as it turns out, her thoughts are simple. She is sorry for distressing the Cursed Duke. But she has no regrets, for she could no more have left Feroluce to die than she could have refused to leave the scullery for the court.

The day, which had only just begun, passes swiftly in sleep.

Feroluce wakes as the sun sets, without seeing anything of it. He unfolds himself and goes to the cave's entrance, which now looks out on a whole sky of stars above a landscape of mountains. The castle is far below, and to the eyes of Rohise as she follows him, invisible. She does not even look for it, for there is something else to be seen.

The great dark shapes of angels are wheeling against the peaks, the stars. And their song begins, up in the starlit spaces. It is a lament, their mourning, pitiless and strong, for Feroluce, who has died in the stone heart of the thing they prey upon.

The tribe of Feroluce do not laugh, but, like a bird or wild beast, they have a kind of equivalent to laughter. This Feroluce now utters, and like a flung lance he launches himself into the air.

Rohise at the cave mouth, abandoned, forgotten, unnoted even by the mass of vampires, watches the winged man as he flies towards his people. She supposes for a moment that she may be able to climb down the tortuous ways of the mountain, undetected. Where then should she go? She does not spend much time on these ideas. They do not interest or involve her. She watches Feroluce and, because she learned long ago the uselessness of weeping, she does not shed tears, though her heart begins to break.

As Feroluce glides, body held motionless, wings outspread on a down-draught, into the midst of the storm of black wings, the red stars of eyes ignite all about him. The great lament dies. The air is very still.

Feroluce waits then. He waits, for the aura of his people is not as he has always known it. It is as if he had come among emptiness. From the silence, therefore, and from nothing else, he learns it all. In the stone he lay and he sang of his death, as the Prince must, dying. And the ritual was completed, and now there is the threnody, the grief, and thereafter the choosing of a new Prince. And none of this is alterable. He is dead. Dead. It cannot and will not be changed.

There is a moment of protest, then, from Feroluce. Perhaps his brief sojourn among men has taught him some of their futility. But as the cry leaves him, all about the huge wings are raised like swords. Talons and teeth and eyes burn against the stars. To protest is to be torn in shreds. He is not of their people now. They can attack and slaughter him as they would any other intruding thing. *Go*, the talons and the teeth and the eyes say to him. *Go far off.*

He is dead. There is nothing left him but to die.

Feroluce retreats. He soars. Bewildered, he feels the power and energy of his strength and the joy of flight, and cannot understand how this is, if he is dead. Yet he *is* dead. He knows it now.

So he closes his eyelids, and his wings. Spear swift he falls. And something shrieks, interrupting the reverie of nihilism. Disturbed, he opens his wings, shudders, turns like a swimmer, finds a ledge against his side and two hands outstretched, holding him by one shoulder, and by his hair.

'No,' says Rohise. (The vampire cloud, wheeling away, have not heard her; she does not think of them.) His eyes stay shut. Holding him, she kisses these eyelids, his forehead, his lips, gently, as she drives her nails into his skin to hold him. The black wings beat, tearing to be free and fall and die. 'No,' says Rohise. 'I love you,' she says. 'My life is your life.' These are the words of the court and of courtly love songs. No matter, she means them. And though he cannot understand her language or her sentiments, yet her passion, purely that, communicates itself, strong and burning as the passions of his kind, who generally love only one thing, which is scarlet. For a second her intensity fills the void which now contains him. But then he dashes himself away from the ledge, to fall again, to seek death again.

Like a ribbon, clinging to him still, Rohise is drawn from the rock and falls with him.

Afraid, she buries her head against his breast, in the shadow of wings and hair. She no longer asks him to reconsider. This is how it must be. *Love* she thinks again, in the instant before they strike the earth. Then that instant comes, and is gone.

Astonished, she finds herself still alive, still in the air. Touching so close feathers have been left on the rocks, Feroluce has swerved away, and upward. Now, conversely, they are whirling towards the very stars. The world seems miles below. Perhaps they will fly into space itself. Perhaps he means to break their bones instead on the cold face of the moon.

He does not attempt to dislodge her, he does not attempt any more to fall and die. But as he flies, he suddenly cries out, terrible lost lunatic cries.

They do not hit the moon. They do not pass through the stars like static rain.

But when the air grows thin and pure there is a peak like a dagger standing in their path. Here, he alights. As Rohise lets go of him, he turns away. He stations himself, sentry-fashion, in the manner of his tribe, at the edge of the pinnacle. But watching for nothing. He has not been able to choose death. His strength and the strong will of another, these have hampered him. His brain has become formless darkness. His eyes glare, seeing nothing.

Rohise, gasping a little in the thin atmosphere, sits at his back, watching for him, in case any harm may come near him.

At last, harm does come. There is a lightening in the east. The frozen, choppy sea of the mountains below and all about, grows visible. It is a marvellous sight, but holds no marvel for Rohise. She averts her eyes from the exquisitely pencilled shapes, looking thin and translucent as paper, the rivers of mist between, the glimmer of nacreous ice. She searches for a blind hole to hide in.

There is a pale yellow wound in the sky when she returns. She grasps Feroluce by the wrist and tugs at him. 'Come,' she says. He looks at her vaguely, as if seeing her from the shore of another country. 'The sun,' she says. 'Quickly.'

The edge of the light runs along his body like a razor. He moves by instinct now, following her down the slippery dagger of the peak, and so eventually into a shallow cave. It is so small it holds him like a coffin. Rohise closes the entrance with her own body. It is the best she can do. She sits facing the sun as it rises, as if prepared to fight. She hates the sun for his sake. Even as the light warms her chilled body, she curses it. Till light and cold and breathlessness fade together.

When she wakes, she looks up into twilight and endless stars, two of which are red. She is lying on the rock by the cave. Feroluce leans over her, and behind Feroluce his quiescent wings fill the sky.

She has never properly understood his nature: Vampire. Yet her own nature, which tells her so much, tells her some vital part of herself is needful to him, and that he is danger, and death. But she loves him, and is not afraid. She would have fallen to die with him. To help him by her death does not seem wrong to her. Thus, she lies still, and smiles at him to reassure him she will not struggle. From lassitude, not fear, she closes her eyes. Presently she feels the soft weight of hair brush by her cheek, and then his cool mouth rests against her throat. But nothing more happens. For some while, they continue in this fashion, she yielding, he kneeling over her, his lips on her skin. Then he moves a little away. He sits, regarding her. She, knowing the unknown act has not been completed, sits up in turn. She beckons to him mutely, telling him with her gestures and her expression *I consent. Whatever is necessary.* But he does not stir. His eyes blaze, but even of these she has no fear. In the end he looks away from her, out across the spaces of the darkness.

He himself does not understand. It is permissible to drink from the body of a pet, the wolf, the eagle. Even to kill the pet, if need demands. Can it be, outlawed from his people, he has lost their composite soul? Therefore, is he soulless now? It does not seem to him he is. Weakened and famished though he is, the vampire is aware of a wild tingling of life. When he stares at the creature which is his food, he finds he sees her differently. He has borne her through the sky, he has avoided death, by some intuitive process, for her sake, and she has led him to safety, guarded him from the blade of the sun. In the beginning it was she who rescued him from the human things which had taken him. She cannot be human, then. Not pet, and not prey. For no, he could not drain her of blood, as he would not seize upon his own kind, even in combat, to drink and feed. He starts to see her as beautiful, not in the way a man beholds a woman, certainly, but as his kind revere the sheen of water in dusk, or flight, or song. There are no words for this. But the life goes on tingling through him. Though he is dead, life.

In the end, the moon does rise, and across the open face of it something wheels by. Feroluce is less swift than was his wont, yet he starts in pursuit, and catches and brings down, killing on the wing, a great night bird. Turning in the air, Feroluce absorbs its liquors. The heat of life now, as well as its assertion, courses through him. He returns to the rock perch, the glorious flaccid bird dangling from his hand. Carefully, he tears the glory of the bird in pieces, plucks the feathers, splits the bones. He wakes the companion (asleep again from weakness) who is not pet or prey, and feeds her morsels of flesh. At first she is unwilling. But her hunger is so enormous and her nature so untamed that quite soon she accepts the slivers of raw fowl.

Strengthened by blood, Feroluce lifts Rohise and bears her gliding down the moon-slit quill-backed land of the mountains, until there is a rocky cistern full of cold, old rains. Here they drink together. Pale white primroses grow in the fissures where the black moss drips. Rohise makes a garland and throws it about the head of her beloved when he does not expect it. Bewildered but disdainful, he touches at the wreath of primroses to see if it is likely to threaten or hamper him. When it does not, he leaves it in place.

Long before dawn this time, they have found a crevice. Because it is so cold, he folds his wings about her. She speaks of her love to him, but he does not hear, only the murmur of her voice, which is musical and does not displease him. And later, she sings him sleepily the little song of the fleur de fur.

VI

There comes a time then, brief, undated, chartless time, when they are together, these two creatures. Not together in any accepted sense, of course, but together in the strange feeling or emotion, instinct or ritual, that can burst to life in an instant or flow to life gradually across half a century, and which men call *Love*.

They are not alike. No, not at all. Their differences are legion and should be unpalatable. He is a supernatural thing and she a human thing, he was a lord and she a scullery sloven. He can fly, she cannot fly. And he is male, she female. What other items are required to make them enemies? Yet they are bound, not merely by love, they are bound by all they are, the very stumbling blocks. Bound, too, because they are doomed. Because the stumbling blocks have doomed them; everything has. Each has been exiled out of their own kind. Together, they cannot even communicate with each other, save by looks, touches, sometimes by sounds, and by songs neither understands, but which each comes to value since the other appears to value them, and since they give expression to that other. Nevertheless, the binding of the doom, the greatest binding, grows, as it holds them fast to each other, mightier and stronger.

Although they do not know it, or not fully, it is the awareness of doom that keeps them there, among the platforms and steps up and down, and the inner cups, of the mountains.

Here it is possible to pursue the airborne hunt, and Feroluce may now and then bring down a bird to sustain them both. But birds are scarce. The richer lower slopes, pastured with goats, wild sheep, and men—they lie far off and far down from this place as a deep of the sea. And Feroluce does not conduct her there, nor does Rohise ask that he should, or try to lead the way, or even dream of such a plan.

But yes, birds are scarce, and the pastures far away, and winter is coming. There are only two seasons in these mountains. High summer, which dies, and the high cold which already treads over the tips of the air and the rock, numbing the sky, making all brittle, as though the whole landscape might snap in pieces, shatter.

How beautiful it is to wake with the dusk, when the silver webs of night begin to form, frost and ice, on everything. Even the ragged dress—once that of a princess—is tinselled and shining with this magic substance, even the mighty wings—once those of a prince—each feather is drawn glittering with thin rime. And oh, the sky, thick as a daisy-field with the white stars. Up there, when they have fed and have strength, they fly, or, Feroluce flies and Rohise flies in his arms, carried by his wings. Up there in the biting chill like a pane of ghostly vitreous, they have become lovers, true blind lovers, embraced and linked, their bodies a bow, coupling on the wing. By the hour that this first happened the girl had forgotten all she had been, and he had forgotten too that she was anything but the essential mate. Sometimes, borne in this way, by wings and by fire, she cries out as she hangs in the ether. These sounds, transmitted through the flawless silence and amplification of the peaks, scatter over tiny half-buried villages countless miles away, where they are heard in fright and taken for the shrieks of malign invisible devils, tiny as bats, and armed with the barbed stings of scorpions. There are always misunderstandings.

After a while, the icy prologues and the stunning starry fields of winter nights give way to the main argument of winter.

The liquid of the pool, where the flowers made garlands, has clouded and closed to stone. Even the volatile waterfalls are stilled, broken cascades of glass. The wind tears through the skin and hair to gnaw the bones. To weep with cold earns no compassion of the cold.

There is no means to make fire. Besides, the one who was Rohise is an animal now, or a bird, and beasts and birds do not make fire, save for the phoenix in the Duke's bestiary. Also, the sun is fire, and the sun is a foe. Eschew fire.

There begin the calendar months of hibernation. The demon lovers too must prepare for just such a measureless winter sleep, that gives no hunger, asks no action. There is a deep cave they have lined with feathers and withered grass. But there are no more flying things to feed them. Long, long ago, the last warm frugal feast, long, long ago the last flight, joining, ecstasy, and song. So, they turn to their cave, to stasis, to sleep. Which each understands, wordlessly, thoughtlessly, is death.

What else? He might drain her of blood, he could persist some while on that, might even escape the mountains, the doom. Or she herself might leave him, attempt to make her way to the places below, and perhaps she could reach them, even now. Others, lost here, have done so. But neither considers these alternatives. The moment for all that is past. Even the death-lament does not need to be voiced again.

Installed, they curl together in their bloodless, icy nest, murmuring a little to each other, but finally still.

Outside, the snow begins to come down. It falls like a curtain. Then the winds take it. Then the night is full of the lashing of whips, and when the sun rises it is white as the snow itself, its flame very distant, giving nothing. The cave mouth is blocked up with snow. In the winter, it seems possible that never again will there be a summer in the world.

Behind the modest door of snow, hidden and secret, sleep is quiet as stars, dense as hardening resin. Feroluce and Rohise turn pure and pale in the amber, in the frigid nest, and the great wings lie like a curious articulated machinery that will not move. And the withered grass and the flowers are crystallized, until the snows shall melt.

At length, the sun deigns to come closer to the earth, and the miracle occurs. The snow shifts, crumbles, crashes off the mountains in rage. The waters hurry after the snow, the air is wrung and racked by splittings and splinterings, by rushes and booms. It is half a year, or it might be a hundred years, later.

Open now, the entry to the cave. Nothing emerges. Then, a flutter, a whisper. Something does emerge. One black feather, and caught in it, the petal of a flower, crumbling like dark charcoal and white, drifting away into the voids below. Gone. Vanished. It might never have been.

But there comes another time (half a year, a hundred years), when an adventurous traveller comes down from the mountains to the pocketed villages the other side of them. He is a swarthy cheerful fellow, you would not take him for herbalist or mystic, but he has in a pot a plant he found high up in the staring crags, which might after all contain anything or nothing. And he shows the plant, which is an unusual one, having slender, dark and velvety leaves, and giving off a pleasant smell like vanilla. 'See, the *Nona Mordica*,' he says. 'The Bite-Me-Not. The flower that repels vampires.'

Then the villagers tell him an odd story, about a castle in another country, besieged by a huge flock, a menace of winged vampires, and how the Duke waited in vain for the magic bush that was in his garden, the Bite-Me-Not, to flower and save them all. But it seems there was a curse on this Duke, who on the very night his daughter was lost, had raped a serving woman, as he had raped others before. But this woman conceived. And bearing the fruit, or flower, of this rape, damaged her, so she lived only a year or two after it. The child grew up unknowing, and in the end betrayed her own father by running away to the vampires, leaving the Duke demoralized. And soon after he went mad, and himself stole out one night, and let the winged fiends into his castle, so all there perished.

'Now if only the bush had flowered in time, as your bush flowers, all would have been well,' the villagers cry.

The traveller smiles. He in turn does not tell them of the heap of peculiar bones, like parts of eagles mingled with those of a woman and a man. Out of the bones, from the heart of them, the bush was rising, but the traveller untangled the roots of it with care; it looks sound enough now in its sturdy pot, all of it twining together. It seems as if two separate plants are growing from a single stem, one with blooms almost black, and one pink-flowered, like a young sunset.

'Flur de fur,' says the traveller, beaming at the marvel, and his luck.

Fleur de feu. Oh flower of fire. That fire is not hate or fear, which makes flowers come, not terror or anger or lust, it is love that is the fire of the Bite-Me-Not, love which cannot abandon, love which cannot harm. Love which never dies.

LUCIUS SHEPARD

THE NIGHT OF WHITE BHAIRAB

Whenever Mr Chatterji went to Delhi on business, twice yearly, he would leave Eliot Blackford in charge of his Katmandu home, and prior to each trip, the transfer of keys and instructions would be made at the Hotel Anapurna. Eliot—an angular, sharp-featured man in his mid-thirties, with thinning blond hair and a perpetually ardent expression—knew Mr Chatterji for a subtle soul, and he suspected that this subtlety had dictated the choice of meeting place. The Anapurna was the Nepalese equivalent of a Hilton, its bar equipped in vinyl and plastic, with a choir-like arrangement of bottles fronting the mirror. Lights were muted, napkins monogrammed. Mr Chatterji, plump and prosperous in a business suit, would consider it an elegant refutation of Kipling's famous couplet ('East is East', etc.) that he was at home here, whereas Eliot, wearing a scruffy robe and sandals was not; he would argue that not only had the twain met, they had actually exchanged places. It was Eliot's own measure of subtlety that restrained him from pointing out what Mr Chatterji could not perceive: that the Anapurna was a skewed version of the American Dream. The carpeting was indoor-outdoor runner; the menu was rife with ludicrous misprints (*Skotch Miss*, *Screwdiver*), and the lounge act—two turbaned, tuxedoed Indians on electric guitar and traps—was managing to turn 'Evergreen' into a doleful raga.

'There will be one important delivery.' Mr Chatterji hailed the waiter and nudged Eliot's shot glass forward. 'It should have been here days ago, but you know these customs people.' He gave an effeminate shudder to express his distaste for the bureaucracy, and cast an expectant eye on Eliot, who did not disappoint.

'What is it?' he asked, certain that it would be an addition to Mr Chatterji's collection: he enjoyed discussing the collection with Americans; it proved that he had an overview of their culture.

'Something delicious!' said Mr Chatterji. He took the tequila bottle from the waiter and—with a fond look—passed it to Eliot. 'Are you familiar with the Carversville Terror?'

'Yeah, sure.' Eliot knocked back another shot. 'There was a book about it.'

'Indeed,' said Mr Chatterji. 'A best seller. The Cousineau mansion was once the most notorious haunted house of your New England. It was torn down several months ago, and I've succeeded in acquiring the fireplace, which'—he sipped his drink—'which was the locus of power. I'm very fortunate to have obtained it.' He fitted his glass into the circle of moisture on the bar and waxed scholarly. 'Aimée Cousineau was a most unusual spirit, capable of a variety of. . . .'

Eliot concentrated on his tequila. These recitals never failed to annoy him, as did—for different reasons—the sleek Western disguise. When Eliot had arrived in Katmandu as a member of the Peace Corps, Mr Chatterji had presented a far less pompous image: a scrawny kid dressed in Levi's that he had wheedled from a tourist. He'd been one of the hangers-on—mostly young Tibetans—who frequented the grubby tea rooms on Freak Street, watching the American hippies giggle over their hash yogurt, lusting after their clothes, their women, their entire culture. The hippies had respected the Tibetans: they were a people of legend, symbols of the occultism then in vogue, and the fact that they liked James Bond movies, fast cars, and Jimi Hendrix had increased the hippies' self-esteem. But they had found laughable the fact that Ranjeesh Chatterji—another Westernized Indian—had liked these same things, and they had treated him with mean condescension. Now, thirteen years later, the roles had been reversed; it was Eliot who had become the hanger-on.

He had settled in Katmandu after his tour was up, his idea being to practise meditation, to achieve enlightenment. But it had not gone well. There was an impediment in his mind—he pictured it as a dark stone, a stone compounded of worldly attachments—that no amount of practice could wear down, and his life had fallen into a futile pattern. He would spend ten months of the year living in a small room near the temple of Swayambhunath, meditating, rubbing away at the stone; and then, during March and September, he would occupy Mr Chatterji's house and debauch himself with liquor and sex and drugs. He was aware that Mr Chatterji considered him a burnout, that the position of caretaker was in effect a form of revenge, a means by which his employer could exercise his own brand of condescension; but Eliot minded neither the label nor the attitude. There were worse things to be than a burnout in Nepal. It was beautiful country, it was inexpensive, it was far from Minnesota (Eliot's home). And the concept of personal failure was meaningless here. You lived, died, and were reborn over and over until at last you attained the ultimate success of non-being: a terrific consolation for failure.

'. . . yet in your country,' Mr Chatterji was saying, 'evil has a sultry character. Sexy! It's as if the spirits were adopting vibrant personalities in order to contend with pop groups and movie stars.'

Eliot thought of a comment, but the tequila backed up on him and he belched instead. Everything about Mr Chatterji—teeth, eyes, hair, gold rings—seemed to be gleaming with extraordinary brilliance. He looked as unstable as a soap bubble, a fat little Hindu illusion.

Mr Chatterji clapped a hand to his forehead. 'I nearly forgot. There will be another American staying at the house. A girl. Very shapely!' He shaped an hourglass in the air. 'I'm quite mad for her, but I don't know if she's trustworthy. Please see she doesn't bring in any strays.'

'Right,' said Eliot. 'No problem.'

'I believe I will gamble now,' said Mr Chatterji, standing and gazing towards the lobby. 'Will you join me?'

'No, I think I'll get drunk. I guess I'll see you in October.'

'You're drunk already, Eliot.' Mr Chatterji patted him on the shoulder. 'Hadn't you noticed?'

Early the next morning, hung over, tongue cleaving to the roof of his mouth, Eliot sat himself down for a final bout of trying to visualize the Avalokitesvara Buddha. All the sounds outside—the buzzing of a motor scooter, birdsong, a girl's laughter—seemed to be repeating the mantra, and the grey stone walls of his room looked at once intensely real and yet incredibly fragile, papery, a painted backdrop he could rip with his hands. He began to feel the same fragility, as if he were being immersed in a liquid that was turning him opaque, filling him with clarity. A breath of wind could float him out the window, drift him across the fields, and he would pass through the trees and mountains, all the phantoms of the material world . . . but then a trickle of panic welled up from the bottom of his soul, from that dark stone. It was beginning to smoulder, to give off poison fumes: a little briquette of anger and lust and fear. Cracks were spreading across the clear substance he had become, and if he didn't move soon, if he didn't break off the meditation, he would shatter.

He toppled out of the lotus position and lay propped on his elbows. His heart raced, his chest heaved, and he felt very much like screaming his frustration. Yeah, that was a temptation. To just say the hell with it and scream, to achieve through chaos what he could not through clarity: to empty himself into the scream. He was trembling, his emotions flowing between self-hate and self-pity. Finally, he struggled up and put on jeans and a cotton shirt. He knew he was close to a breakdown, and he realized that he usually reached this point just before taking up residence at Mr Chatterji's. His life was a frayed thread stretched tight between those two poles of debauchery. One day it would snap.

'The hell with it,' he said. He stuffed the remainder of his clothes into a duffel bag and headed into town.

Walking through Durbar Square—which wasn't really a square but a huge temple complex interspersed with open areas and wound through by cobbled paths—always put Eliot in mind of his brief stint as a tour guide, a career cut short when the agency received complaints about his eccentricity ('. . . As you pick your way among the piles of human waste and fruit rinds, I caution you not to breathe too deeply of the divine afflatus; otherwise, it may forever numb you to the scent of Prairie Cove or Petitpoint Gulch or whatever citadel of gracious living it is that you call home. . . .') It had irked him to have to lecture on the carvings and history of the square, especially to the just-plain-folks who only wanted a Polaroid of Edna or Uncle Jimmy standing next to that weird monkey god on the pedestal. The square was a unique place, and in Eliot's opinion, such unenlightened tourism demeaned it.

Pagoda-style temples of red brick and dark wood towered on all sides, their finials rising into brass lightning bolts. They were alien-looking—you half-expected the sky above them to be of an otherworldly colour and figured by several moons. Their eaves and window screens were ornately carved into the images of gods and demons, and behind a large window screen on the temple of White Bhairab lay the mask of that god. It was almost ten feet high, brass, with a fanciful headdress and long-lobed ears and a mouth full of white fangs; its eyebrows were enamelled red, fiercely arched, but the eyes had the goofy quality common to Newari gods—no matter how wrathful they were, there was something essentially friendly about them, and they reminded Eliot of cartoon germs. Once a year—in fact, a little more than a week from now—the screens would be opened, a pipe would be inserted into the god's mouth, and rice beer would jet out into the mouths of the milling crowds; at some point a fish would be slipped into the pipe, and whoever caught it would be deemed the luckiest soul in the Katmandu Valley for the next year. It was one of Eliot's traditions to make a try for the fish, though he knew that it wasn't luck he needed.

Beyond the square, the streets were narrow, running between long brick buildings three and four storeys tall, each divided into dozens of separate dwellings. The strip of sky between the roofs was bright, burning blue—a void colour—and in the shade the bricks looked purplish. People hung out the windows of the upper storeys, talking back and forth: an exotic tenement life. Small shrines—wooden enclosures containing statuary of stucco or brass—were tucked into wall niches and the mouths of alleys. The gods were everywhere in Katmandu, and there was hardly a corner to which their gaze did not penetrate.

On reaching Mr Chatterji's, which occupied half a block-long building, Eliot made for the first of the interior courtyards; a stair led up from it to Mr Chatterji's apartment, and he thought he would check on what had been left to drink. But as he entered the courtyard—a phalanx of jungly plants arranged around a lozenge of cement—he saw the girl and stopped short. She was sitting in a lawn chair, reading, and she was indeed very shapely. She wore loose cotton trousers, a T-shirt, and a long white scarf shot through with golden threads. The scarf and the trousers were the uniform of the young travellers who generally stayed in the expatriate enclave of Temal: it seemed that they all bought them immediately upon arrival in order to identify themselves to each other. Edging closer, peering between the leaves of a rubber plant, Eliot saw that the girl was doe-eyed, with honey-coloured skin and shoulder-length brown hair interwoven by lighter strands. Her wide mouth had relaxed into a glum expression. Sensing him, she glanced up, startled; then she waved and set down her book.

'I'm Eliot,' he said, walking over.

'I know. Ranjeesh told me.' She stared at him incuriously.

'And you?' He squatted beside her.

'Michaela.' She fingered the book, as if she were eager to get back to it.

'I can see you're new in town.'

'How's that?'

He told her about the clothes, and she shrugged. 'That's what I am,' she said. 'I'll probably always wear them.' She folded her hands on her stomach: it was a nicely rounded stomach, and Eliot—a connoisseur of women's stomachs—felt the beginnings of arousal.

'Always?' he said. 'You plan on being here that long?'

'I don't know.' She ran a finger along the spine of the book. 'Ranjeesh asked me to marry him, and I said maybe.'

Eliot's infant plan of seduction collaped beneath this wrecking ball of a statement, and he failed to hide his incredulity. 'You're in love with Ranjeesh?'

'What's that got to with it?' A wrinkle creased her brow: it was the perfect symptom of her mood, the line a cartoonist might have chosen to express petulant anger.

'Nothing. Not if it doesn't have anything to do with it.' He tried a grin, but to no effect. 'Well,' he said after a pause. 'How do you like Katmandu?'

'I don't get out much,' she said flatly.

She obviously did not want conversation, but Eliot wasn't ready to give up. 'You ought to,' he said. 'The festival of Indra Jatra's about to start. It's pretty wild. Especially on the night of White Bhairab. Buffalo sacrifices, torchlight . . .'

'I don't like crowds,' she said.

Strike two.

Eliot strained to think of an enticing topic, but he had the idea it was a lost cause. There was something inert about her, a veneer of listlessness redolent of Thorazine, of hospital routine. 'Have you seen the Khaa?' he asked.

'The what?'

'The Khaa. It's a spirit . . . though some people will tell you it's partly animal, because over here the animal and spirit worlds overlap. But whatever it is, all the old houses have one, and those that don't are considered unlucky. There's one here.'

'What's it look like?'

'Vaguely anthropomorphic. Black, featureless. Kind of a living shadow. They can stand upright, but they roll instead of walk.'

She laughed. 'No, I haven't seen it. Have you?'

'Maybe,' said Eliot. 'I thought I saw it a couple of times, but I was pretty stoned.'

She sat up straighter and crossed her legs; her breasts jiggled and Eliot fought to keep his eyes centred on her face. 'Ranjeesh tells me you're a little cracked,' she said.

Good ol' Ranjeesh! He might have known that the son of a bitch would have sandbagged him with his new lady. 'I guess I am,' he said, preparing for the brush-off. 'I do a lot of meditation, and sometimes I teeter on the edge.'

But she appeared more intrigued by this admission than by anything else he had told her; a smile melted up from her carefully composed features. 'Tell me some more about the Khaa,' she said.

Eliot congratulated himself. 'They're quirky sorts,' he said. 'Neither good nor evil. They hide in dark corners, though now and then they're seen in the streets or in the fields out near Jyapu. And the oldest ones, the most powerful ones, live in the temples in Durbar Square. There's a story about the one here that's descriptive of how they operate . . . if you're interested.'

'Sure.' Another smile.

'Before Ranjeesh bought this place, it was a guesthouse, and one night a woman with three goitres on her neck came to spend the night. She had two loaves of bread that she was taking home to her family, and she stuck them under her pillow before going to sleep. Around midnight the Khaa rolled into her room and was struck by the sight of her goitres rising and falling as she breathed. He thought they'd make a beautiful necklace, so he took them and put them on his own neck. Then he spotted the loaves sticking out from her pillow. They looked good, so he took them as well and replaced them with two loaves of gold. When the woman woke, she was delighted. She hurried back to her village to tell her family, and on the way, she met a friend, a woman, who was going to market. This woman had four goitres. The first woman told her what had happened, and that night the second woman went to the guesthouse and did exactly the same things. Around midnight the Khaa rolled into her room. He'd grown bored with his necklace, and he gave it to the woman. He'd also decided that bread didn't taste very good, but he still had a loaf and he figured he'd give it another chance. So in exchange for the necklace, he took the woman's appetite for bread. When she woke, she had seven goitres, no gold, and she could never eat bread again the rest of her life.'

Eliot had expected a response of mild amusement, and had hoped that the story would be the opening gambit in a game with a foregone and pleasurable conclusion; but he had not expected her to stand, to become walled off from him again.

'I've got to go,' she said, and with a distracted wave, she made for the front door. She walked with her head down, hands thrust into her pockets, as if counting the steps.

'Where are you going?' called Eliot, taken back.

'I don't know. Freak Street, maybe.'

'Want some company?'

She turned back at the door. 'It's not your fault,' she said, 'but I don't really enjoy your company.'

Shot down!

Trailing smoke, spinning, smacking into the hillside, and blowing up into a fireball.

Eliot didn't understand why it had hit him so hard. It had happened before, and it would again. Ordinarily he would have headed for Temal and found himself another long white scarf and pair of cotton trousers, one less morbidly self-involved (that, in retrospect, was how he characterized Michaela), one who would help him refuel for another bout of trying to visualize Avalokitesvara Buddha. He did, in fact, go to Temal; but he merely sat and drank tea and smoked hashish in a restaurant, and watched the young travellers pairing up for the night. Once he caught the bus to Patan and visited a friend, an old hippie pal named Sam Chipley who ran a medical clinic; once he walked out to Swayambhunath, close enough to see the white dome of the stupa, and atop it, the gilt structure on which the all-seeing eyes of Buddha were painted: they seemed squinty and mean-looking, as if taking unfavourable notice of his approach. But mostly over the next week he wandered through Mr Chatterji's house, carrying a bottle, maintaining a buzz, and keeping an eye on Michaela.

The majority of the rooms were unfurnished, but many bore signs of recent habitation: broken hash pipes, ripped sleeping bags, empty packets of incense. Mr Chatterji let travellers—those he fancied sexually, male and female—use the rooms for up to six months at a time, and to walk through them was to take a historical tour of the American counterculture. The graffiti spoke of concerns as various as Vietnam, the Sex Pistols, women's lib, and the housing shortage in Great Britain, and also conveyed personal messages: 'Ken Finkel please get in touch with me at Am. Ex. in Bangkok . . . love Ruth.' In one of the rooms was a complicated mural depicting Farah Fawcett sitting on the lap of a Tibetan demon, throttling his barbed phallus with her fingers. It all conjured up the image of a mouldering, deranged milieu. Eliot's milieu. At first the tour amused him, but eventually it began to sour him on himself, and he took to spending more and more time on a balcony overlooking the courtyard that was shared with the connecting house, listening to the Newari women sing at their chores and reading books from Mr Chatterji's library. One of the books was titled *The Carversville Terror.*

'. . . bloodcurdling, chilling . . .' said the *New York Times* on the front flap. '. . . the Terror is unrelenting . . .' commented Stephen King. '. . . riveting, gut-wrenching, mind-bending horror . . .' gushed *People* magazine. In neat letters, Eliot appended his own blurb: '. . . piece of crap . . .' The text—written to be read by the marginally literate—was a fictionalized treatment of purportedly real events, dealing with the experiences of the Whitcomb family, who had attempted to renovate the Cousineau mansion during the sixties. Following the usual buildup of apparitions, cold spots, and noisome odours, the family—Papa David, Mama Elaine, young sons Tim and Randy, and teenage Ginny—had met to discuss the situation.

even the kids, thought David, had been aged by the house. Gathered around the dining room table, they looked like a company of the damned—haggard, shadows under their eyes, grim-faced. Even with the windows open and the light streaming in, it seemed there was a pall in the air that no light could dispel. Thank God the damned thing was dormant during the day!

'Well,' he said, 'I guess the floor's open for arguments.'

'I wanna go home!' Tears sprang from Randy's eyes, and on cue, Tim started crying, too.

'It's not that simple,' said David. 'This *is* home, and I don't know how we'll make it if we do leave. The savings account is just about flat.'

'I suppose I could get a job,' said Elaine unenthusiastically.

'I'm not leaving!' Ginny jumped to her feet, knocking over her chair. 'Every time I start to make friends, we have to move!'

'But Ginny!' Elaine reached out a hand to calm her. 'You were the one . . .'

'I've changed my mind!' She backed away, as if she had just recognized them all to be mortal enemies. 'You can do what you want, but I'm staying!' And she ran from the room.

'Oh, God,' said Elaine wearily. 'What's gotten into her?'

What had gotten into Ginny, what was in the process of getting into her and was the only interesting part of the book, was the spirit of Aimée Cousineau. Concerned with his daughter's behaviour, David Whitcomb had researched the house and learned a great deal about the spirit. Aimée Cousineau, née Vuillemont, had been a native of St Berenice, a Swiss village at the foot of the mountain known as the Eiger (its photograph, as well as one of Aimée—a coldly beautiful woman with black hair and cameo features—was included in the central section of the book). Until the age of fifteen, she had been a sweet, unexceptional child; however, in the summer of 1889, while hiking on the slopes of the Eiger, she had become lost in a cave.

The family had all but given up hope, when, to their delight—three weeks later—she had turned up on the steps of her father's store. Their delight was short-lived. This Aimée was far different from the one who had entered the cave. Violent, calculating, slatternly.

Over the next two years, she succeeded in seducing half the men of the village, including the local priest. According to his testimony, he had been admonishing her that sin was not the path to happiness, when she began to undress. 'I'm wed to Happiness,' she told him. 'I've entwined my limbs with the God of Bliss and kissed the scaly thighs of Joy.' Throughout the ensuing affair, she made cryptic comments concerning 'the God below the mountain', whose soul was now forever joined to hers.

At this point the book reverted to the gruesome adventures of the Whitcomb family, and Eliot, bored, realizing it was noon and that Michaela would be sunbathing, climbed to Mr Chatterji's apartment on the fourth floor. He tossed the book on to a shelf and went out onto the balcony. His continued interest in Michaela puzzled him. It occurred to him that he might be falling in love, and he thought that would be nice. Though it would probably lead nowhere, love would be a good kind of energy to have. But he doubted this was the case. Most likely his interest was founded on some fuming product of the dark stone inside him. Simple lust. He looked over the edge of the balcony. She was lying on a blanket—her bikini top beside her—at the bottom of a well of sunlight: thin, pure sunlight like a refinement of honey spreading down and congealing into the mould of a little gold woman. It seemed her heat that was in the air.

That night Eliot broke one of Mr Chatterji's rules and slept in the master bedroom. It was roofed by a large skylight mounted in a ceiling painted midnight blue. The normal display of stars had not been sufficient for Mr Chatterji, and so he'd had the skylight constructed of faceted glass that multiplied the stars, making it appear that you were at the heart of a galaxy, gazing out between the interstices of its blazing core. The walls consisted of a photomural of the Khumbu Glacier and Chomolungma; and bathed in the starlight, the mural had acquired the illusion of depth and chill mountain silence. Lying there, Eliot could hear the faint sounds of Indra Jatra: shouts and cymbals, oboes and drums. He was drawn to the sounds; he wanted to run out into the streets, become an element of the drunken crowds, be whirled through torchlight and delirium to the feet of an idol stained with sacrificial blood. But he felt bound to the house, to Michaela. Marooned in the glow of Mr Chatterji's starlight, floating above Chomolungma and listening to the din of the world below, he could almost believe he was a bodhisattva awaiting a call to action, that his watchfulness had some purpose.

The shipment arrived late in the afternoon of the eighth day. Five enormous crates, each requiring the combined energies of Eliot and three Newari workmen to wrangle up to the third-floor room that housed Mr Chatterji's collection. After tipping the men, Eliot—sweaty, panting—sat down against the wall to catch his breath. The room was about twenty-five feet by fifteen, but looked smaller because of the dozens of curious objects standing around the floor and mounted one above the other on the walls. A brass doorknob, a shattered door, a straight-backed chair whose arms were bound with a velvet rope to prevent anyone from sitting, a discoloured sink, a mirror streaked by a brown stain, a slashed lampshade. They were all relics of some haunting or possession, some grotesque violence, and there were cards affixed to them testifying to the details and referring those who were interested to materials in Mr Chatterji's library. Sitting surrounded by these relics, the crates looked innocuous. Bolted shut, chest-high, branded with customs stamps.

When he had recovered, Eliot strolled around the room, amused by the care that Mr Chatterji had squandered on his hobby; the most amusing thing was that no one except Mr Chatterji was impressed by it: it provided travellers with a footnote for their journals. Nothing more.

A wave of dizziness swept over him—he had stood too soon—and he leaned against one of the crates for support. Jesus, he was in lousy shape! And then, as he blinked away the tangles of opaque cells drifting across his field of vision, the crate shifted. Just a little shift, as if something inside had twitched in its sleep. But palpable, real. He flung himself towards the door, backing away. A chill mapped every knob and articulation of his spine, and his sweat had evaporated, leaving clammy patches on his skin. The crate was motionless. But he was afraid to take his eyes off it, certain that if he did, it would release its pent-up fury. 'Hi,' said Michaela from the doorway.

Her voice electrified Eliot. He let out a squawk and wheeled around, his hands outheld to ward off attack.

'I didn't mean to startle you,' she said. 'I'm sorry.'

'Goddamn!' he said. 'Don't sneak up like that!' He remembered the crate and glanced back at it. 'Listen, I was just locking . . .'

'I'm sorry,' she repeated, and walked past him into the room. 'Ranjeesh is such an idiot about all this,' she said, running her hand over the top of the crate. 'Don't you think?'

Her familiarity with the crate eased Eliot's apprehension. Maybe he had been the one who had twitched: a spasm of overstrained muscles. 'Yeah, I guess.'

She walked over to the straight-backed chair, slipped off the velvet rope, and sat down. She was wearing a pale brown skirt and a plaid blouse that made her look schoolgirlish. 'I want to apologize about the other day,' she said; she bowed her head, and the fall of her hair swung forward to obscure her face. 'I've been having a bad time lately. I have trouble relating to people. To anything. But since we're living here together, I'd like to be friends.' She stood and spread the folds of her skirt. 'See? I even put on different clothes. I could tell the others offended you.'

The innocent sexuality of the pose caused Eliot to have a rush of desire. 'Looks nice,' he said with forced casualness. 'Why've you been having a bad time?'

She wandered to the door and gazed out. 'Do you really want to hear about it?'

'Not if it's painful for you.'

'It doesn't matter,' she said, leaning against the doorframe. 'I was in a band back in the States, and we were doing OK. Cutting an album, talking to record labels. I was living with the guitarist, in love with him. But then I had an affair. Not even an affair. It was stupid. Meaningless. I still don't know why I did it. The heat of the moment, I guess. That's what rock 'n' roll's all about, and maybe I was just acting out the myth. One of the other musicians told my boyfriend. That's the way bands are—you're friends with everyone, but never at the same time. See, I told this guy about the affair. We'd always confided. But one day he got mad at me over something. Something else stupid and meaningless.' Her chin was struggling to stay firm; the breeze from the courtyard drifted fine strands of hair across her face. 'My boyfriend went crazy and beat up my . . .' She gave a dismal laugh. 'I don't know what to call him. My lover. Whatever. My boyfriend killed him. It was an accident, but he tried to run, and the police shot him.'

Eliot wanted to stop her; she was obviously seeing it all again, seeing blood and police flashers and cold white morgue lights. But she was riding a wave of memory, borne along by its energy, and he knew that she had to crest with it, crash with it.

'I was out of it for a while. Dreamy. Nothing touched me. Not the funerals, the angry parents. I went away for months, to the mountains, and I started to feel better. But when I came home, I found that the musician who'd told my boyfriend had written a song about it. The affair, the killings. He'd cut a record. People were buying it, singing the hook when they walked down the street or took a shower. Dancing to it! They were dancing on blood and bones, humming grief, shelling out $5.98 for a jingle about suffering. Looking back, I realize I was crazy, but at the time everything I did seemed normal. More than normal. Directed, inspired. I bought a gun. A ladies' model, the salesman said. I remember thinking how strange it was that there were male and female guns, just like with electric razors. I felt enormous carrying it. I had to be meek and polite or else I was sure people would notice how large and purposeful I was. It wasn't hard to track down Ronnie—that's the guy who wrote the song. He was in Germany, cutting a second album. I couldn't believe it, I wasn't going to be able to kill him! I was so frustrated that one night I went down to a park and started shooting. I missed everything. Out of all the bums and joggers and squirrels, I hit leaves and air. They locked me up after that. A hospital. I think it helped, but . . .' She blinked, waking from a trance. 'But I still feel so disconnected, you know?'

Eliot carefully lifted away the strands of hair that had blown across her face and laid them back in place. Her smile flickered. 'I know,' he said. 'I feel that way sometimes.'

She nodded thoughtfully, as if to verify that she had recognized this quality in him.

They ate dinner in a Tibetan place in Temal; it had no name and was a dump with flyspecked tables and rickety chairs, specializing in water buffalo and barley soup. But it was away from the city centre, which meant they could avoid the worst of the festival crowds. The waiter was a young Tibetan wearing jeans and a T-shirt that bore the legend *Magic Is The Answer*; the earphones of personal stereo dangled about his neck. The walls—visible through a haze of smoke—were covered with snapshots, most featuring the waiter in the company of various tourists, but a few showing an older Tibetan in blue robes and turquoise jewellery, carrying an automatic rifle; this was the owner, one of the Khampa tribesmen who had fought a guerrilla war against the Chinese. He rarely put in an appearance at the restaurant, and when he did, his glowering presence tended to dampen conversation.

Over dinner, Eliot tried to steer clear of topics that might unsettle Michaela. He told her about Sam Chipley's clinic, the time the Dalai Lama had come to Katmandu, the musicians at Swayambhunath. Cheerful, exotic topics. Her listlessness was such an inessential part of her that

Eliot was led to chip away at it, curious to learn what lay beneath; and the more he chipped away, the more animated her gestures, the more luminous her smile became. This was a different sort of smile than she had displayed on their first meeting. It came so suddenly over her face, it seemed an autonomic reaction, like the opening of a sunflower, as if she were facing not you but the principle of light upon which you were grounded. It was aware of you, of course, but it chose to see past the imperfections of the flesh and know the perfected thing you truly were. It boosted your sense of worth to realize that you were its target, and Eliot—whose sense of worth was at low ebb—would have done pratfalls to sustain it. Even when he told his own story, he told it as a joke, a metaphor for American misconceptions of oriental pursuits.

'Why don't you quit it?' she asked. 'The meditation, I mean. If it's not working out, why keep on with it?'

'My life's in perfect suspension,' he said. 'I'm afraid that if I quit practising, if I change anything, I'll either sink to the bottom or fly off.' He tapped his spoon against his cup, signalling for more tea. 'You're not really going to marry Ranjeesh, are you?' he asked, and was surprised at the concern he felt that she actually might.

'Probably not.' The waiter poured their tea, whispery drumbeats issuing from his earphones. 'I was just feeling lost. You see, my parents sued Ronnie over the song, and I ended up with a lot of money—which made me feel even worse . . .'

'Let's not talk about it,' he said.

'It's all right.' She touched his wrist, reassuring, and the skin remained warm after her fingers had withdrawn. 'Anyway,' she went on, 'I decided to travel, and all the strangeness . . . I don't know. I was starting to slip away. Ranjeesh was a kind of sanctuary.'

Eliot was vastly relieved.

Outside, the streets were thronged with festivalgoers, and Michaela took Eliot's arm and let him guide her through the crowds. Newar wearing Nehru hats and white trousers that bagged at the hips and wrapped tightly around the calves; groups of tourists, shouting and waving bottles of rice beer; Indians in white robes and saris. The air was spiced with incense, and the strip of empurpled sky above was so regularly patterned with stars that it looked like a banner draped between the roofs. Near the house, a wild-eyed man in a blue satin robe rushed past, bumping into them, and he was followed by two boys dragging a goat, its forehead smeared with crimson powder: a sacrifice.

'This is crazy!' Michaela laughed.

'It's nothing. Wait till tomorrow night.'

'What happens then?'

'The night of White Bhairab.' Eliot put on a grimace. 'You'll have to watch yourself. Bhairab's a lusty, wrathful sort.'

She laughed again and gave his arm an affectionate squeeze.

Inside the house, the moon—past full, blank and golden—floated dead centre of the square of night sky admitted by the roof. They stood close together in the courtyard, silent, suddenly awkward.

'I enjoyed tonight,' said Michaela; she leaned forward and brushed his cheek with her lips. 'Thank you,' she whispered.

Eliot caught her as she drew back, tipped her chin, and kissed her mouth. Her lips parted, her tongue darted out. Then she pushed him away. 'I'm tired,' she said, her face tightened with anxiety. She walked off a few steps, but stopped and turned back. 'If you want to . . . to be with me, maybe it'll be all right. We could try.'

Eliot went to her and took her hands. 'I want to make love with you,' he said, no longer trying to hide his urgency. And that *was* what he wanted: to make love. Not to ball or bang or screw or any other inelegant version of the act.

But it was not love they made.

Under the starlit blaze of Mr Chatterji's ceiling, she was very beautiful, and at first she was very loving, moving with a genuine involvement; then abruptly, she quit moving altogether and turned her face to the pillow. Her eyes were glistening. Left alone atop her, listening to the animal sound of his breathing, the impact of his flesh against hers, Eliot knew he should stop and comfort her. But the months of abstinence, the eight days of wanting her, all this fused into a bright flare in the small of his back, a reactor core of lust that irradiated his conscience, and he continued to plunge into her, hurrying to completion. She let out a gasp when he withdrew, and curled up, facing away from him.

'God, I'm so sorry,' she said, her voice cracked.

Eliot shut his eyes. He felt sickened, reduced to the bestial. It had been like two mental patients doing nasty on the sly, two fragments of people who together didn't form a whole. He understood now why Mr Chatterji wanted to marry her: he planned to add her to his collection, to enshrine her with the other splinters of violence. And each night he would complete his revenge, substantiate his cultural overview, by making something less than love with this sad, inert girl, this American ghost. Her shoulders shook with muffled sobs. She needed someone to console her, to help her find her own strength and capacity for love. Eliot reached out to her, willing to do his best. But he knew it shouldn't be him.

Several hours later, after she had fallen asleep, unconsolable, Eliot sat in the courtyard, thoughtless, dejected, staring at a rubber plant. It was mired in shadow, its leaves hanging limp. He had been staring for a couple of minutes when he noticed that a shadow in back of the plant was swaying ever so slightly; he tried to make it out, and the swaying subsided. He stood. The chair scraped on the concrete, sounding unnaturally loud. His neck prickled, and he glanced behind him. Nothing. Ye Olde Mental Fatigue, he thought. Ye Olde Emotional Strain. He laughed, and the clarity of the laugh—echoing up through the empty well—alarmed him; it seemed to stir little flickers of motion everywhere in the darkness. What he needed was a drink! The problem was how to get into the bedroom without waking Michaela. Hell, maybe he should wake her. Maybe they should talk more before what had happened hardened into a set of unbreakable attitudes.

He turned towards the stairs . . . and then, yelling out in panic, entangling his feet with the lawn chairs as he leaped backward mid-step, he fell on to his side. A shadow—roughly man-shaped and man-sized—was standing a yard away; it was undulating the way a strand of kelp undulates in a gentle tide. The patch of air around it was rippling, as if the entire image had been badly edited into reality. Eliot scrambled away, coming to his knees. The shadow melted downward, puddling on the cement; it bunched in the middle like a caterpillar, folded over itself, and flowed after him: a rolling sort of motion. Then it reared up, again assuming its manlike shape, looming over him.

Eliot got to his feet, still frightened, but less so. If he had previously been asked to testify as to the existence of the Khaa, he would have rejected the evidence of his bleared senses and come down on the side of hallucination, folktale. But now, though he was tempted to draw that same conclusion, there was too much evidence to the contrary. Staring at the featureless black cowl of the Khaa's head, he had a sense of something staring back. More than a sense. A distinct impression of personality. It was as if the Khaa's undulations were producing a breeze that bore its psychic odour through the air. Eliot began to picture it as a loony, shy old uncle who liked to sit under the basement steps and eat flies and cackle to himself, but who could tell when the first frost was due and knew how to fix the tail on your kite. Weird, yet harmless. The Khaa stretched out an arm: the arm just peeled away from its torso, its hand a thumbless black mitten. Eliot edged back. He wasn't quite prepared to believe it was harmless. But the arm stretched further than he had thought possible and enveloped his wrist. It was soft, ticklish, a river of furry moths crawling over his skin.

In the instant before he jumped away, Eliot heard a whining note inside his skull, and that whining—seeming to flow through his brain with the same suppleness that the Khaa's arm had displayed—was translated into a wordless plea. From it he understood that the Khaa was afraid. Terribly afraid. Suddenly it melted downward and went rolling, bunching, flowing up the stairs; it stopped on the first landing, rolled halfway down, then up again, repeating the process over and over. It came clear to Eliot (*Oh, Jesus! This is nuts!*) that it was trying to convince him to follow. Just like Lassie or some other ridiculous TV animal, it was trying to tell him something, to lead him to where the wounded forest ranger had fallen, where the nest of baby ducks was being threatened by the brush fire. He should walk over, rumple its head, and say, 'What's the matter, girl? Those squirrels been teasing you?' This time his laughter had a sobering effect, acting to settle his thoughts. One likelihood was that his experience with Michaela had been sufficient to snap his frayed connection with consensus reality; but there was no point in buying that. Even if that were the case, he might as well go with it. He crossed to the stairs and climbed towards the rippling shadow on the landing.

'OK, Bongo,' he said. 'Let's see what's got you so excited.'

On the third floor, the Khaa turned down a hallway, moving fast, and Eliot didn't see it again until he was approaching the room that housed Mr. Chatterji's collection. It was standing beside the door, flapping its arms, apparently indicating that he should enter. Eliot remembered the crate.

'No, thanks,' he said. A drop of sweat slid down his rib cage, and he realized that it was unusually warm next to the door.

The Khaa's hand flowed over the doorknob, enveloping it, and when the hand pulled back, it was bulging, oddly deformed, and there was a hole through the wood where the lock mechanism had been. The door swung open a couple of inches. Darkness leaked out of the room, adding an oily essence to the air. Eliot took a backward step. The Khaa dropped the lock mechanism—it materialized from beneath the black, formless hand and clattered to the floor—and latched on to Eliot's arm. Once again he heard the whining, the plea for help, and, since he did not jump away, he had a clearer understanding of the process of translation. He could feel the whining as a cold fluid coursing through his brain, and as the whining died, the message simply appeared—the way an image might appear in a crystal ball. There was an undertone of reassurance to the Khaa's fear, and though Eliot knew this was the mistake people in horror movies were always making, he reached inside the room and fumbled for the wall switch, half-expecting to be snatched up and savaged. He flicked on the light and pushed the door open with his foot.

And wished that he hadn't.

The crates had exploded. Splinters and shards of wood were scattered everywhere, and the bricks had been heaped at the centre of the room. They were dark red, friable bricks like crumbling cakes of dried blood, and each was marked with black letters and numbers that signified its original position in the fireplace. But none were in their proper position now, though they were quite artfully arranged. They had been piled into the shape of a mountain, one that—despite the crudity of its building blocks—duplicated the sheer faces and chimneys and gentle slopes of a real mountain. Eliot recognized it from its photograph. The Eiger. It towered to the ceiling, and under the glare of the lights, it gave off a radiation of ugliness and barbarity. It seemed alive, a fang of dark red meat, and the charred smell of the bricks was like a hum in Eliot's nostrils.

Ignoring the Khaa, who was again flapping its arms, Eliot broke for the landing; there he paused, and after a brief struggle between fear and conscience, he sprinted up the stairs to the bedroom, taking them three at a time. Michaela was gone! He stared at the starlit billows of the sheets. Where the hell . . . her room! He hurtled down the stairs and fell sprawling on the second-floor landing. Pain lanced through his kneecap, but he came to his feet running, certain that something was behind him.

A seam of reddish orange light—not lamplight—edged the bottom of Michaela's door, and he heard a crispy chuckling noise like a fire crackling in a hearth. The wood was warm to the touch. Eliot's hand hovered over the doorknob. His heart seemed to have swelled to the size of a basketball and was doing a fancy dribble against his chest wall. The sensible thing to do would be to get out quick, because whatever lay beyond the door was bound to be too much for him to handle. But instead he did the stupid thing and burst into the room.

His first impression was that the room was burning, but then he saw that though the fire looked real, it did not spread; the flames clung to the outlines of things that were themselves unreal, that had no substance of their own and were made of the ghostly fire: belted drapes, an overstuffed chair and sofa, a carved mantelpiece, all of antique design. The actual furniture—production-line junk—was undamaged. Intense reddish orange light glowed around the bed, and at its heart lay Michaela. Naked, her back arched. Lengths of her hair lifted into the air and tangled, floating in an invisible current; the muscles of her legs and abdomen were coiling, bunching, as if she were shedding her skin. The crackling grew louder, and the light began to rise from the bed, to form into a column of even brighter light; it narrowed at the midpoint, bulged in an approximation of hips and breasts, gradually assuming the shape of a burning woman. She was faceless, a fiery silhouette. Her flickering gown shifted as with the movements of walking, and flames leaped out behind her head like windblown hair.

Eliot was pumped full of terror, too afraid to scream or run. Her aura of heat and power wrapped around him. Though she was within arm's length, she seemed a long way off, inset into a great distance and walking towards him down a tunnel that conformed exactly to her shape. She stretched out a hand, brushing his cheek with a finger. The touch brought more pain than he had ever known. It was luminous, lighting every circuit of his body. He could feel his skin crisping, cracking, fluids leaking forth and sizzling. He heard himself moan: a gush of rotten sound like something trapped in a drain.

Then she jerked back her hand, as if *he* had burned *her*.

Dazed, his nerves screaming, Eliot slumped to the floor and—through blurred eyes—caught sight of a blackness rippling by the door. The Khaa. The burning woman stood facing it a few feet away. It was such an uncanny scene, this confrontation of fire and darkness, of two supernatural systems, that Eliot was shocked to alertness. He had the idea that neither of them knew what to do. Surrounded by its patch of disturbed air, the Khaa undulated; the burning woman crackled and flickered, embedded in her eerie distance. Tentatively, she lifted her hand; but before she could complete the gesture, the Khaa reached with blinding swiftness and its hand enveloped hers.

A shriek like tortured metal issued from them, as if some ironclad principle had been breached. Dark tendrils wound through the burning woman's arm, seams of fire striped the Khaa, and there was a high-pitched humming, a vibration that jarred Eliot's teeth. For a moment he was afraid that spiritual versions of antimatter and matter had been brought into conjunction, that the room would explode. But the hum was sheared off as the Khaa snatched back its hand: a scrap of reddish orange flame glimmered within it. The Khaa melted downward and went rolling out the door. The burning woman—and every bit of flame in the room—shrank to an incandescent point and vanished.

Still dazed, Eliot touched his face. It felt burned, but there was no apparent damage. He hauled himself to his feet, staggered to bed, and collapsed next to Michaela. She was breathing deeply, unconscious. 'Michaela!' He shook her. She moaned, her head rolled from side to side. He heaved her over his shoulder in a fireman's lift and crept out into the hall. Moving stealthily, he eased along the hall to the balcony overlooking the courtyard and peered over the edge . . . and bit his lip to stifle a cry. Clearly visible in the electric blue air of the predawn darkness, standing in the middle of the courtyard, was a tall, pale woman wearing a white nightgown. Her black hair fanned across her back. She snapped her head around to stare at him, her cameo features twisted by a gloating smile, and that smile told Eliot everything he had wanted to know about the possibility of escape. Just try to leave, Aimée Cousineau was saying. Go ahead and try. I'd like that. A shadow sprang erect about a dozen feet away from her, and she turned to it. Suddenly there was a wind in the courtyard: a violent, whirling wind of which she was the calm centre. Plants went flapping up into the well like leathery birds; pots shattered, and the shards flew towards the Khaa. Slowed by Michaela's weight, wanting to get as far as he could from the battle, Eliot headed up the stairs towards Mr Chatterji's bedroom.

It was an hour later, an hour of peeking down into the courtyard, watching the game of hide-and-seek that the Khaa was playing with Aimée Cousineau, realizing that the Khaa was protecting them by keeping her busy . . . it was then that Eliot remembered the book. He retrieved it from the shelf and began to skim through it, hoping to learn something helpful. There was nothing else to do. He picked up at the point of Aimée's rap about her marriage to Happiness, passed over the transformation of Ginny Whitcomb into a teenage monster, and found a second section dealing with Aimée.

In 1895 a wealthy Swiss-American named Armand Cousineau had returned to St Berenice—his birthplace—for a visit. He was smitten with Aimée Vuillemont, and her family, seizing the opportunity to be rid of her, allowed Cousineau to marry Aimée and sail her off to his home in Carversville, New Hampshire. Aimée's taste for seduction had not been curbed by the move. Lawyers, deacons, merchants, farmers: they were all grist for her mill. But in the winter of 1905, she fell in love—obsessively, passionately in love—with a young schoolmaster. She believed that the schoolmaster had saved her from her unholy marriage, and her gratitude knew no bounds. Unfortunately, when the schoolmaster fell in love with another woman, neither did her fury. One night while passing the

Cousineau mansion, the town doctor spotted a woman walking the grounds. '. . . a woman of flame, not burning but composed of flame, her every particular a fiery construct . . .' Smoke was curling from a window; the doctor rushed inside and discovered the schoolmaster wrapped in chains, burning like a log in the vast fireplace. He put out the small blaze spreading from the hearth, and on going back on to the grounds, he stumbled over Aimée's charred corpse.

It was not clear whether Aimée's death had been accidental, a stray spark catching on her nightgown, or the result of suicide; but it *was* clear that thereafter the mansion had been haunted by a spirit who delighted in possessing women and driving them to kill their men. The spirit's supernatural powers were limited by the flesh, but were augmented by immense physical strength. Ginny Whitcomb, for example, had killed her brother Tim by twisting off his arm, and then had gone after her other brother and her father, a harrowing chase that had lasted a day and a night: while in possession of a body, the spirit was not limited to nocturnal activity. . . .

Christ!

The light coming through the skylight was grey.

They were safe!

Eliot went to the bed and began shaking Michaela. She moaned, her eyes blinked open. 'Wake up!' he said. 'We've got to get out!'

'What?' She batted at his hands. 'What are you talking about?'

'Don't you remember?'

'Remember what?' She swung her legs on to the floor, sitting with her head down, stunned by wakefulness; she stood, swayed, and said, 'God, what did you do to me? I feel. . . .' A dull, suspicious expression washed over her face.

'We have to leave.' He walked around the bed to her. 'Ranjeesh hit the jackpot. Those crates of his had an honest-to-God spirit packed in with the bricks. Last night it tried to possess you.' He saw her disbelief. 'You must have blanked out. Here.' He offered the book. 'This'll explain. . . .'

'Oh, God!' she shouted. 'What did you do? I'm all raw inside!' She backed away, eyes wide with fright.

'I didn't do anything.' He held out his palms as if to prove he had no weapons.

'You raped me! While I was asleep!' She looked left, right, in a panic.

'That's ridiculous!'

'You must have drugged me or something! Oh, God! Go away!'

'I won't argue,' he said. 'We have to get out. After that you can turn me in for rape or whatever. But we're leaving, even if I have to drag you.'

Some of her desperation evaporated, her shoulders sagged.

'Look,' he said, moving closer. 'I didn't rape you. What you're feeling is something that goddamn spirit did to you. It was. . . .'

She brought her knee up into his groin.

As he writhed on the floor, curled up around the pain, Eliot heard the door open and her footsteps receding. He caught at the edge of the bed, hauled himself to his knees, and vomited all over the sheets. He fell back and lay there for several minutes, until the pain had dwindled to a powerful throbbing, a throbbing that jolted his heart into the same rhythm; then, gingerly, he stood and shuffled out into the hall. Leaning on the railing, he eased down the stairs to Michaela's room and lowered himself into a sitting position. He let out a shuddering sigh. Actinic flashes burst in front of his eyes.

'Michaela,' he said. 'Listen to me.' His voice sounded feeble: the voice of an old, old man.

'I've got a knife,' she said from just behind the door. 'I'll use it if you try to break in.'

'I wouldn't worry about that,' he said. 'And I sure as hell wouldn't worry about being raped. Now will you listen?'

No response.

He told her everything, and when he was done, she said, 'You're insane. You raped me.'

'I wouldn't hurt you. I. . . .' He had been on the verge of telling her he loved her, but decided it probably wasn't true. He probably just wished that he had a good, clean truth like love. The pain was making him nauseated again, as if the blackish purple stain of his bruises were seeping up into his stomach and filling him with bad gases. He struggled to his feet and leaned against the wall. There was no point in arguing, and there was not much hope that she would leave the house on her own, not if she reacted to Aimée like Ginny Whitcomb. The only solution was to go to the police, accuse her of some crime. Assault. She would accuse him of rape, but with luck they would both be held overnight. And he would have time to wire Mr Chatterji . . . who would believe him. Mr Chatterji was by nature a believer; it simply hadn't fit his notion of sophistication to give credence to his native spirits. He'd be on the first flight from Delhi, eager to document the Terror.

Himself eager to get it over, Eliot negotiated the stairs and hobbled across the courtyard; but the Khaa was waiting, flapping its arms in the shadowed alcove that led to the street. Whether it was an effect of the light or of its battle with Aimée, or, specifically, of the pale scrap of fire visible within its hand, the Khaa looked less substantial. Its blackness was somewhat opaque, and the air around it was blurred, smeary, like waves over a lens: it was as if the Khaa were being submerged more deeply in its own medium. Eliot felt no compunction about allowing it to touch him; he was grateful to it, and his relaxed attitude seemed to intensify the communication. He began to see images in his mind's eye: Michaela's face, Aimée's, and then the two faces were superimposed. He was shown this over and over, and he understood from it that the Khaa wanted the possession to take place. But he didn't understand why. More images. Himself running, Michaela running, Durbar Square, the mask of White Bhairab, the Khaa. Lots of Khaa. Little black hieroglyphs. These images were repeated, too, and after each sequence the Khaa would hold its hand up to his face and display the glimmering scrap of Aimée's fire. Eliot thought he understood, but whenever he tried to convey that he wasn't sure, the Khaa merely repeated the images.

At last, realizing that the Khaa had reached the limits of its ability to communicate, Eliot headed for the street. The Khaa melted down, reared up in the doorway to block his path, and flapped its arms desperately. Once again Eliot had a sense of its weird-old-man-ness. It went against logic to put his trust in such an erratic creature, especially in such a dangerous plan; but logic had little hold on him, and this was a permanent solution. If it worked. If he hadn't misread it. He laughed. The hell with it!

'Take it easy, Bongo,' he said. 'I'll be back as soon as I get my shootin' iron fixed.'

The waiting room of Sam Chipley's clinic was crowded with Newari mothers and children, who giggled as Eliot did a bowlegged shuffle through their midst. Sam's wife led him into the examination room, where Sam—a burly, bearded man, his long hair tied in a ponytail—helped him on to a surgical table.

'Holy shit!' he said after inspecting the injury. 'What you been into, man?' He began rubbing ointment into the bruises.

'Accident,' gritted Eliot, trying not to cry out.

'Yeah, I bet,' said Sam. 'Maybe a sexy little accident who had a change of heart when it come down to strokes. You know, not gettin' it steady might tend to make you a tad intense for some ladies, man. Ever think about that?'

'That's not how it was. Am I all right?'

'Yeah, but you ain't gonna be superstud for a while.' Sam went to the sink and washed his hands. 'Don't gimme that innocent bullshit. You were tryin' to slip it to Chatterji's new squeeze, right?'

'You know her?'

'He brought her over one day, showin' her off. She's a head case, man. You should know better.'

'Will I be able to run?'

Sam laughed. 'Not hardly.'

'Listen, Sam.' Eliot sat up, winced. 'Chatterji's lady. She's in bad trouble, and I'm the only one who can help her. I have to be able to run, and I need something to keep me awake. I haven't slept for a couple of days.'

'I ain't givin' you pills, Eliot. You can stagger through your doper phase without my help.' Sam finished drying his hands and went to sit on a stool beside the window; beyond the window was a brick wall, and atop it a string of prayer flags snapped in the breeze.

'I'm not after a supply, damn it! Just enough to keep me going tonight. This is important, Sam!'

Sam scratched his neck. 'What kind of trouble she in?'

'I can't tell you now,' said Eliot, knowing that Sam would laugh at the idea of something as metaphysically suspect as the Khaa. 'But I will tomorrow. It's not illegal. Come on, man! There's got to be something you can give me.'

'Oh, I can fix you up. I can make you feel like King Shit on Coronation Day.' Sam mulled it over. 'OK, Eliot. But you get your ass back here tomorrow and tell me what's happenin'.' He gave a snort of amusement. 'All I can say is it must be some strange damn trouble for you to be the only one who can save her.'

After wiring Mr Chatterji, urging him to come home at once, Eliot returned to the house and unscrewed the hinges of the front door. He was not certain that Aimée would be able to control the house, to slam doors and make windows stick as she had with her house in New Hampshire, but he didn't want to take any chances. As he lifted the door and set it against the wall of the alcove, he was amazed by its lightness; he felt possessed of a giddy strength, capable of heaving the door up through the well of the courtyard and over the roofs. The cocktail of pain-killers and speed was working wonders. His groin ached, but the ache was distant, far removed from the centre of his consciousness, which was a fount of well-being. When he had finished with the door, he grabbed some fruit juice from the kitchen and went back to the alcove to wait.

In mid-afternoon Michaela came downstairs. Eliot tried to talk to her, to convince her to leave but she warned him to keep away and scuttled back to her room. Then, around five o'clock, the burning woman appeared, floating a few feet above the courtyard floor. The sun had withdrawn to the upper third of the well, and her fiery silhouette was inset into slate-blue shadow, the flames of her hair dancing about her head. Eliot, who had been hitting the pain-killers heavily, was dazzled by her: had she been a hallucination, she would have made his All-Time Top Ten. But even realizing that she was not, he was too drugged to relate to her as a threat. He snickered and shied a piece of broken pot at her. She shrank to an incandescent point, vanished, and that brought home to him his foolhardiness. He took more speed to counteract his euphoria, and did stretching exercises to loosen the kinks and to rid himself of the cramped sensation in his chest.

Twilight blended the shadows in the courtyard, celebrants passed in the street, and he could hear distant drums and cymbals. He felt cut off from the city, the festival. Afraid. Not even the presence of the Khaa, half-merged with the shadows along the wall, served to comfort him. Near dusk, Aimée Cousineau walked into the courtyard and stopped about twenty feet away, staring at him. He had no desire to laugh or throw things. At this distance he could see that her eyes had no whites or pupils or irises. They were dead black. One moment they seemed to be the bulging heads of black screws threaded into her skull; the next they seemed to recede into blackness, into a cave beneath a mountain where something waited to teach the joys of hell to whoever wandered in. Eliot sidled closer to the door. But she turned, climbed the stairs to the second landing, and walked down Michaela's hallway.

Eliot's waiting began in earnest.

An hour passed. He paced between the door and the courtyard. His mouth was cottony; his joints felt brittle, held together by frail wires of speed and adrenaline. This was insane! All he had done was to put them in worse danger. Finally, he heard a door close upstairs. He backed into the street, bumping into two Newari girls, who giggled and skipped away. Crowds of people were moving towards Durbar Square.

'Eliot!'

Michaela's voice. He'd expected a hoarse, demon voice, and when she walked into the alcove, her white scarf glowing palely against the dark air, he was surprised to see that she was unchanged. Her features held no trace of anything other than her usual listlessness.

'I'm sorry I hurt you,' she said, walking towards him. 'I know you didn't do anything. I was just upset about last night.'

Eliot continued to back away.

'What's wrong?' She stopped in the doorway.

It might have been his imagination, the drugs, but Eliot could have sworn that her eyes were much darker than normal. He trotted off a dozen yards or so and stood looking at her.

'Eliot!'

It was a scream of rage and frustration, and he could scarcely believe the speed with which she darted towards him. He ran full tilt at first, leaping sideways to avoid collisions, veering past alarmed, dark-skinned faces; but after a couple of blocks, he found a more efficient rhythm and began to anticipate obstacles, to glide in and out of the crowd. Angry shouts were raised behind him. He glanced back. Michaela was closing the distance, beelining for him, knocking people sprawling with what seemed effortless blows. He ran harder. The crowd grew thicker, and he kept near the walls of the houses, where it was thinnest; but even there it was hard to maintain a good pace. Torches were waved in his face; young men—singing, their arms linked—posed barriers that slowed him further. He could no longer see Michaela, but he could see the wake of her passage. Fists shaking, heads jerking. The entire scene was starting to lose cohesiveness to Eliot. There were screams of torchlight, bright shards of deranged shouts, jostling waves of incense and ordure. He felt like the only solid chunk in a glittering soup that was being poured through a stone trough.

At the edge of Durbar Square, he had a brief glimpse of a shadow standing by the massive gilt doors of Degutale Temple. It was larger and a more anthracitic black than Mr Chatterji's Khaa: one of the old ones, the powerful ones. The sight buoyed his confidence and restored his equilibrium. He had not misread the plan. But he knew that this was the most dangerous part. He had lost track of Michaela, and the crowd was sweeping him along; if she caught up to him now, he would not be able to run. Fighting for elbow room, struggling to keep his feet, he was borne into the temple complex. The pagoda roofs sloped up into darkness like strangely carved mountains, their peaks hidden by a moonless night; the cobbled paths were narrow, barely ten feet across, and the crowd was being squeezed along them, a lava flow of humanity. Torches bobbed everywhere, sending wild licks of shadow and orange light up the walls, revealing scowling faces on the eaves. Atop its pedestal, the gilt statue of Hanuman—the monkey god—looked to be swaying. Clashing cymbals and arrhythmic drumming scattered Eliot's heartbeat; the sinewy wail of oboes seemed to be graphing the fluctuations of his nerves.

As he swept past Hanuman Dhoka Temple, he caught sight of the brass mask of White Bhairab shining over the heads of the crowd like the face of an evil clown. It was less than a hundred feet away, set in a huge niche in a temple wall and illuminated by light bulbs that hung down among strings of prayer flags. The crowd surged faster, knocking him this way and that; but he managed to spot two more Khaa in the doorway of Hanuman Dhoka. Both melted downward, vanishing, and Eliot's hopes soared. They must have located Michaela, they must be attacking! By the time he had been carried to within a few yards of the mask, he was sure that he was safe. They must have finished her exorcism by now. The only problem left was to find her. That, he realized, had been the weak link in the plan. He'd been an idiot not to have foreseen it. Who knows what might happen if she were to fall in the midst of the crowd. Suddenly he was beneath the pipe that stuck out of the god's mouth; the stream of rice beer arching from it looked translucent under the lights, and as it splashed his face (no fish), its coldness acted to wash away his veneer of chemical strength. He was dizzy, his groin throbbed. The great face, with its fierce fangs and goofy, startled eyes, appeared to be swelling and rocking back and forth. He took a deep breath. The thing to do would be to find a place next to a wall where he could wedge himself against the flow of the crowd, wait until it had thinned, and then search for her. He was about to do that very thing when two powerful hands gripped his elbows from behind.

Unable to turn, he craned his neck and peered over his shoulder. Michaela smiled at him: a gloating 'gotcha!' smile. Her eyes were dead-black ovals. She shaped his name with her mouth, her voice inaudible above the music and shouting, and she began to push him ahead of her, using him as a battering ram to forge a path through the crowd. To anyone watching, it might have appeared that he was running interference for her, but his feet were dangling just off the ground. Angry Newar yelled at him as he knocked them aside. He yelled, too. No one noticed. Within seconds they had got clear into a side street, threading between groups of drunkards. People laughed at Eliot's cries for help, and one guy imitated the funny loose-limbed way he was running.

Michaela turned into a doorway, carrying him down a dirt-floored corridor whose walls were carved into ornate screens; the dusky orange lamplight shining through the screens cast a lacework of shadow on the dirt. The corridor widened to a small courtyard, the age-darkened wood of its walls and doors inlaid with intricate mosaics of ivory. Michaela stopped and slammed him against a wall. He was stunned, but he recognized the place to be one of the old Buddhist temples that surrounded the square. Except for a life-sized statue of a golden cow, the courtyard was empty.

'Eliot.' The way she said it, it was more of a curse than a name.

He opened his mouth to scream, but she drew him into an embrace; her grip on his right elbow tightened, and her other hand squeezed the back of his neck, pinching off the scream.

'Don't be afraid,' she said. 'I only want to kiss you.'

Her breasts crushed into his chest, her pelvis ground against him in a mockery of passion, and inch by inch she forced his face down to hers. Her lips parted, and—*oh, Christ Jesus!*—Eliot writhed in her grasp, enlivened by a new horror. The inside of her mouth was as black as her eyes. She wanted him to kiss that blackness, the same she had kissed beneath the Eiger. He kicked and clawed with his free hand, but she was irresistible, her hands like iron. His elbow cracked, and brilliant pain shot through his arm. Something else was cracking in his neck. Yet none of that compared to what he felt as her tongue—a burning black poker—pushed between his lips. His chest was bursting with the need to scream, and everything was going dark. Thinking this was death, he experienced a peevish resentment that death was not—as he'd been led to believe—an end to pain, that it merely added a tickling sensation to all his other pain. Then the searing heat in his mouth diminished, and he thought that death must just have been a bit slower than usual.

Several seconds passed before he realized that he was lying on the ground, several more before he noticed Michaela lying beside him, and—because darkness was tattering the edges of his vision—it was considerably longer before he distinguished the six undulating darknesses that had ringed Aimée Cousineau. They towered over her; their blackness gleamed like thick fur, and the air around them was awash with vibration. In her fluted white nightgown, her cameo face composed in an expression of calm, Aimée looked the antithesis of the vaguely male giants that were menacing her, delicate and finely worked in contrast to their crudity. Her eyes appeared to mirror their negative colour. After a moment, a little wind kicked up, swirling about her. The undulations of the Khaa increased, becoming rhythmic, the movements of boneless dancers, and the wind subsided. Puzzled, she darted between two of them and took a defensive stance next to the golden cow; she lowered her head and stared up through her brows at the Khaa. They melted downward, rolled forward, sprang erect and hemmed her in against the statue. But the stare was doing its damage. Pieces of ivory and wood were splintering, flying off the walls towards the Khaa, and one of them was fading, a mist of black particles accumulating around its body; then, with a shrill noise that reminded Eliot of a jet passing overhead, it misted away.

Five Khaa remained in the courtyard. Aimée smiled and turned her stare on another. Before the stare could take effect, however, the Khaa moved close, blocking Eliot's view of her; and when they pulled back, it was Aimée who showed signs of damage. Rills of blackness were leading from her eyes, webbing her cheeks, making it look as if her face were cracking. Her nightgown caught fire, her hair began to leap. Flames danced on her fingertips, spread to her arms, her breast, and she assumed the form of the burning woman.

As soon as the transformation was complete, she tried to shrink, to dwindle to her vanishing point; but, acting in unison, the Khaa extended their hands and touched her. There was that shriek of tortured metal, lapsing to a high-pitched hum, and to Eliot's amazement, the Khaa were sucked inside her. It was a rapid process. The Khaa faded to a haze, to nothing, and veins of black marbled the burning woman's fire; the blackness coalesced, forming into five tiny stick figures, a hieroglyphic design patterning her gown. With a fuming sound, she expanded again, regaining her normal dimensions, and the Khaa flowed back out, surrounding her. For an instant she stood motionless, dwarfed: a schoolgirl helpless amidst a circle of bullies. Then she clawed at the nearest of them. Though she had no features with which to express emotion, it seemed to Eliot there was desperation in the gesture, in the agitated leaping of her fiery hair. Unperturbed, the Khaa stretched out their enormous mitten hands, hands that spread like oil and enveloped her.

The destruction of the burning woman, of Aimée Cousineau, lasted only a matter of seconds; but to Eliot it occurred within a bubble of slow time, a time during which he achieved a speculative distance. He wondered if—as the Khaa stole portions of her fire and secreted it within their bodies—they were removing disparate elements of her soul, if she consisted of psychologically distinct fragments: the girl who had wandered into the cave, the girl who had returned from it, the betrayed lover. Did she embody gradations of innocence and sinfulness, or was she a contaminated essence, an unfractionated evil? While still involved in this speculation, half a reaction to pain, half to the metallic shriek of her losing battle, he lost consciousness, and when he reopened his eyes, the courtyard was deserted. He could hear music and shouting from Durbar Square. The golden cow stared contentedly into nowhere.

He had the idea that if he moved, he would further break all the broken things inside him; but he inched his left hand across the dirt and rested it on Michaela's breast. It was rising and falling with a steady rhythm. That made him happy, and he kept his hand there, exulting in the hits of her life against his palm. Something shadowy above him. He strained to see it. One of the Khaa . . . No! It was Mr Chatterji's Khaa. Opaquely black, scrap of fire glimmering in its hand. Compared to its big brothers, it had the look of a skinny, sorry mutt. Eliot felt camaraderie towards it.

'Hey, Bongo,' he said weakly. 'We won.'

A tickling at the top of his head, a whining note, and he had an impression not of gratitude—as he might have expected—but of intense curiosity. The tickling stopped, and Eliot suddenly felt clear in his mind. Strange. He was passing out once again, his consciousness whirling, darkening, and yet he was calm and unafraid. A roar came from the direction of the square. Somebody—the luckiest somebody in the Katmandu Valley—had caught the fish. But as Eliot's eyelids fluttered shut, as he had a last glimpse of the Khaa looming above them and felt the warm measure of Michaela's heartbeat, he thought that maybe the crowd was cheering the wrong man.

Three weeks after the night of White Bhairab, Ranjeesh Chatterji divested himself of all worldly possessions (including the gift of a year's free rent at his house to Eliot) and took up residence at Swayambhunath where—according to Sam Chipley, who visited Eliot in the hospital—he was attempting to visualize the Avalokitesvara Buddha. It was then that Eliot understood the nature of his newfound clarity. Just as it had done long ago with the woman's goitres, the Khaa had tried his habituation to meditation on for size, had not cared for it, and had sloughed it off in a handy repository: Ranjeesh Chatterji.

It was such a delicious irony that Eliot had to restrain himself from telling Michaela when she visited that same afternoon; she had no memory of the Khaa, and news of it tended to unsettle her. But otherwise she had been healing right along with Eliot. All her listlessness had eroded over the weeks, her capacity for love was returning and was focused solely on Eliot. 'I guess I needed someone to show me that I was worth an effort,' she told him. 'I'll never stop trying to repay you.' She kissed him. 'I can hardly wait till you come home.' She brought him books and candy and flowers; she sat with him each day until the nurses shooed her away. Yet being the centre of her devotion disturbed him. He was still uncertain whether or not he loved her. Clarity, it seemed, made a man dangerously versatile, his conscience flexible, and instituted a cautious approach to commitment. At least this was the substance of Eliot's clarity. He didn't want to rush into anything.

When at last he did come home, he and Michaela made love beneath the starlight glory of Mr Chatterji's skylight. Because of Eliot's neck brace and cast, they had to manage the act with extreme care, but despite that, despite the ambivalence of his feelings, this time it *was* love they made. Afterward, lying with his good arm around her, he edged nearer to commitment. Whether or not he loved her, there was no way this part of things could be improved by any increment of emotion. Maybe he'd give it a try with her. If it didn't work out, well, he was not going to be responsible for her mental health. She would have to learn to live without him.

'Happy?' he asked, caressing her shoulder.

She nodded and cuddled closer and whispered something that was partially drowned out by the crinkling of the pillow. He was sure he had misheard her, but the mere thought that he hadn't was enough to lodge a nugget of chill between his shoulder blades.

'What did you say?' he asked.

She turned to him and propped herself on an elbow, silhouetted by the starlight, her features obscured. But when she spoke, he realized that Mr Chatterji's Khaa had been true to its erratic traditions on the night of White Bhairab; and he knew that if she were to tip back her head ever so slightly and let the light shine into her eyes, he would be able to resolve all his speculations about the composition of Aimée Cousineau's soul.

'I'm wed to Happiness,' she said.

ROBERT HOLDSTOCK

THORN

At sundown, when the masons and guild carpenters finished their work for the day and trudged wearily back to their village lodgings, Thomas Wyatt remained behind in the half-completed church and listened to the voice of the stone man, calling to him.

The whispered sound was urgent, insistent: 'Hurry! Hurry! I *must* be finished before the others. *Hurry!*'

Thomas, hiding in the darkness below the gallery, felt sure that the ghostly cry could be heard for miles around. But the Watchman, John Tagworthy, was almost completely deaf, now, and the priest was too involved with his own holy rituals to be aware of the way his church was being stolen.

Thomas could hear the priest. He was circling the new church twice, as he always did at sundown, a small, smoking censer in one hand, a book in the other. He walked from right to left. Demons, and the sprites of the old earth, flew before him, birds and bats in the darkening sky. The priest, like all the men who worked on the church—except for Thomas himself—was a stranger to the area. He had long hair and a dark, trimmed beard, an unusual look for a monk.

He talked always about the supreme holiness of the place where his church was being built. He kept a close eye on the work of the craftsmen. He prayed to the north and the south, and constantly was to be seen kneeling at the very apex of the mound, as if exorcizing the ancient spirits buried below.

This was Dancing Hill. Before the stone church there had been a wooden church, and some said that Saint Peter himself had raised the first timbers. And hadn't Joseph, bearing the Grail of Christ, rested on this very spot, and driven out the demons of the earth mound?

But it was Dancing Hill. And sometimes it was referred to by its older name, *Ynys Calidryv*, isle of the old fires. There were other names, too, forgotten now.

'Hurry!' called the stone man from his hidden niche. Thomas felt the cold walls vibrate with the voice of the spectre. He shivered as he felt the power of the earth returning to the carved ragstone pillars, to the neatly positioned blocks. Always at night.

The Watchman's fire crackled and flared in the lee of the south wall. The priest walked away down the hill to the village, stopping just once to stare back at the half-constructed shell of the first stone church in the area. Then he was gone.

Thomas stepped from the darkness and stood, staring up through the empty roof to the clouds and the sky, and the gleaming light that was Jupiter. His heart was beating fast, but a great relief touched his limbs and his mind. And as always, he smiled, then closed his eyes for a moment. He thought of what he was doing. He thought of Beth, of what she would say if she knew his secret work; sweet Beth; with no children to comfort her she was now more alone than ever. But it would not be for much longer. The face was nearly finished . . .

'*Hurry!*'

A few more nights. A few more hours working in darkness, and all the Watchman's best efforts to guard the church would have been in vain.

The church would have been stolen. Thomas would have been the thief!

He moved through the gloom, now, to where a wooden ladder lay against the side wall. He placed the ladder against the high gallery—the leper's gallery—and climbed it. He drew the ladder up behind him and stepped across the debris of wood, stone and leather to the furthest, tightest corner of the place. Bare faces of the coarse ragstone watched the silent church. No mortar joined the stones. Their weight held them secure. They supported nothing but themselves.

At Thomas's muscular insistence, one of them moved, came away from the others.

With twilight gone, but night not yet fully descended, there was enough grey light for him to see the face that was carved there. He stared at the leafy beard, the narrowed, slanting eyes, the wide, flaring nostrils. He saw how the cheeks would look, how the hair would become spiky, how he would include the white and red berries of witch-thorn upon the twigs that clustered round the face . . .

Thomas stared at Thorn, and Thorn watched him by return, a cold smile on cold stone lips. Voices whispered in a sound realm that was neither in the church, nor in another world, but somewhere between the two, a shadowland of voice, movement and memory.

'I must be finished before the others,' the stone man whispered.

'You shall be,' said the mason, selecting chisel and hammer from his leather bag. 'Be patient.'

'I must be finished before the magic ones!' Thorn insisted, and Thomas sighed in irritation.

'You *shall* be finished before the magic ones. No-one has agreed upon the design of their faces, yet.'

The 'magic ones' were what Thomas called the Apostles. The twelve statues were temporarily in place above the altar, bodies completed but faces still smoothly blank.

'To control them I must be here first,' Thorn said.

'I've already opened your eyes. You can see how the other faces are incomplete.'

'Open them better,' said Thorn.

'Very well.'

Thomas reached out to the stone face. He touched the lips, the nose, the eyes. He knew every prominence, every rill, every chisel-mark. The grains of the stone were like pebbles beneath his touch. He could feel the hard-stone intrusion below the right eye, where the rag would not chisel well. There was a hardness, too, in the crown of Thorns, a blemish in the soft rock that would have to be shaped carefully to avoid cracking the whole design. As his fingers ran across the thorn man's lips, cold, old breath tickled him, the woodland man breathing from his time in the long past. As Thomas touched the eyes he felt the eyeballs move, impatient to see better.

I am in a wood grave, and a thousand years lie between us, Thorn had said. *Hurry, hurry. Bring me back.*

In the deepening darkness, working by touch alone, Thomas chiselled the face, bringing back the life of the lost god. The sound of his work was a sequence of shrill notes, stone music in the still church. John Tagworthy, the Watchman, outside by his fire, would be unaware of them. He might see a tallow candle by its glow upon the clouds, he might smell a fart from the distant castle on a still summer's night, but the noises of man and nature had long since ceased to bother his senses.

'Thomas! Thomas Wyatt! Where in God's Name *are* you?'

The voice, hailing him from below, so shocked Thomas that he dropped his chisel, and in desperately trying to catch the tool he cut himself. He stayed silent for a long moment, cursing Jupiter and the sudden band of bright stars for their light. The church was a place of shadows against darkness. As he peered at the north arch he thought he could see a man's shape, but it was only an unfinished timber. He reached for the heavy stone block that would cover the stone face, and as he did so the voice came again.

'God take your gizzard, Thomas Wyatt. It's Simon. Miller's son Simon!'

Thomas crept to the gallery's edge and peered over. The movement drew attention to him. Simon's pale features turned to look at him. 'I heard you working. What are you working on?'

'Nothing,' Thomas lied. 'Practising my craft on good stone with good tools.'

'Show me the face, Thomas,' said the younger man, and Thomas felt the blood drain from his head. *How had he known?* Simon was twenty years old, married for three years and still, like Thomas himself, childless. He was a freeman of course; he worked in his father's mill, but spent a lot of his time in the fields, both his family's strips and the land belonging to the Castle. His great ambition, though, was to be a Guildsman, and masonry was his aspiration.

'What face?'

'Send down the ladder,' Simon urged, and reluctantly Thomas let the wood scaffold down. The miller clambered up to the gallery, breathing hard. He smelled of garlic. He looked eagerly about in the gloom. 'Show me the green man.'

'Explain what you mean.'

'Come on, Thomas! Everybody knows you're shaping the Lord of Wood. I want to see him. I want to know how he looks.'

Thomas could hardly speak. His heart alternately stopped and raced. Simon's words were like stab wounds. *Everybody knew! How could everybody know?*

Thorn had spoken to him and to him alone. He had sworn the mason to silence and secrecy. For thirty days Thomas Wyatt had risked not just a flogging, but almost certain hanging for blasphemy, risked his life for the secret realm. Everybody *knew*?

'If everybody knows, why haven't I been stopped?'

'I don't mean *everybody*,' Simon said, as he felt blindly along the cold walls for a sign of Thomas's work. 'I mean the village. It's spoken in whispers. You're a hero, Thomas. We know what you're doing, and for whom. It's exciting; it's *right*. I've danced with them at the forest cross. I've carried the fire. I *know* how much power remains here. I may take God's name in oath—but that's safe to do. He has no power over me, or any of us. He doesn't belong on Dancing Hill. Don't *worry*, Thomas. We're your friends . . . Ah!'

Simon had found the loose stone. It was heavy and he grunted loudly as he took its weight, letting it down carefully to the floor. His breathing grew soft as he reached for the stone face. But Thomas could see how the young man drew back, fingers extended yet not touching the precious icon.

'There's magic in this, Thomas,' Simon said in awe.

'There's skill—working by night, working with fear—there's skill enough, I'll say that.'

'There's magic in the face,' Simon repeated. 'It's drawing power from the earth below. It's tapping the Dancing Well. There's water in the eyes, Thomas. The dampness of the old well. The face is brilliant.'

He struggled with the covering stone and replaced it. 'I wish it had been me. I wish the green man had chosen me. What an honour, Thomas. Truly.'

Thomas Wyatt watched his friend in astonishment. Was this *really* Simon the miller's son? Was this the young man who had carried the Cross every Resurrection Sunday for ten years? Simon Miller! *I've danced with them at the forest cross.*

'Who have you danced with at the crossroads, Simon?'

'*You* know,' Simon whispered. 'It's alive, Thomas. It's all alive. It's here, around us. It never went away. The Lord of Wood showed us . . .'

'Thorn? Is that who you mean?'

'*Him!*' Simon pointed towards the hidden niche. 'He's been here for years. He came the moment the monks decided to build the church. He came to save us, Thomas. And you're helping. I envy you . . .'

Simon climbed down the ladder. He was a furtive night shape, darting to the high arch where an oak door would soon be fitted, and out across the mud-churned hill, back round the forest, to where the village was a dark place, sleeping.

Thomas followed him down, placing the ladder back against the wall. But on the open hill, almost in sight of the Watchman's fire, he looked to the north, across the forest, to where the ridgeway was a high band of darkness against the pale grey glow of the clouds. Below the ridgeway a fire burned. He knew that he was looking at the forest cross, where the stone road of the Romans crossed the disused track between Woodhurst and Biddenden. He had played there as a child, despite being told never *ever* to follow the broken stone road.

There was a clearing at the deserted crossroads, and years ago he, and

Simon Miller's elder brother Wat, had often found the cold remains of fire and feasts. Outlaws, of course, and the secret baggage trains of the Saxon Knights who journeyed the hidden forest trails. Any other reason for the use of the place would have been unthinkable. Why, there was even an old gibbet, where forest justice was seen to be done . . .

With a shiver he remembered the time when he had come to the clearing and seen the swollen, greyish corpse of a man swinging from that blackened wood. Dark birds had been perched upon its shoulders. The face had had no eyes, no nose, no flesh at all, and the sight of the dead villain had stopped him from ever going back again.

Now, a fire burned at the forest cross. A fire like the fire of thirty nights ago, when Thorn had sent the girl for him . . .

He had woken to the sound of his name being called from outside. His wife, Beth, slept soundly on, turning slightly on the palliasse. It had been a warm night. He had tugged on his britches, and drawn a linen shirt over his shoulders. Stepping outside he had disturbed a hen, which clucked angrily and stalked to another nesting place.

The girl was dressed in dark garments. Her head was covered by a shawl. She was young, though, and the hand that reached for his was soft and pale.

'Who are you?' he said, drawing back. She had tugged at him. His reluctance to go with her was partly fear, partly concern that Beth would see him.

'Iagus goroth! Fiatha! *Fiatha!*' Her words were strange to Thomas. They were *like* the hidden language, but were not of the same tongue.

'Who *are* you?' he insisted, and the girl sighed, still holding his hand. At last she pointed to her bosom. Her eyes were bright beneath the covering of the shawl. Her hair was long and he sensed it to be red, like fire. 'Anuth!' she said. She pointed distantly. 'Thorn. You come with Thorn. With Anuth. Me. *Come.* Thomas. Thomas to Thorn. *Fiatha!*'

She dragged at his hand and he began to run. The grip on his fingers relaxed. She ran ahead of him, skirts swirling, body hunched. He tripped in the darkness, but she seemed able to see every low-hanging branch and proud beechwood root on the track. They entered the wood. He concentrated on her fleeing shape, calling, occasionally, for her to slow down. Each time he went sprawling she came back, making clicking sounds with her mouth, impatient, anxious. She helped him to his feet but immediately took off into the forest depths, heedless of risk to life and limb.

All at once he heard voices, a rhythmic beating, the crackle of fire . . . and the gentle sound of running water. She had brought him to the river. It wound through the forest, and then across downland, towards the Avon.

Through the trees he saw the fire. Anuth took his hand and pulled him, not to the bright glade, but towards the stream. As he walked he stared at the flames. Dark, human shapes passed before the fire. They seemed to be dancing. The heavy rhythm was like the striking of one bone against another. The voices were singing. The language was familiar to him, but incomprehensible.

Anuth dragged him past the firelit glade. He came to the river, and she slipped away. Surprised, he turned, hissing her name; but she had vanished. He looked back at the water, where starlight, and the light of a quarter moon, made the surface seem alive. There was a thick-trunked thorn tree growing from the water's edge. The thorn tree trembled and shifted in the evening wind.

The thorn tree grew before the startled figure of Thomas Wyatt. It rose, it straightened, it stretched. Arms, legs, the gleam of moonlight on eyes and teeth.

'Welcome, Thomas,' said the thorn tree.

He took a step backwards, frightened by the apparition.

'Welcome where?'

In front of him, Thorn laughed. The man's voice rasped, like a child with consumption. 'Look around you, Thomas. Tell me what you see.'

'Darkness. Woodland. A river, stars. Night. Cold night.'

'Take a breath, Thomas. What do you smell?'

'That same night. The river. Leaves and dew. The fire, I can smell the fire. And autumn. All the smells of autumn.'

'When did you last see and smell these things?'

Thomas, confused by the strange midnight encounter, shivered in his clothing. 'Last night. I've always seen and smelled them.'

'Then welcome to a place you know well. Welcome to the always place. Welcome to an autumn night, something that this land has always known, and will always enjoy.'

'But who are you?'

'I have been known by many names.' He came close to the trembling man. His hawthorn crown, with its strange horns, was like a broken tree against the clouds. His beard of leaves and long grass rustled as he spoke. His body quivered where the night breeze touched the clothing of nature that wound around his torso. 'Do you believe in God, Thomas?'

'He died for us. His son. On the cross. He is the Almighty . . .'

Thorn raised his arms. He held them sideways. He was a great cross in the cold night, and his crown of thorns was a beast's antlers. Old fears, forgotten shudders, plagued the villager, Thomas Wyatt. Ancestral cries mocked him. Memories of fire whispered words in the hidden language, confused his mind.

'I am the Cross of God,' said Thorn. 'Touch the wood, touch the sharp thorns . . .'

Thomas reached out. His actions were not his own. His fingers touched the cold flesh of the man's stomach. He felt the ridged muscle in the crossbeam, the bloody points of the thorns that rose from the man's head. He nervously brushed the gnarled wood of the thighs, and the proud branch that rose between them, hot to his fingers, nature's passion, never dying.

'What do you want of me?' Thomas asked quietly.

The cross became a man again. 'To make my image in the new shrine. To make that shrine my own. To make it as mine forever, no matter what manner of worship is performed within its walls . . .'

Thomas stared at the Lord of Wood.

'Tell me what I must do . . .'

Everybody knew, Simon had said. Everybody in the village. It was spoken in whispers. Thomas was a hero. Everybody knew. Everybody but Thomas Wyatt.

'Why have they kept it from me?' he murmured to the night. He had huddled up inside his jacket, and folded his body into the tight shelter of a wall bastion. The encounter with Simon had shaken him badly.

From here he could see north to Biddenden across the gloomy shapelessness of the forest. The castle, and the clustered villages of its demesne, were behind him. He saw only stars, pale clouds, and the flicker of fire, where strange worship occurred.

Why did the fire, in this midnight forest, call to him so much? Why was there such comfort in the thought of the warm glow from the piled branches, and the noisy prattle, and laughter, of those who clustered in its shadowy light? He had danced about a fire often enough: on May eve, at the passing of the day of All Hallows. But those fires were in the village bounds. His soul fluttered, a delighted bird, at the thought of the woodland fire. The smell of autumn, the touch of night's dew, the closeness to the souls of tree and plant; timeless eyes would watch the dancers. They were a shared life with the forest.

Why had he been kept in isolation? *Everybody knew.* The villagers who carried the bleeding, dying Christ through the streets on Resurrection Sunday . . . were they now carrying images of boar and stag and hare about the fire? He—Thomas—was a hero. They spoke of him in whispers. Everybody knew of his work. When had *they* been taken back to the beliefs of old? Had Thorn appeared to each of them as well?

Why didn't he *share* the new belief with them? It was the same belief. He used his craft; they danced for the gods.

As if he were of the same cold stone-stuff upon which he worked, the others kept him distant, watched him from afar. Did Beth know? Thomas shivered. The hours passed. He could feel the gibbet rope around his neck. Only one word out of place, one voice overheard—one whisper to the wrong man, and Thomas Wyatt would be a grey thing, slung by its neck, prey for dark birds. Eyes, nose, the flesh of the face. Every feature that he pecked for Thorn with hammer and chisel would be pecked from him by hard, wet beaks.

From the position of the moon, Thomas realized he had been sitting by the church for several hours. John the Watchman had not walked past. Now that he thought of it, Thomas could hear the man's snoring, coming as if from a far place.

Thomas eased himself to his feet. He lifted his bag gently to his shoulder, over-cautious about the ring and strike of iron tools within the leather. But as he walked towards the path he heard movement in the church. The Watchman snored distantly.

It must be Simon, the miller's son, Thomas thought, back for another look at the face of the woodland god.

Irritated, and still confused, Thomas stepped into the church again, and looked towards the gallery. The ladder was against the balcony. He could hear the stone being moved. There was a time of silence, then the stone was put back. A figure moved to the ladder and began to descend.

Thomas watched in astonishment. He stepped into greater darkness as the priest looked round, then hauled the ladder back to its storage place. All Thomas heard was the sound of the priest's laughter. The man passed through the gloom, long robe swirling through the dust and debris.

Even the priest knew! And that made no sense at all. Thomas slept restlessly, listening to the soft breathing of his wife. Several times the urge to wake her, to speak to her, made him whisper her name and shake her shoulders. But she slumbered on. At sunrise they were up together, but he was so tired he could hardly speak. They ate hard bread, moistened with cold, thin gruel. Thomas tipped the last of their ale into a clay mug. The drink was more meaty than the gruel, but he swallowed the sour liquid and felt its warming tingle.

'The last of the ale,' he said ruefully, tapping the barrel.

'You've been too busy to brew,' Beth said from the table. 'And I'm not skilled.' She was wrapped in a heavy wool cloak. The fire was a dead place in the middle of the small room. Grey ash drifted in the light from the roof hole.

'But no *ale*!' He banged his cup on the barrel in frustration. Beth looked up at him, surprised by his anger.

'We can get ale from the miller. We've done it before and repaid him from our own brewing. It's not the end of the world.'

'I've had no time to brew,' Thomas said, watching Beth through hooded, rimmed eyes. 'I've been working on something of importance. I expect you know what.'

She shrugged. 'Why would I know? You never talk about it.' Her pale face was sweet. She was as pretty now as when he had married her; fuller in body, yes, and wiser in the ways of life. That they were childless had not affected her spirit. She had allowed the wise women to dose her with herbs and bitter spices, to take her to strange stones, and stranger foreigners; she had been seen by apothecaries and doctors, and Thomas had worked in their fields to pay them. And of course, they had prayed. Now Thomas felt too old to care about children. Life was good with Beth, and their sadness had drawn them closer than most couples he knew.

'Everybody knows what I'm working on,' he said bitterly.

'Well, I don't,' she replied. 'But I'd like to . . .'

Perhaps he had been unfair to her. Perhaps she too was kept apart from the village's shared knowledge. He lied to her. 'You must not say a word to anyone. But I'm working on the face of Jesus.'

Beth was delighted. 'Oh Thomas! That's wonderful. I'm so proud of you.' She came round to him and hugged him. Outside, Master mason Tobias Craven called out his name, among others, and he trudged up to the church on Dancing Hill.

His work was uneven and lazy that day. The chisel slipped, the stone splintered, the hammer caught his thumb twice. He was distracted and deeply concerned by what he had seen the night before. When the priest came to the church, to walk among the bustle of activity and inspect the day's progress, Thomas watched him carefully, hoping for some sign of recognition. But the man just smiled, and nodded, then carried the small light of Christ to the altar, and said silent prayers for an hour or more.

At sundown, Thomas felt his body shaking. When the priest called the craftsmen—Thomas included—into the vestry for wine, Thomas stood by the door, staring at the dark features of the Man of God. The priest, handing him his cup, merely said, 'God be with you, Thomas.' It was what he always said.

Tobias Craven came over to him. His face was grey with dust, his clothing heavy with dirt. His dialect was difficult for Thomas to understand, and Thomas was suspicious of the gesture anyway. Would he now discover that the foreigners, too, knew of the face of the woodland deity, half completed behind its door of stone?

'Your work is good, Thomas. Not today, perhaps, but usually. I've watched you.'

'Thank you.'

'At first I was reluctant to allow you to work as a mason among us. It was at the priest's insistence: one local man to work in every craft. It seemed a superstitious idea to me. But now I'm glad. I approve. It's an enlightened gesture, I realize, to allow local men, not of Guilds, to display their skills. And your skill is remarkable.'

Thomas swallowed hard. 'To be a Guildsman would be a great honour.'

Master Tobias looked crestfallen. 'Aye, but alas. I wish I had seen your work when you were twenty, not thirty. But I can write a note for you, to get you better work in the area.'

'Thank you,' Thomas said again.

'Have you travelled, Thomas?'

'Only to Glastonbury. I made a pilgrimage in the third year of my marriage.'

'Glastonbury,' Master Tobias repeated, smiling. 'Now that is a fine Abbey. I've seen it just once. Myself, I worked at York, and at Carlisle, on the Minsters. I was not a Master, of course. But that was cherished work. Now I'm a Guild Master, building tiny churches in remote places. But it gives fulfilment to the soul, and one day I shall die and be buried in the shadow of a place I have built myself. There is satisfaction in the thought.'

'May that not be for many years.'

'Thank you, Thomas.' Tobias drained his cup. 'And now, from God's work to nature's work—'

Thomas paled. Did he mean woodland worship? The Master mason winked at him.

'A good night's sleep!'

When the others had gone, Thomas slipped out of the sheltering woodland and made his way back to the church. The Watchman was fussing with his fire. There was less cloud this evening and the land, though murky, was quite visible for many miles around.

Inside the church, Thomas looked up at the gallery. Uncertainty made him hesitate, then he shook his head. 'Until I understand better . . .' he murmured, and made to turn for home.

'Thomas!' Thorn called. 'Hurry, Thomas.'

Strange green light played off the stone of the church. It darted around him, like will-o'-the-wisp. Fingers prodded him forward, but when he turned there was nothing but shadow.

Again, Thorn called to him.

With a sigh, Thomas placed the ladder against the gallery and climbed up to the half-finished face. Thorn smiled at him. The narrow eyes sparkled with moisture. The leaves and twigs that formed his hair and beard seemed to rustle. The stone strained to move.

'Hurry, Thomas. Open my eyes better.'

'I'm frightened,' the man said. 'Too many people know what I'm doing.'

'Carve me. Shape my face. I must be here before the others. *Hurry!*'

The lips of the forest god twitched with the ghostly figure's anguish. Thomas reached out to the cold stone and felt its stillness. It was just a carving. It had no life. He imagined the voice. It was just a man who told him to make the carving, a man dressed in woodland disguise. Until he knew he was safe, he would not risk discovery. He climbed back down the ladder. Thorn called to him, but Thomas ignored the cry.

At his house a warm fire burned in the middle of the room, and an iron pot of thick vegetable broth steamed above it. There was fresh ale from the miller, and Beth was pleased to see him home so early. She stitched old clothes, seated on a low stool, close to the wood fire. Thomas ate, then drank ale, leaning on the table, his mason's tools spread out before him. The ale was strong and soon went to his head. He felt dizzy, sublimely detached from his body. The warmth, the sensation of drunkenness, his full stomach, all of these things made him drowsy, and slowly his head sank to his arms . . .

A cold blast of air on his neck half roused him. His name was being called. At first he thought it was Beth, but soon, as he surfaced from pleasant oblivion, he recognised the rasping voice of Thorn.

The fire burned high, fanned by the draught from the open door. Beth still sat on her stool, but was motionless and silent, staring at the flames. He spoke her name, but she didn't respond. Thorn called to him again and he looked out at the dark night. He felt a sudden chill of fear. He gathered his tools into his bag and stepped from the house.

Thorn stood in the dark street, a tall figure, his horns of wood black against the sky. There was a strong smell of earth about him. He moved towards Thomas, leaf-clothes rustling.

'The work is unfinished, Thomas.'

'I'm afraid for my life. Too many people know what I'm doing.'

'Only the finishing of the face matters. Your fear is of no consequence. You agreed to work for me. You must go to the church. Now.'

'But if I'm caught!'

'Then another will be found. Go back to the work, Thomas. Open my eyes properly. It *must* be done.'

He turned from Thorn and sighed. There was something wrong with Beth and it worried him, but the persuasive power of the night figure was too strong to counter, and he began to walk wearily towards the church. Soon the village was invisible behind him. Soon the church was a sharp relief against the night sky. The Watchman's fire burned high, and the autumn night was sweet with the smell of woodsmoke. The Watchman himself seemed to be dancing, or so Thomas thought at first. He strained to see better and soon realized that John had fallen asleep and set light to his clothing. He was brushing and beating at his leggings, his grunts of alarm like the evening call of a boar.

The moment's humour passed and a sudden anger took Thomas. Thorn's words were like sharp stab wounds to his pride: his fear was of no consequence. Only the work of carving mattered. He would be caught and it would be of no consequence. He would swing, slowly strangling, from the castle gallows and it would be of no consequence. Another would be found!

'No!' he said aloud. 'No. I will *not* work for Thorn tonight. Tonight is *my* night. Damn Thorn. Damn the face. Tomorrow I will open its eyes, but not now.'

And with a last glance at the Watchman, who had extinguished the fire and settled down again, he turned back to the village.

But as he approached his house, aware of the glow of the fire through the small window, his anger changed to a sudden dread. He began to feel sick. He wanted to cry out, to alert the village. A voice in his head urged him to turn and go back to the night wood. His house, once so welcoming, threatened him deeply. It seemed surrounded by an aura, detached from the real world.

He walked slowly to the small window. He could hear the crackle and spit of the flames. Wood smoke was sweet in the air. Somewhere, at the village bounds, two dogs barked.

The feeling of apprehension in him grew, a strangling weed that made him dizzy. But he looked through the window. And he did not faint, nor cry out, at what he saw within, though a part of his spirit, part of his life, flew away from him then, abandoning him, making him wither and age; making him die a little.

Thorn stood with his back to the fire. His mask of autumn leaves and spiky wood was bright and eerie—dark hair curled from beneath the mask. His arms were wound around with creeper and twine, and twigs of oak, elm and lime were laced upon this binding. Save for these few fragments of nature's clothing he was naked. The black hair on his body gave him the appearance of a burned oak stump, gnarled and weathered by the years. His manhood was a smooth, dark branch, cut to the length of firewood.

Beth was on her knees before him, her weight taken on her elbows. Her skirts were on the floor beside her. The yellow flames cast a flickering glow upon her plump, pale flesh, and Thomas half closed his eyes in despair. He managed to stifle his scream of anguish, but he could not stop himself from watching.

And he uttered no sound, despite the pain, as Thorn dropped down upon the waiting woman.

As he ran to the church the Watchman woke, then stood up, picking up his heavy staff. Thomas Wyatt knocked him down, then drew a flaming wood brand from the brazier. Tool-bag on his shoulder he entered the church, and held the fire high. The ladder was against the balcony. Pale features peered down at him and the ladder began to move. But Simon, the miller's son, was not quite quick enough. Casting the burning wood aside, Thomas leapt for the scaffold and began to ascend.

'I was just looking, Thomas,' Simon cried, then tried to fling the ladder back. Thomas clutched at the balcony, then hauled himself to safety. He said no word to Simon, who backed against the wall where the loose stone was fitted.

'You mustn't touch him, Thomas!'

In the darkness, Simon's eyes were gleaming orbs of fear. Thomas took him by the shoulders and flung him to the balcony, then used a stone to strike him.

'No, Thomas! No!'

The younger man had toppled over the balcony. He held on for dear life, fingers straining to hold his weight.

'Tricked!' screamed Thomas. 'All a trick! Duped! Cuckolded! All of you knew. All of you *knew*!'

'No, Thomas. In the Name of God, it wasn't like that!'

His hammer was heavy. He swung it high. Simon's left hand vanished and the man's scream of pain was deafening. 'She had no other way!' he cried hysterically. 'No, Thomas! No! She chose it! She *chose* it! Thorn's gift to you both.'

The hammer swung. Crushed fingers left bloody marks upon the balcony. Simon crashed to the floor below and was still.

'All of you knew!' Thomas Wyatt cried. He wrenched the loose stone away. Thorn watched him from the blackness through his half-opened eyes. Thomas could see every feature, every line. The mouth stretched in a mocking grin. The eyes narrowed, the nostrils flared.

'Fool. Fool!' whispered the stone man. 'But you cannot stop me now.'

Thomas slapped his hand against the face. The blow stung his flesh. He reached for his chisel, placed the sharp tool against one of the narrow eyes.

'NO!' screeched Thorn. His face twisted and turned. The stone of the church shuddered and groaned. Thomas hesitated. A green glow came from the features of the deity. The eyes were wide with fear, the lips drawn back below the mask. Thomas raised his hammer.

'NO!' screamed the head again. Arms reached from the wall. The light expanded. Thomas backed off, terrified by the spectre which had appeared there, a ghastly green version of Thorn himself, a creature half ghost, half stone, tied to the wall of the church, but reaching out from the cold rock, reaching for Thomas Wyatt, reaching to kill him.

Thomas raised the chisel, raised the hammer. He ran back to the face of Thorn and with a single, vicious blow, drove a gouging furrow through the right eye.

The church shuddered. A block of stone fell from the high wall, striking Thomas on the shoulder. The whole balcony vibrated with Thorn's pain and anger.

Again he struck. The left eye cracked, a great split in the stone. Dampness oozed from the wound. The scream from the wall was deafening. Below the balcony, yellow light glimmered. The Watchman, staring up to where Thomas performed his deed of vengeance.

Then a crack appeared down the whole side of the church. The entire gallery where Thomas had worked dropped by a man's height, and Thomas was flung to the balcony. He struggled to keep his balance, then went over the wall, scrabbling at the air. Thorn's stone-scream was a nightmare sound. Air was cool on the mason's skin. A stone pedestal broke his fall. Broke his back. The village woke to the sound of the priest's terrible scream. He stumbled from the mason's house, hands clutching at his eyes, trying to staunch the flow of blood. He scrabbled at the wood mask, stripping away the thorn, the oak, the crisp brown leaves, exposing dark hair, a thin dark beard.

The priest—Thorn's priest—turned blind eyes to the church. Naked, he began to stagger and stumble towards the hill. Behind him, the villagers followed, torches burning in the night.

Thomas lay across the marble pillar, a few feet from the ground. There was no sensation in his body, though his lungs expanded to draw air into his chest. He lay like a sacrificial victim, arms above his head, legs limp. The Watchman circled him in silence. The church was still.

Soon the priest approached him, hands stretched out before him. The pierced orbs of his eyes glistened as he leaned close to Thomas Wyatt.

'Are you dying, then?'

'I died a few minutes ago,' Thomas whispered. The priest's hands on his face were gentle. Blood dripped from the savaged eyes.

'Another will come,' Thorn said. 'There are many of us. The work will be completed. No church will stand that is not a shrine to the true faith. The spirit of Christ will find few havens in England.'

'Beth . . .' Thomas whispered. He could feel the bird of life struggling to escape him. The Watchman's torch was already dimming.

Thorn raised Thomas's head, a finger across the dry lips. 'You should not have seen,' said the priest. 'It was a gift for a gift. Our skills, the way of ritual, of fertility, for your skill with stone. Another will come to replace me. Another will be found to finish your work. But there will be no child for you, now. No child for Beth.'

'What have I done?' Thomas whispered. 'By all that's holy, what have I done?'

From above him, from a thousand miles away, came the ring of chisel on stone.

'Hurry,' he heard Thorn call into the night. 'Hurry!'

TERRY PRATCHETT

TROLL BRIDGE

The air blew off the mountains, filling the air with fine ice crystals.

It was too cold to snow. In weather like this wolves came down into villages, trees in the heart of the forest exploded when they froze.

In weather like this right-thinking people were indoors, in front of the fire, telling stories about heroes.

It was an old horse. It was an old rider. The horse looked like a shrink-wrapped toast rack; the man looked as though the only reason he wasn't falling off was because he couldn't muster the energy. Despite the bitterly cold wind, he was wearing nothing but a tiny leather kilt and a dirty bandage on one knee.

He took the soggy remnant of a cigarette out of his mouth and stubbed it out on his hand.

'Right,' he said, 'let's do it.'

'That's all very well for you to say,' said the horse. 'But what if you have one of your dizzy spells? And your back is playing up. How shall I feel, being eaten because your back's played you up at the wrong moment?'

'It'll never happen,' said the man. He lowered himself on to the chilly stones, and blew on his fingers. Then, from the horse's pack, he took a sword with an edge like a badly maintained saw and gave a few half-hearted thrusts at the air.

'Still got the old knackcaroony,' he said. He winced, and leaned against a tree.

'I'll swear this bloody sword gets heavier every day.'

'You ought to pack it in, you know,' said the horse. 'Call it a day. This sort of thing at your time of life. It's not right.'

The man rolled his eyes.

'Blast that damn distress auction. This is what comes of buying something that belonged to a wizard,' he said, to the cold world in general. 'I looked at your teeth, I looked at your hooves, it never occurred to me to *listen*.'

'Who did you think was bidding against you?' said the horse.

Cohen the Barbarian stayed leaning against the tree. He was not sure that he could pull himself upright again.

'You must have plenty of treasure stashed away,' said the horse. 'We could go Rimwards. How about it? Nice and warm. Get a nice warm place by a beach somewhere, what do you say?'

'No treasure,' said Cohen. 'Spent it all. Drank it all. Gave it all away. Lost it.'

'You should have saved some for your old age.'

'Never thought I'd *have* an old age.'

'One day you're going to die,' said the horse. 'It might be today.'

'I know. Why do you think I've come here?'

The horse turned and looked down towards the gorge. The road here was pitted and cracked. Young trees were pushing up between the stones. The forest crowded in on either side. In a few years, no one would know there'd even been a road here. By the look of it, no one knew now.

'You've come here to *die*?'

'No. But there's something I've always been meaning to do. Ever since I was a lad.'

'Yeah?'

Cohen tried easing himself upright again. Tendons twanged their red-hot messages down his legs.

'My dad,' he squeaked. He got control again. 'My dad,' he said, 'said to me—' He fought for breath.

'Son,' said the horse, helpfully.

'What?'

'Son,' said the horse. 'No father ever calls his boy "son" unless he's about to impart wisdom. Well-known fact.'

'It's *my* reminiscence.'

'Sorry.'

'He said . . . Son . . . yes, OK . . . Son, when you can face down a troll in single combat, then you can do anything.'

The horse blinked at him. Then it turned and looked down, again, through the tree-jostled road to the gloom of the gorge. There was a stone bridge down there.

A horrible feeling stole over it.

Its hooves jiggled nervously on the ruined road.

'Rimwards,' it said. 'Nice and warm.'

'No.'

'What's the good of killing a troll? What've you got when you've killed a troll?'

'A dead troll. That's the point. Anyway, I don't have to kill it. Just defeat it. One on one. *Mano a* . . . troll. And if I didn't try my father would turn in his mound.'

'You told *me* he drove you out of the tribe when you were eleven.'

'Best day's work he ever did. Taught me to stand on other people's feet. Come over here, will you?'

The horse sidled over. Cohen got a grip on the saddle and heaved himself fully upright.

'And you're going to fight a troll today,' said the horse.

Cohen fumbled in the saddlebag and pulled out his tobacco pouch. The wind whipped at the shreds as he rolled another skinny cigarette in the cup of his hands.

'Yeah,' he said.

'And you've come all the way out here to do it.'

'Got to,' said Cohen. 'When did you last see a bridge with a troll under it? There were hundreds of 'em when I was a lad. Now there's more trolls in the cities than there are in the mountains. Fat as butter, most of 'em. What did we fight all those wars for? Now . . . cross that bridge.'

It was a lonely bridge across a shallow, white, and treacherous river in a deep valley. The sort of place where you got—

A grey shape vaulted over the parapet and landed splayfooted in front of the horse. It waved a club.

'All *right*,' it growled.

'Oh—' the horse began.

The troll blinked. Even the cold and cloudy winter skies seriously reduced the conductivity of a troll's silicon brain, and it had taken it this long to realize that the saddle was unoccupied.

It blinked again, because it could suddenly feel a knife point resting on the back of its neck.

'Hello,' said a voice by its ear.

The troll swallowed. But very carefully.

'Look,' it said desperately, 'It's tradition, OK? A bridge like this, people ort to *expect* a troll . . . 'Ere,' it added, as another thought crawled past, ''ow come I never 'eard you creepin' up on me?'

'Because I'm *good* at it,' said the old man.

'That's right,' said the horse. 'He's crept up on more people than you've had frightened dinners.'

The troll risked a sideways glance.

'Bloody hell,' it whispered. 'You think you're Cohen the Barbarian, do you?'

'What do *you* think?' said Cohen the Barbarian.

'Listen,' said the horse, 'if he hadn't wrapped sacks round his knees you could have told by the clicking.'

It took the troll some time to work this out.

'Oh, *wow*,' it breathed. 'On *my* bridge! Wow!'

'What?' said Cohen.

The troll ducked out of his grip and waved its hands frantically. 'It's all right! It's all right!' it shouted, as Cohen advanced. 'You've got me! You've got me! I'm not arguing! I just want to call the family up, all right? Otherwise no one'll ever believe me. *Cohen the Barbarian*! On *my* bridge!'

Its huge stony chest swelled further. 'My bloody brother-in-law's always swanking about his huge bloody wooden bridge, that's all my wife ever talks about. Hah! I'd like to see the look on his face . . . oh, no! What can you think of me?'

'Good question,' said Cohen.

The troll dropped its club and seized one of Cohen's hands.

'Mica's the name,' it said. 'You don't know what an honour this is!'

He leaned over the parapet. 'Beryl! Get up here! Bring the kids!'

He turned back to Cohen, his face glowing with happiness and pride.

'Beryl's always sayin' we ought to move out, get something better, but I tell her, this bridge has been in our family for generations, there's always been a troll under Death Bridge. It's tradition.'

A huge female troll carrying two babies shuffled up the bank, followed by a tail of smaller trolls. They lined up behind their father, watching Cohen owlishly.

'This is Beryl,' said the troll. His wife glowered at Cohen. 'And this—' he propelled forward a scowling smaller edition of himself, clutching a junior version of his club—'is my lad Scree. A real chip off the old block. Going to take on the bridge when I'm gone, ain't you, Scree. Look, lad, this is Cohen the Barbarian! What d'you think o' that, eh? On *our* bridge! We don't just have rich fat soft ole merchants like your uncle Pyrites gets,' said the troll, still talking to his son but smirking past him to his wife, 'we 'ave proper heroes like they used to in the old days.'

The troll's wife looked Cohen up and down.

'Rich, is he?' she said.

'Rich has got nothing to do with it,' said the troll.

'Are you going to kill our dad?' said Scree suspiciously.

''*Corse* he is,' said Mica severely. 'It's his job. An' then I'll get famed in song an' story. This is Cohen the Barbarian, right, not some bugger from the village with a pitchfork. 'E's a famous hero come all this way to see us, so just you show 'im some respect.

'Sorry about that, sir,' he said to Cohen. 'Kids today. You know how it is.'

The horse started to snigger.

'Now look—' Cohen began.

'I remember my dad tellin' me about you when I was a pebble,' said Mica. ''E bestrides the world like a clossus, he said.'

There was silence. Cohen wondered what a clossus was, and felt Beryl's stony gaze fixed upon him.

'He's just a little old man,' she said. 'He don't look very heroic to me. If he's so good, why ain't he *rich*?'

'Now you listen to me—' Mica began.

'This is what we've been waiting for, is it?' said his wife. 'Sitting under a leaky bridge the whole time? Waiting for people that never come? Waiting for little old bandy-legged old men? I should have listened to my mother! You want me to let our son sit under a bridge waiting for some little old man to kill him? That's what being a troll is all about? Well, it ain't happening!'

'Now you just—'

'Hah! Pyrites doesn't get little old men! He gets big fat merchants! He's *someone*. You should have gone in with him when you had the chance!'

'I'd rather eat worms!'

'Worms? Hah? Since when could we afford to eat worms?'

'Can we have a word?' said Cohen.

He strolled towards the far end of the bridge, swinging his sword from one hand. The troll padded after him.

Cohen fumbled for his tobacco pouch. He looked up at the troll, and held out the bag.

'Smoke?' he said.

'That stuff can kill you,' said the troll.

'Yes. But not today.'

'Don't you hang about talking to your no-good friends!' bellowed Beryl, from her end of the bridge. 'Today's your day for going down to the sawmill! You know Chert said he couldn't go on holding the job open if you weren't taking it seriously!'

Mica gave Cohen a sorrowful little smirk.

'She's very supportive,' he said.

'I'm not climbing all the way down to the river to pull you out again!' Beryl roared. 'You tell him about the billy goats, Mr Big Troll!'

'Billy goats?' said Cohen.

'I don't know *anything* about billy goats,' said Mica. 'She's always going on about billy goats. I have no knowledge whatsoever about billy goats.' He winced.

They watched Beryl usher the young trolls down the bank and into the darkness under the bridge.

'The thing is,' said Cohen, when they were alone, 'I wasn't intending to kill you.'

The troll's face fell.

'You weren't?'

'Just throw you over the bridge and steal whatever treasure you've got.'

'You were?'

Cohen patted him on the back. 'Besides,' he said, 'I like to see people with . . . good memories. That's what the land needs. Good memories.'

The troll stood to attention.

'I try to do my best, sir,' it said. 'My lad wants to go off to work in the city. I've tole him, there's bin a troll under this bridge for nigh on five hundred year—'

'So if you just hand over the treasure,' said Cohen, 'I'll be getting along.'

The troll's face creased in sudden panic.

'Treasure? Haven't got any,' it said.

'Oh, come *on*,' said Cohen. 'Well-set up bridge like this?'

'Yeah, but no one uses this road any more,' said Mica. 'You're the first one along in months, and that's a fact. Beryl says I ought to have gone in with her brother when they built that new road over his bridge, but,' he raised his voice, 'I said, there's been trolls under this bridge—'

'Yeah,' said Cohen.

'The trouble is, the stones keep on falling out,' said the troll. 'And you'd never believe what those masons charge. Bloody dwarfs. You can't trust 'em.' He leaned towards Cohen. 'To tell you the truth, I'm having to work three days a week down at my brother-in-law's lumber mill just to make ends meet.'

'I thought your brother-in-law had a bridge?' said Cohen.

'One of 'em has. But my wife's got brothers like dogs have fleas,' said the troll. He looked gloomily into the torrent. 'One of 'em's a lumber merchant down in Sour Water, one of 'em runs the bridge, and the big fat one is a merchant over on Bitter Pike. Call that a proper job for a troll?'

'One of them's in the bridge business, though,' said Cohen.

'Bridge business? Sitting in a box all day charging people a silver piece to walk across? Half the time he ain't even there! He just pays some dwarf to take the money. And he calls himself a troll! You can't tell him from a human till you're right up close!'

Cohen nodded understandingly.

'D'you know,' said the troll, 'I have to go over and have dinner with them every week? All three of 'em? And listen to 'em go on about moving with the times . . .'

He turned a big, sad face to Cohen.

'What's wrong with being a troll under a bridge?' he said. 'I was brought up to be a troll under a bridge. I want young Scree to be a troll under a bridge after I'm gone. What's wrong with that? You've got to have trolls under bridges. Otherwise, what's it all about? What's it all *for*?'

They leaned morosely on the parapet, looking down into the white water.

'You know,' said Cohen slowly, 'I can remember when a man could ride all the way from here to the Blade Mountains and never see another living thing.' He fingered his sword. 'At least, not for very long.'

He threw the butt of his cigarette into the water. 'It's all farms now. All little farms, run by little people. And *fences* everywhere. Everywhere you look, farms and fences and little people.'

'She's right, of course,' said the troll, continuing some interior conversation. 'There's no future in just jumping out from under a bridge.'

'I mean,' said Cohen, 'I've nothing against farms. Or farmers. You've got to have them. It's just that they used to be a long way off, around the edges. Now *this* is the edge.'

'Pushed back all the time,' said the troll. 'Changing all the time. Like my brother-in-law Chert. A lumber mill! A *troll* running a lumber mill! And you should see the mess he's making of Cutshade Forest!'

Cohen looked up, surprised.

'What, the one with the giant spiders in it?'

'Spiders? There ain't no spiders now. Just stumps.'

'Stumps? *Stumps*? I used to like that forest. It was . . . well, it was darksome. You don't get proper darksome any more. You really knew what terror was, in a forest like that.'

'You want darksome? He's replanting with spruce,' said Mica.

'Spruce!'

'It's not his idea. He wouldn't know one tree from another. That's all down to Clay. He put him up to it.'

Cohen felt dizzy. 'Who's Clay?'

'I said I'd got *three* brothers-in-law, right? He's the merchant. So he said replanting would make the land easier to sell.'

There was a long pause while Cohen digested this.

Then he said, 'You can't sell Cutshade Forest. It doesn't belong to anyone.'

'Yeah. He says that's why you can sell it.'

Cohen brought his fist down on the parapet. A piece of stone detached itself and tumbled down into the gorge.

'Sorry,' he said.

'That's all right. Bits fall off all the time, like I said.'

Cohen turned. 'What's happening? I remember all the big old wars. Don't you? You must have fought.'

'I carried a club, yeah.'

'It was supposed to be for a bright new future and law and stuff. That's what people said.'

'Well, I fought because a big troll with a whip told me to,' said Mica, cautiously. 'But I know what you mean.'

'I mean it wasn't for farms and spruce trees. Was it?'

Mica hung his head. 'And here's me with this apology for a bridge. I feel really bad about it,' he said, 'you coming all this way and everything—'

'And there was some king or other,' said Cohen, vaguely, looking at the water. 'And I think there were some wizards. But there was a king. I'm pretty certain there was a king. Never met him. You know?' He grinned at the troll. 'I can't remember his name. Don't think they ever told me his name.'

About half an hour later Cohen's horse emerged from the gloomy woods on to a bleak, windswept moorland. It plodded on for a while before saying, 'All right . . . how much did you give him?'

'Twelve gold pieces,' said Cohen.

'Why'd you give him twelve gold pieces?'

'I didn't have more than twelve.'

'You must be mad.'

'When I was just starting out in the barbarian hero business,' said Cohen, 'every bridge had a troll under it. And you couldn't go through a forest like we've just gone through without a dozen goblins trying to chop your head off.' He sighed. 'I wonder what happened to 'em all?'

'You,' said the horse.

'Well, yes. But I always thought there'd be some more. I always thought there'd be some more edges.'

'How old are you?' said the horse.

'Dunno.'

'Old enough to know better, then.'

'Yeah. Right.' Cohen lit another cigarette and coughed until his eyes watered.

'Going soft in the head!'

'Yeah.'

'Giving your last dollar to a troll!'

'Yeah.' Cohen wheezed a stream of smoke at the sunset.

'Why?'

Cohen stared at the sky. The red glow was as cold as the slopes of hell. An icy wind blew across the steppes, whipping at what remained of his hair.

'For the sake of the way things should be,' he said.

'Hah!'

'For the sake of things that were.'

'Hah!'

Cohen looked down.

He grinned.

'And for three addresses. One day I'm going to die,' he said, 'but not, I think, today.'

The air blew off the mountains, filling the air with fine ice crystals. It was too cold to snow. In weather like this wolves came down into villages, trees in the heart of the forest exploded when they froze. Except there were fewer and fewer wolves these days, and less and less forest.

In weather like this right-thinking people were indoors, in front of the fire.

Telling stories about heroes.

SELECT BIBLIOGRAPHY

SINGLE- AND SHARED-AUTHOR COLLECTIONS

Because of the overlap between the two genres, some of the collections below contain science fiction as well as fantasy. In a few cases, where successful short story sequences have been followed by novels using the same setting or the same characters, I have cited the novels as well.

Anderson, Poul, *Operation Chaos* (New York, 1971).
Benson, E. F., *The Room in the Tower and other stories* (London, 1912).
—— *Spook Stories* (London, 1928).
Bradley, Marion Zimmer, *Lythande* (New York, 1986).
Brunner, John, *The Compleat Traveller in Black* (London, 1987).
Buchan, John, *The Watcher at the Threshold* (London, 1902).
—— *The Moon Endureth* (London, 1912).
—— *The Runagates Club* (London, 1928).
—— *The Best Short Stories of JB*, ed. David Daniell (2 vols., London, 1980).
Carter, Angela, *Fireworks: Nine Profane Pieces* (London, 1974).
—— *The Bloody Chamber* (London, 1979).
Chambers, Robert W., *The King in Yellow* (New York, 1895).
Davidson, Avram, *Or All the Seas with Oysters* (New York, 1962).
—— *What Strange Seas and Skies* (New York, 1965).
—— *Strange Seas and Shores* (New York, 1971).
—— *Collected Fantasies* (New York, 1982).
—— *The Adventures of Doctor Eszterhazy* (Philadelphia, 1990).
de Camp, L. Sprague, *The Purple Pterodactyls* (New York, 1979).
—— and Pratt, Fletcher, *Tales from Gavagan's Bar* (expanded edn., New York, 1978).
—— *The Complete Compleat Enchanter* (New York, 1989).
del Rey, Lester, *. . . And Some Were Human* (Philadelphia, 1948).
—— *Robots and Changelings* (New York, 1957).
Derleth, August, *The Mask of Cthulhu* (1948).
—— *The Trail of Cthulhu* (1962).
Dunsany, Lord, *At the Edge of the World*, ed. Lin Carter (New York, 1970).
—— *Beyond the Fields We Know*, ed. Lin Carter (New York, 1972).
—— *Over the Hills and Far Away*, ed. Lin Carter (New York, 1974).
Ellison, Harlan, *Deathbird Stories* (New York, 1975).

Forster, E. M., *The Collected Short Stories of EMF* (London, 1947).
Heinlein, Robert A., *Waldo and Magic Inc.* (New York, 1950).
Howard, Robert E., *Skull-Face and others* (Sauk City, Wis., 1946).
—— *The Dark Man and others* (Sauk City, Wis., 1963).
—— 'The Conan Cycle', ed. L. Sprague de Camp and others, in 19 vols. (projected), (New York, 1953 onwards): see further Tuck in 'Works of Reference' below.
Kuttner, Henry, *Ahead of Time* (New York, 1953).
—— *Bypass to Otherness* (New York, 1961).
—— *The Best of HK* (2 vols., London, 1966).
Lanier, Sterling E., *The Peculiar Exploits of Brigadier Ffellowes* (New York, 1971).
—— *The Curious Quests of Brigadier Ffellowes* (New York, 1986).
Lee, Tanith, *Unsilent Night* (Cambridge, Mass., 1981).
—— *Red as Blood, or Tales from the Sisters Grimmer* (New York, 1983).
—— *The Gorgon and other Beastly Tales* (New York, 1985).
—— *Forests of the Night* (London, 1989).
LeGuin, Ursula K., *The Wind's Twelve Quarters* (New York, 1985).
—— *Buffalo Gals and other Animal Presences* (New York, 1985).
Leiber, Fritz, *Swords and Deviltry* (New York, 1968).
—— *Swords in the Mist* (New York, 1968).
—— *The Swords of Lankhmar* (New York, 1968).
—— *Swords against Death* (New York, 1970).
—— *Swords against Wizardry* (New York, 1970).
—— *Swords and Ice-Magic* (New York, 1977).
—— *The Knight and Knave of Swords* (New York, 1988).
Lovecraft, H. P., *The Outsider and others* (Sauk City, Wis., 1939).
—— *The Lurking Fear* (New York, 1947).
—— *The Haunter of the Dark* (London, 1951).
—— *The Dunwich Horror and others* (Sauk City, Wis., 1963).
—— *Dagon and other Macabre Tales* (Sauk City, Wis., 1965).
—— and Derleth, August, *The Watchers out of Time and others* (New York, 1947).
MacDonald, George, *Evenor: Three tales*, ed. Lin Carter (New York, 1972).
Machen, Arthur, *The Great God Pan, and The Inmost Light* (London, 1894).
Moorcock, Michael, *The Stealer of Souls* (London, 1963).
—— *Stormbringer* (London, 1965).
—— *The Singing Citadel* (London, 1970).
Moore, Catherine L., *Jirel of Joiry* (New York, 1969).
Niven, Larry, *All the Myriad Ways* (New York, 1969).
—— *The Flight of the Horse* (New York, 1972).
—— *The Magic Goes Away* (New York, 1978).
—— *Convergent Series* (New York, 1979).
Norton, Andre, *The Many Worlds of AN* (New York, 1974).
Peake, Mervyn, Ballard, J. G., and Aldiss, Brian, *The Inner Landscape* (London, 1969).

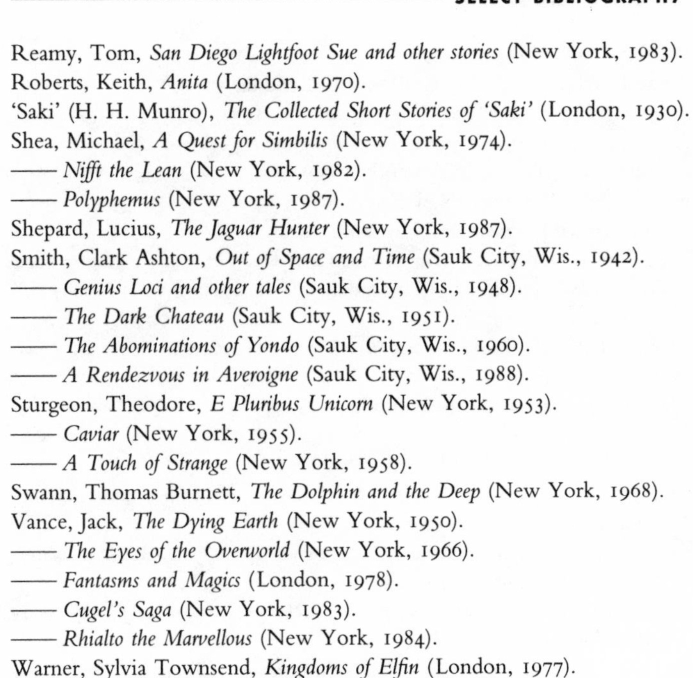

Reamy, Tom, *San Diego Lightfoot Sue and other stories* (New York, 1983).
Roberts, Keith, *Anita* (London, 1970).
'Saki' (H. H. Munro), *The Collected Short Stories of 'Saki'* (London, 1930).
Shea, Michael, *A Quest for Simbilis* (New York, 1974).
—— *Nifft the Lean* (New York, 1982).
—— *Polyphemus* (New York, 1987).
Shepard, Lucius, *The Jaguar Hunter* (New York, 1987).
Smith, Clark Ashton, *Out of Space and Time* (Sauk City, Wis., 1942).
—— *Genius Loci and other tales* (Sauk City, Wis., 1948).
—— *The Dark Chateau* (Sauk City, Wis., 1951).
—— *The Abominations of Yondo* (Sauk City, Wis., 1960).
—— *A Rendezvous in Averoigne* (Sauk City, Wis., 1988).
Sturgeon, Theodore, *E Pluribus Unicorn* (New York, 1953).
—— *Caviar* (New York, 1955).
—— *A Touch of Strange* (New York, 1958).
Swann, Thomas Burnett, *The Dolphin and the Deep* (New York, 1968).
Vance, Jack, *The Dying Earth* (New York, 1950).
—— *The Eyes of the Overworld* (New York, 1966).
—— *Fantasms and Magics* (London, 1978).
—— *Cugel's Saga* (New York, 1983).
—— *Rhialto the Marvellous* (New York, 1984).
Warner, Sylvia Townsend, *Kingdoms of Elfin* (London, 1977).
Wellman, Manly Wade, *John the Balladeer* (New York, 1988).
Wolfe, Gene, *Storeys from the Old Hotel* (London, 1988).
Yolen, Jane, *Tales of Wonder* (New York, 1983).

ANTHOLOGIES

Adrian, Jack (ed.), *Strange Tales from the 'Strand'* (London, 1991).
Baldick, Chris (ed.), *The Oxford Book of Gothic Tales* (Oxford, 1992).
Bensen, D. R. (ed.), *The Unknown* (New York, 1963).
Bradley, Marion Zimmer (ed.), *Sword and Sorceress 1–9* (New York, 1984–92).
Card, Orson Scott (ed.), *Dragons of Light* (New York, 1980).
—— *Dragons of Darkness* (New York, 1981).
Carter, Lin (ed.), *Dragons, Elves and Heroes* (New York, 1969).
—— *The Young Magicians* (New York, 1970).
—— *Flashing Swords, 1–4* (New York, 1973–7).
Datlow, Ellen and Windling, Terri (eds.), *The Best Fantasy and Horror 1–2* (New York, 1988–9).
—— *The Best Fantasy 3–5* (New York, 1990–2).
de Camp, L. Sprague (ed.), *Warlocks and Warriors: An Anthology of Heroic Fantasy* (New York, 1970).
Derleth, August (ed.), *Tales of the Cthulhu Mythos by H. P. Lovecraft and others* (2 vols., Sauk City, Wis., 1969).
Ferman, Edward (ed.), *The Best from 'Fantasy and Science Fiction': A Thirty-Year*

Retrospective (New York, 1980).

Frost, Brian J. (ed.), *The Book of the Werewolf* (London, 1973).

Hill, Douglas (ed.), *Warlocks and Warriors* (London, 1972).

Hoskins, Robert (ed.), *Swords against Tomorrow* (New York, 1970).

Kushner, Ellen (ed.), *Basilisk* (New York, 1980).

Pronzini, Bill (ed.), *Werewolf!* (New York, 1979).

Schmidt, Stanley (ed.), *Unknown* (New York, 1988).

Silverberg, Robert, and Greenberg, Martin H. (eds.), *The Fantasy Hall of Fame* (New York, 1983); published in England as *The Mammoth Book of Fantasy All-Time Greats* (London, 1988).

Wilkins, Cary (ed.), *A Treasury of Fantasy: Heroic Adventures in Imaginary Lands* (New York, 1981).

Yolen, Jane (ed.), *Shapeshifters* (New York, 1978).

WORKS OF CRITICISM

Cornwell, Neil, *The Literary Fantastic: From Gothic to Postmodernism* (Hemel Hempstead, 1990).

Hume, Kathryn, *Fantasy and Mimesis: Responses to Reality in Western Literature* (New York and London, 1984).

Hunter, Lynette, *Modern Allegory and Fantasy: Rhetorical Stances of Contemporary Writing* (London, 1989).

Irwin, W. R., *The Game of the Impossible: The Rhetoric of Fantasy* (Urbana, Ill., 1976).

Jackson, Rosemary, *Fantasy: The Literature of Subversion* (New York and London, 1981).

Kroeber, Karl, *Romantic Fantasy and Science Fiction* (New Haven and London, 1988).

Manlove, C. N., *Modern Fantasy: Five Studies* (Cambridge, 1975).

Moorcock, Michael, *Wizardry and Wild Romance: A Study of Epic Fantasy* (London, 1987).

Morse, Donald E. (ed.), *The Fantastic in World Literature and the Arts* (New York, 1987).

Pesch, Helmut W., *Fantasy: Theorie und Geschichte einer Literarischen Gattung* (Passau, 1982).

Prickett, Stephen, *Victorian Fantasy* (Hassocks, Sussex, 1979).

Rabkin, Eric S., *The Fantastic in Literature* (Princeton, NJ, 1976).

Schlobin, Roger C. (ed.), *The Aesthetics of Fantasy Literature and Art* (Notre Dame, 1982).

Shreffler, Philip A., *The H. P. Lovecraft Companion* (Westport, Conn., 1977).

Todorov, Tzvetan, *The Fantastic: A Structural Approach to a Literary Genre*, trans. Richard Howard (Ithaca, 1975).

WORKS OF REFERENCE

Bleiler, E. F., *The Checklist of Science Fiction and Supernatural Fiction* (London, 1978).

Tuck, Donald H., *The Encyclopaedia of Science Fiction and Fantasy* (3 vols., Chicago 1974, 1978, 1982).

SOURCES

The following notes indicate the place of first publication for all the stories in this anthology, together with convenient reprints in book form where possible. These latter are often useful as a guide to connected stories. Copyright information appears in a separate section of acknowledgements.

'The Demon Pope' by Richard Garnett (1835–1906), appeared in the first edition of *Twilight of the Gods* (Unwin: London, 1888). Later editions in 1903 and 1924 (the latter with an introduction by T. E. Lawrence) added twelve stories to the original sixteen. The Penguin edition of 1947 reprinted the stories in the first edition.

'The Fortress Unvanquishable, Save for Sacnoth' by Edward John Moreton Drax Plunkett, 18th Baron Dunsany (1878–1957) was published first in Dunsany's collection *The Sword of Welleran* (George Allen: London, 1908), and is reprinted in a Dunsany selection edited by Lin Carter, *At the Edge of the World* (Ballantine: New York, 1970).

'Through the Dragon Glass' by Abraham Merritt (1864–1943) was first printed in *All-Story Weekly* for 24 Nov. 1917, and reprinted in Merritt's collection *The Fox Woman* (Avon: New York, 1949) and in *The Young Magicians*, ed. Lin Carter (Ballantine: New York, 1969).

'The Nameless City' by Howard P. Lovecraft (1890–1937) is said to have been written in 1921, but was published in a magazine, *Transatlantic Circular*, too ephemeral to carry a date. It was reprinted in *Weird Tales* for November 1938 and in the Lovecraft collections *The Outsider and Others* (Arkham: Sauk City, Wis., 1939) and *The Lurking Fear* (Avon: New York, 1947). It is accepted as the first story in the 'Cthulhu Mythos', eventually comprising more than 120 stories by more than twenty writers.

'The Wind in the Portico' by John Buchan, 1st Baron Tweedsmuir (1875–1940) appeared first in *Pall Mall Magazine* for March 1928, and was reprinted in his *The Runagates Club* (Hodder and Stoughton: London, 1928).

'The Tower of the Elephant' by Robert E. Howard (1906–36) appeared first in *Weird Tales* for March 1933, and is reprinted in the collaborative collection *Conan*, by Howard, L. Sprague de Camp and Lin Carter (Lancer: New York, 1967).

'Xeethra' by Clark Ashton Smith (1893–1961) was printed first in *Weird Tales* for December 1934, and reprinted in the Smith collections *Lost Worlds* (Arkham: Sauk City, Wis., 1944) and *A Rendezvous in Averoigne* (Arkham: Sauk City, Wis., 1988). The later, corrected version has been used here.

'Jirel Meets Magic' by Catherine L. Moore (b. 1911) appeared in *Weird Tales* for July 1935, and is reprinted with other 'Jirel' stories in Moore's *Jirel of Joiry* (Paperback Library: New York, 1969). Moore married Henry Kuttner (see below) in 1940.

'The Bleak Shore' by Fritz Leiber (1910–92) was the second story in Leiber's cycle of tales about Fafhrd and the Gray Mouser to be printed, in *Unknown* for November 1940. It is reprinted in Leiber's collections *Two Sought Adventure: Exploits of Fafhrd and the Gray Mouser* (Gnome: New York, 1957) and *Swords against Death* (Ace: New York, 1970).

'Homecoming' by Ray Bradbury (b. 1920) appeared first in his collection *Dark Carnival* (Arkham: Sauk City, Wis., 1947), and again in his collection *The October Country* (Ballantine: New York, 1955).

'See You Later' by Henry Kuttner (1914–58) was the second of four 'Hogben' stories, and was first published in *Thrilling Wonder Stories* for June 1949. The Hogben stories have never been collected; this one was reprinted in *The Best of Henry Kuttner Vol. 1* (Mayflower-Dell: London, 1966).

'Liane the Wayfarer' by Jack Vance (b. 1916) is the fourth of six connected stories in Vance's *The Dying Earth* (Hillman: 1950), with later sequels in Vance's *Eyes of the Overworld* (Ace: New York, 1966) and *Cugel's Saga* (Timescape: New York, 1983), and Michael Shea's permitted continuation *A Quest for Simbilis* (Daw: New York, 1974).

'The Desrick on Yandro' by Manly Wade Wellman (b. 1905), first published in *The Magazine of Fantasy and Science Fiction* for June 1952, is the second of what proved to be a long string of stories with the same central character, now collected in *John the Balladeer* (Baen: New York, 1988).

'The Silken-Swift . . .' by Theodore Sturgeon (1918–85) appeared first in *Fantasy and Science Fiction* for November 1953 and virtually simultaneously in the Sturgeon collection *E Pluribus Unicorn* (Abelard: New York, 1953).

'Operation Afreet' by Poul Anderson (b. 1926) was printed first in *Fantasy and Science Fiction* for September 1956, and later incorporated in revised form into a novel made up of connected stories, *Operation Chaos* (Lancer: New York, 1971).

'The Singular Events which Occurred in the Hovel on the Alley off of Eye Street' by Avram Davidson (1923–93) was printed first in *Fantasy and Science Fiction* for February 1962, and reprinted in his collection *What Strange Stars and Skies* (Ace: New York, 1965).

'The Sudden Wings' by Thomas Burnett Swann (1928–76) appeared first in *Science Fantasy* for October 1962. It has not been reprinted to my knowledge.

'Same Time, Same Place' by Mervyn Peake (1911–68) appeared first in *Science Fantasy* for August 1963. It has not been reprinted to my knowledge.

'Timothy' by Keith Roberts (b. 1935) appeared first in *Impulse* for September 1966, and was reprinted in his collection of same-heroine stories *Anita* (Millington: London, 1976).

'The Kings of the Sea' by Sterling E. Lanier (b. 1927) was first printed in *Fantasy and Science Fiction* for November 1968, and reprinted in his collection of linked stories *The Peculiar Exploits of Brigadier Ffellowes* (Walker: New York, 1971).

'Not Long Before the End' by Larry Niven (b. 1938) appeared first in *Fantasy and Science Fiction* for April 1969, and was reprinted in Niven's collection *All the Myriad Ways* (Ballantine: New York, 1971).

'The Wager Lost by Winning' by John Brunner (b. 1934) appeared first in *Fantastic* for April 1970. A later revised version of it was printed in his *The Compleat Traveller in Black* (Methuen: London, 1987).

'Lila the Werewolf' by Peter S. Beagle (b. 1939) was first printed as a chapbook in New York, 1974, and reprinted in *Werewolf*!, ed. Bill Pronzini (Arbor House: New York, 1979).

'Johanna' by Jane Yolen (b. 1939) came out first in *Shape-Shifters*, ed. Jane Yolen (Seabury: New York, 1978), and was reprinted in her collection *Tales of Wonder* (Schocken: New York, 1983).

'The Erl-King' by Angela Carter (1940–92) was printed first in the magazine *Bananas*, and then in slightly altered form in her collection *The Bloody Chamber* (Gollancz: London, 1979). I have not been able to see the earlier version and have reprinted the later one.

'Beyond the Dead Reef' appeared first in *Fantasy and Science Fiction* for January 1983, and was reprinted in *The Year's Best Fantasy Stories*, ed. Donald Wollheim, 1983. 'James Tiptree Jr.' was the usual pen-name of Alice Sheldon (1916–87).

'Subworld' by Phyllis Eisenstein (b. 1946) also appeared first in *Fantasy and Science Fiction* for January 1983, and has not to my knowledge been reprinted.

'Bite-Me-Not or Fleur de Fur' by Tanith Lee (b. 1947) appeared first in *Isaac Asimov's Science Fiction Magazine* for October 1984, and was reprinted in her collection *Forests of the Night* (Unwin Hyman: London, 1989).

'The Night of White Bhairab' by Lucius Shepard (b. 1948) was first printed in *Fantasy and Science Fiction* for October 1984, and reprinted in his collection *The Jaguar Hunter* (Arkham: Sauk City, Wis., 1987).

'Thorn' by Robert Holdstock (b. 1948) was first printed in *Fantasy and Science Fiction* for October 1986, and reprinted in his collection *The Bone Forest* (Grafton: London, 1991).

'Troll Bridge' by Terry Pratchett (b. 1948) appeared first in *After the King: Stories Written in Honour of J. R. R. Tolkien*, ed. Martin H. Greenberg (Pan: London, 1992).

SOURCE ACKNOWLEDGEMENTS

Full source information is given above. The editor and publisher are grateful for permission to include the following copyright stories in this volume.

Poul Anderson, 'Operation Afreet', © 1956 by Fantasy House Inc., © renewed 1984 by Poul Anderson. Reprinted by permission of the author and Scott Meredith Literary Agency Inc., 845 Third Avenue, New York 10022.

Peter S. Beagle, 'Lila the Werewolf', © Peter S. Beagle 1974.

Ray Bradbury, 'Homecoming'. Copyright © Ray Bradbury 1946, renewed 1974 by Ray Bradbury. Reprinted by permission of Don Congdon Associates Inc.

John Brunner, 'The Wager Lost by Winning', copyright © Brunner Fact & Fiction Ltd. 1970. Reprinted by permission of the author.

Angela Carter, 'The Erl-King', © Angela Carter 1979. Reprinted by permission of Victor Gollancz Ltd.

Avram Davidson, 'The Singular Events which occurred in the Hovel on the Alley off of Eye Street'. Copyright © Avram Davidson 1962.

Phyllis Eisenstein, 'Subworld'. Copyright © Phyllis Eisenstein 1983. Reprinted by permission of the author.

Robert Holdstock, 'Thorn', © 1986 Robert Holdstock. Reprinted by permission of the author.

Henry Kuttner, 'See You Later'. Copyright © Standard Magazines, 1949, renewed 1977 by C. L. Moore. Reprinted by permission of Don Congdon Associates Inc.

Sterling E. Lanier, 'The Kings of the Sea'. Copyright © 1968 by Sterling E. Lanier. Reprinted by permission of Curtis Brown Ltd.

Tanith Lee, 'Bite-Me-Not or Fleur de Fur'. Copyright © Tanith Lee 1984. Reprinted by permission of the author.

Fritz Leiber, 'The Bleak Shore'. © Fritz Leiber 1957.

H. P. Lovecraft, 'The Nameless City', reprinted in *The Lurking Fear* (Avon Books: New York, 1947).

Abraham Merritt, 'Through the Dragon Glass', first printed in *All-Story Weekly*, 24 Nov. 1917, reprinted in *The Fox Woman* (Avon Books: New York, 1949).

Catherine L. Moore, 'Jirel Meets Magic'. Copyright © Weird Tales, 1935, 1962, renewed 1990 by Thomas Reggie. Reprinted by permission of Don Congdon Associates Inc.

Larry Niven, 'Not Long Before the End'. Reprinted by permission of Ralph Vicinanza Ltd.

Mervyn Peake, 'Same Time, Same Place'. Reprinted by permission of David Higham Associates.

Edward John Moreton Drax Plunkett (Baron Dunsany), 'The Fortress Unvanquishable, Save for Sacnoth'. Copyright the Estate of Baron Dunsany.

Terry Pratchett, 'Troll Bridge'. Copyright © 1992 Terry Pratchett. Used with permission.

Keith Roberts, 'Timothy'. Copyright © Keith Roberts 1966.

Lucius Shepard, 'The Night of White Bhairab'. Copyright Lucius Shepard. Reprinted by permission of the author.

Clark Ashton Smith, 'Xeethra', first printed in *Weird Tales*, December 1934, reprinted in *Lost Worlds* (1944) and *A Rendezvous In Averoigne* (Arkham House Publishers Inc., 1988).

Theodore Sturgeon, 'The Silken-Swift . . .'. Copyright © the Estate of Theodore Sturgeon 1953.

James Tiptree Jr., 'Beyond the Dead Reef', Copyright © 1983 by James Tiptree Jr., reprinted by permission of the author's estate and the Estate's agent, Virginia Kidd.

Jack Vance, 'Liane the Wayfarer'. Reprinted by permission of Ralph Vicinanza Ltd.

Manly Wade Wellman, 'The Desrick on Yandro'. Reprinted by permission of Ralph Vicinanza Ltd.

Jane Yolen, 'Johanna'. Copyright © 1978 by Jane Yolen. Reprinted by permission of Curtis Brown Ltd.

OUP apologize for any errors or omissions in the above list and would be grateful to be notified of any corrections that should be incorporated in the next edition or reprint of this volume.